# The Impact of Caffeine and Coffee on Human Health

# The Impact of Caffeine and Coffee on Human Health

Special Issue Editors

**Christina Bamia**
**Marilyn Cornelis**

MDPI • Basel • Beijing • Wuhan • Barcelona • Belgrade

**MDPI**

*Special Issue Editors*
Christina Bamia
National and Kapodistrian
University of Athens
Greece

Marilyn Cornelis
Northwestern University
Feinberg School of Medicine
USA

*Editorial Office*
MDPI
St. Alban-Anlage 66
4052 Basel, Switzerland

This is a reprint of articles from the Special Issue published online in the open access journal *Nutrients* (ISSN 2072-6643) from 2018 to 2019 (available at: https://www.mdpi.com/journal/nutrients/special_issues/impact_of_caffeine_and_coffee_on_human_health).

For citation purposes, cite each article independently as indicated on the article page online and as indicated below:

LastName, A.A.; LastName, B.B.; LastName, C.C. Article Title. *Journal Name* **Year**, *Article Number*, Page Range.

ISBN 978-3-03921-834-9 (Pbk)
ISBN 978-3-03921-835-6 (PDF)

# Contents

# About the Special Issue Editors

**Christina Bamia** graduated from the National Metsovio Technical University of Athens. She has an MSc in Statistics with Applications in Medicine from the University of Southampton, a PhD in Medical Statistics from the University of London (LSHTM) and a Diploma in Epidemiology & Population Health from the University of London (LSHTM). Since 2016, she has been Associate Professor of Epidemiology and Medical Statistics in the Athens Medical School. From 2000, she participated as a Research Associate and co-investigator in several EC-funded programs with the EPIC study and CHANCES project. She was a member of the working group organized by the International Agency on Research of Cancer (IARC) Monographs to evaluate the carcinogenicity of drinking coffee, maté, and very hot beverages (May 2016). Her main working areas are nutritional epidemiology, dietary pattern analyses and associated methodological issues. To date, she has 120 peer-reviewed publications and more than 6000 citations.

**Marilyn Cornelis**'s research aims to couple modern high-throughput omic-technologies to traditional clinical and epidemiological methods to enhance biological understanding of how diet and nutrition contributes to chronic disease. She has a special interest in the genetics of coffee consumption, caffeine metabolism, taste preferences and other dietary behaviors.

nutrients

MDPI

*Editorial*

# The Impact of Caffeine and Coffee on Human Health

**Marilyn C. Cornelis**

Department of Preventive Medicine, Northwestern University Feinberg School of Medicine,
Chicago, IL 60611, USA; Marilyn.cornelis@northwestern.edu; Tel.: +1-312-503-4548

Received: 11 February 2019; Accepted: 13 February 2019; Published: 16 February 2019

Coffee is one of the most widely consumed beverages in the world and is also a major source of caffeine for most populations [1]. This special issue of *Nutrients*, "The Impact of Caffeine and Coffee on Human Health" contains nine reviews and 10 original publications of timely human research investigating coffee and caffeine habits and the impact of coffee and caffeine intake on various diseases, conditions, and performance traits.

With increasing interest in the role of coffee in health, general knowledge of population consumption patterns and within the context of the full diet is important for both research and public health. Reyes and Cornelis [1] used 2017 country-level volume sales (proxy for consumption) of caffeine-containing beverages (CCBs) to demonstrate that coffee and tea remain the leading CCBs consumed around the world. In a large coordinated effort spanning 10 European countries, Landais et al. [2] quantified self-reported coffee and tea intakes and assessed their contribution to the intakes of selected nutrients in adults where variation in consumption was mostly driven by geographical region. Overall, coffee and tea contributed to less than 10% of the energy intake. However, the greatest contribution to total sugar intake was observed in Southern Europe (up to ~20%). These works not only emphasize the wide prevalence of coffee and tea drinking, but also the need for data on coffee and tea additives in epidemiological studies of these beverages in certain countries as they may offset any potential benefits these beverages have on health.

Doepker et al. [3] provided a user-friendly synopsis of their systematic review [4] of caffeine safety, which concluded that caffeine doses (400 mg/day for healthy adults, for example) previously determined in 2003 [5] as not to be associated with adverse effects, remained generally appropriate despite new research conducted since then. Further concerning caffeine safety is the systematic review of caffeine-related deaths by Capelletti et al. [6]. Suicide, accidental, and intentional poisoning were the most common causes of death and most cases involved infants, psychiatric patients, and athletes. Both Doepker et al. [3] and Capelletti et al. [6] alluded to the increasing interest in the area of between-person sensitivity resulting from environmental and genetic factors, of which the latter is a topic of additional papers in this special issue and thus reiterates this interest.

Advancements in high-throughput analyses of the human genome, transcriptome, proteome, and metabolome have presented coffee researchers with an unprecedented opportunity to optimize their research approach while acquiring mechanistic and causal insight to their observed associations [7]. Three timely reviews [8–10] and an original report [11] addressed the topic of human genetics and coffee and caffeine consumption. Interest in this area received a boost by the success of genome-wide association studies (GWAS), which identified multiple genetic variants associated with habitual coffee and caffeine consumption as discussed by Cornelis and Munafo [8] in their review of Mendelian randomization (MR) studies on coffee and caffeine consumption. MR is a technique that uses genetic variants as instrumental variables to assess whether an observational association between a risk factor (i.e., coffee) and an outcome aligns with a causal effect. The application of this approach to coffee and health is growing, but has important statistical and conceptual challenges that warrant consideration in the interpretation of the results. Southward et al. [9] and Fulton et al. [10] reviewed the impact of genetics on physiological responses to caffeine. Both emphasized a current clinical interest limited to

*CYP1A2* and *ADORA2A* variations, suggesting opportunities to expand this research to more recent loci identified by GWAS. Despite the advancements in integrating genetics into clinical trials of caffeine, such designs remain susceptible to limitations [9,10,12,13]. Some of these limitations were further highlighted by Shabir et al. [14] in their critical review on the impact of caffeine expectancies on sport, exercise, and cognitive performance. Interestingly, the original findings from a randomized controlled trial of regular coffee, decaffeinated coffee, and placebo suggested the stimulant activity of coffee beyond its caffeine content, raising issues with the use of decaffeinated coffee as a placebo [15].

The impact of coffee intake on gene expression and the lipidome were investigated by Barnung et al. [16] and Kuang et al. [17], respectively. Barnung et al. [16] reported on the results from a population-based whole-blood gene expression analysis of coffee consumption that pointed to metabolic, immune, and inflammation pathways. Using samples from a controlled trial of coffee intake, Kuang et al. [17] reported that coffee intake led to lower levels of specific lysophosphatidylcholines. These two reports provide both novel and confirmatory insight into mechanisms by which coffee might be impacting health and further demonstrate the power of high-throughput omic technologies in the nutrition field.

Heavy coffee and caffeine intake continue to be seen as potentially harmful on pregnancy outcomes [18]. Leviton [19] discussed the biases inherent in studies of coffee consumption during pregnancy and argued that all of the reports of detrimental effects of coffee could be explained by one or more of these biases. The impact of dietary caffeine intake on assisted reproduction technique (ART) outcomes has also garnered interest. An original report by Ricci et al. [20] in this special issue found no relationship between the caffeine intake of subfertile couples and negative ART outcomes.

Van Dijk et al. [21] reviewed the effects of caffeine on myocardial blood flow, which support a significant and clinically relevant influence of recent caffeine intake on cardiac perfusion measurements during adenosine and dipyridamole induced hyperemia. Original observational reports on the association between habitual coffee consumption and liver fibrosis [22], depression [23], hearing [24], and cognition indices [25] have extended the research in these areas to new populations.

Finally, given the widespread availability of caffeine in the diet and the increasing public and scientific interest in the potential health consequences of habitual caffeine intake, Reyes and Cornelis [1] assessed how current caffeine knowledge and concern has been translated into food-based dietary guidelines (FBDG) from around the world; focusing on CCBs. Several themes emerged, but in general, FBDG provided an unfavorable view of CCBs, which was rarely balanced with recent data supporting the potential benefits of specific beverage types.

This collection of original and review papers provides a useful summary of the progress on the topic of caffeine, coffee, and human health. It also points to the research needs and limitations of the study design, which should be considered going forward and when critically evaluating the research findings.

**Conflicts of Interest:** The author declares no conflict of interest.

## References

1.    Reyes, C.M.; Cornelis, M.C. Caffeine in the diet: Country-level consumption and guidelines. *Nutrients* **2018**, *10*, 1772. [CrossRef] [PubMed]
2.    Landais, E.; Moskal, A.; Mullee, A.; Nicolas, G.; Gunter, M.J.; Huybrechts, I.; Overvad, K.; Roswall, N.; Affret, A.; Fagherazzi, G.; et al. Coffee and tea consumption and the contribution of their added ingredients to total energy and nutrient intakes in 10 European countries: Benchmark data from the late 1990s. *Nutrients* **2018**, *10*, 725. [CrossRef] [PubMed]
3.    Doepker, C.; Franke, K.; Myers, E.; Goldberger, J.J.; Lieberman, H.R.; O'Brien, C.; Peck, J.; Tenenbein, M.; Weaver, C.; Wikoff, D. Key findings and implications of a recent systematic review of the potential adverse effects of caffeine consumption in healthy adults, pregnant women, adolescents, and children. *Nutrients* **2018**, *10*, 1536. [CrossRef] [PubMed]

4.  Wikoff, D.; Welsh, B.T.; Henderson, R.; Brorby, G.P.; Britt, J.; Myers, E.; Goldberger, J.; Lieberman, H.R.; O'Brien, C.; Peck, J. Systematic review of the potential adverse effects of caffeine consumption in healthy adults, pregnant women, adolescents, and children. *Food Chem. Toxicol.* **2017**, *109*, 585–648. [CrossRef] [PubMed]

5.  Nawrot, P.; Jordan, S.; Eastwood, J.; Rostein, J.; Hugenholtz, A.; Feeley, M. Effects of caffeine on human health. *Food Addit. Contam.* **2003**, *20*, 1–30. [CrossRef] [PubMed]

6.  Cappelletti, S.; Piacentino, D.; Fineschi, V.; Frati, P.; Cipolloni, L.; Aromatario, M. Caffeine-related deaths: Manner of deaths and categories at risk. *Nutrients* **2018**, *10*, 611. [CrossRef] [PubMed]

7.  Cornelis, M.C. Toward systems epidemiology of coffee and health. *Curr. Opin. Lipidol.* **2015**, *26*, 20–29. [CrossRef] [PubMed]

8.  Cornelis, M.C.; Munafo, M.R. Mendelian randomization studies of coffee and caffeine consumption. *Nutrients* **2018**, *10*, 1343. [CrossRef] [PubMed]

9.  Southward, K.; Rutherfurd-Markwick, K.; Badenhorst, C.; Ali, A. The role of genetics in moderating the inter-individual differences in the ergogenicity of caffeine. *Nutrients* **2018**, *10*, 1352. [CrossRef] [PubMed]

10. Fulton, J.L.; Dinas, P.C.; Carrillo, A.E.; Edsall, J.R.; Ryan, E.J.; Ryan, E.J. Impact of genetic variability on physiological responses to caffeine in humans: A systematic review. *Nutrients* **2018**, *10*, 1373. [CrossRef] [PubMed]

11. Kokaze, A.; Ishikawa, M.; Matsunaga, N.; Karita, K.; Yoshida, M.; Ochiai, H.; Shirasawa, T.; Yoshimoto, T.; Minoura, A.; Oikawa, K.; et al. Nadh dehydrogenase subunit-2 237 leu/met polymorphism influences the association of coffee consumption with serum chloride levels in male Japanese health checkup examinees: An exploratory cross-sectional analysis. *Nutrients* **2018**, *10*, 1344. [CrossRef] [PubMed]

12. Southward, K.; Rutherfurd-Markwick, K.; Badenhorst, C.; Ali, A. Response to "are there non-responders to the ergogenic 3 effects of caffeine ingestion on exercise performance?". *Nutrients* **2018**, *10*, 1175. [CrossRef] [PubMed]

13. Grgic, J. Are there non-responders to the ergogenic effects of caffeine ingestion on exercise performance? *Nutrients* **2018**, *10*, 1736. [CrossRef] [PubMed]

14. Shabir, A.; Hooton, A.; Tallis, J.; Higgins, M. The influence of caffeine expectancies on sport, exercise, and cognitive performance. *Nutrients* **2018**, *10*, 1528. [CrossRef] [PubMed]

15. Haskell-Ramsay, C.F.; Jackson, P.A.; Forster, J.S.; Dodd, F.L.; Bowerbank, S.L.; Kennedy, D.O. The acute effects of caffeinated black coffee on cognition and mood in healthy young and older adults. *Nutrients* **2018**, *10*, 1386. [CrossRef] [PubMed]

16. Barnung, R.; Nøst, T.; Ulven, S.M.; Skeie, G.; Olsen, K. Coffee consumption and whole-blood gene expression in the norwegian women and cancer post-genome cohort. *Nutrients* **2018**, *10*, 1047. [CrossRef]

17. Kuang, A.; Erlund, I.; Herder, C.; Westerhuis, J.A.; Tuomilehto, J.; Cornelis, M.C. Lipidomic response to coffee consumption. *Nutrients* **2018**, *10*, 1851. [CrossRef]

18. Poole, R.; Kennedy, O.J.; Roderick, P.; Fallowfield, J.A.; Hayes, P.C.; Parkes, J. Coffee consumption and health: Umbrella review of meta-analyses of multiple health outcomes. *BMJ* **2017**, *359*, j5024. [CrossRef]

19. Leviton, A. Biases inherent in studies of coffee consumption in early pregnancy and the risks of subsequent events. *Nutrients* **2018**, *10*, 1152. [CrossRef]

20. Ricci, E.; Noli, S.; Cipriani, S.; La Vecchia, I.; Chiaffarino, F.; Ferrari, S.; Mauri, P.A.; Reschini, M.; Fedele, L.; Parazzini, F. Maternal and paternal caffeine intake and art outcomes in couples referring to an italian fertility clinic: A prospective cohort. *Nutrients* **2018**, *10*, 1116. [CrossRef]

21. Van Dijk, R.; Ties, D.; Kuijpers, D.; van der Harst, P.; Oudkerk, M. Effects of caffeine on myocardial blood flow: A systematic review. *Nutrients* **2018**, *10*, 1083. [CrossRef] [PubMed]

22. Yaya, I.; Marcellin, F.; Costa, M.; Morlat, P.; Protopopescu, C.; Pialoux, G.; Santos, M.E.; Wittkop, L.; Esterle, L.; Gervais, A.; et al. Impact of alcohol and coffee intake on the risk of advanced liver fibrosis: A longitudinal analysis in hiv-hcv coinfected patients (anrs hepavih co-13 cohort). *Nutrients* **2018**, *10*, 705. [CrossRef] [PubMed]

23. Navarro, A.M.; Abasheva, D.; Martinez-Gonzalez, M.A.; Ruiz-Estigarribia, L.; Martin-Calvo, N.; Sanchez-Villegas, A.; Toledo, E. Coffee consumption and the risk of depression in a middle-aged cohort: The sun project. *Nutrients* **2018**, *10*, 1333. [CrossRef] [PubMed]

24. Lee, S.Y.; Jung, G.; Jang, M.J.; Suh, M.W.; Lee, J.H.; Oh, S.H.; Park, M.K. Association of coffee consumption with hearing and tinnitus based on a national population-based survey. *Nutrients* **2018**, *10*, 1429. [CrossRef]
25. Haller, S.; Montandon, M.L.; Rodriguez, C.; Herrmann, F.R.; Giannakopoulos, P. Impact of coffee, wine, and chocolate consumption on cognitive outcome and MRI parameters in old age. *Nutrients* **2018**, *10*, 1391. [CrossRef] [PubMed]

*Article*

# Lipidomic Response to Coffee Consumption

Alan Kuang [1], Iris Erlund [2], Christian Herder [3,4], Johan A. Westerhuis [5,6], Jaakko Tuomilehto [7,8,9] and Marilyn C. Cornelis [1,*]

[1] Department of Preventive Medicine, Northwestern University Feinberg School of Medicine, 680 North Lake Shore Drive, Suite 1400, Chicago, IL 60611, USA; alan.kuang@northwestern.edu
[2] Genomics and Biomarkers Unit, National Institute for Health and Welfare, P.O. Box 30, 00271 Helsinki, Finland; iris.erlund@thl.fi
[3] Institute for Clinical Diabetology, German Diabetes Center, Leibniz Center for Diabetes Research at Heinrich Heine University Düsseldorf, 40225 Düsseldorf, Germany; Christian.Herder@DDZ.UNI-DUESSELDORF.DE
[4] German Center for Diabetes Research (DZD), 85764 München-Neuherberg, Germany
[5] Biosystems Data Analysis, Swammerdam Institute for Life Sciences, University of Amsterdam, Science Park 904, 1098 XH Amsterdam, The Netherlands; j.a.westerhuis@uva.nl
[6] Centre for Human Metabolomics, Faculty of Natural Sciences, North-West University (Potchefstroom Campus), Private Bag X6001, Potchefstroom 2520, South Africa
[7] Disease Risk Unit, National Institute for Health and Welfare, 00271 Helsinki, Finland; jaakko.tuomilehto@thl.fi
[8] Department of Public Health, University of Helsinki, 00014 Helsinki, Finland
[9] Saudi Diabetes Research Group, King Abdulaziz University, 21589 Jeddah, Saudi Arabia
[*] Correspondence: marilyn.cornelis@northwestern.edu; Tel.: +1-312-503-4548; Fax: +1-312-908-9588

Received: 25 October 2018; Accepted: 22 November 2018; Published: 1 December 2018

**Abstract:** Coffee is widely consumed and contains many bioactive compounds, any of which may impact pathways related to disease development. Our objective was to identify individual lipid changes in response to coffee drinking. We profiled the lipidome of fasting serum samples collected from a previously reported single blinded, three-stage clinical trial. Forty-seven habitual coffee consumers refrained from drinking coffee for 1 month, consumed 4 cups of coffee/day in the second month and 8 cups/day in the third month. Samples collected after each coffee stage were subject to quantitative lipidomic profiling using ion-mobility spectrometry–mass spectrometry. A total of 853 lipid species mapping to 14 lipid classes were included for univariate analysis. Three lysophosphatidylcholine (LPC) species including LPC (20:4), LPC (22:1) and LPC (22:2), significantly decreased after coffee intake ($p < 0.05$ and $q < 0.05$). An additional 72 species mapping to the LPC, free fatty acid, phosphatidylcholine, cholesteryl ester and triacylglycerol classes of lipids were nominally associated with coffee intake ($p < 0.05$ and $q > 0.05$); 58 of these decreased after coffee intake. In conclusion, coffee intake leads to lower levels of specific LPC species with potential impacts on glycerophospholipid metabolism more generally.

**Keywords:** coffee; caffeine; lipids; biomarkers; trial; lysophosphatidylcholine; lipidomics

## 1. Introduction

Coffee is one of the most widely consumed beverages in the world and has been implicated in numerous diseases such as type 2 diabetes (T2D) and cardiovascular disease [1–4]. The causal and precise molecular mechanisms that underlie the beneficial and adverse effects of coffee remain unclear. Coffee is the major source of caffeine for many populations [5], but it also contains hundreds of other compounds, many of which might impact pathways related to disease development or prevention [6].

High-throughput omic profiling techniques enable thorough studies of an individual's response to coffee intake and provide potentially new mechanistic insight to the role coffee plays in health [7,8].

We recently performed a comprehensive metabolomics study of coffee consumption leveraging serum samples collected during a coffee trial [8,9]. Over 100 metabolites were significantly associated with coffee intake; several mapping to xanthine, benzoate, steroid, endocannabinoid and fatty acid (acylcholine) metabolism. We extend this work to comprehensive lipid profiling for the first time. Lipid molecules are a subset of the metabolome and serve as ubiquitous and essential multifunctional metabolites [10]. Lipids are directly exposed to intracellular and extracellular biochemical changes and as a result undergo various modifications themselves [10]. Our objective was to identify individual lipid changes in response to coffee in order to gain more insight into biological mechanisms by which coffee may impact health.

## 2. Subjects and Methods

### 2.1. Coffee Trial

Serum samples analyzed in the current study were obtained from participants completing an investigator-blinded, three-stage controlled trial in 2009–2010 that lasted for 3 months (Supplementary Note 1, ISRCTN registry: ISRCTN12547806) [9]. Briefly, habitual coffee consumers <65 years of age, residing in Finland, free of T2D, but with an elevated risk of T2D were eligible for participation. The participants received packages of coffee and brewed the coffee daily at home with their own coffee machine using paper filters. During the first month, participants refrained from drinking coffee, whereas in the second month they were instructed to consume 4 cups coffee/day (1 cup = 150 mL, Juhla Mokka brand) and in the third month 8 cups/day. Of the 49 participants recruited, 47 completed the trial. Baseline characteristics of these 47 participants are shown in Table S1. Several clinical biomarkers were measured and analyzed as part of the initial report as previously described [9]. Serum concentrations of total cholesterol, High-density lipoprotein (HDL) cholesterol, apo A-I and adiponectin increased significantly in response to coffee intake, whereas interleukin-18, 8-isoprostane, and the ratios of low-density lipoprotein (LDL) to HDL cholesterol and of apo B and apo A-I decreased significantly. The trial was conducted in accordance with the Declaration of Helsinki (1964), as amended in South Africa (1996), and approved by Joint Authority for the Hospital District of Helsinki and Uusimaa Ethics Committee, Department of Medicine, Helsinki, Finland. Written informed consent was obtained from all participants.

### 2.2. Lipidomics Assay, Data Acquisition and Processing

Lipid species were measured in fasting serum samples collected after each coffee stage (True Mass Complex Lipid Panel, Metabolon, Research Triangle Park, NC, USA). Lipids were extracted from samples using dichloromethane and methanol in a modified Bligh-Dyer extraction in the presence of internal standards with the lower, organic, phase being used for analysis. The extracts were concentrated under nitrogen and reconstituted in 0.25 mL of dichloromethane:methanol (50:50) containing 10 mM ammonium acetate. The extracts were placed in vials for infusion-mass spectrometry (MS) analyses, performed on a SelexION equipped Sciex 5500 QTRAP using both positive and negative mode electrospray. Each sample was subjected to two analyses, with ion mobility spectrometry (IMS)-MS conditions optimized for lipid classes monitored in each analysis. The 5500 QTRAP was operated in MRM mode to monitor the transitions for over 1100 lipids from up to 14 lipid classes including cholesteryl esters (CE), triacylglycerols (TAG), diacylglycerols (DAG), free fatty acids (FFA), phosphatidylcholines (PC), phosphatidylethanolamines (PE), phosphatidylinositols (PI), lysophosphatidylcholines (LPC), lysophosphatidylethanolamines (LPE), sphingomyelin (SM), ceramide (CER), hexosylceramides (HCER), lactosylceramides (LCER), dihydroceramides (DCER). Individual lipid species were quantified based on the ratio of signal intensity for target compounds to the signal intensity for an assigned internal standard of known concentration. Missing values were imputed with the observed minimum value. Individual lipid species that contained more than 20% missing values across the first (0 cups/day) and third (8 cups/day) trial stages were not included for statistical analysis (120 lipid species, Table S2) leaving a total of 853 lipid species for analysis. The same data, but with missing values treated as 0, were also expressed as mole% determined by calculating the proportion of individual species within each class. In secondary analysis, lipid species data were used to derive additional and biologically meaningful lipid traits. Lipid class concentrations were calculated from the sum of all molecular species within a class. For lipid classes containing more than one fatty acid (FA) per species (i.e., DAG, PC, PE, PI, and TAG) we also determined FA concentrations by calculating the sum of individual FAs within each of these classes. These traits were derived prior to excluding the lipids in Table S2 (see above). The final set of lipid species (primary traits) and derived lipid traits (secondary) analyzed in the current study are listed in Table S3.

*2.3. Statistical Analysis*

All statistical analyses were performed using R, SAS version 9.4 (SAS Institute Inc, Cary, NC, USA) or Matlab. To explore the data and identify any outlier samples we first performed standard principal component analysis (PCA) and multilevel PCA [11]. For the latter, we generated a data matrix of the within-person variation by subtracting individual lipid values from the mean lipid value of all three coffee stages, per participant, per lipid. Repeated measures ANOVA was used to test the relationship between coffee treatment and each individual lipid species. P-values were further adjusted for multiple comparisons by the Benjamini–Hochberg procedure and the false discovery rate (FDR)-adjusted P-values, expressed as $q$-values, are reported [12]. All nominal ($p < 0.05$) associations are presented but only those with a $q$-value $< 0.05$ are defined as statistically significant. We computed ordinary Pearson correlations to explore the latent relationships of changes in identified coffee lipids across treatments. These analyses were additionally supplemented with data for metabolites and clinical biomarkers that previously changed in response to coffee in this coffee trial (Table S4) [8,9]. Formal cross-platform integration analysis will be a focus of another report. Correlation networks were constructed using Cytoscape [13]. In secondary analysis, lipid class and fatty acid concentrations were also subject to univariate analysis. A multivariate approach was also pursued as traditionally done with high-throughput data and is presented in Supplementary Note 2 and Figure S5.

## 3. Results

PCA or multilevel-PCA demonstrated no clear separation of samples by coffee stage (Figure S1). As a result, no clear outliers were detected and thus all samples were included for our primary analysis.

Serum lipid class concentrations (data not shown) or distributions (Figure 1a) did not significantly change in response to coffee intake. A total of 75 lipid species were at least nominally associated with coffee intake and these mapped to 8 lipid classes ($p < 0.05$, Table 1, Figure 2a and Figure S2). When applying an FDR correction, LPC 20:4, 22:1 and 22:2 remained significantly associated with coffee intake (Figure 2b). Similar results were observed when lipid species concentrations were expressed as mole% (Figure 1b and data not shown). When FA concentrations of DAG, PC, PE, PI, and TAG were examined, no associations met statistical significance (data not shown).

Results of correlation analysis of changes among previously identified clinical [9] and metabolite [8] markers of coffee response and the 75 nominal to significant lipid species identified here (Table 1) are presented in Figure S3. Generally lipid species of the same class or sharing fatty acid chains clustered together. Changes in TAGs that increased in response to coffee, however, did not correlate with changes in TAGs that decreased in response to coffee. Changes in lipid species generally correlated with metabolites that also decreased in response to coffee and thus unlikely originated from the coffee beverage itself. These metabolites were also lipid derivatives; particularly those of the acyl choline and endocannabinoid pathways. Besides kynurenine and xanthines, few other aqueous metabolites were consistently represented among correlations with either clinical makers or lipid species. No changes in lipids or metabolites were consistently correlated with clinical markers that responded to coffee.

**Table 1.** Significant lipid markers of coffee consumption *.

| Lipid Class † | Lipid Species | Group Effect | | Fold of Change § | | |
|---|---|---|---|---|---|---|
| | | *p*-Value | *q*-Value | 4 Cups/0 Cup | 8 Cups/0 Cup | 8 Cups/4 Cups |
| CE | CE(20:4) | 0.0296 | 0.4529 | 0.9 | 0.92 | 1.02 |
| FFA | FFA(20:3) | 0.0021 | 0.297 | 0.9 | 0.87 | 0.96 |
| | FFA(20:4) | 0.0012 | 0.2492 | 0.95 | 0.87 | 0.91 |
| | FFA(22:2) | 0.0481 | 0.4529 | 0.95 | 0.89 | 0.94 |
| | FFA(22:6) | 0.0415 | 0.4529 | 0.98 | 0.89 | 0.91 |
| TAG | TAG47:1-FA17:0 | 0.0483 | 0.4529 | 1.26 | 1.4 | 1.11 |
| | TAG51:3-FA15:0 | 0.0401 | 0.4529 | 0.82 | 0.91 | 1.11 |
| | TAG52:4-FA16:1 | 0.0317 | 0.4529 | 0.8 | 0.92 | 1.15 |
| | TAG52:5-FA16:1 | 0.0329 | 0.4529 | 0.77 | 0.89 | 1.16 |
| | TAG52:5-FA20:5 | 0.05 | 0.4529 | 1.07 | 1.25 | 1.18 |
| | TAG52:6-FA16:1 | 0.041 | 0.4529 | 0.78 | 0.9 | 1.14 |
| | TAG53:3-FA16:0 | 0.0211 | 0.4529 | 0.88 | 0.87 | 1 |
| | TAG53:3-FA18:1 | 0.0242 | 0.4529 | 0.9 | 0.93 | 1.03 |
| | TAG53:4-FA16:0 | 0.0229 | 0.4529 | 0.84 | 0.89 | 1.07 |
| | TAG53:4-FA18:2 | 0.0289 | 0.4529 | 0.82 | 0.88 | 1.08 |
| | TAG53:5-FA18:3 | 0.048 | 0.4529 | 0.87 | 0.92 | 1.06 |
| | TAG54:3-FA18:1 | 0.0354 | 0.4529 | 0.82 | 0.9 | 1.09 |
| | TAG54:3-FA20:1 | 0.0368 | 0.4529 | 0.84 | 0.94 | 1.13 |
| | TAG54:4-FA20:1 | 0.0306 | 0.4529 | 0.82 | 0.94 | 1.14 |
| | TAG55:3-FA18:1 | 0.0353 | 0.4529 | 0.82 | 0.86 | 1.05 |
| | TAG55:4-FA18:1 | 0.0198 | 0.4529 | 0.82 | 0.85 | 1.04 |
| | TAG55:5-FA18:1 | 0.0208 | 0.4529 | 0.77 | 0.83 | 1.08 |
| | TAG56:3-FA18:1 | 0.0103 | 0.4529 | 0.81 | 0.87 | 1.07 |
| | TAG56:3-FA20:1 | 0.0155 | 0.4529 | 0.79 | 0.86 | 1.09 |
| | TAG56:4-FA18:1 | 0.0124 | 0.4529 | 0.8 | 0.87 | 1.08 |
| | TAG56:4-FA20:1 | 0.0314 | 0.4529 | 0.71 | 0.81 | 1.14 |
| | TAG56:4-FA20:2 | 0.0141 | 0.4529 | 0.84 | 0.88 | 1.05 |
| | TAG56:5-FA18:1 | 0.0221 | 0.4529 | 0.83 | 0.9 | 1.09 |
| | TAG56:5-FA20:2 | 0.0051 | 0.4529 | 0.77 | 0.84 | 1.08 |
| | TAG56:5-FA20:3 | 0.0215 | 0.4529 | 0.83 | 0.89 | 1.08 |
| | TAG56:5-FA20:4 | 0.0447 | 0.4529 | 0.84 | 0.91 | 1.07 |
| | TAG56:6-FA18:2 | 0.0132 | 0.4529 | 0.77 | 0.88 | 1.13 |
| | TAG56:6-FA20:2 | 0.0206 | 0.4529 | 0.76 | 0.84 | 1.11 |
| | TAG56:6-FA20:3 | 0.0077 | 0.4529 | 0.77 | 0.85 | 1.1 |
| | TAG56:6-FA20:4 | 0.0306 | 0.4529 | 0.81 | 0.88 | 1.08 |
| | TAG56:7-FA18:2 | 0.0457 | 0.4529 | 0.8 | 0.91 | 1.14 |
| | TAG56:7-FA20:3 | 0.042 | 0.4529 | 0.79 | 0.85 | 1.07 |
| | TAG56:7-FA22:4 | 0.0484 | 0.4529 | 0.87 | 0.92 | 1.06 |
| | TAG56:7-FA22:5 | 0.0384 | 0.4529 | 0.85 | 0.95 | 1.12 |
| | TAG56:9-FA20:4 | 0.0458 | 0.4529 | 0.83 | 0.92 | 1.11 |
| | TAG56:9-FA22:6 | 0.0229 | 0.4529 | 0.85 | 0.92 | 1.08 |
| | TAG57:8-FA22:6 | 0.0093 | 0.4529 | 0.87 | 0.91 | 1.04 |
| | TAG58:10-FA20:5 | 0.0161 | 0.4529 | 0.86 | 0.94 | 1.09 |
| | TAG58:10-FA22:5 | 0.0391 | 0.4529 | 0.74 | 0.84 | 1.14 |
| | TAG58:10-FA22:6 | 0.0388 | 0.4529 | 0.72 | 0.8 | 1.11 |
| | TAG58:7-FA22:4 | 0.0294 | 0.4529 | 0.81 | 0.89 | 1.11 |
| | TAG58:7-FA22:5 | 0.0109 | 0.4529 | 0.79 | 0.85 | 1.07 |
| | TAG58:8-FA20:4 | 0.0324 | 0.4529 | 0.85 | 0.9 | 1.06 |
| | TAG58:8-FA22:5 | 0.0386 | 0.4529 | 0.79 | 0.85 | 1.08 |
| | TAG58:9-FA22:5 | 0.0478 | 0.4529 | 0.78 | 0.86 | 1.1 |
| | TAG60:10-FA22:5 | 0.0349 | 0.4529 | 0.85 | 0.9 | 1.06 |
| | TAG60:10-FA22:6 | 0.0357 | 0.4529 | 0.82 | 0.92 | 1.13 |
| | TAG60:11-FA22:5 | 0.0038 | 0.4529 | 0.8 | 0.92 | 1.16 |

**Table 1.** *Cont.*

| Lipid Class † | Lipid Species | Group Effect | | Fold of Change § | | |
|---|---|---|---|---|---|---|
| | | *p*-Value | *q*-Value | 4 Cups/0 Cup | 8 Cups/0 Cup | 8 Cups/4 Cups |
| LPC | LPC(15:0) | 0.0142 | 0.4529 | 0.95 | 0.92 | 0.97 |
| | LPC(17:0) | 0.0017 | 0.2886 | 0.96 | 0.9 | 0.93 |
| | LPC(18:1) | 0.0423 | 0.4529 | 0.98 | 0.93 | 0.95 |
| | LPC(20:2) | 0.0094 | 0.4529 | 0.95 | 0.89 | 0.93 |
| | LPC(20:3) | 0.0362 | 0.4529 | 0.94 | 0.91 | 0.96 |
| | LPC(20:4) | <0.0001 | 0.0088 | 0.94 | 0.87 | 0.93 |
| | LPC(22:1) | <0.0001 | 0.0313 | 0.91 | 0.78 | 0.86 |
| | LPC(22:2) | <0.0001 | 0.0051 | 0.94 | 0.79 | 0.84 |
| PC | PC(17:0/20:4) | 0.0183 | 0.4529 | 0.91 | 0.91 | 1 |
| | PC(18:0/16:1) | 0.0274 | 0.4529 | 1.09 | 1.3 | 1.19 |
| | PC(18:0/18:3) | 0.0375 | 0.4529 | 1.13 | 1.24 | 1.1 |
| | PC(18:0/20:2) | 0.0143 | 0.4529 | 1 | 1.11 | 1.11 |
| | PC(18:0/20:3) | 0.0361 | 0.4529 | 0.96 | 1.08 | 1.12 |
| | PC(18:1/20:4) | 0.0152 | 0.4529 | 0.92 | 0.91 | 0.99 |
| PE | PE(18:0/20:1) | 0.0203 | 0.4529 | 0.97 | 1.12 | 1.16 |
| | PE(O-16:0/18:2) | 0.0301 | 0.4529 | 1.08 | 1.19 | 1.11 |
| | PE(O-18:0/20:3) | 0.0458 | 0.4529 | 0.98 | 1.12 | 1.15 |
| | PE(P-16:0/18:2) | 0.0246 | 0.4529 | 1.07 | 1.18 | 1.1 |
| | PE(P-16:0/22:4) | 0.025 | 0.4529 | 0.89 | 1.01 | 1.14 |
| | PE(P-18:0/18:2) | 0.0406 | 0.4529 | 1.04 | 1.15 | 1.1 |
| DCER | DCER(24:0) | 0.0475 | 0.4529 | 1 | 1.1 | 1.1 |
| LCER | LCER(26:1) | 0.0097 | 0.4529 | 0.95 | 1.08 | 1.13 |

CE: cholesteryl ester; FFA: free fatty acid; TAG: triacylglycerol; LPC: lysophosphatidylcholine; PC: phosphatidylcholine; PE: phosphatidylethanolamine; DCER: dihydroceramide; LCER: lactosylceramide. * Shown are results from RMA that meet nominal significance ($p < 0.05$, column 3). Bold-faced lipid species meet significance threshold of $p < 0.05$ (column 3) and $q < 0.05$ (column 4). † neutral lipids: CE, FFA, TAG; phospholipids: LPC, PC, PE; sphingolipids: DCER, LCER. § ANOVA contrasts: lipid levels that increase in response to coffee are shaded red ($p < 0.05$) or pink ($0.05 < p < 0.10$) and lipid levels that decrease are colored green ($p < 0.05$) or light green ($0.05 < p < 0.10$).

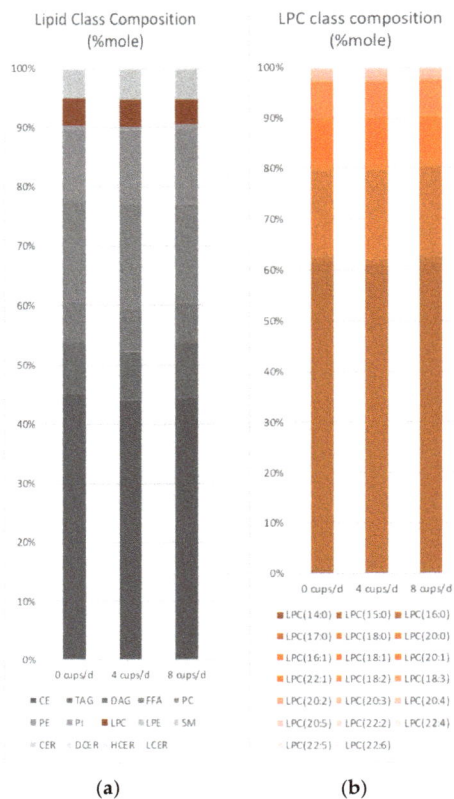

**Figure 1.** Lipid class (**a**) and LPC (**b**) composition response to coffee intake.

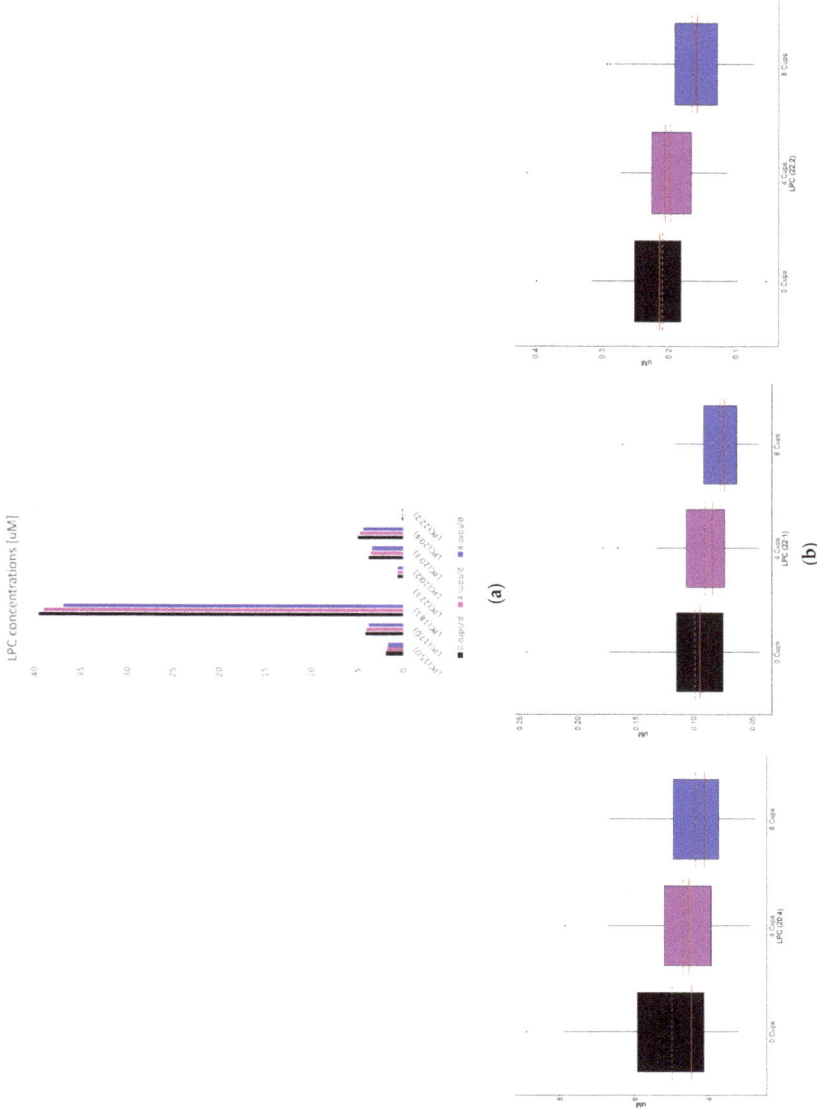

**Figure 2.** LPC concentration response to coffee intake. Shown are all nominally (**a**) to significant (**a**,**b**) LPCs that changed in response to coffee.

## 4. Discussion

The current study is the first controlled trial-based lipidomic assessment of coffee intake. We found three LPC species (LPC (20:4), LPC (22:1) and LPC (22:2)) that significantly decreased after 4 and 8 cups per day. Several other species mapping to the LPC, FFA, PC and CE classes showed nominal but plausible changes. Although the current lipid species analysis is unique from that of our previous metabolomic analysis [8] of the same samples the findings taken together suggest that coffee drinking has more of an immediate impact on non-lipid than lipid metabolites over the duration of the coffee trial examined here.

The lipidomics platform was unable to distinguish between fatty acid isoforms, their position on a glycerol backbone (i.e., sn-1 vs. sn-2) or define their bond type (acyl- or alkyl-). Several lipid species at least nominally associated with coffee response contained FA(20:4). In our previous metabolomics report [8], arachidonic acid (AA, 20:4n6) and LPC (20:4n6) were specifically measured and decreased in response to coffee ($p < 0.05$, $q < 0.05$ for AA and $p < 0.05$, $q > 0.05$ for LPC (20:4n6)). These findings, along with the correlation patterns among these lipid variables (Figure S3) suggest most contain the n6 form of FA(20:4). The only relevant isoforms for FA(22:1) and FA(22:2) are 22:1n9 (erucic acid) and 22:2n6 (docosadienoic acid), respectively.

Figure S4 shows the biological relationships among the neutral and phospholipid lipid classes measured in the current study. LPC is a bioactive phospholipid synthesized primarily from plasma membrane- and lipoprotein-PC by phospholipase A1 (PLA1) or PLA2 that cleave the PC sn-1 or sn-2 ester bond, respectively [14–16]. LPC can also be formed by lecithin cholesterol acyltransferase in HDL, from oxidation of LDL and by endothelial lipase. LPC transports glycerophospholipid components between tissues but is also a ligand for specific signaling receptors and activates several second messengers [17]. Much experimental data have implicated LPC in atherosclerosis and acute and chronic inflammation but results support both beneficial and adverse properties [18,19]. The conflicting biological properties of LPC might be due to their fatty acyl composition, with saturated or monounsaturated LPC presenting with greater pro-atherogenic properties than polyunsaturated LPCs [20–23]. In the current study, 18 of the 20 LPCs examined tended to decrease with coffee intake but none of these shared a particular fatty acyl composition pattern (i.e., saturated or polyunsaturated fatty acids) (Table 1 and data not shown). LPC(20:4n6) sn-2, a potential isoform of LPC (20:4), is particularly interesting because it intersects several metabolic pathways that lead to the production of potent signaling molecules such as 2-arachidonoyl-lysophosphatidic acid, and specific eicosanoids and endocannabinoids [24–26]. Metabolites of the endocannabinoid system as well as choline (a product of LPC metabolism) and glycerol-3-phosphate (a product of LPC metabolism and endocannabinoid synthesis) significantly decreased in response to coffee intake in our previous metabolomic study [8,18,27]. Interestingly, PLA2 also contributes to endocannabinoid synthesis [27]. Taken together, decreased LPC levels in response to coffee align with decreased levels of downstream metabolites in similar biological pathways, most notably glycerophospholipid metabolism.

Although the caffeine component of coffee is known to stimulate lipolysis in the acute setting [28–31], the mechanisms and constituents of habitual coffee drinking leading to decreased LPC in the current study are unclear. The resistance of LDL to oxidative damage (a source of LPC) in humans increases after consumption of coffee and this might be explained by the incorporation of coffee's phenolic acids into LDL [32]. Indeed, polyphenols, including those in coffee, decreased LPC production induced by oxidation [33].

Population-based or cross-sectional metabolomic/lipidomic studies of self-reported habitual coffee intake have also reported specific lipid species associated with coffee intake (Table S5). Direct comparison with the current report is difficult given the study designs, different lipidomic platforms used, and limitations in lipid species quantification (i.e., detected signals are usually a sum of several isobaric/isomeric lipids). Interestingly, however, Miranda et al. [34] focused exclusively on LPC species and found an inverse association between the plasma levels of LPC(16:1 a), LPC(18:1 a) and LPC(20:4 a) and habitual coffee intake, particularly when comparing intakes > 100 mL/day to 0 mL/day.

The uncertain biological implications of lower LPC levels also extends to human studies of diseases or conditions of which are potentially modified by coffee consumption (Table S6). Most are cross-sectional and include small sample sizes and generate some significant findings that are not confirmed in other studies. Applicable to the original motivation of the data from our coffee trial examined here is a recent meta-analysis of metabolite changes and risk of T2D [35]; only three lipid markers were significantly associated with risk: LPC(18:0), SM(16:0) and FFA(18:1) [35]. None of these lipid species changed in response to coffee in the current study.

All lipid species that potentially increased in response to coffee did so only after the period of 8 cups per day. These lipid species also tended to correlate among each other rather than with lipid species that decreased in response to coffee, suggesting distinct lipid pathways altered by low and high coffee intake. PC species were a notable exception. PC species that increased in response to 8 cups of coffee all contained FA(18:0) and their changes directly correlated with changes in LPCs, TAGs, and acylcholines that decreased in response to 4 and 8 cups. This might suggest a shared lipid pathway impacted by coffee intake and the increase in PC observed only after 8 cups per day is a delayed dose response.

In the initial report of the current coffee trial, several clinical lipid and inflammatory biomarkers changed in response to coffee [9]. None of these were convincingly correlated with lipids or metabolites measured in the current or recent report and underscores additional information accessible via high-throughput or more precise omic analysis. Triglycerides, for example, did not significantly change in the trial, yet when analyzing TAG species we observed TAGs that potentially increased and decreased in response to coffee. Nevertheless, a special complication in the analysis and clinical interpretation of TAGs is the large number of isobars resulting from presence of different combinations of the three acyl moieties and their regioisomers.

The application of lipidomics to a clinical study of coffee intake with repeated measures, large contrasts in coffee intake, excellent participant compliance and standardized protocols for sample handling and storage are major strengths of the current study. As a clinical trial it addresses many of the limitations of observational studies. In addition, the composition of brewed coffee varies as a function of bean type, roast and preparation methods; factors for which detailed information is rarely collected in population-based studies of coffee. Participants of our clinical trial were all provided the same coffee: a medium roast, 100% Arabica blend of Brazilian, Columbian, Central American and African coffee which is a popular type of coffee in Finland. Despite these strengths, several weaknesses of the study should be acknowledged. Our one-group study design without randomization, lack of blinding of participants and placebo control were limitations. We cannot rule out an impact of time-varying factors that may induce significant associations due to correlations with coffee. No specific guidelines were provided on coffee additives (i.e., sugar, cream) or beverages to consume in the place of coffee during the month of coffee abstinence. The very low levels of xanthine metabolites in the first month suggest participants largely refrained from consuming any caffeine-containing beverages [8,9]. Our previous report presented no obvious overlap with potential metabolite markers of dairy or tea consumption or lifestyle factors [8]. Body weight, a proxy for energy balance, remained stable throughout the trial. All participants for the current study were Finnish habitual coffee drinkers at increased risk of T2D which may limit the generalizability of our findings to other groups.

## 5. Conclusions

Our study provides the first thorough analysis of the lipidomic changes in response to controlled coffee consumption. The new findings suggest coffee alters glycerophospholipid metabolism and build on our previous metabolomic results that yield novel candidate pathways that offer insight to the mechanisms by which coffee may be exerting its health effects.

**Supplementary Materials:** The following are available online at http://www.mdpi.com/2072-6643/10/12/1851/s1, Supplementary Notes, Figure S1: PCA and multilevel PCA analysis, Figure S2: Serum concentrations of lipid species nominally associated with changes in response to coffee intake, Figure S3: Pearson correlations of changes in variable (lipid/metabolite/biomarker) levels after a) 4 cups/d compared to 0 cups/d b) 8 cups/d compared to 0 cups/d and c) 8 cups/d compared to 4 cups/d, Figure S4: Lipid pathway including neutral and phospholipid lipid classes measured in the current study, Figure S5: Multilevel PCA, Table S1: Baseline characteristics of coffee trial participants ($N = 47$), Table S2: Lipid species excluded from the current analysis, Table S3: Lipids analyzed in the current study, Table S4: Metabolite markers of coffee response reported by Cornelis et al, 2016, Table S5: Population-based lipidomic studies of habitual coffee consumption, Table S6: Circulating lysophosphatidylcholines and coffee-implicated disease or conditions.

**Author Contributions:** A.K. analyzed the data. I.E., C.H. and J.T. lead the coffee trial (Kempf et al., 2010) and provided samples for the current study. M.C.C. and J.A.W. supervised the statistical analysis. M.C.C. acquired the lipidomics data and was responsible for the current study concept, study design and final content. M.C.C. and A.K. wrote the paper. All authors critically revised for important intellectual content and approved the final manuscript.

**Funding:** This work was supported by the American Diabetes Association (ADA, 7-13-JF-15 to MCC). The original trial was supported by a grant from the Institute of Scientific Information on Coffee, which is a consortium of major European Coffee Companies (JT). The German Diabetes Center was supported by the Ministry of Culture and Science of the State of North Rhine-Westphalia (MKW NRW), the German Federal Ministry of Health (BMG) and in part by a grant from the German Federal Ministry of Education and Research (BMBF) to the German Center for Diabetes Research (DZD).

**Acknowledgments:** We thank Paulig Oy, Helsinki, Finland for the donation of coffee for this trial. Matlab computations in this paper were run on the Quest cluster supported in part through the computational resources and staff contributions provided for the Quest high performance computing facility at Northwestern University, which is jointly supported by the Office of the Provost, the Office for Research, and Northwestern University Information Technology.

**Conflicts of Interest:** The authors declare no conflict of interest

## References

1. Reyes, C.M.; Cornelis, M.C. Caffeine in the diet: Country-level consumption and guidelines. *Nutrients* **2018**, *10*, 1772. [CrossRef] [PubMed]

2. Cornelis, M. Gene-coffee interactions and health. *Curr. Nutr. Rep.* **2014**, *3*, 178–195. [CrossRef]

3. Higdon, J.V.; Frei, B. Coffee and health: A review of recent human research. *Crit. Rev. Food Sci. Nutr.* **2006**, *46*, 101–123. [CrossRef] [PubMed]

4. Cowan, T.E.; Palmnas, M.S.; Yang, J.; Bomhof, M.R.; Ardell, K.L.; Reimer, R.A.; Vogel, H.J.; Shearer, J. Chronic coffee consumption in the diet-induced obese rat: Impact on gut microbiota and serum metabolomics. *J. Nutr. Biochem.* **2014**, *25*, 489–495. [CrossRef] [PubMed]

5. Fredholm, B.B.; Battig, K.; Holmen, J.; Nehlig, A.; Zvartau, E.E. Actions of caffeine in the brain with special reference to factors that contribute to its widespread use. *Pharmacol. Rev.* **1999**, *51*, 83–133. [PubMed]

6. Gilbert, R.M. Caffeine consumption. In *The Methylxanthine Beverages and Foods: Chemistry, Consumption, and Health Effects*; Spiller, G.A., Ed.; Alan R. Liss Inc.: New York, NY, USA, 1984; pp. 185–213.

7. Cornelis, M.C.; Hu, F.B. Systems epidemiology: A new direction in nutrition and metabolic disease research. *Curr. Nutr. Rep.* **2013**, *2*. [CrossRef] [PubMed]

8. Cornelis, M.C.; Erlund, I.; Michelotti, G.A.; Herder, C.; Westerhuis, J.A.; Tuomilehto, J. Metabolomic response to coffee consumption: Application to a three-stage clinical trial. *J. Intern. Med.* **2018**, *283*, 544–557. [CrossRef] [PubMed]

9. Kempf, K.; Herder, C.; Erlund, I.; Kolb, H.; Martin, S.; Carstensen, M.; Koenig, W.; Sundvall, J.; Bidel, S.; Kuha, S.; et al. Effects of coffee consumption on subclinical inflammation and other risk factors for type 2 diabetes: A clinical trial. *Am. J. Clin. Nutr.* **2010**, *91*, 950–957. [CrossRef] [PubMed]

10. Stephenson, D.J.; Hoeferlin, L.A.; Chalfant, C.E. Lipidomics in translational research and the clinical significance of lipid-based biomarkers. *Transl. Res.* **2017**, *189*, 13–29. [CrossRef] [PubMed]

11. Farnell, D.J.; Popat, H.; Richmond, S. Multilevel principal component analysis (MPCA) in shape analysis: A feasibility study in medical and dental imaging. *Comput. Methods Programs Biomed.* **2016**, *129*, 149–159. [CrossRef] [PubMed]

12. Benjamini, Y.; Hochberg, Y. Controlling the false ciscovery rate: A practical and powerful approach to multiple testing. *J. R. Stat. Soc. Ser. B (Methodol.)* **1995**, *57*, 289–300.

13. Shannon, P.; Markiel, A.; Ozier, O.; Baliga, N.S.; Wang, J.T.; Ramage, D.; Amin, N.; Schwikowski, B.; Ideker, T. Cytoscape: A software environment for integrated models of biomolecular interaction networks. *Genome Res.* **2003**, *13*, 2498–2504. [CrossRef] [PubMed]

14. Dennis, E.A.; Cao, J.; Hsu, Y.-H.; Magrioti, V.; Kokotos, G. Phospholipase a2 enzymes: Physical structure, biological function, disease implication, chemical inhibition, and therapeutic intervention. *Chem. Rev.* **2011**, *111*, 6130–6185. [CrossRef] [PubMed]

15. Richmond, G.S.; Smith, T.K. Phospholipases a(1). *Int. J. Mol. Sci.* **2011**, *12*, 588–612. [CrossRef] [PubMed]

16. Yamashita, A.; Hayashi, Y.; Nemoto-Sasaki, Y.; Ito, M.; Oka, S.; Tanikawa, T.; Waku, K.; Sugiura, T. Acyltransferases and transacylases that determine the fatty acid composition of glycerolipids and the metabolism of bioactive lipid mediators in mammalian cells and model organisms. *Prog. Lipid Res.* **2014**, *53*, 18–81. [CrossRef] [PubMed]

17. Tomura, H.; Mogi, C.; Sato, K.; Okajima, F. Proton-sensing and lysolipid-sensitive g-protein-coupled receptors: A novel type of multi-functional receptors. *Cell Signal.* **2005**, *17*, 1466–1476. [CrossRef] [PubMed]

18. Schmitz, G.; Ruebsaamen, K. Metabolism and atherogenic disease association of lysophosphatidylcholine. *Atherosclerosis* **2010**, *208*, 10–18. [CrossRef] [PubMed]

19. Matsumoto, T.; Kobayashi, T.; Kamata, K. Role of lysophosphatidylcholine (LPC) in atherosclerosis. *Curr. Med. Chem.* **2007**, *14*, 3209–3220. [CrossRef] [PubMed]

20. Hung, N.D.; Sok, D.-E.; Kim, M.R. Prevention of 1-palmitoyl lysophosphatidylcholine-induced inflammation by polyunsaturated acyl lysophosphatidylcholine. *Inflamm. Res.* **2012**, *61*, 473–483. [CrossRef] [PubMed]

21. Akerele, O.; Cheema, S. Fatty acyl composition of lysophosphatidylcholine is important in atherosclerosis. *Med. Hypotheses* **2015**, *85*, 754–760. [CrossRef] [PubMed]

22. Ojala, P.; Hirvonen, T.; Hermansson, M.; Somerharju, P.; Parkkinen, J. Acyl chain-dependent effect of lysophosphatidylcholine on human neutrophils. *J. Leukoc. Boil.* **2007**, *82*, 1501–1509. [CrossRef] [PubMed]

23. Aiyar, N.; Disa, J.; Ao, Z.; Ju, H.; Nerurkar, S.; Willette, R.N.; Macphee, C.H.; Johns, D.G.; Douglas, S.A. Lysophosphatidylcholine induces inflammatory activation of human coronary artery smooth muscle cells. *Mol. Cell. Biochem.* **2007**, *295*, 113–120. [CrossRef] [PubMed]

24. Pete, M.J.; Exton, J.H. Purification of a lysophospholipase from bovine brain that selectively deacylates arachidonoyl-substituted lysophosphatidylcholine. *J. Boil. Chem.* **1996**, *271*, 18114–18121. [CrossRef]

25. Di Marzo, V. 'Endocannabinoids' and other fatty acid derivatives with cannabimimetic properties: Biochemistry and possible physiopathological relevance. *Biochim. Biophys. Acta (BBA)-Lipids Lipid Metab.* **1998**, *1392*, 153–175. [CrossRef]

26. Aoki, J.; Inoue, A.; Okudaira, S. Two pathways for lysophosphatidic acid production. *Biochim. Biophys. Acta (BBA)-Mol. Cell Boil. Lipids* **2008**, *1781*, 513–518. [CrossRef] [PubMed]

27. Maccarrone, M. Metabolism of the endocannabinoid anandamide: Open questions after 25 years. *Front. Mol. Neurosci.* **2017**, *10*, 166. [CrossRef] [PubMed]

28. Beaudoin, M.-S.; Robinson, L.E.; Graham, T.E. An oral lipid challenge and acute intake of caffeinated coffee additively decrease glucose tolerance in healthy men. *J. Nutr.* **2011**, *141*, 574–581. [CrossRef] [PubMed]

29. Mougios, V.; Ring, S.; Petridou, A.; Nikolaidis, M.G. Duration of coffee-and exercise-induced changes in the fatty acid profile of human serum. *J. Appl. Physiol.* **2003**, *94*, 476–484. [CrossRef] [PubMed]

30. Bellet, S.; Kershbaum, A.; Finck, E.M. Response of free fatty acids to coffee and caffeine. *Metabolism* **1968**, *17*, 702–707. [CrossRef]

31. Hodgson, A.B.; Randell, R.K.; Jeukendrup, A.E. The metabolic and performance effects of caffeine compared to coffee during endurance exercise. *PLoS ONE* **2013**, *8*, e59561. [CrossRef] [PubMed]

32. Natella, F.; Nardini, M.; Belelli, F.; Scaccini, C. Coffee drinking induces incorporation of phenolic acids into LDL and increases the resistance of LDL to ex vivo oxidation in humans. *Am. J. Clin. Nutr.* **2007**, *86*, 604–609. [CrossRef] [PubMed]

33. Cartron, E.; Carbonneau, M.-A.; Fouret, G.; Descomps, B.; Léger, C.L. Specific antioxidant activity of caffeoyl derivatives and other natural phenolic compounds: LDL protection against oxidation and decrease in the proinflammatory lysophosphatidylcholine production. *J. Nat. Prod.* **2001**, *64*, 480–486. [CrossRef] [PubMed]

34. Miranda, A.M.; Carioca, A.A.F.; Steluti, J.; da Silva, I.; Fisberg, R.M.; Marchioni, D.M. The effect of coffee intake on lysophosphatidylcholines: A targeted metabolomic approach. *Clin. Nutr. (Edinburgh Scotland)* **2017**, *36*, 1635–1641. [CrossRef] [PubMed]

35. Park, J.-E.; Lim, H.R.; Kim, J.W.; Shin, K.-H. Metabolite changes in risk of type 2 diabetes mellitus in cohort studies: A systematic review and meta-analysis. *Diabetes Res. Clin. Pract.* **2018**, *140*, 216–227. [CrossRef] [PubMed]

*nutrients*

MDPI

*Article*

# Association of Coffee Consumption with Hearing and Tinnitus Based on a National Population-Based Survey

Sang-Youp Lee [1], Gucheol Jung [2], Myoung-jin Jang [2], Myung-Whan Suh [1], Jun Ho Lee [1,3], Seung Ha Oh [1,3] and Moo Kyun Park [1,3,*]

[1] Department of Otorhinolaryngology-Head and Neck Surgery, Seoul National University College of Medicine, Seoul 03080, Korea; lsy738@hanmail.net (S.-Y.L.); drmung@naver.com (M.-W.S.); junlee@snu.ac.kr (J.H.L.); shaoh@snu.ac.kr (S.H.O.)
[2] Medical Research Collaborating Center, Seoul National University Hospital, Seoul 03080, Korea; 9E887@snuh.org (G.J.); mjjang2014@naver.com (M.-j.J.)
[3] Sensory Organ Research Institute, Seoul National University Medical Research Center Seoul, Seoul 03080, Korea
* Correspondence: aseptic@snu.ac.kr or entpmk@gmail.com; Tel.: +82-2-2072-2446; Fax: +82-2-745-2387

Received: 21 August 2018; Accepted: 30 September 2018; Published: 4 October 2018

**Abstract:** Coffee is the one of the most common beverages worldwide and has received considerable attention for its beneficial health effects. However, the association of coffee with hearing and tinnitus has not been well studied. The aim of this study was to investigate the association of coffee with hearing and tinnitus based on a national population-based survey. We evaluated hearing and tinnitus data from the 2009–2012 Korean National Health and Nutrition Examination Survey and their relationship with a coffee consumption survey. All patients underwent a medical interview, physical examination, hearing test, tinnitus questionnaire and nutrition examination. Multivariable logistic regression models were used to examine the associations between coffee and hearing loss or tinnitus. We evaluated 13,448 participants (≥19 years) participants. The frequency of coffee consumption had a statistically significant inverse correlation with bilateral hearing loss in the 40–64 years age group. Daily coffee consumers had 50–70% less hearing loss than rare coffee consumers, which tended to be a dose-dependent relationship. In addition, the frequency of coffee consumption had an inverse correlation with tinnitus in the 19–64 years age group but its association was related with hearing. Brewed coffee had more of an association than instant or canned coffee in the 40–64 years age group. These results suggest a protective effect of coffee on hearing loss and tinnitus.

**Keywords:** adult; coffee; hearing; protection; tinnitus

---

## 1. Introduction

Coffee is the most commonly consumed beverage, apart from water, in the world [1]. Coffee and its compounds have various effects on human health [1,2]. Coffee consumption has been associated with a decreased risk of cancers [3], diabetes [4], Parkinson's disease [5], liver disease [6] and cardiovascular disease [7]. It has also been associated with low birth weight, preterm birth [8] and fractures in women [9,10]. Dementia and depression are related to coffee consumption [11,12]. However, its effect could be related to coffee consumption behavior, social relationships and culture. The health effects of coffee have not received much consideration. Actually, about half of consumers believe drinking coffee is bad for their health [13].

Coffee contains over 1000 bioactive compounds [14,15] with functions, including antioxidant, anti-inflammatory, anti-fibrotic and anticancer effects [1]. In addition, coffee contains polyphenols, such as caffeic acid and caffeic acid phenethyl ester, which have antioxidant effects and protect against hearing loss in vivo and in vitro [16,17]. Caffeine is an important component of coffee that varies according to the preparation method [1].

Hearing loss is a major public health problem [18,19]. One-fifth of adults suffer from hearing loss if mild and unilateral hearing loss are included [20]. Hearing loss is third in terms of disease burden [21]. Common causes of hearing loss are age and noise exposure. Hearing loss affects communication and relationships with people. In particular, it affects talking and has been associated with depression and anxiety [22]. Hearing loss is associated with decreased cognitive performance and dementia [23].

Tinnitus is not a single disease entity. Actually, it is a symptom that decreases quality of life and is related with hearing loss and aging. The prevalence of tinnitus is 12–30% worldwide [24,25]. In addition, coffee and caffeine are often blamed as a cause of tinnitus [26,27]. However, the effect of caffeine on tinnitus remains controversial [28,29]. Few large population-based studies have investigated the effect of coffee consumption on hearing and tinnitus [30].

The aim of this study was to investigate the association of coffee with hearing and tinnitus in adult and elderly participants based on a national population-based survey. We compared the consumption frequency and type of coffee and the prevalence of hearing loss and tinnitus.

## 2. Materials and Methods

### 2.1. Study Population

This study used data from the Korean National Health and Nutrition Examination Survey (KNHANES). This survey collects information, such as health and nutritional status, from a representative sample of the general Korean population to assess the health-related behavior, health condition and nutritional state of Koreans.

The subjects were asked about their hearing, symptoms of tinnitus, health behavior and nutrition by questionnaire. The participants were asked about the annoyance of tinnitus measured by the following answers: "No," "slightly annoying" and "very annoying and difficult to sleep." Information about the subjects included sleep time, stress, education level (less than middle school or beyond high school), education level of the parents, income (<25%, 25–50%, 50–75%, or >75% according to the equivalized household income per month), current smoking status and alcohol drinking status (social drinker, heavy drinker, or problem drinker). Health status (hypertension, diabetes, anemia, renal failure, thyroid disease, osteoporosis and menopause) was also checked. Duration of occupational exposure to noise and earphone and headphone use time were measured.

Physical examinations were conducted by a physician to assess any problems with the tympanic membrane or other ear, nose and throat problems, including perforation or retraction of the tympanic membrane, otitis media with effusion and cholesteatoma. Pure tone audiometry was performed at 0.5, 1, 2, 3, 4 and 6 kHz in a soundproof room. The severity of hearing loss was based on a lower threshold of unilateral hearing loss and a higher threshold for bilateral hearing loss. The pure tone average was the average of the hearing levels at 0.5, 1, 2 and 3 kHz or 0.5, 1, 2 and 4 kHz, whereas the high frequency hearing level was the average of the hearing levels at 3, 4 and 6 kHz. Blood samples were collected and analyzed in a single laboratory (Neodin Medical Institute, Seoul, Korea).

In total, 36,067 individuals participated in the 2009–2012 KNHANES. Individuals with ear disease (external ear problem, middle ear problem, inner ear problem, retrocochlear problem, congenital hearing loss and systemic disease) were not included here. Of them, 27,492 participants were age ≥19 years. Among 27,492 participants aged ≥19, 9294 participants were excluded because they did not complete all three component surveys (health interview, health examination, and nutrition surveys) (n = 4480) or examined from January 1 to July 20 in 2009 (n = 3299, auditory test data were not available) or aged ≥ 65 in 2012 (n = 1515, FFQ was surveyed for subjects aged 19–64 years in 2012). Of the remaining 18,198 subjects, additional 4750 were excluded because they did not receive hearing threshold testing nor respond to tinnitus-related questions (n = 975) or did not respond for coffee consumption frequency (n = 629) or have missing values for covariates considered in this study (n = 3146). Finally, 13,448 subjects (4633 subjects aged 19–39, 6631 aged 19–39 and 2184 aged ≥65 years) were included in the analysis for the present study (Figure 1). This study was approved by the Institutional Review Board of the Seoul National University Hospital (IRB number: E-1808-064-965).

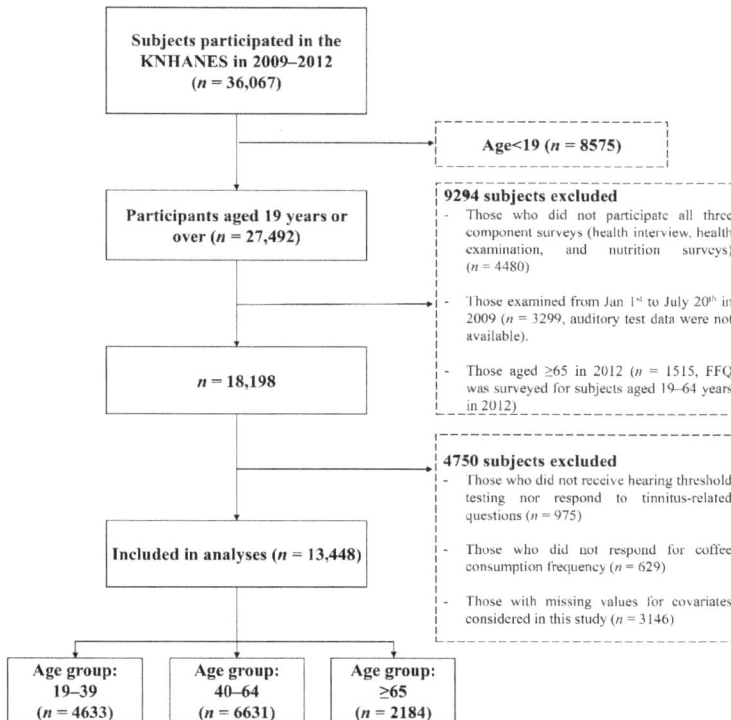

**Figure 1.** Flow chart of the selection process. KNHANES, Korean National Health and Nutrition Examination Survey.

## 2.2. Assessment of Coffee Consumption

Coffee consumption frequency was assessed using the food-frequency questionnaire (FFQ). Participants were asked to indicate how frequently they consumed coffee over the previous year based on ten categories (none, 6–11 times per year, once per month, two to three times per month, once per week, two to three times per week, four to six times per week, once per day, twice per day and three times per day) in 2009–2011 and on nine categories (never or seldom, once per month, two to three times per month, once per week, two to four times per week, five to six times per week, once per day, twice per day and three times per day) in 2012, in which the first two categories in the previous FFQ version were combined into "never or seldom" and four times per week was grouped into two or three times per week.

Coffee consumption frequency was categorized into rarely, monthly, weekly and daily using the FFQ data as follows: rarely, less than once per month; monthly, one to three times per month; weekly, one to six times per week; and daily, once or more per day.

The information on the types and amount of all coffee that participants consumed over the past 24 h was collected by trained dietitians 1 week after the health interview. The type of coffee was grouped into brewed, instant, or canned coffee using the 24-hour dietary recall method.

## 2.3. Statistical Analysis

The subjects' characteristics according to coffee consumption frequency are presented as median (interquartile range) or number (proportion) and compared using the Fisher exact test (binary covariates), the chi-square test (more than three categories), or the Wilcoxon rank-sum test (continuous covariates). Multivariable logistic regression models were used to examine the associations between coffee consumption and hearing loss or tinnitus. The analyses were adjusted for the following potential confounders: age, sex, education, parents' education, perceived stress, exposure to indoor secondhand smoke, current smoking, heavy drinking, drinking-related problems, menopause, history of hypertension, diabetes mellitus, anemia, kidney failure, thyroid disorder, tympanic membrane perforation, cholesteatoma and otitis media with effusion. The multivariable models for tinnitus or annoyance related to tinnitus included hearing loss as well as the potential covariates described above. To examine the association of the type of coffee consumed with hearing loss and tinnitus, multivariable logistic regression analyses were performed for the adjusted associations between coffee type consumed and hearing loss or tinnitus. All statistical analyses were performed using SAS software (version 9.2; SAS Institute, Cary, NC, USA).

## 3. Results

The prevalence rates of unilateral and bilateral hearing loss in the study population were 1.19% and 0.17% for subjects in the 19–39 years age group, 5.01% and 2.9% for subjects in the 40–64 years age group and 14.24% and 20.97% for subjects in the ≥65 years age group. The prevalence rates of tinnitus and tinnitus-related annoyance were 18.07% and 3.86% for subjects in the 19–39 years age group, 19.92% and 6.24% for subjects in the 40–64 years age group and 27.98% and 12.82% for subjects in the ≥65 years age group (Table 1). The participants' characteristics according to age group and the frequency of coffee consumption showed that there were differences in the covariates according to coffee consumption: age, sex, educational level, house income, sleeping duration, stress, exposure to indoor secondhand smoke, current smoking, heavy drinking, difficulties controlling alcohol use, menopause, hypertension, diabetes, kidney failure and thyroid disorder (Table S1).

**Table 1.** Characteristics of study subjects in Korean National Health and Nutrition Examination Survey KNHANES (2009–2012) by frequency of coffee consumption.

| Symptoms by Group | Frequency of Coffee Consumption | | | | |
|---|---|---|---|---|---|
| | Total | Rarely | Monthly | Weekly | Daily |
| Age group (19–39) | *n* = 4633 | *n* = 634 | *n* = 304 | *n* = 873 | *n* = 2822 |
| Hearing loss *, *n* (%) | | | | | |
| Unilateral | 55 (1.19%) | 11 (1.74%) | 2 (0.66%) | 6 (0.69%) | 36 (1.28%) |
| Bilateral | 8 (0.17%) | 1 (0.16%) | 0 (0.00%) | 3 (0.34%) | 4 (0.14%) |
| Tinnitus, *n* (%) | 837 (18.07%) | 130 (20.50%) | 62 (20.39%) | 177 (20.27%) | 468 (16.58%) |
| Tinnitus-related annoyance, *n* (%) | 179 (3.86%) | 33 (5.21%) | 13 (4.28%) | 26 (2.98%) | 107 (3.79%) |
| Age group (40–64) | *n* = 6631 | *n* = 656 | *n* = 308 | *n* = 899 | *n* = 4768 |
| Hearing loss *, *n* (%) | | | | | |
| Unilateral | 332 (5.01%) | 40 (6.1%) | 15 (4.87%) | 47 (5.23%) | 230 (4.82%) |
| Bilateral | 192 (2.90%) | 31 (4.73%) | 18 (5.84%) | 33 (3.67%) | 110 (2.31%) |
| Tinnitus, *n* (%) | 1321 (19.92%) | 149 (22.71%) | 64 (20.78%) | 191 (21.25%) | 917 (19.23%) |
| Tinnitus-related annoyance, *n* (%) | 414 (6.24%) | 50 (7.62%) | 18 (5.84%) | 66 (7.34%) | 280 (5.87%) |
| Age group (≥65) | *n* = 2184 | *n* = 429 | *n* = 122 | *n* = 383 | *n* = 1250 |
| Hearing loss *, *n* (%) | | | | | |
| Unilateral | 311 (14.24%) | 71 (16.55%) | 18 (14.75%) | 58 (15.14%) | 164 (13.12%) |
| Bilateral | 458 (20.97%) | 99 (23.08%) | 20 (16.39%) | 79 (20.63%) | 260 (20.80%) |
| Tinnitus, *n* (%) | 611 (27.98%) | 135 (31.47%) | 38 (31.15%) | 95 (24.80%) | 343 (27.44%) |
| Tinnitus-related annoyance, *n* (%) | 280 (12.82%) | 73 (17.02%) | 18 (14.75%) | 47 (12.27%) | 142 (11.36%) |

* Hearing loss ≥ 41 dB for four-frequency average of pure-tone thresholds at 500, 1000, 2000 and 4000 Hz.

Table 2 shows the results of the association between coffee consumption and hearing loss. No significant correlation was detected between coffee consumption frequency and unilateral hearing loss across all age groups. No significant correlation was detected between bilateral hearing loss and coffee consumption frequency in the 19–39 and ≥65 years age groups. However, daily coffee consumption resulted in a significantly decreased risk of bilateral hearing loss in the 40–64 years age group, compared with the rare consumption group (adjusted odds ratio (aOR), 0.50; 95% confidence interval (CI), 0.33–0.78; $p$ = 0.0021), whereas monthly or weekly consumers did not show a significant difference relative to rare consumers. In the 40–64 years age group, odds ratio of mild and moderate hearing loss in daily coffee consumers and mild hearing loss in weekly coffee consumers were significantly lower than those of rare coffee consumers (Table S7). In addition, as the frequency of coffee consumption increased there tended to be a decrease in bilateral hearing loss in the 40–64 years age group.

Table 3 shows the results of the association between coffee consumption and tinnitus and tinnitus-related annoyance. In the univariable analysis, the prevalence of tinnitus in daily coffee consumers was lower than that in the rare coffee consumers in the 19–39 years (unadjusted OR, 0.77; 95% CI, 0.62–0.96; $p$ = 0.0186) and 40–64 years (unadjusted odds ratio (OR), 0.81; 95% CI, 0.67–0.99; $p$ = 0.0357) age groups. However, in the multivariable models adjusted for potential confounders, the relationships between daily coffee consumers and rare consumers were not significant in the 19–39 year (aOR, 0.80; 95% CI, 0.63–1.00; $p$ = 0.0548) and the 40–64 years age groups (aOR, 0.90; 95% CI, 0.73–1.10; $p$ = 0.3066). An inverse association was observed between tinnitus-related annoyance and coffee consumption in weekly coffee consumers aged 19–39 years (unadjusted OR, 0.56; 95% CI, 0.33-0.95; $p$ = 0.0298) and daily coffee consumers aged ≥ 65 years (unadjusted OR, 0.63; 95% CI, 0.46–0.85; $p$ = 0.0026). However, the associations in weekly coffee consumers aged 19–39 (aOR, 0.58; 95% CI, 0.34–1.01; $p$ = 0.0529) and daily coffee consumers aged ≥ 65 years (aOR, 0.77; 95% CI, 0.54–1.09; $p$ = 0.1355) were not significant in the multivariable analysis.

We investigated associations between types of coffee and hearing loss, tinnitus and tinnitus-related annoyance using multivariable analysis. Table 4 shows "adjusted" odds ratio of hearing loss, tinnitus and tinnitus-related annoyance for three types of coffee. The odds of unilateral hearing loss or tinnitus-related annoyance did not reach statistical significance for all age groups. However, the odds ratio of bilateral hearing loss for brewed coffee in 40–64 years age group is significantly lower than 1 (aOR, 0.61; 95% CI, 0.44–0.84; $p$ = 0.0028). And the odds ratio of tinnitus for brewed coffee in 19–39 years age group is significantly lower than 1 (aOR, 0.82; 95% CI, 0.70–0.97; $p$ = 0.0175). Contrary, the odds ratio of tinnitus for canned coffee in 40–64 years age group is significantly higher than 1 (aOR, 1.49; 95% CI, 1.17–1.90; $p$ = 0.0011).

Table 2. Odds ratios and 95% confidence intervals of coffee consumption for hearing loss.

| Frequency of Coffee Consumption | Univariable Analysis | | | | Multivariable Analysis * | | | |
|---|---|---|---|---|---|---|---|---|
| | Hearing Loss (Unilateral) | | Hearing Loss (Bilateral) | | Hearing Loss (Unilateral) | | Hearing Loss (Bilateral) | |
| | OR (95% CI) | p-Value | OR (95% CI) | p-Value | OR (95% CI) | p-Value | OR (95% CI) | p-Value |
| **Age 19–39** | | | | | | | | |
| Rarely | Reference | | Reference | | Reference | | Reference | |
| Monthly | 0.38 (0.08, 1.70) | 0.2039 | 0.69 (0.03, 17.15) | 0.8231 | 0.42 (0.09, 1.99) | 0.2768 | 0.59 (0.06, 6.35) | 0.6656 |
| Weekly | 0.39 (0.14, 1.07) | 0.0664 | 1.70 (0.25, 11.55) | 0.5880 | 0.45 (0.16, 1.28) | 0.1338 | 1.53 (0.37, 6.32) | 0.5607 |
| Daily | 0.73 (0.37, 1.45) | 0.3688 | 0.67 (0.11, 4.29) | 0.6765 | 0.76 (0.36, 1.59) | 0.4595 | 0.74 (0.19, 2.94) | 0.6715 |
| **Age 40–64** | | | | | | | | |
| Rarely | Reference | | Reference | | Reference | | Reference | |
| Monthly | 0.79 (0.43, 1.45) | 0.4442 | 1.25 (0.69, 2.28) | 0.4604 | 0.80 (0.42, 1.52) | 0.4891 | 1.19 (0.62, 2.26) | 0.6051 |
| Weekly | 0.85 (0.55, 1.31) | 0.4611 | 0.77 (0.47, 1.27) | 0.3023 | 0.86 (0.54, 1.36) | 0.5141 | 0.74 (0.43, 1.26) | 0.2624 |
| Daily | 0.78 (0.55, 1.10) | 0.1603 | 0.48 (0.32, 0.72) | 0.0004 | 0.85 (0.59, 1.23) | 0.3865 | 0.50 (0.33, 0.78) | 0.0021 |
| **Age ≥ 65** | | | | | | | | |
| Rarely | Reference | | Reference | | Reference | | Reference | |
| Monthly | 0.87 (0.50, 1.53) | 0.6344 | 0.65 (0.39, 1.11) | 0.1155 | 0.89 (0.48, 1.66) | 0.7070 | 0.72 (0.40, 1.29) | 0.2720 |
| Weekly | 0.90 (0.62, 1.31) | 0.5841 | 0.87 (0.62, 1.21) | 0.3997 | 0.97 (0.64, 1.47) | 0.8765 | 0.90 (0.62, 1.31) | 0.5953 |
| Daily | 0.76 (0.56, 1.03) | 0.0778 | 0.88 (0.67, 1.14) | 0.3212 | 0.85 (0.60, 1.19) | 0.3360 | 0.84 (0.62, 1.14) | 0.2669 |

* Adjusted for age, sex, education, parents' education, perceived stress, exposure to indoor secondhand smoke, current smoking, heavy drinking, drinking-related problem, menopause, history of hypertension, diabetes mellitus, anemia, kidney failure, thyroid disorder, tympanic membrane perforation, cholesteatoma and otitis media with effusion. OR, odds ratio. CI, confidence interval.

Table 3. Odds ratios and 95% confidence intervals of coffee consumption for tinnitus.

| Frequency of Coffee Consumption | Univariable Analysis | | | | Multivariable Analysis * | | | |
|---|---|---|---|---|---|---|---|---|
| | Tinnitus | | Tinnitus-Related Annoyance | | Tinnitus | | Tinnitus-Related Annoyance | |
| | OR (95% CI) | p-Value | OR (95% CI) | p-Value | OR (95% CI) | p-Value | OR (95% CI) | p-Value |
| **Age 19–39** | | | | | | | | |
| Rarely | Reference | | Reference | | Reference | | Reference | |
| Monthly | 0.99 (0.71, 1.39) | 0.9688 | 0.81 (0.42, 1.57) | 0.5382 | 1.09 (0.77, 1.54) | 0.6366 | 0.94 (0.48, 1.84) | 0.8541 |
| Weekly | 0.99 (0.77, 1.27) | 0.9129 | 0.56 (0.33, 0.95) | 0.0298 | 1.05 (0.81, 1.37) | 0.7013 | 0.58 (0.34, 1.01) | 0.0529 |
| Daily | 0.77 (0.62, 0.96) | 0.0186 | 0.72 (0.48, 1.07) | 0.1043 | 0.80 (0.63, 1.00) | 0.0548 | 0.76 (0.50, 1.16) | 0.2035 |
| **Age 40–64** | | | | | | | | |
| Rarely | Reference | | Reference | | Reference | | Reference | |
| Monthly | 0.89 (0.64, 1.24) | 0.4998 | 0.75 (0.43, 1.31) | 0.3159 | 0.92 (0.65, 1.29) | 0.6210 | 0.75 (0.42, 1.35) | 0.3376 |
| Weekly | 0.92 (0.72, 1.17) | 0.4892 | 0.96 (0.66, 1.41) | 0.8349 | 0.97 (0.75, 1.25) | 0.8249 | 1.07 (0.72, 1.60) | 0.7322 |
| Daily | 0.81 (0.67, 0.99) | 0.0357 | 0.76 (0.55, 1.03) | 0.0795 | 0.90 (0.73, 1.10) | 0.3066 | 0.92 (0.66, 1.29) | 0.6464 |
| **Age ≥ 65** | | | | | | | | |
| Rarely | Reference | | Reference | | Reference | | Reference | |
| Monthly | 0.99 (0.64, 1.52) | 0.9463 | 0.84 (0.48, 1.48) | 0.5530 | 1.02 (0.63, 1.65) | 0.9417 | 0.94 (0.49, 1.78) | 0.8446 |
| Weekly | 0.72 (0.53, 0.98) | 0.0357 | 0.68 (0.46, 1.01) | 0.0582 | 0.71 (0.51, 1.00) | 0.0514 | 0.74 (0.48, 1.16) | 0.1881 |
| Daily | 0.82 (0.65, 1.05) | 0.1109 | 0.63 (0.46, 0.85) | 0.0026 | 0.95 (0.72, 1.24) | 0.6899 | 0.77 (0.54, 1.09) | 0.1355 |

* Adjusted for age, sex, education, parents' education, perceived stress, exposure to indoor secondhand smoke, current smoking, heavy drinking, drinking-related problem, menopause, history of hypertension, diabetes mellitus, anemia, kidney failure, thyroid disorder, tympanic membrane perforation, cholesteatoma, otitis media with effusion, and hearing loss.

**Table 4.** Odds ratios and 95% confidence intervals of coffee type.

| Coffee Type | Hearing Loss (Unilateral) | | Hearing Loss (Bilateral) | |
|---|---|---|---|---|
| | OR * (95% CI) | *p*-Value | OR * (95% CI) | *p*-Value |
| Age group: 19–39 | | | | |
| Brewed coffee (yes vs. no) | 0.95 (0.53, 1.69) | 0.8599 | 0.46 (0.16, 1.27) | 0.1333 |
| Instant coffee (yes vs. no) | 0.65 (0.22, 1.91) | 0.4369 | 0.64 (0.08, 4.88) | 0.6683 |
| Canned coffee (yes vs. no) | 1.02 (0.39, 2.64) | 0.9752 | 1.28 (0.32, 5.08) | 0.7223 |
| Age group: 40–64 | | | | |
| Brewed coffee (yes vs. no) | 1.04 (0.81, 1.34) | 0.7414 | 0.61 (0.44, 0.84) | 0.0028 |
| Instant coffee (yes vs. no) | 1.11 (0.81, 1.51) | 0.5216 | 0.70 (0.45, 1.09) | 0.1133 |
| Canned coffee (yes vs. no) | 1.41 (0.86, 2.30) | 0.1694 | 0.63 (0.23, 1.73) | 0.3672 |
| Age group: ≥65 | | | | |
| Brewed coffee (yes vs. no) | 0.84 (0.64, 1.10) | 0.2083 | 1.02 (0.80, 1.30) | 0.8886 |
| Instant coffee (yes vs. no) | 0.87 (0.60, 1.26) | 0.4472 | 0.74 (0.53, 1.04) | 0.0842 |
| Canned coffee (yes vs. no) | 1.38 (0.51, 3.74) | 0.5233 | 0.89 (0.30, 2.64) | 0.8294 |
| | Tinnitus | | Tinnitus-Related Annoyance | |
| | OR † (95% CI) | *p*-Value | OR † (95% CI) | *p*-Value |
| Age group: 19–39 | | | | |
| Brewed coffee (yes vs. no) | 0.82 (0.70, 0.97) | 0.0175 | 1.09 (0.80, 1.50) | 0.5821 |
| Instant coffee (yes vs. no) | 1.08 (0.84, 1.41) | 0.5439 | 1.04 (0.62, 1.74) | 0.8812 |
| Canned coffee (yes vs. no) | 0.95 (0.74, 1.23) | 0.7163 | 1.13 (0.68, 1.86) | 0.6454 |
| Age group: 40–64 | | | | |
| Brewed coffee (yes vs. no) | 1.00 (0.87, 1.14) | 0.9359 | 0.96 (0.77, 1.19) | 0.6790 |
| Instant coffee (yes vs. no) | 0.96 (0.81, 1.14) | 0.6779 | 0.96 (0.72, 1.27) | 0.7660 |
| Canned coffee (yes vs. no) | 1.49 (1.17, 1.90) | 0.0011 | 1.28 (0.83, 1.95) | 0.2619 |
| Age group: ≥65 | | | | |
| Brewed coffee (yes vs. no) | 0.92 (0.74, 1.14) | 0.4318 | 0.89 (0.66, 1.20) | 0.4441 |
| Instant coffee (yes vs. no) | 1.19 (0.90, 1.57) | 0.2263 | 0.95 (0.64, 1.42) | 0.7967 |
| Canned coffee (yes vs. no) | 0.85 (0.37, 1.97) | 0.7051 | 0.63 (0.14, 2.74) | 0.5328 |

* Adjusted for age, sex, education, parents' education, perceived stress, exposure to indoor secondhand smoke, current smoking, heavy drinking, drinking-related problem, menopause, history of hypertension, diabetes mellitus, anemia, kidney failure, thyroid disorder, tympanic membrane perforation, cholesteatoma and otitis media with effusion. † Adjusted for age, sex, education, parents' education, perceived stress, exposure to indoor secondhand smoke, current smoking, heavy drinking, drinking-related problem, menopause, history of hypertension, diabetes mellitus, anemia, kidney failure, thyroid disorder, tympanic membrane perforation, cholesteatoma, otitis media with effusion and hearing loss.

## 4. Discussion

This study demonstrated the inverse correlation of the frequency of coffee consumption with hearing loss in middle aged Koreans. The prevalence of bilateral hearing loss in daily coffee consumers was significantly lower in the 40–64 years age group compared to the other age groups. However, no significant correlation was observed between unilateral hearing loss and coffee consumption in any other age group. Tinnitus and tinnitus-related annoyance were not related with coffee consumption. Instant coffee consumers aged 40–64 years had less hearing loss and less tinnitus-related annoyance than those in the ≥65 years age group and for any type of coffee.

No previous large-scale study has demonstrated the effects of coffee on hearing loss and tinnitus. Many people believe that coffee has a harmful effect on hearing. Actually, there is a report that caffeine in coffee has detrimental effects on recovery from acoustic overstimulation events [31,32]. Caffeine, a major ingredient of coffee, is proved to be an aggravating factor of Meniere's disease [33]. Many otologists prescribe a coffee restriction for cases of hearing loss from Meniere's disease [34]. Some studies have indicated that coffee has preventive effects on hearing loss but most studies were conducted on a specific type of hearing loss or situation. Chang et al. reported that noise-induced hearing loss in workers is less frequent in tea or coffee consumers (OR = 0.03,

95% CI: 0.01–0.51) [35]. Caffeine improves transmission in the peripheral and central brain auditory pathways [36]. Caffeine improves auditory processing in preterm infants, resulting in improved neurodevelopmental outcomes [37]. Hong et al. reported that coffee ameliorates the hearing threshold shift and delayed latency of auditory evoked potentials in patients with diabetic neuropathy [38]. Coffee improves auditory neuropathy in diabetic mice. In addition, trigonelline—the main active compound in coffee extracts—facilitates recovery from pyridoxine-induced auditory neuropathy in a mouse model [39].

Our univariable analyses determined that the prevalence of tinnitus in daily coffee consumers in the 19–39 and 40–64 years age groups was lower than that in rare coffee consumers of the same age groups. The prevalence of tinnitus in weekly coffee consumers age ≥65 years was lower than that in rare coffee consumers in the same age group. However, this tendency disappeared in the multivariable analysis, suggesting that coffee consumption itself does not have a direct correlation with tinnitus. Some covariates, such as bilateral hearing loss, stress and sleep, can indirectly affect tinnitus.

The relationship between coffee and tinnitus is controversial. There is an opinion that caffeine in coffee stimulates ascending auditory pathways or reduces the suppressive effect on the central nervous system, which evokes tinnitus [40]. Other studies have argued that the stimulation increases the detection of tinnitus through increased arousal or anxiety [41]. Caffeine in coffee is known to have deleterious effects on sleep [42,43] and it can aggravate tinnitus-associated distress [44]. However, one study reported that stopping caffeine intake does not improve tinnitus symptoms [29]. Another study reported that higher caffeine intake is associated with a lower risk of tinnitus in women [45]. McComack et al. reported that persistent tinnitus decreases with caffeinated coffee consumption (OR = 0.99 per cup/day) and consumption of caffeinated coffee appears to be associated with lower levels of reported transient tinnitus [30].

The relationship between the frequency of coffee consumption and the occurrence of tinnitus in the 19–39 and ≥65 years age groups appeared to be quite different from that of the 40–64 years age group. The abatement of tinnitus in the 40–64 years age group can be explained by a decrease in bilateral hearing loss. However, no significant decreases in bilateral hearing loss were observed in the 19–39 years and ≥65 years age groups. One of the covariates, such as perceived stress, can reduce tinnitus. In fact, an inverse correlation has been reported between perceived stress and tinnitus [46–48]. Coffee consumption is associated with social activity; thus, it is highly probable that socially active people have a relatively lower level of perceived stress and low stress can lower the incidence of tinnitus.

Types of coffee have association with hearing loss and tinnitus. Our results suggest that brewed coffee can have preventive effects on bilateral hearing loss and tinnitus but canned coffee can have inducing effect on tinnitus for some age groups. Difference in preparation method, heat treatment (freeze-drying or high temperature sterilization) and expiration date seem to have affected on bioactive constituents in each type of coffee [49,50]. However, since the details of coffee type are very diverse, it is difficult to make uniform conclusion.

We have some limitations in this study. First, this is an observational study, so it is difficult to generalize the result of this study to the causal relationship from coffee consumption to hearing loss and tinnitus. To confirm causality, well-controlled experimental design will be needed. Second, we could not analyze unilateral and bilateral tinnitus separately, because KNHANES data did not discern tinnitus side. However, bilateral tinnitus is different from unilateral one in the point of heritability and prognosis [51]. Therefore, it is desirable to discern the side of tinnitus. Third, the frequency of coffee consumption was estimated by questionnaires of subjects. It depends on the memory of the subjects, so it is possible that there is a measurement error.

## 5. Conclusions

According to the results of KNHANES analyses, coffee consumers had a low prevalence of bilateral hearing loss. However, the path of lower bilateral hearing loss and tinnitus varied according to age group. The incidence of bilateral hearing loss was low in coffee consumers aged 40–64 years, which influenced the low prevalence of tinnitus. However, other covariates, such as the low perceived stress of coffee consumers, seemed to be the main cause for the low prevalence of tinnitus in the 19–39 and ≥65 years age groups. In addition, brewed coffee consumers had lower rate of bilateral hearing loss and tinnitus.

**Supplementary Materials:** The following are available online at http://www.mdpi.com/2072-6643/10/10/1429/s1, Table S1: Characteristics of study subjects in KNHANES (2009–2012) by coffee consumption and age group (19–39), Table S2: Characteristics of study subjects in KNHANES (2009–2012) by coffee consumption and age group (40–64), Table S3: Characteristics of study subjects in KNHANES (2009–2012) by coffee consumption and age group (≥65), Table S4: Degree of unilateral hearing loss by coffee consumption, Table S5: Degree of bilateral hearing loss by coffee consumption, Table S6: Odds ratio and 95% confidence intervals by coffee consumption for degree of unilateral hearing loss, Table S7: Odds ratio and 95% confidence intervals by coffee consumption for degree of bilateral hearing loss.

**Author Contributions:** Conceptualization, M.K.P.; Investigation, M.K.P., M.-j.J., S.-Y.L., G.J.; Methodology, M.K.P., M.-j.J., S.-Y.L., G.J., S.H.O. J.H.L., M.-W.S.; Data analysis, M.-j.J., G.J.; Original draft preparation, M.K.P., S.-Y.L.; Review and editing, M.K.P., M.-j.J., S.-Y.L., G.J., S.H.O. J.H.L., M.-W.S.; Supervision, S.H.O. J.H.L., M.-W.S.; Project administration, S.-Y.L., G.J.

**Funding:** This research was supported by Basic Science Research Program through the National Research Foundation of Korea (NRF) funded by the Ministry of Science, ICT & Future Planning (NRF- 2017R1D1A 1B03034832).

**Conflicts of Interest:** The authors declare no conflict of interest.

## References

1. Ludwig, I.A.; Clifford, M.N.; Lean, M.E.; Ashihara, H.; Crozier, A. Coffee: Biochemistry and potential impact on health. *Food Funct.* **2014**, *5*, 1695–1717. [CrossRef] [PubMed]
2. Butt, M.S.; Sultan, M.T. Coffee and its consumption: Benefits and risks. *Crit. Rev. Food Sci. Nutr.* **2011**, *51*, 363–373. [CrossRef] [PubMed]
3. Bohn, S.K.; Blomhoff, R.; Paur, I. Coffee and cancer risk, epidemiological evidence and molecular mechanisms. *Mol. Nutr. Food Res.* **2014**, *58*, 915–930. [CrossRef] [PubMed]
4. Santos, R.M.; Lima, D.R. Coffee consumption, obesity and type 2 diabetes: A mini-review. *Eur. J. Nutr.* **2016**, *55*, 1345–1358. [CrossRef] [PubMed]
5. Qi, H.; Li, S. Dose-response meta-analysis on coffee, tea and caffeine consumption with risk of Parkinson's disease. *Geriatr. Gerontol. Int.* **2014**, *14*, 430–439. [CrossRef] [PubMed]
6. Salomone, F.; Galvano, F.; Li Volti, G. Molecular Bases Underlying the Hepatoprotective Effects of Coffee. *Nutrients* **2017**, *9*, 85. [CrossRef] [PubMed]
7. Cano-Marquina, A.; Tarin, J.J.; Cano, A. The impact of coffee on health. *Maturitas* **2013**, *75*, 7–21. [CrossRef] [PubMed]
8. Rhee, J.; Kim, R.; Kim, Y.; Tam, M.; Lai, Y.; Keum, N.; Oldenburg, C.E. Maternal Caffeine Consumption during Pregnancy and Risk of Low Birth Weight: A Dose-Response Meta-Analysis of Observational Studies. *PLoS ONE* **2015**, *10*, e0132334. [CrossRef] [PubMed]
9. Lee, D.R.; Lee, J.; Rota, M.; Lee, J.; Ahn, H.S.; Park, S.M.; Shin, D. Coffee consumption and risk of fractures: A systematic review and dose-response meta-analysis. *Bone* **2014**, *63*, 20–28. [CrossRef] [PubMed]
10. Poole, R.; Kennedy, O.J.; Roderick, P.; Fallowfield, J.A.; Hayes, P.C.; Parkes, J. Coffee consumption and health: Umbrella review of meta-analyses of multiple health outcomes. *BMJ* **2017**, *359*, j5024. [CrossRef] [PubMed]
11. Liu, Q.P.; Wu, Y.F.; Cheng, H.Y.; Xia, T.; Ding, H.; Wang, H.; Wang, Z.M.; Xu, Y. Habitual coffee consumption and risk of cognitive decline/dementia: A systematic review and meta-analysis of prospective cohort studies. *Nutrition* **2016**, *32*, 628–636. [CrossRef] [PubMed]
12. Grosso, G.; Micek, A.; Castellano, S.; Pajak, A.; Galvano, F. Coffee, tea, caffeine and risk of depression: A systematic review and dose-response meta-analysis of observational studies. *Mol. Nutr. Food Res.* **2016**, *60*, 223–234. [CrossRef] [PubMed]
13. Samoggia, A.; Riedel, B. Coffee consumption and purchasing behavior review: Insights for further research. *Appetite* **2018**, *129*, 70–81. [CrossRef] [PubMed]
14. Jeszka-Skowron, M.; Zgola-Grzeskowiak, A.; Waskiewicz, A.; Stepien, L.; Stanisz, E. Positive and negative aspects of green coffee consumption–antioxidant activity versus mycotoxins. *J. Sci. Food. Agric.* **2017**, *97*, 4022–4028. [CrossRef] [PubMed]

15. Moreira, A.S.P.; Nunes, F.M.; Simoes, C.; Maciel, E.; Domingues, P.; Domingues, M.R.M.; Coimbra, M.A. Data on coffee composition and mass spectrometry analysis of mixtures of coffee related carbohydrates, phenolic compounds and peptides. *Data Brief* **2017**, *13*, 145–161. [CrossRef] [PubMed]

16. Park, M.K.; Im, G.J.; Chang, J.; Chae, S.W.; Yoo, J.; Han, W.G.; Hwang, G.H.; Jung, J.Y.; Choi, J.; Jung, H.H.; et al. Protective effects of caffeic acid phenethyl ester (CAPE) against neomycin-induced hair cell damage in zebrafish. *Int. J. Pediatr. Otorhinolaryngol.* **2014**, *78*, 1311–1315. [CrossRef] [PubMed]

17. Choi, J.; Kim, S.H.; Rah, Y.C.; Chae, S.W.; Lee, J.D.; Md, B.D.; Park, M.K. Effects of caffeic acid on cisplatin-induced hair cell damage in HEI-OC1 auditory cells. *Int. J. Pediatr. Otorhinolaryngol.* **2014**, *78*, 2198–2204. [CrossRef] [PubMed]

18. Chadha, S.; Cieza, A. Promoting global action on hearing loss: World hearing day. *Int. J. Audiol.* **2017**, *56*, 145–147. [CrossRef] [PubMed]

19. Graydon, K.; Waterworth, C.; Miller, H.; Gunasekera, H. Global burden of hearing impairment and ear disease. *J. Laryngol. Otol.* **2018**, 1–8. [CrossRef] [PubMed]

20. Stevens, G.; Flaxman, S.; Brunskill, E.; Mascarenhas, M.; Mathers, C.D.; Finucane, M.; Global Burden of Disease Hearing Loss Expert Group. Global and regional hearing impairment prevalence: An analysis of 42 studies in 29 countries. *Eur. J. Public Health* **2013**, *23*, 146–152. [CrossRef] [PubMed]

21. WHO. *The Global Burden of Disease: 2004 Update*; World Health Organization: Geneva, Switzerland, 2008.

22. Gomaa, M.A.; Elmagd, M.H.; Elbadry, M.M.; Kader, R.M. Depression, Anxiety and Stress Scale in Patients with Tinnitus and Hearing Loss. *Eur. Arch. Otorhinolaryngol.* **2014**, *271*, 2177–2184. [CrossRef] [PubMed]

23. Peracino, A. Hearing loss and dementia in the aging population. *Audiol. Neurootol.* **2014**, *19*, 6–9. [CrossRef] [PubMed]

24. Kim, H.J.; Lee, H.J.; An, S.Y.; Sim, S.; Park, B.; Kim, S.W.; Lee, J.S.; Hong, S.K.; Choi, H.G. Analysis of the prevalence and associated risk factors of tinnitus in adults. *PLoS ONE* **2015**, *10*, e0127578. [CrossRef] [PubMed]

25. McCormack, A.; Edmondson-Jones, M.; Somerset, S.; Hall, D. A systematic review of the reporting of tinnitus prevalence and severity. *Hear Res.* **2016**, *337*, 70–79. [CrossRef] [PubMed]

26. Schleuning, A.J.; Shi, Y.J.; Martin, W.H. Tinnitus. In *Head and Neck Surgery—Otolaryngology*; Bailey, B.J., Johnson, J.T., Newlands, S.D., Eds.; Lippincott Williams & Wilkins: Philadelphia, PA, USA, 2006; p. 2241.

27. Crummer, R.W.; Hassan, G.A. Diagnostic approach to tinnitus. *Am. Fam. Physician* **2004**, *69*, 120–126. [PubMed]

28. Figueiredo, R.R.; Rates, M.J.; Azevedo, A.A.; Moreira, R.K.; Penido Nde, O. Effects of the reduction of caffeine consumption on tinnitus perception. *Braz. J. Otorhinolaryngol.* **2014**, *80*, 416–421. [CrossRef] [PubMed]

29. Claire, L.S.; Stothart, G.; McKenna, L.; Rogers, P.J. Caffeine abstinence: An ineffective and potentially distressing tinnitus therapy. *Int. J. Audiol.* **2010**, *49*, 24–29. [CrossRef] [PubMed]

30. McCormack, A.; Edmondson-Jones, M.; Mellor, D.; Dawes, P.; Munro, K.J.; Moore, D.R.; Fortnum, H. Association of dietary factors with presence and severity of tinnitus in a middle-aged UK population. *PLoS ONE* **2014**, *9*, e114711. [CrossRef] [PubMed]

31. Zawawi, F.; Bezdjian, A.; Mujica-Mota, M.; Rappaport, J.; Daniel, S.J. Association of Caffeine and Hearing Recovery after Acoustic Overstimulation Events in a Guinea Pig Model. *JAMA Otolaryngol. Head Neck Surg.* **2016**, *142*, 383–388. [CrossRef] [PubMed]

32. Mujica-Mota, M.A.; Gasbarrino, K.; Rappaport, J.M.; Shapiro, R.S.; Daniel, S.J. The effect of caffeine on hearing in a guinea pig model of acoustic trauma. *Am. J. Otolaryngol.* **2014**, *35*, 99–105. [CrossRef] [PubMed]

33. Sanchez-Sellero, I.; San-Roman-Rodriguez, E.; Santos-Perez, S.; Rossi-Izquierdo, M.; Soto-Varela, A. Caffeine intake and Meniere's disease: Is there relationship? *Nutr. Neurosci.* **2017**, 1–8. [CrossRef] [PubMed]

34. Sharon, J.D.; Trevino, C.; Schubert, M.C.; Carey, J.P. Treatment of Meniere's Disease. *Curr. Treat. Options Neurol.* **2015**, *17*, 341. [CrossRef] [PubMed]

35. Chang, S.J.; Chang, C.K. Prevalence and risk factors of noise-induced hearing loss among liquefied petroleum gas (LPG) cylinder infusion workers in Taiwan. *Ind. Health* **2009**, *47*, 603–610. [CrossRef] [PubMed]

36. Dixit, A.; Vaney, N.; Tandon, O.P. Effect of caffeine on central auditory pathways: An evoked potential study. *Hear Res.* **2006**, *220*, 61–66. [CrossRef] [PubMed]

37. Maitre, N.L.; Chan, J.; Stark, A.R.; Lambert, W.E.; Aschner, J.L.; Key, A.P. Effects of caffeine treatment for apnea of prematurity on cortical speech-sound differentiation in preterm infants. *J. Child Neurol.* **2015**, *30*, 307–313. [CrossRef] [PubMed]

38. Hong, B.N.; Yi, T.H.; Park, R.; Kim, S.Y.; Kang, T.H. Coffee improves auditory neuropathy in diabetic mice. *Neurosci. Lett.* **2008**, *441*, 302–306. [CrossRef] [PubMed]

39. Hong, B.N.; Yi, T.H.; Kim, S.Y.; Kang, T.H. High-dosage pyridoxine-induced auditory neuropathy and protection with coffee in mice. *Biol. Pharm. Bull.* **2009**, *32*, 597–603. [CrossRef] [PubMed]

40. Nehlig, A.; Daval, J.L.; Debry, G. Caffeine and the central nervous system: Mechanisms of action, biochemical, metabolic and psychostimulant effects. *Brain Res. Brain Res. Rev.* **1992**, *17*, 139–170. [CrossRef]

41. Bagueley D, A.G.; McFerran, D.; McKenna, L. *Tinnitus: Consequences and Moderating Factors*; Wiley-Blackwell: West Sussex, UK, 2013; p. 100.

42. Miller, B.; O'Connor, H.; Orr, R.; Ruell, P.; Cheng, H.L.; Chow, C.M. Combined caffeine and carbohydrate ingestion: Effects on nocturnal sleep and exercise performance in athletes. *Eur. J. Appl. Physiol.* **2014**, *114*, 2529–2537. [CrossRef] [PubMed]

43. Park, J.; Han, J.W.; Lee, J.R.; Byun, S.; Suh, S.W.; Kim, T.; Yoon, I.Y.; Kim, K.W. Lifetime Coffee Consumption, Pineal Gland Volume and Sleep Quality in Late Life. *Sleep* **2018**. [CrossRef] [PubMed]

44. Miguel, G.S.; Yaremchuk, K.; Roth, T.; Peterson, E. The effect of insomnia on tinnitus. *Ann. Otol. Rhinol. Laryngol.* **2014**, *123*, 696–700. [CrossRef] [PubMed]

45. Glicksman, J.T.; Curhan, S.G.; Curhan, G.C. A prospective study of caffeine intake and risk of incident tinnitus. *Am. J. Med.* **2014**, *127*, 739–743. [CrossRef] [PubMed]

46. Mazurek, B.; Szczepek, A.J.; Hebert, S. Stress and tinnitus. *HNO* **2015**, *63*, 258–265. [CrossRef] [PubMed]

47. Schaaf, H.; Flohre, S.; Hesse, G.; Gieler, U. Chronic stress as an influencing factor in tinnitus patients. *HNO* **2014**, *62*, 108–114. [CrossRef] [PubMed]

48. Roland, L.T.; Lenze, E.J.; Hardin, F.M.; Kallogjeri, D.; Nicklaus, J.; Wineland, A.M.; Fendell, G.; Peelle, J.E.; Piccirillo, J.F. Effects of mindfulness based stress reduction therapy on subjective bother and neural connectivity in chronic tinnitus. *Otolaryngol. Head Neck Surg.* **2015**, *152*, 919–926. [CrossRef] [PubMed]

49. Caprioli, G.; Cortese, M.; Sagratini, G.; Vittori, S. The influence of different types of preparation (espresso and brew) on coffee aroma and main bioactive constituents. *Int J. Food Sci. Nutr.* **2015**, *66*, 505–513. [CrossRef] [PubMed]

50. Narita, Y.; Inouye, K. Decrease in the acrylamide content in canned coffee by heat treatment with the addition of cysteine. *J. Agric. Food Chem.* **2014**, *62*, 12218–12222. [CrossRef] [PubMed]

51. Maas, I.L.; Bruggemann, P.; Requena, T.; Bulla, J.; Edvall, N.K.; Hjelmborg, J.V.B.; Szczepek, A.J.; Canlon, B.; Mazurek, B.; Lopez-Escamez, J.A.; et al. Genetic susceptibility to bilateral tinnitus in a Swedish twin cohort. *Genet. Med.* **2017**, *19*, 1007–1012. [CrossRef] [PubMed]

![nutrients logo] *nutrients*

MDPI

*Article*

# Impact of Coffee, Wine, and Chocolate Consumption on Cognitive Outcome and MRI Parameters in Old Age

Sven Haller [1,2,3,*], Marie-Louise Montandon [4], Cristelle Rodriguez [5], François R. Herrmann [6] and Panteleimon Giannakopoulos [4,5]

1   CIRD—Centre d'Imagerie Rive Droite, Rue Chantepoulet 21, 1201 Genève, Switzerland
2   Department of Surgical Sciences, Radiology, Uppsala University, 751 85 Uppsala, Sweden
3   Faculty of Medicine, University of Geneva, 1205 Genève, Switzerland
4   Department of Psychiatry, Faculty of Medicine, University of Geneva, 1205 Genève, Switzerland; mlmontandon@hotmail.com (M.-L.M.); Panteleimon.Giannakopoulos@hcuge.ch (P.G.)
5   Division of Institutional Measures, Medical Direction, University Hospitals of Geneva, Rue Gabrielle-Perret-Gentil 4, 1205 Genève, Switzerland; Cristelle.Rodriguez@hcuge.ch
6   Division of Geriatrics, Department of Internal Medicine, Rehabilitation and Geriatrics, University of Geneva, 1205 Genève, Switzerland; Francois.Herrmann@hcuge.ch
*   Correspondence: sven.haller@me.com; Tel.: +41-7-6482-1754

Received: 20 August 2018; Accepted: 26 September 2018; Published: 1 October 2018

**Abstract:** Coffee, wine and chocolate are three frequently consumed substances with a significant impact on cognition. In order to define the structural and cerebral blood flow correlates of self-reported consumption of coffee, wine and chocolate in old age, we assessed cognition and brain MRI measures in 145 community-based elderly individuals with preserved cognition (69 to 86 years). Based on two neuropsychological assessments during a 3-year follow-up, individuals were classified into stable-stable (52 sCON), intermediate (61 iCON) and deteriorating-deteriorating (32 dCON). MR imaging included voxel-based morphometry (VBM), tract-based spatial statistics (TBSS) and arterial spin labelling (ASL). Concerning behavior, moderate consumption of caffeine was related to better cognitive outcome. In contrast, increased consumption of wine was related to an unfavorable cognitive evolution. Concerning MRI, we observed a negative correlation of wine and VBM in bilateral deep white matter (WM) regions across all individuals, indicating less WM lesions. Only in sCON individuals, we observed a similar yet weaker association with caffeine. Moreover, again only in sCON individuals, we observed a significant positive correlation between ASL and wine in overlapping left parietal WM indicating better baseline brain perfusion. In conclusion, the present observations demonstrate an inverse association of wine and coffee consumption with cognitive performances. Moreover, low consumption of wine but also moderate to heavy coffee drinking was associated with better WM preservation and cerebral blood-flow notably in cognitively stable elders.

**Keywords:** caffeine; wine; chocolate; aging; cognition

---

## 1. Introduction

Coffee, wine and chocolate are three frequently consumed substances with a significant impact on cognitive performances.

Early studies in community-based samples suggested that moderate caffeine consumption is associated with decreased incidence of both mild cognitive impairment (MCI) and clinically overt AD [1–4]. More recently, a case-control study revealed that plasma caffeine levels greater than 1200 ng/mL in MCI subjects were associated with no conversion to dementia during a 2–4-year follow-up [5]. Importantly, in the Italian Longitudinal study of aging, moderate caffeine consumption over time (from 1 to 2 cups of coffee/day) was associated with lower incidence rate of MCI in cognitively intact older individuals. However, an inverse association was found for those who increased their daily caffeine consumption [6,7].

A U-shape relationship between cognitive performance and wine consumption has been postulated with a marked detrimental effect of heavy drinking but a decrease of Alzheimer disease (AD) and dementia risk among light to moderate drinkers. However, this latter association has been challenged due to confounding by socioeconomic class and intelligence (for review see References [8–10]).

Recent lines of evidence suggest that regular consumption of cocoa is associated with dose-dependent improvements in general cognition, attention, processing speed, and working memory that have been documented in animal models of normal aging but also in a limited series of healthy elders (for review see [11,12]).

The impact of these substances on resting state brain function and AD pathology has been intensively explored. A limited number of randomized controlled trials explored the acute effects of caffeine, cocoa flavonoids and alcohol in brain function and perfusion [13–15]. Overall, caffeine intake was associated with a significant reduction of ASL-measured gray matter cerebral blood flow, increased load-related activation compared to placebo in the left and right dorsolateral prefrontal cortex during working memory encoding, but decreased load-related activation in the left thalamus during working memory maintenance. Alcohol intake led to increased cerebral blood flow in a dose-dependent manner (for review see Joris et al. [16]). Chronic caffeine intake has been shown to reduce Aβ-induced cell death in vitro, decrease brain amyloid levels [6,17–21], reduce hippocampal tau phosphorylation and proteolytic fragments but also mitigate several proinflammatory and oxidative stress markers in AD transgenic models [22]. Several studies pointed to a caffeine-mediated decrease of resting-state connectivity across the brain in healthy controls. More recently, it was shown that although this is true in respect to visual and motor areas, the blood oxygenation level dependent (BOLD) functional connectivity of the default mode network (DMN) might increase via the recruitment of attentional networks partly explaining the caffeine-mediated elevated alertness [23–26]. Low concentrations of ethanol have been shown to protect against toxicity induced by Aβ oligomers [27]. In alcohol drinkers (without misuse or dependence), resting state functional connectivity is reduced in posterior cortical areas as precuneus, postcentral gyrus, insula, right fusiform and lingual gyri and visual cortex [28] but also in the sub-callosal cortex, in left temporal fusiform cortex and left inferior temporal gyrus [29]. In the same line, cocoa extracts reduce oligomerization of beta amyloid and modulates the brain neurotrophic-derived factor signalling pathway in AD animal models [30,31]. At the cellular level, chocolate and other flavonoids interact with signalization cascades involving protein and lipid kinases that lead to the inhibition of neuronal death by apoptosis induced by neurotoxicants such as oxygen radicals and promote neuronal survival and synaptic plasticity (for review see [32]).

Contrasting with the substantial amount of data on resting state fMRI effects of wine, coffee and chocolate intake, a surprisingly low number of studies addressed the consequences of their chronic consumption on structural MRI parameters in healthy controls (without any misuse or addiction-related behaviors). Most of them concerned alcohol beverages and remain highly controversial. Linear decrease of grey matter (GM) volumes were reported with weekly alcohol consumption mainly in men whereas white matter (WM) volume analysis led to conflicting data [33–36]. Regular caffeine use is known to reduce arterial spin labelling (ASL)-assessed cerebral blood flow (CBF) [37,38] in healthy controls. To our knowledge, there were no studies investigating the relationship between chocolate consumption and structural MRI parameters as well as ASL-assessed CBF.

In order to define the structural and cerebral blood flow correlates of regular consumption of coffee, wine and chocolate in old age, we performed voxel-based morphometry (VBM), tract-based spatial statistics (TBSS) that detect changes in grey and white matter microstructure and arterial spin labelling (ASL) perfusion imaging in a community-based series of 145 elderly individuals aged from

69.3 to 85.8 who were cognitively preserved at inclusion and underwent two neuropsychological assessments during a subsequent 3-year period.

## 2. Materials and Methods

### 2.1. Participants

The data engaged in this article was retrieved from an ongoing large population-based longitudinal study on healthy aging that is still ongoing in the Geneva and Lausanne counties. The cohort included 526 elderly Caucasian white individuals living in Geneva and Lausanne catchment area. Due to the need for excellent French knowledge (in order to participate in detailed neuropsychological testing) the vast majority of the participants were Swiss (or born in French-speaking European countries, 92%). At baseline, all individuals were evaluated with an extensive neuropsychological battery, including the Mini-Mental State Examination (MMSE) [39], the Hospital Anxiety and Depression Scale (HAD [40]), and the Lawton Instrumental Activities of Daily Living (IADL, [41]). Cognitive assessment included (a) attention (Digit-Symbol-Coding [42], Trail Making Test A [43]), (b) working memory (verbal: Digit Span Forward [44]), visuo-spatial: Visual Memory Span (Corsi) [45], (c) episodic memory (verbal: RI-48 Cued Recall Test [46]), visual: Shapes Test [47], (d) executive functions (Trail Making Test B [43], Wisconsin Card Sorting Test and Phonemic Verbal Fluency Test), (e) language (Boston Naming [48]), (f) visual gnosis (Ghent Overlapping Figures), (g) praxis: ideomotor [49], reflexive [50], and constructional (Consortium to Establish a Registry for Alzheimer's Disease (CERAD), Figures copy [51]). All individuals were also evaluated with the Clinical Dementia Rating scale (CDR) [52]. In agreement with the criteria of Petersen et al. [53], participants with a CDR of 0.5 but no dementia and a score exceeding 1.5 standard deviations below the age-appropriate mean in any of the cognitive tests were classified as MCI and were excluded. Participants with neither dementia nor MCI were classified as cognitively healthy controls and underwent full neuropsychological assessment at follow-ups, on average 18 and 36 months later. Exclusion criteria included psychiatric or neurologic disorders, sustained head injury, history of major medical disorders (neoplasm or cardiac illness), alcohol or drug abuse, regular use of neuroleptics, antidepressants or psychostimulants and contraindications to MR imaging. To control for the confounding effect of cardiovascular diseases, individuals with subtle cardiovascular symptoms and a history of stroke, severe hypertension and transient ischemic episodes were also excluded from the present study.

At follow-up, which took place 18 months after inclusion, the cognitively healthy individuals underwent full neuropsychological assessment. Individuals who obtained stable cognitive scores over the baseline and follow-up evaluation were classified as stable controls. The progressive control group obtained a follow-up evaluation of at least 0.5 standard deviations (SD) lower than measured at baseline, on a minimum of two cognitive tests. Two neuropsychologists clinically assessed all individuals independently. The final classification was determined by a trained neuropsychologist considering both the results of the neuropsychological tests and overall clinical assessment [54]. All of the case's individuals were assessed once again 18 months later with the same neuropsychological battery. The participants were subsequently grouped as described above (−0.5 SD in at least two cognitive tests), with comparison of the scores of the latest assessment. Stable individuals showing no changes in the second assessment were classified in the stable-stable (sCON) group and progressive individuals demonstrating a further decline as deteriorating-deteriorating (dCON). The intermediate group (iCON) refers to participants demonstrating a fluctuating scoring pattern, incorporating stable-progressive, progressive-stable or progressive-improved individuals.

The final sample consisted of 52 sCON (mean age 73 ± 3 years; 32 women), 61 iCON (mean age 73 ± 3 years; 30 women) and 32 dCON (mean age 74 ± 4.0 years; 18 women). All participants gave informed written consent after formal approval by the local Ethics Committee.

The timeline of neuropsychological assessment, MR imaging and questionnaire is illustrated online in Figure S1.

## 2.2. Substance Questionnaire

Usual caffeinated foods and beverages (coffee, chocolate) consumption as well as wine intake were assessed by a self-administered questionnaire. Participants were asked to complete the questionnaire entering the amount consumed by day, month and year (see online Supplementary Material). After reception of the questionnaire and in case of doubt, additional information was obtained by phone calls in order to obtain a global estimation of the consumption. In contrast, the type of coffee preparation or wine was not explored further since no lines of evidence indicate a differential impact of these preparations (or type of wine) in the human brain. The caffeine questionnaire was derived from Reference [55] and related caffeine content can be found in References [56,57].

## 2.3. MRI Data Acquisition

Imaging data were acquired on a 3T MRI scanner (TRIO SIEMENS Medical Systems, Erlangen, Germany) Essential data include: a high-resolution T1-weighted anatomical scan (magnetization prepared rapid gradient echo (MPRAGE), 256 × 256 matrix, 176 slices 1 mm isotropic, TR = 2.27 ms), a pulsed ASL sequence (64 × 64 matrix, 24 slices, voxel size 3.44 × 3.44 × 5 mm$^3$, TE = 12 ms, TR = 4000 ms, inversion time (TI) 1600 ms) and a diffusion tensor imaging DTI sequence (b = 0 and 30 diffusion directions with $b$ = 1000 s/mm$^2$, 128 × 128 matrix, voxel size 2.0 × 2.0 × 2.0 mm$^3$, TE = 82.4 ms, TR = 7900 ms and 1 average).

Additional sequences included axial fast spin-echo T2w imaging (4000/105, 30 sections, 4-mm section thickness), susceptibility weighted imaging (28/20, 208 × 256 × 128 matrix, 1 mm × 1mm × 1 mm voxel size) were performed to exclude brain disease, such as ischemic stroke, subdural hematomas, or space-occupying lesions.

## 2.4. Statistical Analysis of Demographic and Substance Data

Comparison among the three groups were performed with Fisher exact test, Kruskal-Wallis test or one way ANOVA according to the distribution of the variables. Caffeinated foods and beverages were considered as continuous variables, z-scores and also as tertile (light, moderate, heavy consumers). Consumption of coffee was divided in tertile as follows: light (0–28 cups/month), moderate (29–60 cups/month), heavy (61–168 cups/month). Light drinkers for wine corresponded to a consumption of 0–8 units /month, moderate to a consumption of 9–28 units /month, and heavy to a consumption of 29–200 units/month. Consumption of chocolate was divided in tertile as follows: light (0–20 serving/month), moderate: 20–80 serving/month, heavy: 81–226 serving/month). Unadjusted, adjusted and multiple ordered logistic regression models were used to predict group membership (see results section for details) from the different type of consumptions (chocolate, coffee and wine).

## 2.5. MR Data Analysis

### 2.5.1. Whole-Brain Voxel-Based Morphometry (VBM)

The voxel-based morphometry analysis was carried out using the FSL software package [58], according to the standard procedure. The essential processing steps included brain extraction using Brain Extraction Tool [59], tissue-type segmentation using FMRIB's Automated Segmentation Tool [60], nonlinear transformation into Montreal Neurological Institute (MNI) reference space, and creation of a study-specific GM template to which the native GM images were then nonlinearly re-registered. The modulated segmented images were then smoothed with an isotropic Gaussian kernel with a sigma of 2 mm. Finally, the voxel-wise FSL General Linear Model was applied by using permutation-based non-parametric testing with the FSL Randomize Tool with the threshold-free cluster enhancement (TFCE) correction for multiple comparisons [61], considering fully corrected $p$ values < 0.05 as significant. The analysis was performed twice. First, the analysis was performed across all participants across the entire brain using coffee, wine or chocolate as dependent variables- and age, gender, education and MMSE score as potential confounders. Second, the analysis was performed as separate models for the groups sCON, iCON and dCON using only one explanatory variable (coffee, wine or chocolate) and again age, gender, education and MMSE score as non-explanatory variables.

### 2.5.2. Arterial Spin Labelling (ASL)

The reconstructed relCBF (relative cerebral blood flow) ASL perfusion images were spatially normalized using a linear spatial alignment from ASL raw data to the individual high-resolution 3DT1 image, followed by the application of the non-linear spatial registration determined in the pre-processing of the 3DT1 data. The spatial transformations were then applied to the relCBF maps calculated directly on the MRI scanner, this two-steps approach results in a non-linear spatial registration of the ASL relCBF map into the MNI space. We then calculated the whole brain average relCBF, which was compared between groups with caffeine, wine and chocolate as dependent variables with age, gender, education and MMSE score as potential confounders. Moreover, we applied a voxel-wise local permutation-based, with threshold-free cluster enhancement (TFCE) correction for multiple comparisons, considering fully corrected $p$ values < 0.05 as significant. The statistical models were performed similar to VBM described above.

### 2.5.3. Diffusion Tensor Imaging (DTI) Tract Based Spatial Statistics (TBSS)

The TBSS analysis of the DTI data was done implementing the FSL software package [58], according to the standard procedure described in detail [62]. All subjects' FA data were projected onto a mean FA skeleton using a non-linear spatial registration. The tract skeleton is the basis for voxel-wise cross-subject statistics and reduces potential misregistrations as the source for false-positive or false-negative analysis results. The other DTI-derived parameters—longitudinal, radial, and mean diffusivity were analyzed in the same way using spatial transformation parameters that were estimated in the initial FA analysis. Similar to the VBM analysis above, the TBSS was analyzed using voxel-wise statistical analysis was performed TFCE correction for multiple comparisons, considering fully corrected $p$ values < 0.05 as significant. We used the John Hopkins University DTI-based white matter tractography atlas, which is distributed in the FSL package, for anatomic labeling of the supra-threshold voxels. The statistical models were performed similar to VBM described above.

### 2.5.4. GM Region of Interest (ROI) Analysis

In addition to the voxel-wise whole-brain analysis described above, we additionally performed a region of interest (ROI) analysis. The whole was parcellated into 133 regions using the Combinostics cMRI software package [63]. We performed bivariate linear regression models to predict each MRI regional parameters from group and each substance entered either as $z$-score or as an ordinal variable (tertile).

## 3. Results

### 3.1. Clinical, Demographic and Substance Data

The clinical and demographic data are summarized in Table 1. There were no statistically significant differences in age, gender and education among the groups sCON, iCON and dCON.

When including one type of consumption as z-score in ordered logistic regression model to predict group membership without and while adjusting for age, sex, education level and MMS, only wine was associated with an increased risk of adverse evolution ($OR_{unadjusted}$ 1.012, 95% CI 1.002–1.023; $p = 0.017$ unadjusted), ($OR_{adjusted}$ 1.012, 95% CI 1.001–1.022; $p = 0.028$ adjusted). In a multiple ordered logistic regression model adjusted for the same confounders as above and all type of consumptions, wine consumption remained significantly associated with the dCON status ($OR_{adjusted}$ 1.401, 95% CI 1.003–1.955; $p = 0.048$).

When analyzing the consumption data as tertile, moderate coffee drinkers are less likely to be classified as dCON ($OR_{unadjusted}$ 0.451, 95% CI 0.214–0.950; $p = 0.036$) ($OR_{adjusted}$ 0.447, 95% CI 0.210–0.952; $p = 0.037$). This observation persists after adjusting for wine and chocolate consumption $OR_{adjusted} = 0.455$; 95% CI 0.208–0.995; $p = 0.048$.

**Table 1.** Clinical, demographic and substance data by evolution groups.

| | sCON (Stable-Stable/ Stable-Improved) | iCON (Stable-Progressed/ Progressed-Stable/ Progressed-Improved) | dCON (Progressed-Progressed) | Total | *p* Value |
|---|---|---|---|---|---|
| N | 52 | 61 | 32 | 145 | |
| Age | 73.6 ± 3.4 | 73.9 ± 3.3 | 74.0 ± 3.8 | 73.8 ± 3.5 | 0.898 |
| Gender | | | | | 0.321 |
| Female | 33 (63.5%) | 30 (49.2%) | 18 (56.3%) | 81 (55.9%) | |
| Male | 19 (36.5%) | 31 (50.8%) | 14 (43.8%) | 64 (44.1%) | |
| Education (year) | | | | | 0.315 |
| <9 | 10 (19.2%) | 5 (8.2%) | 6 (18.8%) | 21 (14.5%) | |
| 9–12 | 20 (38.5%) | 29 (47.5%) | 16 (50.0%) | 65 (44.8%) | |
| >12 | 22 (42.3%) | 27 (44.3%) | 10 (31.3%) | 59 (40.7%) | |
| MMSE | 28.6 ± 1.2 | 28.3 ± 1.3 | 28.5 ± 1.7 | 28.5 ± 1.4 | 0.534 |
| Chocolate (serving/month) | 61.3 ± 58.5 | 56.0 ± 49.2 | 46.4 ± 44.4 | 55.8 ± 51.7 | 0.443 |
| Coffee (cup/month) | 56.3 ± 32.6 | 50.6 ± 36.1 | 58.7 ± 43.2 | 54.4 ± 36.5 | 0.535 |
| Wine (glass/month) | 18.6 ± 18.3 | 28.1 ± 29.9 | 34.5 ± 43.7 | 26.1 ± 30.7 | 0.054 |
| Chocolate (tertile) | | | | | 0.689 |
| Light | 18 (34.6%) | 20 (32.8%) | 15 (46.9%) | 53 (36.6%) | |
| Moderate | 17 (32.7%) | 22 (36.1%) | 7 (21.9%) | 46 (31.7%) | |
| Heavy | 17 (32.7%) | 19 (31.1%) | 10 (31.3%) | 46 (31.7%) | |
| Coffee (tertile) | | | | | 0.228 |
| Light | 12 (23.1%) | 25 (41.0%) | 13 (40.6%) | 50 (34.5%) | |
| Moderate | 21 (40.4%) | 19 (31.1%) | 7 (21.9%) | 47 (32.4%) | |
| Heavy | 19 (36.5%) | 17 (27.9%) | 12 (37.5%) | 48 (33.1%) | |
| Wine (tertile) | | | | | 0.154 |
| Light | 24 (46.2%) | 17 (27.9%) | 12 (37.5%) | 53 (36.6%) | |
| Moderate | 19 (36.5%) | 30 (49.2%) | 8 (25.0%) | 57 (39.3%) | |
| Heavy | 9 (17.3%) | 14 (23.0%) | 12 (37.5%) | 35 (24.1%) | |

### 3.2. MRI Analysis across the Entire Group

Across all participants, we observed a negative correlation in VBM with wine notably in bilateral deep white matter regions (Figure 1).

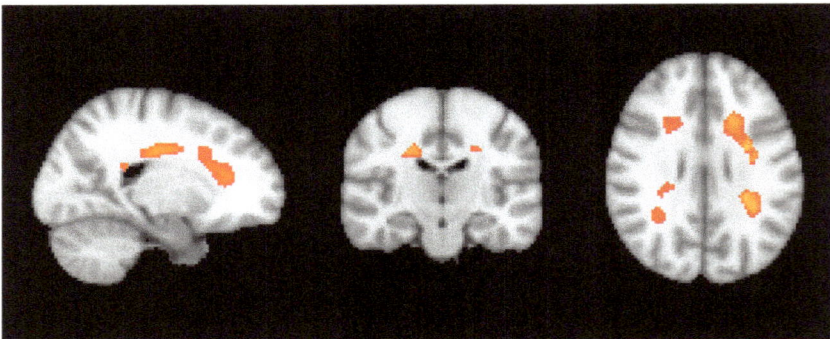

**Figure 1.** Negative correlation between wine and VBM across all individuals. $p < 0.05$ TFCE corrected.

In contrast, no significant differences were observed for ASL or TBSS measures as a function of the substances studied.

### 3.3. Group MRI Analysis

In sCON cases, we observed a significant positive correlation between ASL measures and wine in left parietal white matter (Figure 2), overlapping with the results of the VBM correlation of all individuals reported above.

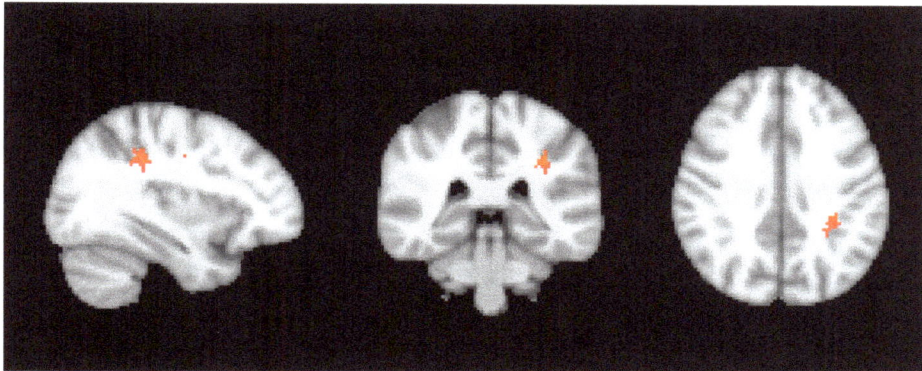

**Figure 2.** Positive correlation between wine and ASL for only sCON individuals. $p < 0.05$ TFCE corrected.

Moreover, we observed a negative correlation between VBM and caffeine only in sCON individuals notably in the white matter that was more pronounced in left parietal and right frontal regions (Figure 3).

Importantly, there were no significant associations between these substances and MRI findings in both iCON and dCON groups.

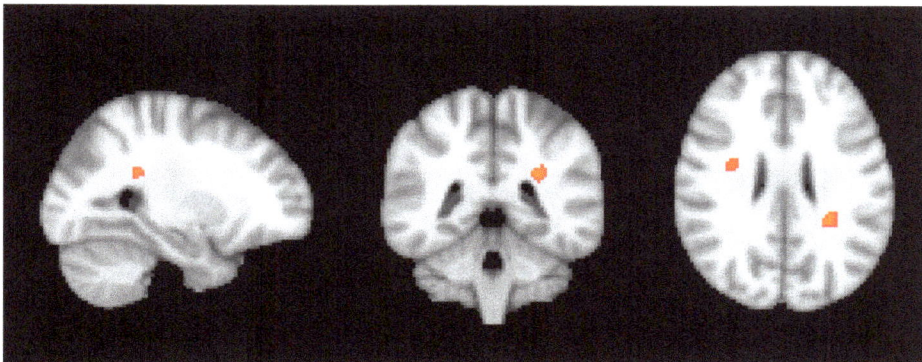

**Figure 3.** Negative correlation between caffeine and VBM for only sCON individuals. $p < 0.05$ TFCE corrected.

## 4. Discussion

We demonstrate an inverse association of wine and coffee consumption with cognitive performances. In addition, low consumption of wine but also moderate to heavy coffee drinking was associated with better WM preservation and cerebral blood-flow notably in cognitively stable elders.

At the behavioral level, the present study reveals that moderate consumption of caffeine is related to better cognitive outcome in a community-based sample of 145 elderly controls that undergo two detailed neuropsychological follow-ups in a 3-year period. Importantly, this association is limited to low quantities and did not persist in cases with very subtle signs of cognitive instability (iCON) or early phases of cognitive decline (dCON).

In contrast, increased consumption of wine is related to unfavorable cognitive evolution. The relationship between drinking and cognitive performances in old age remains a highly controversial issue. The deleterious effect of heavy wine consumption on cognitive evolution over time in elderly controls has been already documented [8–10,64,65]. Several lines of evidence have suggested that moderate drinking could have a slight positive impact on memory and verbal abilities [66,67] but negative data have been also reported [64,68]. In our highly selected cases that mostly consumed very low levels of alcohol (more than 75% among them consumed less than one unit per day and almost one third less than eight units per month), we failed to document a positive association between moderate wine drinking and cognition. In contrast, we found a negative relationship between increased wine consumption and neuropsychological performances as already suggested previously (for review see Reference [64]). It should, however, be noted that this finding was obtained when using z-scores but not tertiles indicating that the heavy consumption of a limited number of elderly controls led to this result. In contrast to wine, moderate caffeine consumption (up to two cups of coffee/day) was associated with better cognitive outcome in our 3-year follow-up. This observation parallels several previous reports on the protection conferred by moderate caffeine consumption in cognitive aging [1–5]. Not surprisingly, chronic chocolate consumption was not associated with cognition in our elderly controls. A positive effect of cocoa products seems to be confined to acute consumption as previously reported [11,12].

Concerning brain MRI, we first assessed the entire dataset of healthy elderly controls and observed a negative correlation between wine consumption and VBM in bilateral fronto-parietal white matter (WM). This result may appear contra-intuitive at first glance, as VBM is usually used to assess modifications in grey matter (GM) concentration. However, it should be noted that microvascular WM lesion are very frequent in the elderly population. They appear as hypersignal on T2w/FLAIR (fluid attenuated inversion recovery) sequences, and are usually reported on those sequences, e.g., using the Fazekas score. Although less evident and consequently usually less frequently assessed, those microvascular WM lesions also appear as a hypointense signal on T1w images, which is the basis of the VBM analysis. The negative correlation between wine and VBM in WM indicates less hypointense signal on T1 and consequently a reduced severity of WM lesions with increasing wine intake. Interestingly, the additionally performed TBSS analysis of the WM skeleton did not reveal significant differences in FA (fractional anisotropy), which is considered as a microstructural marker of axonal integrity. Taking together the results of VBM and TBSS, this indicates that increased wine intake may reduce microvascular lesions of the fronto-parietal WM, while association between this consumption and microstructural integrity of the WM seems more difficult to establish. Interestingly, an increasing number of studies point to the positive association between low to moderate wine consumption and WM integrity. In particular, Verbaten reported less white matter damage in elderly light and moderate drinkers [33]. Similar results were reported by Mukamal for elders consuming less than six units per week [69] for the vast majority of the present cases. Interestingly and unlike cognitive performances, we did not detect a negative association between heavy drinking and WM integrity. The absence of a U-shape association here may be related to the limited number of heavy drinkers in this sample and low exposure to cardiovascular risk factors due to the exclusion criteria.

A separate set of findings concerned with the association between consumption and brain structure as a function of the cognitive fate in this longitudinal series. We built regression models for each subgroup. Based on repeated neurocognitive testing, the healthy controls were sub-classified into sCON, iCON and dCON. It is important to emphasize that even for the dCON participants, the cognitive profile remains within the normal limits at follow-up, however, the individual cognitive profile slightly decreased two times at 18 and 36 months follow-up. In contrast, the cognitive profile remains constant twice for the sCON participants, and is intermediate for the iCON participants. Only in the sCON individuals, we observed a positive correlation between wine and ASL in the WM, overlapping with the regions of the VBM results across all participants reported above. This indicates that wine does not only reduce the WM lesion load, but also improves brain perfusion at baseline; however, this effect is limited in cases who remained cognitively stable over time. It is noteworthy that among sCON cases, only six cases corresponded to the classical definition of heavy drinking ($\geq 8$ units for women and 15 for men), the mean consumption being less than one unit/day. In the same line, we found a negative association between caffeine consumption and VBM only for sCON participants in the right frontal and left parietal WM regions, without a significant association with TBSS parameters. Similar to the argumentat above, this might indicate that caffeine reduces WM lesion load only in sCON participants, without having a significant effect on WM microstructural integrity. Interestingly, and in contrast to wine, most of the sCON cases were of moderate or heavy consumption of caffeine, not supporting the idea of a U-shaped association between caffeine consumption and WM lesions. Moreover, the positive association between caffeine consumption and cognition was present only in sCON participants consistent with the view that caffeine is a cognitive normalizer rather than a cognitive enhancer [70,71]. As for cognitive outcome, chocolate consumption was not associated with the MRI parameters studied in the present series suggesting that the chronic consumption of chocolate is not beneficial nor deleterious for brain integrity or cognitive performances in old age.

## 5. Conclusions

In conclusion, the present observations confirm the opposite associations between wine and coffee consumption on cognitive performances, suggesting a detrimental effect of heavy drinking and benefits of chronic consumption of moderate quantities of coffee. The low consumption of wine but also moderate to heavy coffee drinking is associated with better WM preservation and cerebral blood-flow in cognitively stable elders without significant cerebrovascular pathologies. Strengths of the present study include the longitudinal follow-up with detained neuropsychological battery in all of our community-dwelling cases and absence of health-related confounders such as neurological, psychiatric and cerebrovascular pathologies. Several limitations should however be considered when interpreting these data. First, our cohort of healthy controls was without significant vascular pathology and a high level of daily functioning without any symptom of substance abuse is not representative of the entire spectrum of old age. Second, current consumption was assessed with a food questionnaire based on self-reporting, leading to possible underestimation of wine consumption. Third, no data on lifetime consumption were obtained, so the possible deleterious or beneficial effect of wine and coffee use at midlife cannot be assessed. Finally, MRI assessment was performed at baseline and thus we cannot comment on the association between MRI structural parameter changes and wine and coffee consumption over time. Future studies in large community-based samples combining self and proxy-reports, lifetime assessment of wine and coffee consumption and repeated MRI scans are needed to shed additional light into the complex relationships between these substances and structural MRI parameters in old age.

**Supplementary Materials:** The following are available online at http://www.mdpi.com/2072-6643/10/10/1391/s1, Figure S1: Timeline of imaging and neuropsychological testing.

**Author Contributions:** Conceptualization, S.H. and P.G.; Methodology, S.H., P.G.; Formal Analysis, S.H., M.-L.M., F.R.H.; Investigation, C.R.; Patient recruitment and neuropsychological testing, C.R.; Writing-Review & Editing, S.H., M.-L.M., P.G.; Funding Acquisition, C.R., P.G.

**Funding:** This work is supported by Swiss National Foundation grants SNF 3200B0-1161193 and SPUM 33CM30-124111 and an unrestricted grant from the Assocation Suisse pour la Recherche Alzheimer.

**Acknowledgments:** We thank all volunteers for participating in this study.

**Conflicts of Interest:** The authors declare no conflict of interest. The funders had no role in the design of the study; in the collection, analyses, or interpretation of data; in the writing of the manuscript, and in the decision to publish the results.

## References

1. Lindsay, J.; Laurin, D.; Verreault, R.; Hébert, R.; Helliwell, B.; Hill, G.B.; McDowell, I. Risk factors for Alzheimer's disease: A prospective analysis from the Canadian Study of Health and Aging. *Am. J. Epidemiol.* **2002**, *156*, 445–453. [CrossRef] [PubMed]

2. Maia, L.; de Mendonca, A. Does caffeine intake protect from Alzheimer's disease? *Eur. J. Neurol.* **2002**, *9*, 377–382. [CrossRef] [PubMed]

3. Van Gelder, B.M.; Buijsse, B.; Tijhuis, M.; Kalmijn, S.; Giampaoli, S.; Nissinen, A.; Kromhout, D. Coffee consumption is inversely associated with cognitive decline in elderly european men: The fine study. *Eur. J. Clin. Nutr.* **2006**, *61*, 226–232. [CrossRef] [PubMed]

4. Eskelinen, M.H.; Ngandu, T.; Tuomilehto, J.; Soininen, H.; Kivipelto, M. Midlife coffee and tea drinking and the risk of late-life dementia: A population-based CAIDE study. *J. Alzheimers Dis.* **2009**, *16*, 85–91. [CrossRef] [PubMed]

5. Ritchie, K.; Artero, S.; Portet, F.; Brickman, A.; Muraskin, J.; Beanino, E.; Ancelin, M.; Carrière, I. Caffeine, cognitive functioning, and white matter lesions in the elderly: Establishing causality from epidemiological evidence. *J. Alzheimers Dis.* **2010**, *20* (Suppl. 1), S161–S166. [CrossRef] [PubMed]

6. Chu, Y.-F.; Chang, W.-H.; Black, R.M.; Liu, J.-R.; Sompol, P.; Chen, Y.; Wei, H.; Zhao, Q.; Cheng, I.H. Crude caffeine reduces memory impairment and amyloid β1–42 levels in an alzheimer's mouse model. *Food Chem.* **2012**, *135*, 2095–2102. [CrossRef] [PubMed]

7. Solfrizzi, V.; Panza, F.; Imbimbo, B.P.; D'Introno, A.; Galluzzo, L.; Gandin, C.; Misciagna, G.; Guerra, V.; Osella, A.; Baldereschi, M.; et al. Coffee consumption habits and the risk of mild cognitive impairment: The Italian longitudinal study on aging. *J. Alzheimers Dis.* **2015**, *47*, 889–899. [CrossRef] [PubMed]

8. Ilomaki, J.; Jokanovic, N.; Tan, E.C.; Lonnroos, E. Alcohol consumption, dementia and cognitive decline: An overview of systematic reviews. *Curr. Clin. Pharmacol.* **2015**, *10*, 204–212. [CrossRef] [PubMed]

9. Panza, F.; Frisardi, V.; Seripa, D.; Logroscino, G.; Santamato, A.; Imbimbo, B.P.; Scafato, E.; Pilotto, A.; Solfrizzi, S. Alcohol consumption in mild cognitive impairment and dementia: Harmful or neuroprotective. *Int. J. Geriatr. Psychiatry* **2012**, *27*, 1218–1238. [CrossRef] [PubMed]

10. Topiwala, A.; Ebmeier, K.P. Effects of drinking on late-life brain and cognition. *Evid. Based Ment. Health* **2018**, *21*, 12–15. [CrossRef] [PubMed]

11. Sokolov, A.N.; Pavlova, M.A.; Klosterhalfen, S.; Enck, P. Chocolate and the brain: Neurobiological impact of cocoa flavanols on cognition and behavior. *Neurosci. Biobehav. Rev.* **2013**, *37*, 2445–2453. [CrossRef] [PubMed]

12. Socci, V.; Tempesta, D.; Desideri, G.; De Gennaro, L.; Ferrara, M. Enhancing human cognition with cocoa flavonoids. *Front. Nutr.* **2017**, *4*. [CrossRef] [PubMed]

13. Decroix, L.; Tonoli, C.; Soares, D.D.; Tagougui, S.; Heyman, E.; Meeusen, R. Acute cocoa flavanol improves cerebral oxygenation without enhancing executive function at rest or after exercise. *Appl. Physiol. Nutr. Metab.* **2016**, *41*, 1225–1232. [CrossRef] [PubMed]

14. Klaassen, E.B.; de Groot, R.H.M.; Evers, E.A.T.; Snel, J.; Veerman, E.C.I.; Ligtenberg, A.J.M.; Jolles, J.; Veltman, D.J. The effect of caffeine on working memory load-related brain activation in middle-aged males. *Neuropharmacology* **2013**, *64*, 160–167. [CrossRef] [PubMed]

15. Vidyasagar, R.; Greyling, A.; Draijer, R.; Corfield, D.R.; Parkes, L.M. The effect of black tea and caffeine on regional cerebral blood flow measured with arterial spin labeling. *J. Cereb. Blood Flow Metab.* **2013**, *33*, 963–968. [CrossRef] [PubMed]

16. Joris, P.J.; Mensink, R.P.; Adam, T.C.; Liu, T.T. Cerebral blood flow measurements in adults: A review on the effects of dietary factors and exercise. *Nutrients* **2018**, *10*. [CrossRef] [PubMed]

17. Arendash, G.W.; Schleif, W.; Rezai-Zadeh, K.; Jackson, E.K.; Zacharia, L.C.; Cracchiolo, J.R.; Shippy, D.; Tan, J. Caffeine protects alzheimer's mice against cognitive impairment and reduces brain β-amyloid production. *Neuroscience* **2006**, *142*, 941–952. [CrossRef] [PubMed]

18. Arendash, G.W.; Mori, T.; Cao, C.; Mamcarz, M.; Runfeldt, M.; Dickson, A.; Rezai-Zadeh, K.; Tan, J.; Citron, B.A.; Lin, X.; et al. Caffeine reverses cognitive impairment and decreases brain amyloid-beta levels in aged Alzheimer's disease mice. *J. Alzheimers Dis.* **2009**, *17*, 661–680. [CrossRef] [PubMed]

19. Dall'Igna, O.P.; Fett, P.; Gomes, M.W.; Souza, D.O.; Cunha, R.A.; Lara, D.R. Caffeine and adenosine A2a receptor antagonists prevent β-amyloid (25–35)-induced cognitive deficits in mice. *Exp. Neurol.* **2007**, *203*, 241–245. [CrossRef] [PubMed]

20. Espinosa, J.; Rocha, A.; Nunes, F.; Costa, M.S.; Schein, V.; Kazlauckas, V.; Kalinine, E.; Souza, D.O.; Cunha, R.A.; Porciúncula, L.O.; et al. Caffeine consumption prevents memory impairment, neuronal damage, and adenosine A2A receptors upregulation in the hippocampus of a rat model of sporadic dementia. *J. Alzheimers Dis.* **2013**, *34*, 509–518. [CrossRef] [PubMed]

21. Han, K.; Jia, N.; Li, J.; Yang, L.; Min, L.Q. Chronic caffeine treatment reverses memory impairment and the expression of brain BNDF and TrkB in the PS1/APP double transgenic mouse model of Alzheimer's disease. *Mol. Med. Rep.* **2013**, *8*, 737–740. [CrossRef] [PubMed]

22. Cao, C.; Wang, L.; Lin, X.; Mamcarz, M.; Zhang, C.; Bai, G.; Nong, J.; Sussman, S.; Arendash, G. Caffeine synergizes with another coffee component to increase plasma GCSF: Linkage to cognitive benefits in Alzheimer's mice. *J. Alzheimers Dis.* **2011**, *25*, 323–335. [CrossRef] [PubMed]

23. Laurienti, P.J.; Field, A.S.; Burdette, J.H.; Maldjian, J.A.; Yen, Y.-F.; Moody, D.M. Dietary caffeine consumption modulates fmri measures. *NeuroImage* **2002**, *17*, 751–757. [CrossRef] [PubMed]

24. Rack-Gomer, A.L.; Liau, J.; Liu, T.T. Caffeine reduces resting-state bold functional connectivity in the motor cortex. *NeuroImage* **2009**, *46*, 56–63. [CrossRef] [PubMed]

25. Wong, C.W.; Olafsson, V.; Tal, O.; Liu, T.T. Anti-correlated networks, global signal regression, and the effects of caffeine in resting-state functional mri. *NeuroImage* **2012**, *63*, 356–364. [CrossRef] [PubMed]

26. Tal, O.; Diwakar, M.; Wong, C.-W.; Olafsson, V.; Lee, R.; Huang, M.-X.; Liu, T.T. Caffeine-induced global reductions in resting-state bold connectivity reflect widespread decreases in meg connectivity. *Front. Hum. Neurosci.* **2013**, *7*. [CrossRef] [PubMed]

27. Muñoz, G.; Urrutia, J.C.; Burgos, C.F.; Silva, V.; Aguilar, F.; Sama, M.; Yeh, H.H.; Opazo, C.; Aguayo, L.G. Low concentrations of ethanol protect against synaptotoxicity induced by aβ in hippocampal neurons. *Neurobiol. Aging* **2015**, *36*, 845–856. [CrossRef] [PubMed]

28. Vergara, V.M.; Liu, J.; Claus, E.D.; Hutchison, K.; Calhoun, V. Alterations of resting state functional network connectivity in the brain of nicotine and alcohol users. *Neuroimage* **2017**, *151*, 45–54. [CrossRef] [PubMed]

29. Spagnolli, F.; Cerini, R.; Cardobi, N.; Barillari, M.; Manganotti, P.; Storti, S.; Mucelli, R.P. Brain modifications after acute alcohol consumption analyzed by resting state fMRI. *Magn. Reson. Imaging* **2013**, *31*, 1325–1330. [CrossRef] [PubMed]

30. Wang, J.; Varghese, M.; Ono, K.; Yamada, M.; Levine, S.; Tzavaras, N.; Gong, B.; Hurst, W.J.; Blitzer, R.D.; Pasinetti, G.M. Cocoa extracts reduce oligomerization of amyloid-β: Implications for cognitive improvement in Alzheimer's disease. *J. Alzheimers Dis.* **2014**, *41*, 643–650. [CrossRef] [PubMed]

31. Cimini, A.; Gentile, R.; D'Angelo, B.; Benedetti, E.; Cristiano, L.; Avantaggiati, M.L.; Giordano, A.; Ferri, C.; Desideri, G. Cocoa powder triggers neuroprotective and preventive effects in a human alzheimer's disease model by modulating bdnf signaling pathway. *J. Cell. Biochem.* **2013**, *114*, 2209–2220. [CrossRef] [PubMed]

32. Nehlig, A. The neuroprotective effects of cocoa flavanol and its influence on cognitive performance. *Br. J. Clin. Pharmacol.* **2013**, *75*, 716–727. [CrossRef] [PubMed]

33. Verbaten, M.N. Chronic effects of low to moderate alcohol consumption on structural and functional properties of the brain: Beneficial or not? *Hum. Psychopharmacol. Clin. Exp.* **2009**, *24*, 199–205. [CrossRef] [PubMed]

34. Anstey, K.J.; Jorm, A.F.; Réglade-Meslin, C.; Maller, J.; Kumar, R.; von Sanden, C.; Windsor, T.D.; Rodgers, B.; Wen, W.; Sachdev, P. Weekly alcohol consumption, brain atrophy, and white matter hyperintensities in a community-based sample aged 60 to 64 years. *Psychosom. Med.* **2006**, *68*, 778–785. [CrossRef] [PubMed]

35. Debruin, E.; Hulshoffpol, H.; Schnack, H.; Janssen, J.; Bijl, S.; Evans, A.; Leonkenemans, J.; Kahn, R.; Verbaten, M. Focal brain matter differences associated with lifetime alcohol intake and visual attention in male but not in female non-alcohol-dependent drinkers. *NeuroImage* **2005**, *26*, 536–545. [CrossRef] [PubMed]

36. Sachdev, P.S.; Chen, X.; Wen, W.; Anstey, K.J.; Anstry, K.J. Light to moderate alcohol use is associated with increased cortical gray matter in middle-aged men: A voxel-based morphometric study. *Psychiatry Res.* **2008**, *163*, 61–69. [CrossRef] [PubMed]

37. Addicott, M.A.; Yang, L.L.; Peiffer, A.M.; Burnett, L.R.; Burdette, J.H.; Chen, M.Y.; Hayasaka, S.; Kraft, R.A.; Maldjian, J.A.; Laurienti, P.J. The effect of daily caffeine use on cerebral blood flow: How much caffeine can we tolerate? *Hum. Brain Mapp.* **2009**, *30*, 3102–3114. [CrossRef] [PubMed]

38. Pelligrino, D.A.; Xu, H.L.; Vetri, F. Caffeine and the control of cerebral hemodynamics. *J. Alzheimers Dis.* **2010**, *20* (Suppl. 1), S51–S62. [CrossRef] [PubMed]

39. Folstein, M.F.; Folstein, S.E.; McHugh, P.R. "Mini-mental state". A practical method for grading the cognitive state of patients for the clinician. *J. Psychiatr. Res.* **1975**, *12*, 189–198. [CrossRef]

40. Zigmond, A.S.; Snaith, R.P. The hospital anxiety and depression scale. *Acta Psychiatr. Scand.* **1983**, *67*, 361–370. [CrossRef] [PubMed]

41. Barberger-Gateau, P.; Commenges, D.; Gagnon, M.; Letenneur, L.; Sauvel, C.; Dartigues, J.-F. Instrumental activities of daily living as a screening tool for cognitive impairment and dementia in elderly community dwellers. *J. Am. Geriatr. Soc.* **1992**, *40*, 1129–1134. [CrossRef] [PubMed]

42. Wechsler, D.A. *Wechsler Memory Scale*, 3rd ed.; Psychological Corporation: San Antonio, TX, USA, 1997.

43. REITAN, R.M. Validity of the trail making test as an indicator of organic brain damage. *Percept. Mot. Ski.* **1958**, *8*, 271–276. [CrossRef]

44. Wechsler, D. *Manual for the Wechsler Adult Intelligence Scale*; Psychological Corporation: New York, NY, USA, 1955.

45. Milner, B. Interhemispheric differences in the localization of psychological processes in man. *Columbia Méd. Bull.* **1971**, *27*, 272–277. [CrossRef]

46. Buschke, H.; Sliwinski, M.J.; Kuslansky, G.; Lipton, R.B. Diagnosis of early dementia by the double memory test: Encoding specificity improves diagnostic sensitivity and specificity. *Neurology* **1997**, *48*, 989–996. [CrossRef] [PubMed]

47. Baddley, A.; Emslie, H.; Nimmo-Smith, I. *Doors and People: A Test of Visual and Verbal Recall and Recognition*; Bury St Edmunds: St Edmundsbury, UK, 1994.

48. Kaplan, E.F.; Goodglass, H.; Weintraub, S. *The Boston Naming Test*, 2nd ed.; Lea & Febiger: Philadelphia, PA, USA, 1983.

49. Schnider, A.; Hanlon, R.E.; Alexander, D.N.; Benson, D.F. Ideomotor apraxia: Behavioral dimensions and neuroanatomical basis. *Brain Lang.* **1997**, *58*, 125–136. [CrossRef] [PubMed]

50. Poeck, K. Clues to the Nature of disruption to limb Praxis. In *Neuropsychological Studies of Apraxia and Related Disorders*; Elsevier: Amsterdam, The Netherlands, 1985; pp. 99–109.

51. Welsh, K.A.; Butters, N.; Mohs, R.C.; Beekly, D.; Edland, S.; Fillenbaum, G.; Heyman, A. The consortium to establish a registry for alzheimer's disease (cerad). Part V. A normative study of the neuropsychological battery. *Neurology* **1994**, *44*, 609–614. [CrossRef] [PubMed]

52. Hughes, C.P.; Berg, L.; Danziger, W.L.; Coben, L.A.; Martin, R.L. A new clinical scale for the staging of dementia. *Columbia J. Psychiatry* **1982**, *140*, 566–572. [CrossRef]

53. Petersen, R.C.; Doody, R.; Kurz, A.; Mohs, R.C.; Morris, J.C.; Rabins, P.V.; Ritchie, K.; Rossor, M.; Thal, L.; Winblad, B. Current concepts in mild cognitive impairment. *Arch. Neurol.* **2001**, *58*, 1985–1992. [CrossRef] [PubMed]

54. Xekardaki, A.; Rodriguez, C.; Montandon, M.-L.; Toma, S.; Tombeur, E.; Herrmann, F.R.; Zekry, D.; Lovblad, K.-O.; Barkhof, F.; Giannakopoulos, P.; et al. Arterial spin labeling may contribute to the prediction of cognitive deterioration in healthy elderly individuals. *Radiology* **2015**, *274*, 490–499. [CrossRef] [PubMed]

55. Bolca, S.; Huybrechts, I.; Verschraegen, M.; De Henauw, S.; Van de Wiele, T. Validity and reproducibility of a self-administered semi-quantitative food-frequency questionnaire for estimating usual daily fat, fibre, alcohol, caffeine and theobromine intakes among belgian post-menopausal women. *Int. J. Environ. Res. Public Heal.* **2009**, *6*, 121–150. [CrossRef] [PubMed]

56. Harland, B.F. Caffeine and nutrition. *Nutrition* **2000**, *16*, 522–526. [CrossRef]

57. Heckman, M.A.; Weil, J.; de Mejia, E.G. Caffeine (1, 3, 7-trimethylxanthine) in foods: A comprehensive review on consumption, functionality, safety, and regulatory matters. *J. Food Sci.* **2010**, *75*, R77–R87. [CrossRef] [PubMed]

58. FSL Software Package. Available online: http://www.fmrib.ox.ac.uk/fsl/ (accessed on 20 February 2018).

59. Brain Extraction Tool. Available online: http://www.fmrib.ox.ac.uk/fsl/fslwiki/BET (accessed on 20 February 2018).

60. FMRIB's Automated Segmentation Tool. Available online: http://www.fmrib.ox.ac.uk/fsl/fslwiki/fast (accessed on 20 February 2018).

61. Smith, S.; Nichols, T. Threshold-free cluster enhancement: Addressing problems of smoothing, threshold dependence and localisation in cluster inference. *NeuroImage* **2009**, *44*, 83–98. [CrossRef] [PubMed]

62. Smith, S.M.; Jenkinson, M.; Johansen-Berg, H.; Rueckert, D.; Nichols, T.E.; Mackay, C.E.; Watkins, K.E.; Ciccarelli, O.; Cader, M.Z.; Matthews, P.M.; et al. Tract-based spatial statistics: Voxelwise analysis of multi-subject diffusion data. *Neuroimage* **2006**, *31*, 1487–1505. [CrossRef] [PubMed]

63. Combinostics cMRI Software Package. Available online: https://www.cneuro.com (accessed on 20 February 2018).

64. Sabia, S.; Elbaz, A.; Britton, A.; Bell, S.; Dugravot, A.; Shipley, M.; Kivimaki, M.; Singh-Manoux, A. Alcohol consumption and cognitive decline in early old age. *Neurology* **2014**, *82*, 332–339. [CrossRef] [PubMed]

65. Xu, G.; Liu, X.; Yin, Q.; Zhu, W.; Zhang, R.; Fan, X. Alcohol consumption and transition of mild cognitive impairment to dementia. *Psychiatry Clin. Neurosci.* **2009**, *63*, 43–49. [CrossRef] [PubMed]

66. Corley, J.; Jia, X.; Brett, C.E.; Gow, A.J.; Starr, J.M.; Kyle, J.A.M.; McNeill, G.; Deary, I.J. Alcohol intake and cognitive abilities in old age: The lothian birth cohort 1936 study. *Neuropsychology* **2011**, *25*, 166–175. [CrossRef] [PubMed]

67. Huntley, J.; Corbett, A.; Wesnes, K.; Brooker, H.; Stenton, R.; Hampshire, A.; Ballard, C. Online assessment of risk factors for dementia and cognitive function in healthy adults. *Int. J. Geriatr. Psychiatry* **2018**, *33*, e286–e293. [CrossRef] [PubMed]

68. Lobo, E.; Dufouil, C.; Marcos, G.; Quetglas, B.; Saz, P.; Guallar, E.; Lobo, A. Is There an Association Between Low-to-Moderate Alcohol Consumption and Risk of Cognitive Decline? *Am. J. Epidemiol.* **2010**, *172*, 708–716. [CrossRef] [PubMed]

69. Mukamal, K.J. Alcohol consumption and abnormalities of brain structure and vasculature. *Am. J. Geriatr. Cardiol.* **2004**, *13*, 22–28. [CrossRef] [PubMed]

70. Haller, S.; Montandon, M.L.; Rodriguez, C.; Moser, D.; Toma, S.; Hofmeister, J.; Sinanaj, I.; Lovblad, K.O.; Giannakopoulos, P. Acute caffeine administration effect on brain activation patterns in mild cognitive impairment. *J. Alzheimers Dis.* **2014**, *41*, 101–112. [CrossRef] [PubMed]

71. Cunha, R.A.; Agostinho, P.M. Chronic caffeine consumption prevents memory disturbance in different animal models of memory decline. *J. Alzheimers Dis.* **2010**, *20* (Suppl. 1), S95–S116. [CrossRef] [PubMed]

*nutrients*

MDPI

*Article*

# The Acute Effects of Caffeinated Black Coffee on Cognition and Mood in Healthy Young and Older Adults

Crystal F. Haskell-Ramsay [1,*], Philippa A. Jackson [1], Joanne S. Forster [1], Fiona L. Dodd [1], Samantha L. Bowerbank [2] and David O. Kennedy [1]

[1]   Brain, Performance and Nutrition Research Centre, Northumbria University, Newcastle Upon-Tyne NE1 8ST, UK; philippa.jackson@northumbria.ac.uk (P.A.J.); jo.forster@northumbria.ac.uk (J.S.F.); f.dodd@northumbria.ac.uk (F.L.D.); david.kennedy@northumbria.ac.uk (D.O.K.)
[2]   Faculty of Health and Life Sciences, Northumbria University, Newcastle Upon-Tyne NE1 8ST, UK; samantha.bowerbank@northumbria.ac.uk
*   Correspondence: crystal.haskell-ramsay@northumbria.ac.uk; Tel.: +44-191-2274875

Received: 20 August 2018; Accepted: 24 September 2018; Published: 30 September 2018

**Abstract:** Cognitive and mood benefits of coffee are often attributed to caffeine. However, emerging evidence indicates behavioural effects of non-caffeine components within coffee, suggesting the potential for direct or synergistic effects of these compounds when consumed with caffeine in regular brewed coffee. The current randomised, placebo-controlled, double-blind, counterbalanced-crossover study compared the effects of regular coffee, decaffeinated coffee, and placebo on measures of cognition and mood. Age and sex effects were explored by comparing responses of older (61–80 years, $N = 30$) and young (20–34 years, $N = 29$) males and females. Computerised measures of episodic memory, working memory, attention, and subjective state were completed at baseline and 30 min post-drink. Regular coffee produced the expected effects of decreased reaction time and increased alertness when compared to placebo. When compared to decaffeinated coffee, increased digit vigilance accuracy and decreased tiredness and headache ratings were observed. Decaffeinated coffee also increased alertness when compared to placebo. Higher jittery ratings following regular coffee in young females and older males represented the only interaction of sex and age with treatment. These findings suggest behavioural activity of coffee beyond its caffeine content, raising issues with the use of decaffeinated coffee as a placebo and highlighting the need for further research into its psychoactive effects.

**Keywords:** coffee; caffeine; chlorogenic acids; phenolic; cognition; cognitive; mood; age; sex

---

## 1. Introduction

Coffee consumption is associated with a number of health benefits in elderly men and women including reduced risk of cardiovascular disease (CVD) [1], lower incidence of type 2 diabetes mellitus [2], and decreased death from inflammatory diseases [3], CVD [4,5], and all-cause mortality [6,7]. A number of physiological factors associated with these conditions are relevant to cognitive function in healthy ageing, as well as pathological ageing conditions such as dementia or Alzheimer's disease (AD). Indeed, a number of epidemiological studies have demonstrated an association between higher coffee consumption and better performance on cognitive tests in older adults [8,9], as well as an inverse relationship between coffee consumption and risk of dementia/AD [10–14].

Cognitive benefits from coffee consumption are typically attributed to caffeine, which exerts its effects through non-selective antagonism of adenosine $A_1$ and $A_{2A}$ receptors [15]. In support of this, a number of studies have demonstrated the ability of caffeine to improve measures of attention and increase ratings of alertness [16–18]. However, coffee contains more than 1000 different compounds including phenolics, diterpenes, and melanoidins [19], all of which have the potential to affect behaviour either directly or indirectly through interaction with caffeine. This is demonstrated by studies showing direct psychoactive effects and modulation of caffeine's effects by the amino acid l-theanine, present in tea [20–22]. Similarly, lengthened startled blink onset latency has been shown following decaffeinated coffee as compared to caffeinated coffee, caffeinated juice, and non-caffeinated juice [23]. Chlorogenic acids (CGA) are a group of phenolic compounds representing the principal non-caffeine components in coffee [24] and have been explored in relation to mood and cognition in healthy, elderly participants [25]. In comparison to regular CGA decaffeinated coffee (224 mg CGA, 5 mg caffeine), high CGA decaffeinated coffee (521 mg CGA, 11 mg caffeine) increased alertness and decreased negative emotional processing, whereas caffeinated coffee (244 mg CGA, 167 mg caffeine) increased accuracy on a sustained attention task and improved mood. These results indicate that the addition of CGA to regular decaffeinated coffee can modulate its effects on behaviour. The effects of CGA were explored further in a study comparing 540 mg isolated CGA, 6 g decaffeinated green blend coffee (532 mg CGA), and placebo [26]. Whilst positive effects on mood were observed following decaffeinated green blend coffee, these effects were not evident following CGA in isolation, which also led to detrimental effects to cognition at 120 min post-drink. This provides further evidence for behavioural effects of decaffeinated coffee and highlights the need to consider the synergistic contribution of non-caffeine compounds in coffee.

Coffee is one of the most widely consumed beverages in the world, yet intervention trials examining the specific impact of consuming regular, brewed coffee on cognition and mood are lacking. Given the potential for non-caffeine components within coffee to exert psychoactive effects or to interact synergistically with caffeine, it is important that the effects of regular coffee and decaffeinated coffee are compared to placebo. In addition, despite physiological differences between men and women, including in their nutrient needs and in cognitive performance [27,28], sex differences are rarely considered in nutritional intervention trials. This is particularly important here, as studies of the relationship between coffee and cognitive decline have indicated that whilst reduced risk is related to coffee consumption in men [29], the effect is more pronounced in women [30,31]. This suggests that greater effects of coffee consumption may be observed in older adults as a consequence of cognitive decline, and that these beneficial effects may be enhanced in females. Furthermore, given the impact of the menstrual cycle on resting metabolic rate [32] and systemic clearance of caffeine [33], it is also possible that sex differences in response will be moderated by age. In order to explore this further, the current study compared the behavioural effects of regular coffee, decaffeinated coffee and placebo in elderly participants (61–80 years) to those in a younger (20–34 years) adult group and examined differential responses in men and women. As debate continues as to whether caffeine's effects are modulated by habitual consumption [34,35] only those who regularly consumed coffee and tea were included.

## 2. Materials and Methods

### 2.1. Design

A randomised, placebo-controlled, double-blind, counterbalanced-crossover design was employed. The study was approved by Northumbria University's Faculty of Health and Life Sciences Ethics Committee (reference: SUB057_Forster_090216; approved: 26 February 2016) and was conducted in accordance with the Declaration of Helsinki.

## 2.2. Participants

Seventy-two participants were drawn through an opportunity sample within Newcastle upon Tyne and the surrounding areas. Thirty-six of these represented an older group aged 61 to 80 years (18 male), and 36 represented a younger comparator group aged 20 to 34 years (18 male). Sample size was determined from a power calculation based upon previous data showing improvements to cognition and mood in habitual caffeine consumers following 150 mg caffeine [16]. An effect size of $d = 0.6$ indicated that a total of 72 participants would allow detection of significant effects with a power of 0.8. All participants were healthy non-smokers for whom English was their first language. Participants were not currently taking medication with the exception of contraception in young female participants and those used in the treatment of arthritis, high blood pressure, high cholesterol, and reflux-related conditions in the older participant group. Due to the potential impact of habitual caffeine intake on response, only those who regularly consumed more than two cups of coffee or three cups of tea (equating to $\geq$150 mg caffeine/day) were included. Participants were paid £60 for taking part.

Thirteen participants were excluded from the per protocol analysis (12 based on high (>1 μg/mL) pre-dose caffeine salivary levels, and one due to under-consumption of the drink provided). The population for analysis (see Figure 1) consisted of 30 older adults (14 male) and 29 young (16 males). Participant characteristics for each age group by sex can be found in Table 1.

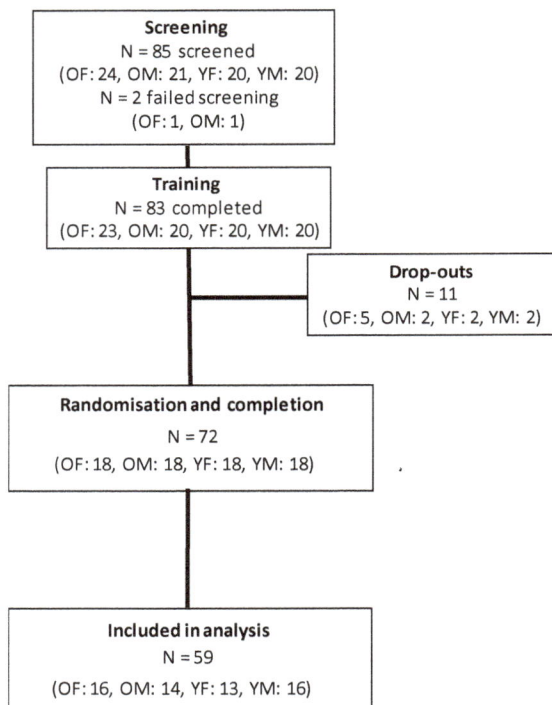

**Figure 1.** Final participant disposition. N = Number of participants; OF = Older Female; OM = Older Male; YF = Younger Female; YM = Younger Male.

**Table 1.** Participant demographics (SD = Standard Deviation).

| | Young | | | | Older | | | |
|---|---|---|---|---|---|---|---|---|
| | Male | | Female | | Male | | Female | |
| | Mean | SD | Mean | SD | Mean | SD | Mean | SD |
| Age | 26.3 | 4.4 | 26.2 | 3.6 | 67.7 | 6.3 | 67.1 | 3.4 |
| Years in education | 18 | 3 | 17 | 3 | 16 | 5 | 14 | 4 |
| Body Mass Index (BMI) | 25.7 | 3.8 | 23.8 | 3.6 | 25.9 | 3.4 | 26.1 | 3.9 |
| Caffeine consumption (mg/day) | 327 | 88.2 | 351 | 110.4 | 426 | 74.4 | 394 | 87.8 |
| Coffee consumption (cups/day) | 2.88 | 1.54 | 2.54 | 1.45 | 2.64 | 1.13 | 3.59 | 0.93 |
| Fruit and vegetables (portions/day) | 4.3 | 1.5 | 4.1 | 1.4 | 4.4 | 1.9 | 5.4 | 1.8 |

### 2.3. Treatment

At each study visit, one of the following drinks was administered by an independent third party with no further involvement in the study.

- 220 mL water mixed with 2.5 g coffee flavouring (placebo)
- 220 mL regular coffee (without milk and sugar) containing 100 mg caffeine
- 220 mL decaffeinated coffee (without milk and sugar) containing ~5 mg caffeine

The order in which participants received each drink was determined by computer-generated random allocation (Latin square) for each sex (male, female) by group (older and younger age comparators). Regular and decaffeinated coffee were brewed using two separate drip filter coffee makers following a standardised brewing procedure, including the use of filter papers to minimise cafestol and kahweol levels. Placebo consisted of 2.5 g flavouring (maltodextrin 2.26 g, dark roast 0.1 g, mild roast 0.1 g, and coffee natural 0.04 g—Firmenich SA, Meyrin, Satigny, Switzerland) added to boiling water. Drinks were matched for temperature (58 °C) and served in an opaque thermal beaker with a black opaque straw with 5 min allowed for drinking.

### 2.4. Salivary Caffeine Levels

Saliva samples were obtained using salivettes (Sarstedt, Leicester, UK). Samples were taken immediately prior to baseline assessments in order to confirm compliance to abstinence and following post-drink assessments to confirm effective caffeine absorption. The saliva samples were immediately frozen at $-20$ °C until thawing. Once thawed, salivette tubes were centrifuged at $15,000\times g$ for 10 min. Stock solutions of caffeine, paraxanthine, and benzotriazole (internal standard) were prepared in type I ultra-pure water at a concentration of 100 µg/mL. Calibration standards for caffeine and paraxanthine were prepared between 0.05 and 5.00 µg/mL. Quality control samples were also prepared at a concentration of 2.5 µg/mL. Internal standard (50 µL at 5.0 µg/mL) was added to 50 µL of each standard and sample in duplicate. To extract the compounds 2 mL of ethyl acetate was added and solutions were vortex mixed for 3 min following by centrifugation at $4000\times g$ for 10 min. The organic layer was transferred to a clean tube and dried under a stream of nitrogen at 45 °C. The residue was reconstituted in 100 µL of mobile phase and 50 µL injected onto the column.

Saliva samples were analysed with high-performance liquid chromatography. The HPLC system was an Agilent 1260 Infinity™ (Cheadle, Greater Manchester, UK) consisting of an Infinity™ quaternary pump, an Infinity™ Autosampler with integrated column oven and an Infinity™ multi-wavelength detector set at 280 nm. Instrument control and data processing was performed using Agilent OpenLab™ CDS (Agilent Technologies Ltd., Cheadle, UK). Chromatographic separation was achieved on a Kinetex $C_{18}$ column (4.6 × 250 mm i.d., particle size 5 µm; Phenomenex Ltd., Macclesfield, UK). The mobile phase consisted of acetonitrile, acetic acid and type I ultra-pure water (5:1:95, $v\%:v\%:v\%$) delivered at a flow rate of 1.00 mL/min.

*2.5. Cognitive and Mood Measures*

With the exception of driving ability, all cognitive and mood measures were delivered using the Computerised Mental Performance Assessment System (COMPASS, Northumbria University, Newcastle upon Tyne, UK), a purpose-designed software application for the flexible delivery of randomly generated parallel versions of standard and novel cognitive assessment tasks. This assessment system has previously been shown to be sensitive to nutritional interventions [36,37] including caffeine [20]. The tasks and mood scales were chosen based on their known sensitivity to caffeine or susceptibility to ageing. Tasks were presented in the same order on each occasion and, with the exception of the paper and pencil tasks (immediate and delayed word recall and verbal fluency), responses were made using a response pad. The entire selection of tasks took approximately 25 min to complete. See Table 2 for order and scoring of tasks completed at baseline. Due to the potential for interference from repeat completions, computerised location learning and driving simulation were only completed post-dose on study visits with the final session from their training day used as the statistical baseline in the analyses.

**Table 2.** Cognitive tasks completed at baseline and 30 min post-dose in order of presentation (computerised location learning and driving ability are described below).

| Task | Descriptor | Scoring | Domain |
|---|---|---|---|
| Word presentation | A series of words is displayed on the screen, one word at a time. In this case, 15 words were presented with a display time of 1 s and interstimulus interval of 1 s | - | |
| Immediate word recall | Participants are instructed to write down the words that were presented. In this case, 60 s were given to complete the task | Number correct and number of errors | Episodic memory |
| Picture presentation | A series of photographic images are displayed on the screen, one at a time. In this case, 15 images were presented with a display time of 2 s and an interstimulus interval of 1 s | - | |
| Simple reaction time | An upwards pointing arrow is displayed on the screen at irregular intervals. Participants must respond as quickly as they can as soon as they see the arrow appear. In this case, 50 stimuli were presented | Reaction time (ms) | Attention |
| Digit vigilance | A fixed number appears on the right of the screen and a series of changing numbers appear on the left of the screen at the rate of 150 per minute. Participants are required to make a response when the number on the left matches the number on the right. In this case the task lasted for 3 min | Accuracy (%), reaction time for the correct responses (ms) and false alarms (number) | Attention |
| Numeric working memory | Five single target numbers are displayed on the screen, one at a time. Participants are required to memorise these numbers as they appear. Once the target series has been presented, numbers are displayed one at a time and participants are required to indicate if each number was presented in the previous list or not. In this case, three trials were completed | Accuracy (%) and reaction time for the correct responses (ms) | Working memory |
| Verbal fluency | Participants are presented with a letter on screen and asked to write down as many words as they can, beginning with that letter. In this case, the letters presented were A, T, C, F, M, and S and 60 s were given to complete the task | Number correct permitted words, with names and perseverations discounted from the total score | Language |
| Delayed word recall | Participants are instructed to write down the words that were presented to them at the beginning of the assessment. In this case, 60 s were given to complete the task | Number correct and number of errors | Episodic memory |

Table 2. *Cont.*

| Task | Descriptor | Scoring | Domain |
|---|---|---|---|
| Rapid visual information processing | A continuous series of single digits are presented in the centre of the screen at the rate of 100 per minute. Participants are required to make a response when three consecutive odd or three consecutive even digits are displayed. In this case, the task lasted for 5 min, with eight correct target strings presented in each minute. | Accuracy (%), reaction time for the correct responses (ms) and false alarms (number) | Attention |
| Delayed word recognition | All target words that were shown during Word presentation plus an equal number of decoys are displayed on the screen one at a time. Participants indicate if they remember seeing the word earlier or not. | Accuracy (%) and reaction time for the correct responses (ms) | Episodic memory |
| Delayed picture recognition | All target pictures shown during Picture presentation plus an equal number of decoys are displayed on the screen one at a time. Participants indicate if they remember seeing the picture earlier or not. | Accuracy (%) and reaction time for the correct responses (ms) | Episodic memory |

## 2.6. Caffeine Research Visual Analogue Scales

Prior to cognitive assessment, subjective state was assessed with the Caffeine Research Visual Analogue Scales [38], which have previously been used in caffeine research [16,21,39]. The following descriptors are presented on-screen: 'relaxed', 'alert', 'jittery', 'tired', 'tense', 'headache', 'overall mood', and 'mentally fatigued'. Participants are asked to rate how much these descriptors match their current state by placing an 'x' on a line with the end points labelled 'not at all' (left hand end) and 'extremely' (right hand end); with the exception of 'headache', which is labelled 'no headache' and 'extreme headache'; and 'overall mood', which is labelled 'very bad' and 'very good'. Ratings are scored as % along the line from left to right.

## 2.7. Computerised Location Learning—Learning Phase

Location learning was assessed with a computerised task modified from Kessels et al. [40]. Participants are shown a grid containing pictures of objects. Following a timed delay they are shown an empty grid and asked to relocate the objects to the correct location shown to them previously. In the current study, this was repeated five times during the learning phase, with objects presented for 15 s, a gap of 10 s before the empty grid was shown, and a pause of 5 s between each trial. For each of the five learning trials, a displacement score is calculated as the sum of the errors made for each object (calculated by counting the number of cells the object had to be moved both horizontally and vertically in order to be in the correct location) from each trial. A learning index is also calculated as the average relative difference in performance between trials $[((A - B)/A + (B - C)/B + (C - D)/C + (D - E)/D)/4]$.

## 2.8. Computerised Location Learning—Delayed Trial

During the delayed trial, which took place 30 min after completion of the learning phase, participants are again asked to place the objects in the correct location on the empty grid with no further prompting. The delayed trial is scored for displacement and delayed recall, which is calculated as the difference between displacement score on the final learning trial and the delayed trial.

## 2.9. Driving Ability

A PC based driving simulation (Driving Simulator 2013, Excalibur Publishing Limited, Banbury Oxfordshire, UK) was used to assess driving ability. Driving was controlled via a steering wheel and pedals with gears set to fully automatic. The task lasted for 3 min and is scored on the basis of adhering to road rules and driving ability. Specifically, the task is scored for errors, which are given when deviating too much from the track; deviating too much from the instructed directions; not indicating; speeding; colliding. If the drive ended (either because of collision or because of exceeding 10 errors) the task was restarted but no more than two restarts (three drives in total) were allowed.

*2.10. Procedure*

Potential participants attended the Brain Performance and Nutrition Research Centre at Northumbria University for an initial screening session where they gave informed consent prior to participation. Their eligibility was assessed in accordance with the criteria outlined in the 'Participants' section and training on the computerised tasks was provided. This consisted of five completions of cognitive tasks and took place on a single day. Participants attended three study visits, separated by at least seven days to allow for washout and to prevent confound due to caffeine abstention instructions. These instructions required abstention from caffeine from noon the day before study visits but this did not exceed 24 h in order to minimise any potential withdrawal effects. Consumption of alcohol and over-the-counter medication was also restricted for 24 h (48 h in the case of systemic antihistamines). On the morning of study visits, participants ate their usual breakfast at least 1 h prior to arrival at the laboratory with the time and composition of breakfast standardised across visits. Participants attended the laboratory at 9:45 a.m. and were screened to ensure eligibility for testing that day, this included checking they were in good health and had adhered to instructions regarding breakfast consumption, and caffeine, alcohol, and medication restrictions. A food diary was used to aid with breakfast standardisation and a saliva sample was obtained to confirm adherence to caffeine abstention instructions. All testing took place in a suite of dedicated temperature-controlled university laboratories with participants visually isolated from each other and wearing noise-reduction headphones to decrease the impact of any auditory distractions. Baseline assessments of cognition and mood were completed and participants were then given their drink for that day. After 30 min of rest in the laboratory, the learning phase of a computerised Location Learning Test (cLLT) was completed before parallel versions of the tasks completed at baseline. This was followed by the delayed trial of the cLLT, a driving simulation task and a final saliva sample for assessment of caffeine levels. At the end of the final visit only, participants were asked to guess which drink they believed they had consumed that day. See Figure 2 for a schematic depicting the study visit running order.

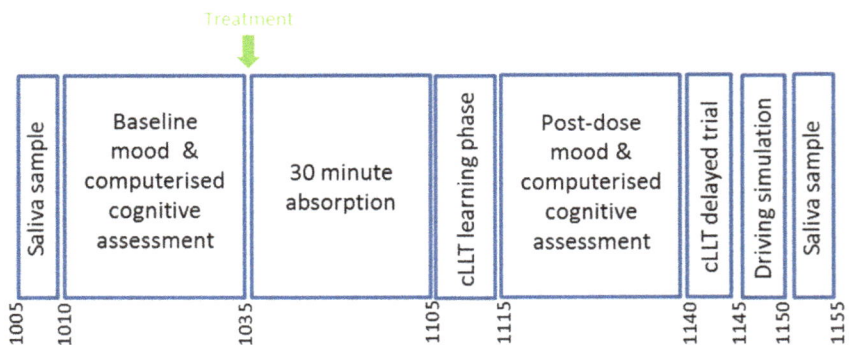

**Figure 2.** Study visit timeline.

*2.11. Statistics*

The post-dose outcome measures were modelled using the MIXED procedure in SPSS (version 24.0, IBM Corp., Armonk, NY, USA) which included the respective baseline values and the terms treatment, age, sex, treatment × age, treatment × sex and treatment × age × sex as fixed factors. In the case of computerised location learning and driving simulation, baseline values were taken from the final training session. Significant effects were followed up with Bonferroni corrected pairwise comparisons.

## 3. Results

*3.1. Treatment-Related Effects*

### 3.1.1. Salivary Caffeine

Baseline salivary caffeine values were 0.17 μg/mL, confirming adherence to caffeine abstention instructions. A significant main effect of treatment was observed on post-dose salivary caffeine ($F(2, 101.1) = 155.6$, $p < 0.0001$). Pairwise comparisons revealed significantly greater levels following caffeinated coffee compared to placebo ($p < 0.0001$) and decaffeinated coffee ($p < 0.0001$). See Figure 3.

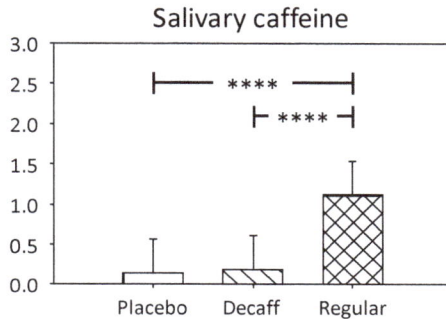

**Figure 3.** Adjusted means + standard error for salivary caffeine measured in μg/mL. Significant treatment effect **** $p < 0.001$.

### 3.1.2. Digit Vigilance

A significant main effect of treatment was observed for digit vigilance accuracy ($F(2, 101.1) = 4.44$, $p = 0.014$). Pairwise comparisons revealed significantly greater accuracy following regular coffee compared to decaffeinated coffee ($p = 0.01$). See Figure 4a.

Digit vigilance reaction time was also significantly affected by treatment ($F(2, 71.3) = 5.07$, $p = 0.009$). Pairwise comparisons revealed significantly faster responses following regular coffee compared to placebo ($p = 0.009$). See Figure 4b.

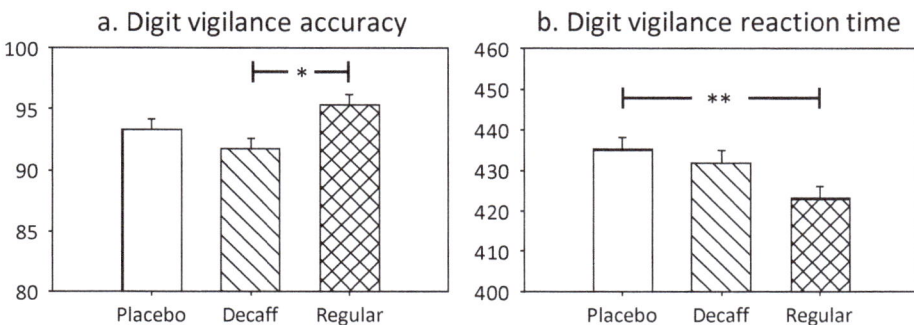

**Figure 4.** *Cont.*

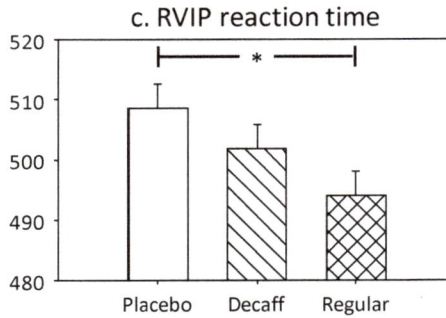

**Figure 4.** Adjusted means + standard error for those cognitive measures showing significant effects of treatment. (**a**) Digit vigilance accuracy; (**b**) Digit vigilance reaction time; (**c**) Rapid Visual Information Processing (RVIP) reaction time. Accuracy is measured as % and reaction time in milliseconds. * $p < 0.05$; ** $p < 0.01$.

### 3.1.3. Rapid Visual Information Processing

Rapid visual information processing reaction time showed a significant effect of treatment ($F(2, 102.9) = 3.77$, $p = 0.026$). This was due to significantly faster responses following regular coffee compared to placebo ($p = 0.02$). See Figure 4c. A significant treatment x sex interaction was also observed for false alarms ($F(2, 93.3) = 4.55$, $p = 0.013$) but pairwise comparisons revealed no significant effects.

### 3.1.4. Computerised Location Learning Delayed Trial

Computerised location learning recall showed a significant treatment x sex interaction ($F(2, 104) = 3.46$, $p = 0.035$). However, pairwise comparisons revealed no significant effects.

### 3.1.5. Alert

Ratings of 'alertness' were significantly affected by treatment ($F(2, 106) = 9.86$, $p < 0.0001$). This was due to significantly higher ratings following regular coffee ($p < 0.0005$) and decaffeinated coffee ($p = 0.0048$) compared to placebo. See Figure 5a.

### 3.1.6. Tired

A significant main effect of treatment on 'tired' ratings was also observed ($F(2, 101.4) = 12.31$, $p = 0.0001$). Pairwise comparisons revealed this was due to significantly lower ratings following regular coffee compared to decaffeinated coffee ($p = 0.003$) and placebo ($p < 0.0001$). See Figure 5b.

### 3.1.7. Headache

A significant main effect of treatment on headache ratings ($F(2, 92.9) = 6.31$, $p = 0.003$) was due to significantly lower ratings following regular coffee compared to decaffeinated coffee ($p = 0.0049$) and placebo ($p = 0.015$). See Figure 5c.

### 3.1.8. Overall Mood

'Overall mood' was significantly affected by treatment ($F(2, 105.8) = 5.56$, $p < 0.005$). This was due to significantly higher ratings following regular coffee compared to placebo ($p = 0.004$). See Figure 5d.

### 3.1.9. Mental Fatigue

A significant main effect of treatment on 'mental fatigue' ratings was also observed (F(2, 97.5) = 4.43, *p* = 0.014). Pairwise comparisons revealed this was due to significantly lower ratings following regular coffee compared to placebo (*p* = 0.01). See Figure 5e.

### 3.1.10. Jittery

A significant treatment × age × sex interaction was observed on jittery ratings (F(3, 76.2) = 3.01, *p* = 0.035). Pairwise comparisons revealed significantly higher ratings following regular coffee compared to placebo in young females (*p* = 0.046) and compared to decaffeinated coffee in older males (0.045). See Figure 5f.

Unadjusted means, standard deviations, and *F* and *p* values for all factors (treatment, age, sex) and their interactions can be found in Tables S1–S3.

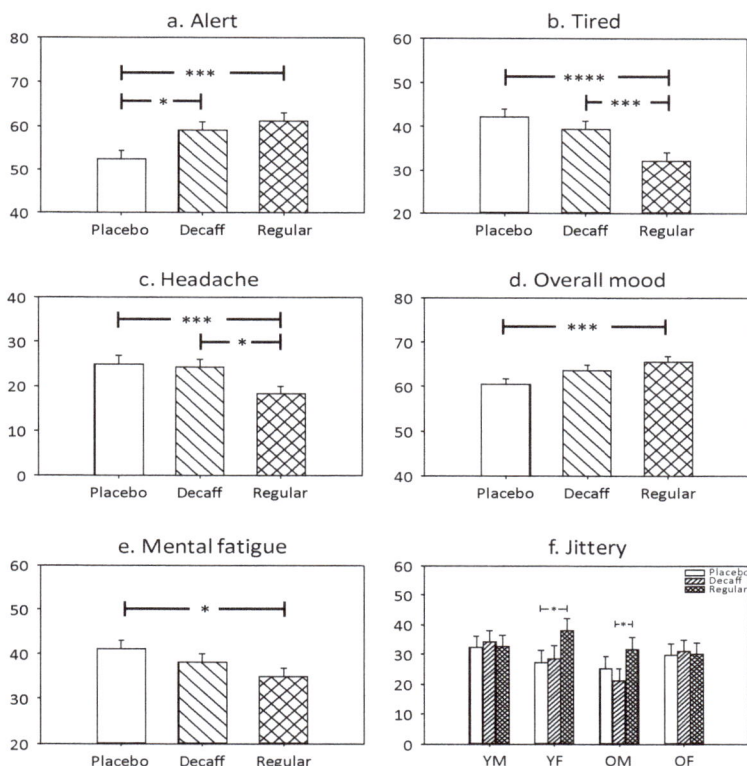

**Figure 5.** Adjusted means + standard error for those mood measures showing significant treatment-related effects. (**a**) Alert; (**b**) Tired; (**c**) Headache; (**d**) Overall mood; (**e**) Mental fatigue; (**f**) Jittery. Ratings are measured as % along a visual analogue scale with higher values indicating greater response. YM = young male; YF = young female; OM = older male; OF = older female; * *p* < 0.05; *** *p* < 0.005; **** *p* < 0.001.

*3.2. Treatment Guess*

Seventy-one percent of participants correctly guessed which drink they had received at the final visit. Eighty-one percent correctly guessed they had received regular coffee, 72% correctly identified decaffeinated coffee, and 58% were able to correctly identify placebo as their final drink.

## 4. Discussion

Consumption of 220 mL of regular coffee containing 100 mg caffeine led to faster responses during digit vigilance and rapid visual information processing tasks when compared to placebo, and to increased digit vigilance accuracy when compared to decaffeinated coffee. In terms of mood effects, ratings of alertness and overall mood were higher and mental fatigue ratings lower following regular coffee compared to placebo. Tiredness and headache ratings were lower following regular coffee compared to placebo and decaffeinated coffee. Rating of jitteriness was the only outcome to show an interaction with sex and age indicating higher ratings following regular coffee when compared to placebo in young females and when compared to decaffeinated coffee in older males. Decaffeinated coffee also engendered an increase in subjective alertness, compared to placebo, whereas accuracy of digit vigilance and tired and headache ratings were impaired in comparison to regular coffee. A beneficial effect of decaffeinated coffee was also observed following a treatment × age × sex interaction, which indicated that ratings of jitteriness were significantly lower following decaffeinated compared to regular coffee in older men. The pattern of response to decaffeinated coffee generally fell between responses to regular coffee and placebo. Specifically, numeric working memory accuracy and reaction time, reaction time for attention tasks and mood ratings, all followed the order of placebo > decaffeinated > regular coffee (see Supplementary Tables), with the exception of relaxed and tense ratings which showed a preferential effect of decaffeinated coffee.

The findings with regards regular coffee are largely in line with the reported effects of caffeine, which only has a consistent beneficial effect on attention task performance and subjective alertness/arousal [16–18]. Whilst this could be taken as support for the notion that caffeine is the sole contributor to the effects, the finding of psychoactive effects of decaffeinated coffee, in terms of increased alertness when compared to placebo and a pattern of lower effects than regular coffee in comparison to placebo, supports the suggestion of a modulatory role for the non-caffeine compounds within coffee. The effects of decaffeinated coffee presented here are broadly in line with previous results showing impairment to accuracy of a sustained attention task in comparison to regular coffee [25] and increases in alertness when compared to placebo and the phenolic acid CGA in isolation [26]. Previous studies have highlighted CGA as a potentially important component of coffee. However, whilst there is some evidence for beneficial modulation of coffee's effects by increasing CGA content [25], the effects of CGA in isolation were largely negative [26]. This potentially highlights an issue in applying a reductionist approach to nutritional interventions where complex interactions between many different components may be required to see optimum results. Although composition is varied depending on roasting and brewing techniques, caffeine generally only accounts for ~1% and CGA ~10% of the weight of coffee beans, this leaves almost 90% of the constituents unaccounted for. It is also important to note that the CGA profile may be altered as part of the decaffeination process and therefore any analysis of effects must take account of the impact of decaffeination on other constituents [41].

The observed benefits for regular coffee are expected due to the known effects of caffeine in antagonising adenosine $A_1$ and $A_{2A}$ receptors thereby, increasing oxygen metabolism [42] and upregulating various neurotransmitters including noradrenaline, dopamine, serotonin, acetylcholine, and GABA [15]. Caffeine and its metabolites also have a number of mechanistic properties that make them liable to have a modulatory or interactive effect when caffeine is co-consumed with other bioactive compounds. These include the inhibition of enzymes involved in the breakdown of neurotransmitters (e.g., acetylcholinesterase and monoamine oxidase) and cellular signalling molecules (e.g., phosphodiesterase and PARP (poly(ADP-ribose)polymerase)) [43,44] and a role as a competitive substrate for a number of cytochrome P450 (CYP) enzymes (CYP2A1, CYP2E1, and CYP1A1) that metabolise endogenous and exogenous chemicals in the human body [45–47]. Of particular relevance here, low-doses of caffeine have therefore been shown to increase the bioavailability of phenolic compounds [48–50] and have a synergistic effect in terms of the cardiovascular benefits of polyphenols.

Coffee also has the potential to impact glucose metabolism as evidenced by an increase in insulin sensitivity observed following decaffeinated coffee when compared to placebo [51]. Interestingly, this effect was not apparent following regular coffee, which may be due to counteractive effects of caffeine and non-caffeine components within regular coffee. Support for this comes from data showing decreased insulin sensitivity following caffeine [52]. Moreover, area under the curve (AUC) profiles for serum insulin indicate that caffeine increases AUC when compared to decaffeinated coffee and placebo, whilst regular coffee produced a trend towards the same when compared to decaffeinated coffee, with similar profiles evinced for glucose AUC [53]. CGA derivatives have been shown to increase insulin sensitivity in rats [54], and further support for the role of phenolic compounds in this effect comes from data showing modulation of glucose and insulin response following phenolic-rich berries [55–57] as well as a reduction in the postprandial blood glucose response following grape seed extract [58]. Similarly, caffeine is known to have a vasoconstrictive effect, including reduced cerebral blood flow (CBF) [39], whereas phenolic-rich foods have demonstrated the opposite effect. Of particular relevance is the ability of phenolic-rich cocoa to increase CBF when compared to a phenolic-poor control matched for methylxanthine content [59,60]. These findings indicate the ability of coffee components to counteract the negative effects of caffeine and a potential synergy whereby phenolic compounds increase CBF, and therefore oxygen supply, whilst caffeine increases brain activity and subsequent oxygen metabolism. It is also possible that caffeine increases absorption of phenolics as has been shown following consumption of cocoa [48] but is as yet untested following coffee.

A further consideration is that due to a focus on psychoactive effects of caffeine, the cognitive and mood effects of coffee have typically been measured at 30 to 120 min post-dose coinciding with a peak in caffeine levels at around 40-min post-ingestion [61]. However, analysis of the fate of CGA following coffee consumption shows that whilst a number of phenolic acids and their derivatives peak between 30 and 60 min, others do not appear until between 4- and 6-h post-ingestion [62]. It is therefore necessary to extend the testing period in order to fully examine the impact of these metabolites. This is also true in relation to caffeine, which has a half-life of around 5 h [63], and has demonstrated behavioural effects up to 8 h post-dose [64]. It is therefore probable that any effects of coffee observed at 6-h would represent an interaction between phenolic acids and caffeine and the measurement of biomarkers would aid in elucidating the role of each.

Learning and episodic memory tasks showed the expected effect in terms of significantly lower performance in the older cohort when compared to young (see Supplementary Tables). However, no interactions between age and treatment were observed on any cognitive measure. Whilst learning and memory are not typically susceptible to caffeine, it has been proposed that these tasks may show sensitivity in low arousal situations as is expected in the elderly as energetic resources diminish [65]. However, in the current study there was no evidence of higher arousal in the young sample when compared to the older participants on subjective measures of 'arousal' or psychomotor tasks. This may suggest that the older cohort studied here were relatively high functioning, as is supported by their status being higher than national averages both in relation to fruit and vegetable consumption and education level [66]. It has also been suggested that cognitive benefits of caffeine consumption may be more pronounced in those aged over 80 years [31,67]. Therefore, the findings reported here do not preclude interaction effects in an older sample with poorer nutritional status and/or lower education level.

Similarly, although sex differences were observed in the current study, these did not interact with treatment as may have been expected from data showing greater benefits of coffee consumption in women than men [30]. However, the potential mechanisms underlying sex differences following habitual consumption, including sex steroid levels [68,69], haemodynamic mechanisms [70], and uric acid responses [71,72], are unlikely to exert effects over a 30-min time period. Furthermore, given the impact of the menstrual cycle and hormonal contraception on metabolism, it is possible that any differential sex effects in the younger cohort were obscured by the lack of control for menstrual cycle phase and the inclusion of four hormonal contraceptive users in this study. This also potentially explains large variations in salivary caffeine following regular coffee in young females that were not observed in the young male group. Similar large variations in response were shown for older men and women indicating individual differences in response to caffeine. This variability is largely due to differences in CYP1A2 activity, which is influenced by a number of factors including sex and genetic polymorphisms [73].

Polymorphisms of the ADORA2A gene may only explain the age, sex, treatment interaction, which indicated that older men and younger women experienced greater feelings of jitteriness following regular coffee. It has previously been reported that those who are T/T homozygous at nucleotide positions 1976C > T and 2592Tins experience increases in anxiety after caffeine administration that are not observed in the other genotypic groups [74,75]. Moreover, sex differences in response have been noted for 1976TT homozygotes, whereby females are more susceptible to anxiogenic effects of caffeine than males [76,77]. Interestingly, whilst there is evidence for reduced caffeine intake in 1976TT homozygotes [78], others have shown that intake of coffee, but not other sources of caffeine, is increased. It was also shown that increased habitual consumption moderated anxiogenic effects of caffeine such that they were only observed in non/light consumers of caffeine, irrespective of genotype [79]. This indicates that even in those with a genetic predisposition, tolerance to anxiogenic effects can occur with habitual consumption. As the older men in the current study consumed less coffee than their female counterparts, this may in part explain the specificity of 'jittery' effects observed here. However, there are a multitude of factors that impact on interindividual differences [80], which require further exploration before definitive conclusions can be reached.

In the current study, although the addition of a true placebo builds on research previously limited to comparing effects following caffeinated and decaffeinated coffee, one important limitation is the omission of a caffeine-only arm. The inclusion of a caffeine-only condition would have allowed a direct comparison of any synergistic effects between caffeine and the other bioactive compounds in coffee, including the phenolic compounds. It may also have facilitated in blinding of drinks. Although the placebo drink in the current study was somewhat effective in that 42% incorrectly identified it, only 1 of 19 participants mistook it for regular coffee. It is important to note that this measure was only included at the end of the final visit when all three drinks had been consumed and, therefore, does not rule out the blinding of participants at earlier visits. However, as the stimulant effects of caffeine are easily detected, the inclusion of a caffeine-containing 'placebo' would provide an active control for regular coffee and reduce the ability of participants to correctly identify regular coffee.

The findings presented here suggest behavioural activity of coffee beyond its caffeine content. In fact, only one cognitive measure and two subjective measures showed significant differences between regular and decaffeinated coffee in favour of regular coffee. This highlights two key issues with studies which compare regular and decaffeinated coffee. Firstly, these studies attribute any differential effects to caffeine without considering the potential for interaction with other components. Secondly, any synergistic effects of caffeine and other coffee components within regular coffee are likely to be underestimated due to the potential for behavioural effects of decaffeinated coffee used as the control. If the effects of regular coffee are to be fully understood, it is important that future research compares these to the equivalent dose of caffeine, decaffeinated coffee, and placebo. Furthermore, research in this area must include plasma levels of potentially important compounds, including phenolic compounds. This would allow assessment of the impact of caffeine on the pharmacodynamic profile of other components in coffee. An extended testing period is also recommended in order to capture effects of colonic metabolites of phenolics appearing at ~8 h. Further research is also required in which cognition is measured alongside potential underlying mechanisms including, but not limited to, glucoregulation and modulation of cerebral haemodynamics. Finally, in order to capture the impact of interindividual differences in metabolism of caffeine and other components of coffee on behavioural outcomes, genetic factors should also be considered.

**Supplementary Materials:** The following are available online at http://www.mdpi.com/2072-6643/10/10/1386/s1, Table S1: saliva; Table S2: cognition; Table S3: mood.

**Author Contributions:** C.F.H-R., P.A.J., J.S.F., F.L.D. and D.O.K. contributed to the design of the study; J.S.F. and F.L.D. collected the data; S.L.B. conducted the salivary analysis; C.F.H.-R. conducted the data analysis; all authors contributed to and reviewed the final publication.

**Funding:** This research and APC were funded by the Institute for Scientific Information on Coffee (ISIC) who collaborated on aspects of the design but had no further involvement in the study.

**Conflicts of Interest:** The authors declare no conflicts of interest.

## References

1. Rodriguez-Artalejo, F.; Lopez-Garcia, E. Coffee consumption and cardiovascular disease: A condensed review of epidemiological evidence and mechanisms. *J. Agric. Food Chem.* **2018**, *66*, 5257–5263. [CrossRef] [PubMed]
2. Jiang, X.B.; Zhang, D.F.; Jiang, W.J. Coffee and caffeine intake and incidence of type 2 diabetes mellitus: A meta-analysis of prospective studies. *Eur. J. Nutr.* **2014**, *53*, 25–38. [CrossRef] [PubMed]
3. Andersen, L.F.; Jacobs, D.R., Jr.; Carlsen, M.H.; Blomhoff, R. Consumption of coffee is associated with reduced risk of death attributed to inflammatory and cardiovascular diseases in the Iowa women's health study. *Am. J. Clin. Nutr.* **2006**, *83*, 1039–1046. [CrossRef] [PubMed]
4. Greenberg, J.A.; Dunbar, C.C.; Schnoll, R.; Kokolis, R.; Kokolis, S.; Kassotis, J. Caffeinated beverage intake and the risk of heart disease mortality in the elderly: A prospective analysis. *Am. J. Clin. Nutr.* **2007**, *85*, 392–398. [CrossRef] [PubMed]
5. Freedman, N.D.; Park, Y.; Abnet, C.C.; Hollenbeck, A.R.; Sinha, R. Association of coffee drinking with total and cause-specific mortality. *N. Engl. J. Med.* **2012**, *366*, 1891–1904. [CrossRef] [PubMed]
6. Gunter, M.J.; Murphy, N.; Cross, A.J.; Dossus, L.; Dartois, L.; Fagherazzi, G.; Kaaks, R.; Kuhn, T.; Boeing, H.; Aleksandrova, K.; et al. Coffee drinking and mortality in 10 european countries: A multinational cohort study. *Ann. Intern. Med.* **2017**, *167*, 236–247. [CrossRef] [PubMed]
7. Park, S.Y.; Freedman, N.D.; Haiman, C.A.; Le Marchand, L.; Wilkens, L.R.; Setiawan, V.W. Association of coffee consumption with total and cause-specific mortality among nonwhite populations. *Ann. Intern. Med.* **2017**, *167*, 228–235. [CrossRef] [PubMed]
8. Jarvis, M.J. Does caffeine intake enhance absolute levels of cognitive performance? *Psychopharmacology* **1993**, *110*, 45–52. [CrossRef] [PubMed]
9. Araujo, L.F.; Giatti, L.; Reis, R.C.; Goulart, A.C.; Schmidt, M.I.; Duncan, B.B.; Ikram, M.A.; Barreto, S.M. Inconsistency of association between coffee consumption and cognitive function in adults and elderly in a cross-sectional study (elsa-brasil). *Nutrients* **2015**, *7*, 9590–9601. [CrossRef] [PubMed]
10. Eskelinen, M.H.; Kivipelto, M. Caffeine as a protective factor in dementia and Alzheimer's disease. *J. Alzheimers Dis.* **2010**, *20*, S167–S174. [CrossRef] [PubMed]
11. Eskelinen, M.H.; Ngandu, T.; Tuomilehto, J.; Soininen, H.; Kivipelto, M. Midlife coffee and tea drinking and the risk of late-life dementia: A population-based caide study. *J. Alzheimers Dis.* **2009**, *16*, 85–91. [CrossRef] [PubMed]
12. Maia, L.; de Mendonca, A. Does caffeine intake protect from alzheimer's disease? *Eur. J. Neurol.* **2002**, *9*, 377–382. [CrossRef] [PubMed]
13. Lindsay, J.; Laurin, D.; Verreault, R.; Hebert, R.; Helliwell, B.; Hill, G.B.; McDowell, I. Risk factors for alzheimer's disease: A prospective analysis from the canadian study of health and aging. *Am. J. Epidemiol.* **2002**, *156*, 445–453. [CrossRef] [PubMed]
14. Liu, Q.P.; Wu, Y.F.; Cheng, H.Y.; Xia, T.; Ding, H.; Wang, H.; Wang, Z.M.; Xu, Y. Habitual coffee consumption and risk of cognitive decline/dementia: A systematic review and meta-analysis of prospective cohort studies. *Nutrition* **2016**, *32*, 628–636. [CrossRef] [PubMed]
15. Fredholm, B.B.; Battig, K.; Holmen, J.; Nehlig, A.; Zvartau, E.E. Actions of caffeine in the brain with special reference to factors that contribute to its widespread use. *Pharmacol. Rev.* **1999**, *51*, 83–133. [PubMed]

16. Haskell, C.F.; Kennedy, D.O.; Wesnes, K.A.; Scholey, A.B. Cognitive and mood improvements of caffeine in habitual consumers and habitual non-consumers of caffeine. *Psychopharmacology* **2005**, *179*, 813–825. [CrossRef] [PubMed]

17. Smit, H.J.; Rogers, P.J. Effects of low doses of caffeine on cognitive performance, mood and thirst in low and higher caffeine consumers. *Psychopharmacology* **2000**, *152*, 167–173. [CrossRef] [PubMed]

18. Childs, E.; de Wit, H. Subjective, behavioral, and physiological effects of acute caffeine in light, nondependent caffeine users. *Psychopharmacology* **2006**, *185*, 514–523. [CrossRef] [PubMed]

19. Renouf, M.; Marmet, C.; Giuffrida, F.; Lepage, M.; Barron, D.; Beaumont, M.; Williamson, G.; Dionisi, F. Dose-response plasma appearance of coffee chlorogenic and phenolic acids in adults. *Mol. Nutr. Food Res.* **2014**, *58*, 301–309. [CrossRef] [PubMed]

20. Dodd, F.L.; Kennedy, D.O.; Riby, L.M.; Haskell-Ramsay, C.F. A double-blind, placebo-controlled study evaluating the effects of caffeine and l-theanine both alone and in combination on cerebral blood flow, cognition and mood. *Psychopharmacology* **2015**, *232*, 2563–2576. [CrossRef] [PubMed]

21. Haskell, C.F.; Kennedy, D.O.; Milne, A.L.; Wesnes, K.A.; Scholey, A.B. The effects of l-theanine, caffeine and their combination on cognition and mood. *Biol. Psychol.* **2008**, *77*, 113–122. [CrossRef] [PubMed]

22. Giles, G.E.; Mahoney, C.R.; Brunye, T.T.; Taylor, H.A.; Kanarek, R.B. Caffeine and theanine exert opposite effects on attention under emotional arousal. *Can. J. Physiol. Pharmacol.* **2017**, *95*, 93–100. [CrossRef] [PubMed]

23. Andrews, S.E.; Blumenthal, T.D.; Flaten, M.A. Effects of caffeine and caffeine-associated stimuli on the human startle eyeblink reflex. *Pharmacol. Biochem. Behav.* **1998**, *59*, 39–44. [CrossRef]

24. Ferruzzi, M.G. The influence of beverage composition on delivery of phenolic compounds from coffee and tea. *Physiol. Behav.* **2010**, *100*, 33–41. [CrossRef] [PubMed]

25. Cropley, V.; Croft, R.; Silber, B.; Neale, C.; Scholey, A.; Stough, C.; Schmitt, J. Does coffee enriched with chlorogenic acids improve mood and cognition after acute administration in healthy elderly? A pilot study. *Psychopharmacology* **2012**, *219*, 737–749. [CrossRef] [PubMed]

26. Camfield, D.A.; Silber, B.Y.; Scholey, A.B.; Nolidin, K.; Goh, A.; Stough, C. A randomised placebo-controlled trial to differentiate the acute cognitive and mood effects of chlorogenic acid from decaffeinated coffee. *PLoS ONE* **2013**, *8*, e82897. [CrossRef] [PubMed]

27. Duff, S.J.; Hampson, E. A sex difference on a novel spatial working memory task in humans. *Brain Cognit.* **2001**, *47*, 470–493. [CrossRef] [PubMed]

28. Herlitz, A.; Rehnman, J. Sex differences in episodic memory. *Curr. Dir. Psychol. Sci.* **2008**, *17*, 52–56. [CrossRef]

29. van Gelder, B.M.; Buijsse, B.; Tijhuis, M.; Kalmijn, S.; Giampaoli, S.; Nissinen, A.; Kromhout, D. Coffee consumption is inversely associated with cognitive decline in elderly european men: The fine study. *Eur. J. Clin. Nutr.* **2007**, *61*, 226–232. [CrossRef] [PubMed]

30. Arab, L.; Biggs, M.L.; O'Meara, E.S.; Longstreth, W.T.; Crane, P.K.; Fitzpatrick, A.L. Gender differences in tea, coffee, and cognitive decline in the elderly: The cardiovascular health study. *J. Alzheimers Dis.* **2011**, *27*, 553–566. [CrossRef] [PubMed]

31. Ritchie, K.; Carriere, I.; de Mendonca, A.; Portet, F.; Dartigues, J.F.; Rouaud, O.; Barberger-Gateau, P.; Ancelin, M.L. The neuroprotective effects of caffeine: A prospective population study (the three city study). *Neurology* **2007**, *69*, 536–545. [CrossRef] [PubMed]

32. Henry, C.J.; Lightowler, H.J.; Marchini, J. Intra-individual variation in resting metabolic rate during the menstrual cycle. *Br. J. Nutr.* **2003**, *89*, 811–817. [CrossRef] [PubMed]

33. Lane, J.D.; Steege, J.F.; Rupp, S.L.; Kuhn, C.M. Menstrual cycle effects on caffeine elimination in the human female. *Eur. J. Clin. Pharmacol.* **1992**, *43*, 543–546. [CrossRef] [PubMed]

34. James, J.E.; Rogers, P.J. Effects of caffeine on performance and mood: Withdrawal reversal is the most plausible explanation. *Psychopharmacology* **2005**, *182*, 1–8. [CrossRef] [PubMed]

35. Smith, A.P.; Christopher, G.; Sutherland, D. Acute effects of caffeine on attention: A comparison of non-consumers and withdrawn consumers. *J. Psychopharmacol.* **2013**, *27*, 77–83. [CrossRef] [PubMed]

36. Kennedy, D.O.; Wightman, E.L.; Reay, J.L.; Lietz, G.; Okello, E.J.; Wilde, A.; Haskell, C.F. Effects of resveratrol on cerebral blood flow variables and cognitive performance in humans: A double-blind, placebo-controlled, crossover investigation. *Am. J. Clin. Nutr.* **2010**, *91*, 1590–1597. [CrossRef] [PubMed]

37. Stonehouse, W.; Conlon, C.A.; Podd, J.; Hill, S.R.; Minihane, A.M.; Haskell, C.; Kennedy, D. Dha supplementation improved both memory and reaction time in healthy young adults: A randomized controlled trial. *Am. J. Clin. Nutr.* **2013**, *97*, 1134–1143. [CrossRef] [PubMed]

38. Rogers, P.J.; Martin, J.; Smith, C.; Heatherley, S.V.; Smit, H.J. Absence of reinforcing, mood and psychomotor performance effects of caffeine in habitual non-consumers of caffeine. *Psychopharmacology* **2003**, *167*, 54–62. [CrossRef] [PubMed]

39. Kennedy, D.O.; Haskell, C.F. Cerebral blood flow and behavioural effects of caffeine in habitual and non-habitual consumers of caffeine: A near infrared spectroscopy study. *Biol. Psychol.* **2011**, *86*, 298–306. [CrossRef] [PubMed]

40. Kessels, R.P.; Nys, G.M.; Brands, A.M.; van den Berg, E.; Van Zandvoort, M.J. The modified location learning test: Norms for the assessment of spatial memory function in neuropsychological patients. *Arch. Clin. Neuropsychol.* **2006**, *21*, 841–846. [CrossRef] [PubMed]

41. Fujioka, K.; Shibamoto, T. Chlorogenic acid and caffeine contents in various commercial brewed coffees. *Food Chem.* **2008**, *106*, 217–221. [CrossRef]

42. Griffeth, V.E.; Perthen, J.E.; Buxton, R.B. Prospects for quantitative fmri: Investigating the effects of caffeine on baseline oxygen metabolism and the response to a visual stimulus in humans. *Neuroimage* **2011**, *57*, 809–816. [CrossRef] [PubMed]

43. Geraets, L.; Moonen, H.J.; Wouters, E.F.; Bast, A.; Hageman, G.J. Caffeine metabolites are inhibitors of the nuclear enzyme poly(adp-ribose)polymerase-1 at physiological concentrations. *Biochem. Pharmacol.* **2006**, *72*, 902–910. [CrossRef] [PubMed]

44. Zulli, A.; Smith, R.M.; Kubatka, P.; Novak, J.; Uehara, Y.; Loftus, H.; Qaradakhi, T.; Pohanka, M.; Kobyliak, N.; Zagatina, A.; et al. Caffeine and cardiovascular diseases: Critical review of current research. *Eur. J. Nutr.* **2016**, *55*, 1331–1343. [CrossRef] [PubMed]

45. Carrillo, J.A.; Benitez, J. Clinically significant pharmacokinetic interactions between dietary caffeine and medications. *Clin. Pharmacokinet.* **2000**, *39*, 127–153. [CrossRef] [PubMed]

46. Gunes, A.; Dahl, M.L. Variation in cyp1a2 activity and its clinical implications: Influence of environmental factors and genetic polymorphisms. *Pharmacogenomics* **2008**, *9*, 625–637. [CrossRef] [PubMed]

47. Gambaro, S.E.; Moretti, R.; Tiribelli, C.; Gazzin, S. Brain cytochrome p450 enzymes: A possible therapeutic targets for neurological diseases. *Ther. Targets Neurol. Dis.* **2015**, *2*, e598.

48. Sansone, R.; Ottaviani, J.I.; Rodriguez-Mateos, A.; Heinen, Y.; Noske, D.; Spencer, J.P.; Crozier, A.; Merx, M.W.; Kelm, M.; Schroeter, H.; et al. Methylxanthines enhance the effects of cocoa flavanols on cardiovascular function: Randomized, double-masked controlled studies. *Am. J. Clin. Nutr.* **2017**, *105*, 352–360. [CrossRef] [PubMed]

49. Dulloo, A.G.; Duret, C.; Rohrer, D.; Girardier, L.; Mensi, N.; Fathi, M.; Chantre, P.; Vandermander, J. Efficacy of a green tea extract rich in catechin polyphenols and caffeine in increasing 24-h energy expenditure and fat oxidation in humans. *Am. J. Clin. Nutr.* **1999**, *70*, 1040–1045. [CrossRef] [PubMed]

50. Nakagawa, K.; Nakayama, K.; Nakamura, M.; Sookwong, P.; Tsuduki, T.; Niino, H.; Kimura, F.; Miyazawa, T. Effects of co-administration of tea epigallocatechin-3-gallate (EGCG) and caffeine on absorption and metabolism of EGCG in humans. *Biosci. Biotechnol. Biochem.* **2009**, *73*, 2014–2017. [CrossRef] [PubMed]

51. Reis, C.E.G.; Paiva, C.; Amato, A.A.; Lofrano-Porto, A.; Wassell, S.; Bluck, L.J.C.; Dorea, J.G.; da Costa, T.H.M. Decaffeinated coffee improves insulin sensitivity in healthy men. *Br. J. Nutr.* **2018**, *119*, 1029–1038. [CrossRef] [PubMed]

52. MacKenzie, T.; Comi, R.; Sluss, P.; Keisari, R.; Manwar, S.; Kim, J.; Larson, R.; Baron, J.A. Metabolic and hormonal effects of caffeine: Randomized, double-blind, placebo-controlled crossover trial. *Metabolism* **2007**, *56*, 1694–1698. [CrossRef] [PubMed]

53. Battram, D.S.; Arthur, R.; Weekes, A.; Graham, T.E. The glucose intolerance induced by caffeinated coffee ingestion is less pronounced than that due to alkaloid caffeine in men. *J. Nutr.* **2006**, *136*, 1276–1280. [CrossRef] [PubMed]

54. Adisakwattana, S.; Moonsan, P.; Yibchok-Anun, S. Insulin-releasing properties of a series of cinnamic acid derivatives in vitro and in vivo. *J. Agric. Food Chem.* **2008**, *56*, 7838–7844. [CrossRef] [PubMed]

55. Torronen, R.; Kolehmainen, M.; Sarkkinen, E.; Mykkanen, H.; Niskanen, L. Postprandial glucose, insulin, and free fatty acid responses to sucrose consumed with blackcurrants and lingonberries in healthy women. *Am. J. Clin. Nutr.* **2012**, *96*, 527–533. [CrossRef] [PubMed]

56. Torronen, R.; Kolehmainen, M.; Sarkkinen, E.; Poutanen, K.; Mykkanen, H.; Niskanen, L. Berries reduce postprandial insulin responses to wheat and rye breads in healthy women. *J. Nutr.* **2013**, *143*, 430–436. [CrossRef] [PubMed]

57. Torronen, R.; Sarkkinen, E.; Niskanen, T.; Tapola, N.; Kilpi, K.; Niskanen, L. Postprandial glucose, insulin and glucagon-like peptide 1 responses to sucrose ingested with berries in healthy subjects. *Br. J. Nutr.* **2012**, *107*, 1445–1451. [CrossRef] [PubMed]

58. Sapwarobol, S.; Adisakwattana, S.; Changpeng, S.; Ratanawachirin, W.; Tanruttanawong, K.; Boonyarit, W. Postprandial blood glucose response to grape seed extract in healthy participants: A pilot study. *Pharmacogn. Mag.* **2012**, *8*, 192–196. [CrossRef] [PubMed]

59. Francis, S.T.; Head, K.; Morris, P.G.; Macdonald, I.A. The effect of flavanol-rich cocoa on the fMRI response to a cognitive task in healthy young people. *J. Cardiovasc. Pharmacol.* **2006**, *47*, S215–S220. [CrossRef] [PubMed]

60. Lamport, D.J.; Pal, D.; Moutsiana, C.; Field, D.T.; Williams, C.M.; Spencer, J.P.; Butler, L.T. The effect of flavanol-rich cocoa on cerebral perfusion in healthy older adults during conscious resting state: A placebo controlled, crossover, acute trial. *Psychopharmacology* **2015**, *232*, 3227–3234. [CrossRef] [PubMed]

61. Liguori, A.; Hughes, J.R.; Grass, J.A. Absorption and subjective effects of caffeine from coffee, cola and capsules. *Pharmacol. Biochem. Behav.* **1997**, *58*, 721–726. [CrossRef]

62. Stalmach, A.; Williamson, G.; Crozier, A. Impact of dose on the bioavailability of coffee chlorogenic acids in humans. *Food Funct.* **2014**, *5*, 1727–1737. [CrossRef] [PubMed]

63. Blanchard, J.; Sawers, S.J. Comparative pharmacokinetics of caffeine in young and elderly men. *J. Pharmacokinet. Biopharm.* **1983**, *11*, 109–126. [CrossRef] [PubMed]

64. Haskell-Ramsay, C.F.; Jackson, P.A.; Forster, J.S.; Robertson, B.C.; Kennedy, D.O. Acute effects of three doses of caffeine on attention, motor speed and mood over an 8-hour period. *Appetite* **2018**, in press. [CrossRef]

65. van Boxtel, M.; Schmitt, J. *Coffee, Tea, Chocolate, and the Brain*, 1st ed.; Nehlig, A., Ed.; CRC Press: Boca Raton, FL, USA, 2004.

66. Office for National Statistics (ONS). 2011 Census: Key Statistics for England and Wales. March 2011. Available online: http://www.ons.gov.uk/ons/dcp171778_290685.pdf (accessed on 28 September 2018).

67. Johnson-Kozlow, M.; Kritz-Silverstein, D.; Barrett-Connor, E.; Morton, D. Coffee consumption and cognitive function among older adults. *Am. J. Epidemiol.* **2002**, *156*, 842–850. [CrossRef] [PubMed]

68. Ascherio, A.; Weisskopf, M.G.; O'Reilly, E.J.; McCullough, M.L.; Calle, E.E.; Rodriguez, C.; Thun, M.J. Coffee consumption, gender, and parkinson's disease mortality in the cancer prevention study II cohort: The modifying effects of estrogen. *Am. J. Epidemiol.* **2004**, *160*, 977–984. [CrossRef] [PubMed]

69. Ferrini, R.L.; Barrett-Connor, E. Caffeine intake and endogenous sex steroid levels in postmenopausal women. The rancho bernardo study. *Am. J. Epidemiol.* **1996**, *144*, 642–644. [CrossRef] [PubMed]

70. Hartley, T.R.; Lovallo, W.R.; Whitsett, T.L. Cardiovascular effects of caffeine in men and women. *Am. J. Cardiol.* **2004**, *93*, 1022–1026. [CrossRef] [PubMed]

71. Kiyohara, C.; Kono, S.; Honjo, S.; Todoroki, I.; Sakurai, Y.; Nishiwaki, M.; Hamada, H.; Nishikawa, H.; Koga, H.; Ogawa, S.; et al. Inverse association between coffee drinking and serum uric acid concentrations in middle-aged japanese males. *Br. J. Nutr.* **1999**, *82*, 125–130. [PubMed]

72. Perna, L.; Mons, U.; Schottker, B.; Brenner, H. Association of cognitive function and serum uric acid: Are cardiovascular diseases a mediator among women? *Exp. Gerontol.* **2016**, *81*, 37–41. [CrossRef] [PubMed]

73. Rasmussen, B.B.; Brix, T.H.; Kyvik, K.O.; Brosen, K. The interindividual differences in the 3-demthylation of caffeine alias cyp1a2 is determined by both genetic and environmental factors. *Pharmacogenetics* **2002**, *12*, 473–478. [CrossRef] [PubMed]

74. Alsene, K.; Deckert, J.; Sand, P.; de Wit, H. Association between A2a receptor gene polymorphisms and caffeine-induced anxiety. *Neuropsychopharmacology* **2003**, *28*, 1694–1702. [CrossRef] [PubMed]

75. Childs, E.; Hohoff, C.; Deckert, J.; Xu, K.; Badner, J.; de Wit, H. Association between adora2a and drd2 polymorphisms and caffeine-induced anxiety. *Neuropsychopharmacology* **2008**, *33*, 2791–2800. [CrossRef] [PubMed]

76. Gajewska, A.; Blumenthal, T.D.; Winter, B.; Herrmann, M.J.; Conzelmann, A.; Muhlberger, A.; Warrings, B.; Jacob, C.; Arolt, V.; Reif, A.; et al. Effects of ADORA2A gene variation and caffeine on prepulse inhibition: A multi-level risk model of anxiety. *Prog. Neuropsychopharmacol. Biol. Psychiatry* **2013**, *40*, 115–121. [CrossRef] [PubMed]

77. Domschke, K.; Gajewska, A.; Winter, B.; Herrmann, M.J.; Warrings, B.; Muhlberger, A.; Wosnitza, K.; Glotzbach, E.; Conzelmann, A.; Dlugos, A.; et al. ADORA2A gene variation, caffeine, and emotional processing: A multi-level interaction on startle reflex. *Neuropsychopharmacology* **2012**, *37*, 759–769. [CrossRef] [PubMed]

78. Cornelis, M.C.; El-Sohemy, A.; Campos, H. Genetic polymorphism of the adenosine A2A receptor is associated with habitual caffeine consumption. *Am. J. Clin. Nutr.* **2007**, *86*, 240–244. [CrossRef] [PubMed]

79. Rogers, P.J.; Hohoff, C.; Heatherley, S.V.; Mullings, E.L.; Maxfield, P.J.; Evershed, R.P.; Deckert, J.; Nutt, D.J. Association of the anxiogenic and alerting effects of caffeine with ADORA2A and adora1 polymorphisms and habitual level of caffeine consumption. *Neuropsychopharmacology* **2010**, *35*, 1973–1983. [CrossRef] [PubMed]

80. Nehlig, A. Interindividual differences in caffeine metabolism and factors driving caffeine consumption. *Pharmacol. Rev.* **2018**, *70*, 384–411. [CrossRef] [PubMed]

*nutrients*

MDPI

*Article*

# NADH Dehydrogenase Subunit-2 237 Leu/Met Polymorphism Influences the Association of Coffee Consumption with Serum Chloride Levels in Male Japanese Health Checkup Examinees: An Exploratory Cross-Sectional Analysis

Akatsuki Kokaze [1,*], Mamoru Ishikawa [2,3], Naomi Matsunaga [2], Kanae Karita [2], Masao Yoshida [2], Hirotaka Ochiai [1], Takako Shirasawa [1], Takahiko Yoshimoto [1], Akira Minoura [1], Kosuke Oikawa [1], Masao Satoh [4], Hiromi Hoshino [1] and Yutaka Takashima [2]

[1]  Department of Hygiene, Public Health and Preventive Medicine, Showa University School of Medicine, 1-5-8 Hatanodai, Shinagawa-ku, Tokyo 142-8555, Japan; h-ochiai@med.showa-u.ac.jp (H.O.); shirasawa@med.showa-u.ac.jp (T.S.); yoshimotot@med.showa-u.ac.jp (T.Y.); minoaki@med.showa-u.ac.jp (A.M.); k-oikawa@med.showa-u.ac.jp (K.O.); hhiromi@med.showa-u.ac.jp (H.H.)
[2]  Department of Public Health, Kyorin University School of Medicine, 6-20-2 Shinkawa, Mitaka-shi, Tokyo 181-8611, Japan; ishikawa-m@genkiplaze.or.jp (M.I.); naomim@ks.kyorin-u.ac.jp (N.M.); kanae@ks.kyorin-u.ac.jp (K.K.); yohhy@ks.kyorin-u.ac.jp (M.Y.); yutakat@kyorin-u.ac.jp (Y.T.)
[3]  Mito Red Cross Hospital, 3-12-48 Sannomaru, Mito-shi, Ibaraki 310-0011, Japan
[4]  School of Medical Technology and Health, Faculty of Health and Medical Care, Saitama Medical University, 1397-1 Yamane, Hidaka-shi, Saitama 350-1241, Japan; satoma@saitama-med.ac.jp
*  Correspondence: akokaze@med.showa-u.ac.jp; Tel.: +81-(3)3784-8133; Fax: +81-(3)3784-7733

Received: 6 August 2018; Accepted: 19 September 2018; Published: 20 September 2018

**Abstract:** Background: Nicotinamide adenine dinucleotide (NADH) dehydrogenase subunit-2 237 leucine/methionine (ND2-237 Leu/Met) polymorphism has been shown to modify the association of coffee consumption with the risk of hypertension, dyslipidemia, and abnormal glucose tolerance, and low serum chloride levels have been shown to be associated with all-cause and cardiovascular disease mortality. Therefore, the purpose of the present study was to investigate whether ND2-237 Leu/Met polymorphism influences the association of coffee consumption with serum chloride levels in male Japanese health checkup examinees. Methods: From among individuals visiting the hospital for a regular medical checkup, 402 men (mean age ± standard deviation, 53.9 ± 7.8 years) were selected for inclusion in the study. After ND2-237 Leu/Met genotyping, we conducted an exploratory cross-sectional study to examine the combined association of ND2-237 Leu/Met polymorphism and coffee consumption with serum electrolyte levels. Results: After adjusting for age, body mass index, habitual smoking, alcohol consumption, green tea consumption, and antihypertensive medication, coffee consumption significantly increased serum chloride levels ($p$ for trend = 0.001) in men with the ND2-237Leu genotype. After these adjustments, the odds ratios (ORs) for low levels of serum chloride, defined as <100 mEq/L, were found to be dependent on coffee consumption ($p$ for trend = 0.001). In addition, the OR for low levels of serum chloride was significantly lower in men with the ND2-237Leu genotype who consumed ≥4 compared with <1 cup of coffee per day (OR = 0.096, 95% confidence interval = 0.010–0.934; $p$ = 0.044). However, neither serum chloride levels nor risk of low levels of serum chloride appeared to be dependent on coffee consumption. Conclusions: The results suggest that ND2-237 Leu/Met polymorphism modifies the association of coffee consumption with serum chloride levels in middle-aged Japanese men.

**Keywords:** cardiovascular disease; coffee consumption; gene-diet interaction; longevity; NADH dehydrogenase; polymorphism; serum chloride levels

## 1. Background

Coffee intake has been shown to be a favorable health behavior [1–3], and habitual consumption (3–4 cups per day) is more likely to benefit than to harm health [3]. Several recent meta-analyses of prospective studies have reported finding an inverse relationship between habitual coffee consumption and both all-cause [3–6] and cardiovascular disease (CVD) [4,5] mortality. In addition, Rodríguez-Artalejo and López-García reported that habitual consumption of 3–5 cups of coffee per day reduces the risk of CVD by 15%, and that habitual consumption of >3–5 cups per day does not elevate the risk of CVD [7].

A population-based cohort study reported finding an association between lower serum chloride levels and both all-cause and CVD mortality [8]. That epidemiological study found that the risk ratio for CVD mortality associated with low serum chloride levels was equivalent to or higher than that observed for well-known risk factors of CVD, including hypertension, diabetes, and habitual smoking [8]. Furthermore, low levels of serum chloride have been reported to be associated with increased mortality and risk of CVD in patients with pre-dialysis chronic kidney disease [9], and serum chloride levels have been shown to be negatively associated with mortality in patients with a history of heart failure [10,11]. By contrast, in a large-scale follow-up study involving hypertensive adults, a 1 mEq/L increase in serum chloride levels was found to reduce all-cause mortality by 1.5% after adjusting for confounding factors [12].

Nicotinamide adenine dinucleotide (NADH) dehydrogenase subunit-2 237 leucine/methionine (ND2-237 Leu/Met) polymorphism has been reported to be associated with longevity in the Japanese population [13]. The frequency of the ND2-237Met genotype has been found to be substantially higher in Japanese centenarians compared with the general Japanese population [13], and individuals with the ND2-237Met genotype have been shown to be less likely than those with the ND2-237Leu genotype to develop lifestyle-related diseases [14–18].

ND2-237 Leu/Met polymorphism has been reported to modify the effect of coffee consumption on the risks of hypertension [19], glucose tolerance abnormality [20], dyslipidemia [21], liver damage [22], and anemia [23]. We previously reported that serum chloride levels were significantly lower in obese men with the ND2-237Met genotype than in those with the ND2-237Leu genotype [24]. However, to our knowledge, no studies have been conducted to investigate the joint association of ND2-237 Leu/Met polymorphism and coffee consumption with serum chloride levels.

Therefore, the purpose of the present exploratory cross-sectional study was to investigate whether ND2-237 Leu/Met polymorphism modifies the association of coffee consumption with serum chloride levels in male Japanese health checkup examinees.

## 2. Subjects and Methods

### 2.1. Study Participants

The study participants were recruited from among individuals visiting the Mito Red Cross Hospital for a regular medical checkup between August 1999 and August 2000. This study was conducted in accordance with the Declaration of Helsinki. The Ethics Committee of Kyorin University School of Medicine approved the study protocol. Written informed consent was obtained from all 602 volunteers before participation. Because of the insufficient number of women available for categorization into groups based on the ND2-237 Leu/Met genotype and coffee consumption, females were excluded, as were males with unclear or incomplete data. Finally, 402 Japanese men (mean age ± standard deviation, 53.9 ± 7.8 years) were included in the analysis.

## 2.2. Data Collection

### 2.2.1. Clinical Measurements

The participants' anthropometric, biophysical, and biochemical data were obtained from the results of regular medical checkups. Serum electrolyte levels—namely serum sodium, chloride, potassium, or calcium levels—were determined using an auto-analyzer (HITACHI 7600-110S; Hitachi High Technology Corp., Tokyo, Japan). Body mass index (BMI) was calculated as weight (kg) divided by the square of height (m). Information on antihypertensive treatment was derived from the participants' health records.

### 2.2.2. Self-Administered Questionnaire

A survey regarding coffee intake, habitual smoking, alcohol consumption, and green tea intake was conducted on the participants via a self-administered questionnaire. Similar to previous reports [19–23], coffee consumption was categorized based on the number of cups of coffee consumed per day (<1, 1–3, and ≥4 cups). Habitual smoking was categorized as non- or ex-smokers and current smokers. Alcohol consumption was categorized based on drinking frequency (daily drinkers; occasional drinkers [those who drink several times per week or month]; and non- or ex-drinkers). Green tea consumption was categorized based on the number of cups of green tea consumed per day (≤1, 2-4, and ≥5 cups).

## 2.3. Genotyping

ND2-237 genotyping methods have been described previously [23]. Briefly, DNA was extracted from white blood cells. ND2-237 Leu/Met genotype was determined using the polymerase chain reaction-restriction fragment length polymorphism test. The absence or presence of an *Alu*I site was designated as ND2-237Met or ND2-237Leu, respectively.

## 2.4. Statistical Analyses

All statistical analyses were performed using SAS statistical software (version 9.2 for Windows; SAS Institute, Inc., Cary, NC, USA). Multiple logistic regression analysis was conducted to calculate odds ratios (ORs) for the risk of low levels of serum chloride. In accordance with previous epidemiological studies [8,12], low levels of serum chloride were defined as <100 mEq/L. For the multiple logistic regression analysis and analysis of covariance, habitual smoking (non- or ex-smokers = 0, current smokers = 1), alcohol consumption (non- or ex-drinkers = 0, occasional drinkers = 1, daily drinkers = 2), green tea consumption (≤1 cup per day = 1, 2–4 cups per day = 2, ≥5 cups per day = 3) and antihypertensive treatment (not receiving antihypertensive treatment = 0, receiving antihypertensive treatment = 1) were numerically coded. Two-tailed $p$ values <0.05 were considered statistically significant.

## 3. Results

No statistically significant differences in serum electrolyte levels—namely serum sodium, chloride, potassium, or calcium levels—were observed between the ND2-237Leu and ND2-237Met genotypes (Table 1).

A significant positive association was observed between coffee consumption and serum chloride levels in the men with the ND2-237Leu genotype ($p$ for trend = 0.001) (Table 2). Moreover, serum chloride levels were significantly higher in the participants who consumed ≥4 compared with <1 or 1–3 cups of coffee per day ($p$ = 0.001 and $p$ = 0.026, respectively). After adjusting for age, BMI, habitual smoking, alcohol consumption, green tea consumption, and antihypertensive medication, a significant positive association was found between coffee consumption and serum chloride levels ($p$ for trend = 0.002). In addition, serum chloride levels were significantly higher in the participants who consumed $\geq 4$ compared with <1 cup of coffee per day ($p = 0.010$). Moreover, a significant positive association was observed between coffee consumption and serum sodium levels in men with the ND2-237Leu genotype ($p$ for trend = 0.033); a significant positive association was also found between coffee consumption and serum sodium levels ($p$ for trend = 0.044). By contrast, no statistically significant association was observed between coffee consumption and serum electrolyte levels in the men with the ND2-237Met genotype.

**Table 1.** Clinical characteristics of the study participants by ND2-237 Leu/Met genotype.

| | ND2-237Leu | ND2-237Met | *p* Value |
|---|---|---|---|
| | *N* = 245 | *N* = 157 | |
| Age (years) * | 54.4 (7.8) | 53.2 (7.8) | 0.142 |
| Body mass index (kg/m$^2$) * | 23.3 (2.8) | 23.5 (2.6) | 0.366 |
| Systolic blood pressure (mmHg) * | 125.8 (15.8) | 125.7 (14.1) | 0.934 |
| Diastolic blood pressure (mmHg) ** welch | 73.9 (10.6) | 73.7 (9.1) | 0.817 |
| Low-density lipoprotein cholesterol (mg/dL) * | 121.3 (34.4) | 118.0 (30.8) | 0.319 |
| High-density lipoprotein cholesterol (mg/dL) ** welch | 54.5 (13.5) | 56.2 (16.1) | 0.285 |
| Triglyceride (mg/dL) *** | 115 (84–160) | 112 (84–158) | 0.948 |
| Uric acid (mg/dL) * | 5.94 (1.24) | 5.94 (1.22) | 0.970 |
| Serum sodium (mEq/L) * | 140.3 (2.0) | 140.1 (1.9) | 0.200 |
| Serum chloride (mEq/L) * | 101.3 (2.5) | 100.8 (2.2) | 0.062 |
| Serum potassium (mEq/L) * | 4.19 (0.28) | 4.18 (0.26) | 0.712 |
| Serum calcium (mEq/L) * | 9.33 (0.37) | 9.38 (0.38) | 0.180 |
| Coffee consumption (<1 cup per day/1–3 cups per day/$\geq$4 cups per day) (%) **** | 44.5/46.2/9.3 | 36.3/51.6/12.1 | 0.237 |
| Current smokers (%) **** | 41.7 | 40.8 | 0.852 |
| Alcohol consumption (non- or ex-/occasionally/ daily) (%) **** | 18.2/35.2/46.6 | 13.4/38.2/48.4 | 0.431 |
| Green tea consumption (<1 cup per day/1–4 cups per day/$\geq$5 cups per day) (%) **** | 21.9/41.7/36.4 | 19.8/47.1/33.1 | 0.562 |
| Antihypertensive (%) **** | 19.4 | 13.4 | 0.115 |

Age, body mass index, systolic blood pressure, diastolic blood pressure, low-density lipoprotein cholesterol, high-density lipoprotein cholesterol, uric acid, serum sodium, serum chloride, serum potassium, serum calcium are expressed as means (standard deviation). Triglyceride is expressed as the median (interquartile range). * Student's *t*-test, ** Welch's *t*-test, *** Mann-Whitney test, **** chi-square test. All *p* values depict significant differences between ND2-237Leu and ND2-237Met.

**Table 2.** Serum electrolyte levels by coffee consumption status and ND2-237 Leu/Met genotype.

| | Coffee Consumption | | | *p* for Trend |
|---|---|---|---|---|
| | <1 Cup Per Day | 1–3 Cups Per Day | $\geq$4 Cups Per Day | |
| **ND2-237Leu** | *N* = 109 | *N* = 113 | *N* = 23 | |
| Serum sodium levels (mEq/L) | 140.1 (0.2) | 140.4 (0.2) | 141.2 (0.4) | 0.033 |
| Serum sodium levels (mEq/L) † | 140.1 (0.2) | 140.5 (0.2) | 141.1 (0.4) | 0.044 |
| Serum chloride levels (mEq/L) | 100.9 (0.2) | 101.4 (0.2) | 102.9 (0.5) **,*** | 0.001 |
| Serum chloride levels (mEq/L) † | 100.7 (0.3) | 101.4 (0.3) | 102.4 (0.5) * | 0.002 |
| Serum potassium levels (mEq/L) | 4.19 (0.03) | 4.19 (0.03) | 4.20 (0.05) | 0.904 |
| Serum potassium levels (mEq/L) † | 4.18 (0.03) | 4.20 (0.03) | 4.22 (0.06) | 0.439 |
| Serum calcium levels (mEq/L) | 9.33 (0.04) | 9.33 (0.04) | 9.32 (0.08) | 0.867 |
| Serum calcium levels (mEq/L) † | 9.40 (0.04) | 9.35 (0.04) | 9.33 (0.08) | 0.306 |
| **ND2-237Met** | *N* = 57 | *N* = 81 | *N* = 19 | |
| Serum sodium levels (mEq/L) | 140.1 (0.2) | 140.1 (0.2) | 139.8 (0.4) | 0.679 |
| Serum sodium levels (mEq/L) † | 140.7 (0.3) | 140.8 (0.3) | 140.5 (0.5) | 0.700 |
| Serum chloride levels (mEq/L) | 100.8 (0.3) | 100.9 (0.2) | 100.8 (0.5) | 0.935 |
| Serum chloride levels (mEq/L) † | 101.3 (0.4) | 101.2 (0.3) | 101.3 (0.5) | 0.965 |
| Serum potassium levels (mEq/L) | 4.17 (0.03) | 4.18 (0.03) | 4.18 (0.06) | 0.905 |
| Serum potassium levels (mEq/L) † | 4.14 (0.05) | 4.15 (0.04) | 4.16 (0.07) | 0.836 |
| Serum calcium levels (mEq/L) | 9.39 (0.05) | 9.40 (0.04) | 9.30 (0.09) | 0.507 |
| Serum calcium levels (mEq/L) † | 9.41 (0.07) | 9.39 (0.06) | 9.24 (0.10) | 0.175 |

† Serum sodium levels, † serum chloride levels, † serum potassium levels, and † serum calcium levels are expressed as least-square means (standard error) adjusted for age, body mass index, alcohol consumption, habitual smoking, green tea consumption, and antihypertensive medication. The Bonferroni correction for multiple comparisons was applied; * $p < 0.05$, ** $p < 0.005$ vs. <1 cup of coffee per day, *** $p < 0.05$ vs. 1–3 cups of coffee per day.

A significant negative association was observed between coffee consumption and the risk of low levels of serum chloride among men with the ND2-237Leu genotype ($p$ for trend = 0.032) (Table 3). Moreover, the *OR* for low levels of serum chloride was significantly lower among men with the

ND2-237Leu genotype who consumed ≥4 compared with <1 cup of coffee per day (*OR* = 0.125, 95% confidence interval [CI] = 0.016–0.973; *p* = 0.047). After adjusting for age, BMI, habitual smoking, alcohol consumption, green tea consumption, and antihypertensive medication, the risk of low levels of serum chloride was found to be dependent on coffee consumption (*p* for trend = 0.028). In addition, the *OR* for low levels of serum chloride was found to be significantly lower for men with the ND2-237Leu genotype who consumed ≥4 compared with <1 cup of coffee per day (*OR* = 0.096, 95% CI = 0.010–0.934; *p* = 0.044). However, the association between the ND2-237Met genotype and the risk of low levels of serum chloride did not appear to be statistically dependent on the amount of daily coffee consumption.

**Table 3.** Odds ratios (ORs) and 95% confidence intervals (CIs) for low levels of serum chloride (serum chloride levels <100 mEq/L) by ND2-237 Leu/Met genotype and coffee consumption.

| Genotype and Coffee Consumption | Frequency (%) | | OR (95% CI) | Adjusted OR [†] (95% CI) |
|---|---|---|---|---|
| | Normal Levels of Serum Chloride (Serum Chloride Levels ≥100 mEq/L) | Low Levels of Serum Chloride (Serum Chloride Levels <100 mEq/L) | | |
| | | ND2-237Leu | | |
| <1 cup per day | 80 (73.4) | 29 (26.6) | 1 (reference) | 1 (reference) |
| 1–3 cups per day | 89 (78.8) | 24 (21.2) | 0.744 (0.400–1.382) | 0.615 (0.308–1.226) |
| ≥4 cups per day | 22 (95.7) | 1 (4.3) | 0.125 (0.016–0.973) * | 0.096 (0.010–0.934) * |
| | | | *p* for trend = 0.032 | *p* for trend = 0.028 |
| | | ND2-237Met | | |
| <1 cup per day | 41 (71.9) | 16 (28.1) | 1 (reference) | 1 (reference) |
| 1–3 cups per day | 60 (74.1) | 21 (25.9) | 0.897 (0.419–1.922) | 0.803 (0.339–1.902) |
| ≥4 cups per day | 16 (84.2) | 3 (15.8) | 0.480 (0.123–1.875) | 0.361 (0.076–1.718) |
| | | | *p* for trend = 0.353 | *p* for trend = 0.264 |

[†] OR adjusted for age, body mass index, habitual alcohol consumption, habitual smoking, green tea consumption, and antihypertensive medication. * *p* < 0.05.

## 4. Discussion

Although somewhat limited by the small sample size, the results of the present study suggest that ND2-237 Leu/Met polymorphism and coffee consumption exert a joint association with serum chloride levels in middle-age male Japanese health checkup examinees; namely, serum chloride levels were positively associated with the amount of daily coffee consumption in men with the ND2-237Leu genotype. Moreover, the risk of low levels of serum chloride, which has been shown to be associated with a higher risk of CVD and mortality [8–12], was significantly lower in men with the ND2-237Leu genotype who consumed ≥4 compared with <1 cup of coffee per day. However, coffee consumption did not appear to affect serum chloride levels in men with the ND2-237Met genotype.

In regard to the role of the interplay between caffeine and mitochondria in the cardiovascular system, Ale-Agha, et al. recently established experimentally that caffeine acting jointly with mitochondrial cyclin-dependent kinase inhibitor 1B exerts protective effects against CVD [25]. However, although likely related to biochemical differences between ND2-237Leu and ND2-237Met in response to some compounds in coffee, the physiological mechanisms underlying the joint association of ND2-237 Leu/Met polymorphism and coffee consumption with serum chloride levels remains unknown. NADH dehydrogenase is known as a major site of the generation of reactive oxygen species (ROS) in mitochondria; it is also known itself to be a target of attack by ROS [26]. Some animal experiments have demonstrated that ND2-237Met protects NADH dehydrogenase from ROS attack and/or suppresses ROS production [27,28]. Moreover, coffee consumption has been shown to exert antioxidant potential in humans [29]. The results of our previous research suggest that moderate coffee intake exerts antioxidant effects in men with the ND2-237Leu but not the ND2-237Met genotype [19,20,22]. To elucidate the mechanisms underlying the joint association of ND2-237 Leu/Met polymorphism and coffee consumption with serum chloride levels, further biophysical and biochemical investigations are needed.

Previous cross-sectional studies have reported finding an interactive association of ND2-237 Leu/Met polymorphism and coffee consumption with the risks of hypertension [19], glucose tolerance abnormality [20], dyslipidemia [21], liver damage [22], and erythrocytic parameters [23]. The risk of hypertension was significantly lower in men with the ND2-237Leu genotype who consumed ≥2 compared with ≤1 cup of coffee per day [19], and that of glucose tolerance abnormality was also significantly lower in those who consumed ≥4 compared with <1 cup of coffee per day [20]. Consequently, the risk of abnormally elevated levels of serum liver enzymes was significantly lower in men who consumed ≥3 compared with <1 cup of coffee per day [22]. However, the risk of anemia was significantly higher in those who consumed ≥4 compared with <1 cup of coffee per day [23]. Meanwhile, the risk of dyslipidemia was significantly higher among men with the ND2-237Met genotype who consumed ≥1 compared with <1 cup of coffee per day [21]. Taken together, other than the risk of anemia, moderate levels of coffee intake appear to be more beneficial for health in men with the ND2-237Leu compared with the ND2-237Met genotype.

Although the focus of the present study was on serum chloride levels, serum sodium levels were also significantly and positively associated with coffee consumption in men with the ND2-237Leu genotype. In addition, in a population-based cohort study, De Bacquer, et al. reported finding a remarkably positive association between serum chloride and serum sodium levels [8]. In patients with acute decompensated heart failure, compared with that of serum chloride levels, the prognostic value of serum sodium levels was diminished [10]. Moreover, in patients with post-myocardial infarction accompanied by systolic dysfunction and heart failure, low levels of serum chloride were found to be associated with CVD mortality accompanied by low levels of serum sodium [30]. Recently, a large-scale follow-up study reported finding an association between low serum sodium levels and an increased risk of CVD mortality [31]; therefore, serum sodium levels may affect the pathophysiology of serum chloride levels for CVD. However, to verify whether serum chloride levels act jointly with serum sodium levels in regard to CVD in the general population, further epidemiological studies are needed.

A potential contradiction may exist in relation to these findings. Although low serum chloride levels have been shown to be associated with an increased risk of both all-cause and CVD mortality [8–12], the results of our previous study showed that serum chloride levels were significantly lower in men who had the ND2-237Met genotype—who have been reported to have a genetic tendency toward longevity [13] and resistance against life-threatening diseases [14–18]—compared with those with the ND2-237Leu genotype [24]. Therefore, the genetic advantage of having the ND2-237Met genotype may surpass the physiological disadvantage of having low serum chloride levels.

The present study did have several important limitations. First, the data were collected 18 years ago. Second, the sample size was relatively small. Third, the participants consisted of only men. Fourth, only a single population was analyzed; to prevent errors in genetic epidemiological studies, several independent data sets need to be analyzed. Fifth, although cross-sectional studies can suggest causal associations, they cannot prove valid causality. To overcome this limitation, a follow-up study involving a larger study sample that includes multiple populations is needed. Sixth, we did not obtain any data on salt intake or other dietary factors. Although no significant associations have been found between serum chloride levels and dietary sodium intake [32], higher sodium intake has been found to be a risk factor for CVD in Japan [33]. Potential correlations between additional dietary factors and coffee consumption have been reported [34]. Therefore, a food frequency questionnaire survey will be required in future studies. Finally, we based the categorization of habitual coffee consumption on the number of cups consumed per day. Whether any interaction exists between ND2-237 Leu/Met polymorphism and levels of caffeine, chlorogenic acids, or other unknown compounds in coffee on serum chloride levels remains unclear and therefore warrants further investigation.

## 5. Conclusions

The results of the present exploratory cross-sectional analysis suggest a joint association of ND2-237 Leu/Met polymorphism and coffee consumption with serum chloride levels among male Japanese health checkup examinees. For men with the ND2-237Leu genotype, higher coffee consumption may reduce the risk of low levels of serum chloride. Therefore, daily coffee intake is recommended for men with the ND2-237Leu genotype to reduce the risk of CVD. To the best of our knowledge, this is the first report of the effects of gene–diet interaction on serum chloride levels. These findings may contribute to individualized prevention strategies for CVD.

**Author Contributions:** A.K. designed the study, carried out the epidemiological survey, carried out genotyping, analyzed the data, and drafted the manuscript; M.I. collected the samples; K.K. and M.Y. carried out the epidemiological survey; N.M. assisted with genotyping; H.O., T.S., T.Y., A.M., K.O., M.S., and H.H. assisted in the data analysis and the interpretation of the results; Y.T. designed the study and carried out the epidemiological survey. All authors have read and approved the final manuscript.

**Funding:** This study was supported in part by Grants-in-Aid from the Ministry of Education, Culture, Sports, Science and Technology of Japan (No. 14570355, No. 18590572, No. 23500859, and No. 26350908) and the Chiyoda Mutual Life Foundation.

**Conflicts of Interest:** The authors declare no conflict of interest.

## References

1.  Cano-Marquina, A.; Tarin, J.J.; Cano, A. The impact of coffee on health. *Maturitas* **2013**, *75*, 7–21. [CrossRef] [PubMed]
2.  Cornelis, M.C. Coffee intake. *Prog. Mol. Biol. Transl. Sci.* **2012**, *108*, 293–322. [CrossRef] [PubMed]
3.  Poole, R.; Kennedy, O.J.; Roderick, P.; Fallowfield, J.A.; Hayes, P.C.; Parkes, J. Coffee comsumption and health: Umbrella review of meta-analyses of multiple health outcomes. *BMJ* **2017**, *358*, j5024. [CrossRef] [PubMed]
4.  Malerba, S.; Turati, F.; Galeone, C.; Pelucchi, C.; Verga, F.; La Vecchia, C.; Tavani, A. A meta-analysis of prospective studies of coffee consumption and mortality for all causes, cancers and cardiovascular diseases. *Eur. J. Epidemiol.* **2013**, *28*, 527–539. [CrossRef] [PubMed]
5.  Crippa, A.; Discacciati, A.; Larsson, S.C.; Wolk, A.; Orsini, N. Coffee consumption and mortality from all causes, cardiovascular disease, and cancer: A dose-response meta-analysis. *Am. J. Epidemiol.* **2014**, *180*, 763–775. [CrossRef] [PubMed]
6.  Je, Y.; Giovannucci, E. Coffee consumption and total mortality: A meta-analysis of twenty prospective cohort studies. *Br. J. Nutr.* **2014**, *111*, 1162–1173. [CrossRef] [PubMed]
7.  Rodríguez-Artalejo, F.; López-García, E. Coffee consumption and cardiovascular disease: A condensed review of epidemiological evidence and mechanisms. *J. Agric. Food Chem.* **2018**, *66*, 5257–5263. [CrossRef] [PubMed]
8.  De Bacquer, D.; De Backer, G.; De Buyzere, M.; Kornitzer, M. Is low serum chloride level a risk factor for cardiovascular mortality? *J. Cardiovasc. Risk* **1998**, *5*, 177–184. [CrossRef] [PubMed]
9.  Mandai, S.; Kanda, E.; Iimori, S.; Naito, S.; Noda, Y.; Kikuchi, H.; Akazawa, M.; Oi, K.; Toda, T.; Sohara, E.; et al. Association of serum chloride level with mortality and cardiovascular events in chronic kidney disease: The CKD-ROUTE study. *Clin. Exp. Nephrol.* **2017**, *21*, 104–111. [CrossRef] [PubMed]
10. Grodin, J.L.; Simon, J.; Hachamovitch, R.; Wu, Y.; Jackson, G.; Halkar, M.; Starling, R.C.; Testani, J.M.; Tang, W.H. Prognostic role of serum chloride levels in acute decompensated heart failure. *J. Am. Coll. Cardiol.* **2015**, *66*, 659–666. [CrossRef] [PubMed]
11. Zhang, Y.; Peng, R.; Li, X.; Yu, J.; Chen, X.; Zhou, Z. Serum chloride as a novel marker for adding prognostic information of mortality in chronic heart failure. *Clin. Chim. Acta* **2018**, *483*, 112–118. [CrossRef] [PubMed]
12. McCallum, L.; Jeemon, P.; Hastie, C.E.; Patel, R.K.; Williamson, C.; Redzuan, A.M.; Dawson, J.; Sloan, W.; Muir, S.; Morrison, D.; et al. Serum chloride is an independent predictor of mortality in hypertensive patients. *Hypertension* **2013**, *62*, 836–843. [CrossRef] [PubMed]
13. Tanaka, M.; Gong, J.S.; Zhang, J.; Yoneda, M.; Yagi, K. Mitochondrial genotype associated with longevity. *Lancet* **1998**, *351*, 185–186. [CrossRef]
14. Kokaze, A.; Ishikawa, M.; Matsunaga, N.; Yoshida, M.; Satoh, M.; Teruya, K.; Masuda, Y.; Honmyo, R.; Uchida, Y.; Takashima, Y. NADH dehydrogenase subunit-2 237 Leu/Met polymorphism modifies the effects of alcohol consumption on risk for hypertension in middle-aged Japanese men. *Hypertens. Res.* **2007**, *30*, 213–218. [CrossRef] [PubMed]

15. Wang, D.; Taniyama, M.; Suzuki, Y.; Katagiri, T.; Ban, Y. Association of the mitochondrial DNA 5178A/C polymorphism with maternal inheritance and onset of type 2 diabetes in Japanese patients. *Exp. Clin. Endocrinol. Diabetes* **2001**, *109*, 361–364. [CrossRef] [PubMed]

16. Mukae, S.; Aoki, S.; Itoh, S.; Sato, R.; Nishio, K.; Iwata, T.; Katagiri, T. Mitochondrial 5178A/C genotype is associated with acute myocardial infarction. *Circ. J.* **2003**, *67*, 16–20. [CrossRef] [PubMed]

17. Takagi, K.; Yamada, Y.; Gong, J.S.; Sone, T.; Yokota, M.; Tanaka, M. Association of a 5178C→A (Leu237Met) polymorphism in the mitochondrial DNA with a low prevalence of myocardial infarction in Japanese individuals. *Atherosclerosis* **2004**, *175*, 281–286. [CrossRef] [PubMed]

18. Ohkubo, R.; Nakagawa, M.; Ikeda, K.; Kodama, T.; Arimura, K.; Akiba, S.; Saito, M.; Ookatsu, Y.; Atsuchi, Y.; Yamano, Y.; et al. Cerebrovascular disorders and genetic polymorphisms: Mitochondrial DNA5178C is predominant in cerebrovascular disorders. *J. Neurol. Sci.* **2002**, *198*, 31–35. [CrossRef]

19. Kokaze, A.; Ishikawa, M.; Matsunaga, N.; Karita, K.; Yoshida, M.; Ohtsu, T.; Shirasawa, T.; Sekii, H.; Ito, T.; Kawamoto, T.; et al. NADH dehydrogenase subunit-2 237 Leu/Met polymorphism modulates the effects of coffee consumption on the risk of hypertension in middle-aged Japanese men. *J. Epidemiol.* **2009**, *19*, 231–236. [CrossRef] [PubMed]

20. Kokaze, A.; Ishikawa, M.; Matsunaga, N.; Karita, K.; Yoshida, M.; Ohtsu, T.; Shirasawa, T.; Haseba, Y.; Satoh, M.; Teruya, K.; et al. Longevity-associated mitochondrial DNA 5178 C/A polymorphism modifies the effect of coffee consumption on glucose tolerance in middle-aged Japanese men. In *Handbook on Longevity: Genetics, Diet and Disease*; Bentely, J.V., Keller, M.A., Eds.; Nova Science Publishers: New York, NY, USA, 2009; pp. 139–160.

21. Kokaze, A.; Ishikawa, M.; Matsunaga, N.; Karita, K.; Yoshida, M.; Shimada, N.; Ohtsu, T.; Shirasawa, T.; Ochiai, H.; Kawamoto, T.; et al. Combined effect of longevity-associated mitochondrial DNA 5178 C/A polymorphism and coffee consumption on the risk of hyper-LDL cholesterolemia in middle-aged Japanese men. *J. Hum. Genet.* **2010**, *55*, 577–581. [CrossRef] [PubMed]

22. Kokaze, A.; Yoshida, M.; Ishikawa, M.; Matsunaga, N.; Karita, K.; Ochiai, H.; Shirasawa, T.; Nanri, H.; Mitsui, K.; Hoshino, H.; et al. Mitochondrial DNA 5178 C/A polymorphism modulates the effects of coffee consumption on elevated levels of serum liver enzymes in male Japanese health check-up examinees: An exploratory cross-sectional study. *J. Physiol. Anthropol.* **2016**, *35*, 15. [CrossRef] [PubMed]

23. Kokaze, A.; Ishikawa, M.; Matsunaga, N.; Karita, K.; Yoshida, M.; Ohtsu, T.; Ochiai, H.; Shirasawa, T.; Nanri, H.; Saga, N.; et al. Longevity-associated mitochondrial DNA 5178 C/A polymorphism modulates the effects of coffee consumption on erythrocytic parameters in Japanese men: An exploratory cross-sectional analysis. *J. Physiol. Anthropol.* **2014**, *33*, 37. [CrossRef] [PubMed]

24. Kokaze, A.; Ishikawa, M.; Matsunaga, N.; Yoshida, M.; Makita, R.; Satoh, M.; Teruya, K.; Sekiguchi, K.; Masuda, Y.; Harada, M.; et al. Longevity-associated NADH dehydrogenase subunit-2 polymorphism and serum electrolyte levels in middle-aged obese Japanese men. *Mech. Ageing Dev.* **2005**, *126*, 705–709. [CrossRef] [PubMed]

25. Ale-Agha, N.; Goy, C.; Jakobs, P.; Spyridopoulos, I.; Gonnissen, S.; Dyballa-Rukes, N.; Aufenvenne, K.; von Ameln, F.; Zurek, M.; Spannbrucker, T.; et al. CDKN1B/p27 is localized in mitochondria and improves respiration-dependent processes in the cardiovascular system—New mode of action for caffeine. *PLoS Biol.* **2018**, *16*, e2004408. [CrossRef] [PubMed]

26. Madamanchi, N.R.; Runge, M.S. Mitochondrial dysfunction in atherosclerosis. *Circ. Res.* **2007**, *100*, 460–473. [CrossRef] [PubMed]

27. Gusdon, A.M.; Votyakova, T.V.; Mathews, C.E. mt-Nd2a suppresses reactive oxygen species production by mitochondrial complexes I. and III. *J. Biol. Chem.* **2008**, *283*, 10690–10697. [CrossRef] [PubMed]

28. Stadtman, E.R.; Moskovitz, J.; Berlett, B.S.; Levine, R.L. Cyclic oxidation and reduction of protein methionine residues is an important antioxidant mechanism. *Mol. Cell. Biochem.* **2002**, *234*, 3–9. [CrossRef] [PubMed]

29. Ishizaka, Y.; Yamakado, M.; Toda, A.; Tani, M.; Ishizaka, N. Relationship between coffee consumption, oxidant status, and antioxidant potential in the Japanese general population. *Clin. Chem. Lab. Med.* **2013**, *51*, 1951–1959. [CrossRef] [PubMed]

30. Ferreira, J.P.; Girerd, N.; Duarte, K.; Coiro, S.; McMurray, J.J.; Dargie, H.J.; Pitt, B.; Dickstein, K.; Testani, J.M.; Zannad, F.; et al. Serum chloride and sodium interplay in patients with acute myocardial infarction and heart failure with reduced ejection fraction: An analysis from the high-risk myocardial infarction database initiative. *Circ. Heart Fail.* **2017**, *10*, e003500. [CrossRef] [PubMed]

31. He, X.; Liu, C.; Chen, Y.; He, J.; Dong, Y. Risk of cardiovascular mortality associated with serum sodium and chloride in the general population. *Can. J. Cardiol.* **2018**, *34*, 999–1003. [CrossRef] [PubMed]

32. Van Berge-Landry, H.; James, G.D. Serum electrolyte, serum protein, serum fat and renal responses to a dietary sodium challenge: Allostasis and allostatic load. *Ann. Hum. Biol.* **2004**, *31*, 477–487. [CrossRef] [PubMed]

33. Iso, H. Lifestyle and cardiovascular disease in Japan. *J. Atheroscler. Thromb.* **2011**, *18*, 83–88. [CrossRef] [PubMed]

34. Solvoll, K.; Selmer, R.; Løken, E.B.; Foss, O.P.; Trygg, K. Coffee, dietary habits, and serum cholesterol among men and women 35–49 years of age. *Am. J. Epidemiol.* **1989**, *129*, 1277–1288. [CrossRef] [PubMed]

*Article*

# Coffee Consumption and the Risk of Depression in a Middle-Aged Cohort: The SUN Project

Adela M. Navarro [1,2], Daria Abasheva [1], Miguel Á. Martínez-González [1,3,4,5], Liz Ruiz-Estigarribia [1,4], Nerea Martín-Calvo [1,3,4], Almudena Sánchez-Villegas [6] and Estefanía Toledo [1,3,4,*]

[1] Department of Preventive Medicine and Public Health, School of Medicine, University of Navarra, 31008 Pamplona, Spain; adela.navarro.e@gmail.com (A.M.N.); dabasheva@alumni.unav.es (D.A.); mamartinez@unav.es (M.Á.M.-G.); lruiz.29@alumni.unav.es (L.R.-E.); nmartincalvo@unav.es (N.M.-C.)

[2] Department of Cardiology, Complejo Hospitalario de Navarra, Servicio Navarro de Salud Osasunbidea, 31008 Pamplona, Spain

[3] IdiSNA, Navarra Institute for Health Research, 31008 Pamplona, Spain

[4] Centro de Investigación Biomédica en Red Área de Fisiopatología de la Obesidad y la Nutrición (CIBEROBN), 28029 Madrid, Spain

[5] Department of Nutrition, Harvard TH Chan School of Public Health, Boston, MA 02115, USA

[6] Nutrition Research Group, Research Institute of Biomedical and Health Sciences, University of Las Palmas de Gran Canaria, 35016 Las Palmas de Gran Canaria, Spain; almudena.sanchez@ulpgc.es

* Correspondence: etoledo@unav.es; Tel.: +34-948425600 (ext. 806224)

Received: 30 August 2018; Accepted: 16 September 2018; Published: 19 September 2018

**Abstract:** Coffee is one of the most widely consumed drinks around the world, while depression is considered the major contributor to the overall global burden of disease. However, the investigation on coffee consumption and depression is limited and results may be confounded by the overall dietary pattern. We assessed the relationship between coffee intake and the risk of depression, controlling for adherence to the Mediterranean diet. We studied 14,413 university graduates of the 'Seguimiento Universidad de Navarra' (SUN) cohort, initially free of depression. We evaluated coffee consumption using a validated food-frequency questionnaire (FFQ). Incident depression cases were adjudicated only if the participant met two criteria simultaneously: (a) validated physician-diagnosed depression together with (b) new onset of habitual antidepressant use. Both criteria were needed; participants meeting only one of them were not classified as cases. Participants who drank at least four cups of coffee per day showed a significantly lower risk of depression than participants who drank less than one cup of coffee per day (HR: 0.37 (95% CI 0.15–0.95)). However, overall, we did not observe an inverse linear dose–response association between coffee consumption and the incidence of depression ($p$ for trend = 0.22).

**Keywords:** coffee; depression; cohort study

---

## 1. Introduction

Depression is considered the major contributor to the overall global burden of disease and a common cause of disability worldwide, with more than 300 million people affected [1]. Severe forms of depression can lead to suicide, which is the second leading cause of death in people aged 15–29 years, accounting for 800,000 deaths every year [2]. The lifetime prevalence of depression and the distribution of suicide rates are not uniform. Within Europe, both depression prevalence and suicide rates are higher in northern countries than in southern ones [3]. Nowadays, the prevention of depression represents a public health priority due to its huge social and economic burden.

Some investigations suggest that underlying pathophysiological mechanisms in depression are also present in metabolic syndrome (MetS), obesity, and cardiovascular disease (CVD) [4]. Endothelial dysfunction and an increased production of proinflammatory cytokines may explain the link between depression and CVD [5,6].

On the other hand, coffee is one of the most widely consumed beverages around the world. It is known that coffee contains antioxidant substances with potentially beneficial properties; e.g., chlorogenic acid, flavonoids, melanoidins, and various lipid-soluble compounds such as furans, pyrroles, and maltol [7].

Two recent meta-analyses including three longitudinal and five cross-sectional studies found an inverse association between coffee consumption and depression [8,9]. It is noteworthy that none of the longitudinal studies adjusted their estimates for an overall dietary pattern. Given that coffee consumption may be associated with a high-quality overall dietary pattern and that a healthy dietary pattern, such as the traditional Mediterranean diet (MedDiet), has been associated with a lower risk of depression [10], the overall dietary pattern may be a potential confounder in the association between coffee consumption and depression. Therefore, it is interesting to assess the association between coffee consumption and the risk of depression once adherence to an overall healthy dietary pattern has been accounted for in the analysis. This seems especially relevant when the association is assessed in a Mediterranean setting.

To our knowledge, the effect of coffee on the risk of depression has not been assessed in a Mediterranean cohort and it has neither been assessed if coffee consumption can show an inverse association with depression incidence once adherence to the traditional MedDiet has been accounted for. Thus, the aim of this study was to evaluate whether coffee consumption is independently associated with the risk of depression in the SUN project, a prospective cohort of Spanish graduates, after controlling for adherence to the traditional MedDiet.

## 2. Materials and Methods

### 2.1. Study Population

The "Seguimiento Universidad de Navarra" (SUN) project is a prospective multipurpose cohort of Spanish university graduates. The study methods have been described in more detail elsewhere [11]. Briefly, the SUN project is a dynamic cohort assessing the relationship between diet and chronic diseases. It was developed inspired by the models of the Nurses' Health Study and the Health Professionals Follow-Up Study. Recruitment started in December 1999 and is permanently open. After the initial questionnaire, follow-up questionnaires are mailed every other year to participants to update information on diet and lifestyle and collect information on health outcomes which might have happened in the previous two years. For participants lost to follow-up, the National Death Index is consulted periodically to assess their vital status. Participants are middle-aged university graduates from different Spanish regions.

By 2017, 22,564 participants were recruited. In order to allow the minimal follow-up of two years, we included only those participants who were recruited before March 2014 (2.75 years before the database closing date). Out of 22,279 eligible subjects, we excluded 1990 participants with no follow-up information (retention rate 91%); 1910 participants with total energy intake out of predefined limits (<500 or >3500 kcal/day for women and <800 or >4000 kcal/day for men); participants with previously diagnosed cardiovascular diseases, cancer, or diabetes (*n* = 1798); participants who died before returning their first follow-up questionnaire (*n* = 39); participants with baseline depression, regular antidepressant use, or implausible date or depression diagnosis (*n* = 1811); as well as patients with diagnosed depression during the first 2 years of follow-up or regular antidepressant use at 2 years of follow-up (*n* = 318). The final sample consisted of 14,413 participants who answered at least 1 follow-up questionnaire.

## 2.2. Assessment of Coffee Consumption

The baseline questionnaire included a previously validated 136-item food-frequency questionnaire (FFQ) [12–14]. The serving size for coffee was 50 cc. Information about the consumption of regular and decaffeinated coffee was gathered separately. The FFQ assessed regular food consumption over the previous 12 months and included nine categories of response for the frequency of consumption, ranging from 'never/seldom' to 'more than six times per day'. Then, participants were grouped in four categories according to their level of coffee consumption (<1 cup/day, 1 cup/day, >1–<4 cups/day, ≥4 cups/day).

## 2.3. Case Ascertainment

We adjudicated an incident case of major depressive disorder during follow-up in a participant initially free of any history of depression only if he or she met 2 criteria simultaneously: (a) a validated [15] self-reported new physician-made diagnosis of depression together with (b) new-onset habitual use of antidepressants (in the previous 2 years). Both criteria were needed; participants meeting only one of them were not classified as cases.

## 2.4. Assessment of Covariates

Sociodemographic, anthropometric, lifestyle, and comorbidity information were also collected at baseline and updated every two years through the follow-up questionnaires. The adherence to the MedDiet was established based on the information in the FFQ according to the index defined by Trichopoulou et al. [16]. The latest available information on food composition tables for Spain was utilized by trained dietitians to update the nutrient dataset from the information collected with the FFQ. The baseline questionnaire also included three questions on self-perceived personality traits with scores ranging from 0 to 10. More concretely, these questions assessed self-perceived psychological dependence (0—autonomous to 10—dependent), competitiveness (0—conformist to 10—competitive) and anxiety (0—relaxed to 10—tense) [17].

## 2.5. Statistical Analysis

Baseline quantitative traits of participants were described as the mean and standard deviation according to categories of coffee consumption and baseline qualitative traits, and as the percentage across the same categories. We calculated $p$ values for comparisons across categories of coffee consumption with ANOVA for quantitative variables and with chi-squared tests for qualitative variables.

Cox regression models were fit to assess the association between coffee intake and the risk of clinical depression development. We used age as the underlying time variable in all the analyses. Models were stratified by age and period of completion of baseline questionnaire. Participants contributed to the person-years of follow-up from the study inception until diagnosis of depression, death, or last follow-up questionnaire; whichever occurred first.

In our main analysis, we used total coffee intake as the exposure variable. The group in the lowest level of coffee consumption was used as the reference category in all the analyses. For the linear trend test, the median in each category of coffee consumption was calculated to generate a new quantitative variable. As a sensitivity analysis, we also fit models for regular and decaffeinated coffee separately.

The final model was adjusted for potential confounders such as sex, body-mass index (BMI; 3 categories), physical activity (continuous), alcohol intake (linear and quadratic), smoking status (never/former/current/missing) and package-years of smoking (continuous), total energy intake (continuous), adherence to the traditional MedDiet (continuous), years of university studies (continuous), marital status (3 categories), TV-watching hours (continuous), snacking, following any special diet, baseline hypertension and baseline hypercholesterolemia, self-perception of competitiveness, anxiety, and psychological dependence (continuous), and use of tranquilizers or anxiolytic drugs, and was stratified for age (decades) and recruitment period.

The interactions of coffee consumption (4 categories) with sex, age (2 categories), and smoking status (4 categories) were studied by introducing an interaction term in the model and calculating the likelihood ratio test between the model with the interaction and the model without it.

All analyses were performed with Stata SE 15.0. A two-sided $p$ value below 0.05 was deemed as statistically significant.

## 3. Results

We followed 14,413 participants, 5765 (40%) men and 8648 women, for a mean follow-up time of 10 years (standard deviation (SD): 4). Mean age of participants at recruitment was 36.4 years (SD: 11.5). Among 144,029 person-years follow-up, we identified 199 incident cases of depression. The incidence rate of depression was 1.3/1000 person-years of follow-up in the lowest category of coffee consumption and 1.5, 1.5, and 0.8/1000 persons-years of follow-up in the subsequent categories.

Participants' baseline characteristics by category of coffee consumption are shown in Table 1. On average, participants in the highest category of coffee consumption were older, had a higher average BMI, and reported higher mean total energy intake, lower physical activity, and being more tense compared to participants in the lower coffee consumption categories. Those participants were also more likely to be male, married, current smokers, and to consume more alcohol. At baseline, they also reported higher blood cholesterol levels and were more prone to be following any special diet than their peers in the other categories of coffee consumption.

**Table 1.** Baseline characteristics of participants according to total coffee consumption.

| Cups/Day | Total Coffee Consumption | | | | *p* Value |
|---|---|---|---|---|---|
| | <1 | 1 | >1 and <4 | ≥4 | |
| *N* | 5253 | 2667 | 5928 | 565 | |
| Age at recruitment | 34.5 (11.8) | 37.7 (11.9) | 37.1 (10.8) | 39.5 (11.1) | <0.001 |
| Body-mass index (kg/m²) | 23.2 (3.4) | 23.4 (3.3) | 23.5 (3.4) | 24.1 (3.7) | 0.002 |
| Physical activity in METS | 28.7 (26.1) | 26.8 (23.9) | 25.7 (21.5) | 26.4 (24.9) | <0.001 |
| Total energy in kcal/day | 2292 (630) | 2352 (593) | 2406 (598) | 2479 (653) | <0.001 |
| Adherence to Mediterranean diet (0–9 score) | 4.04 (1.79) | 4.36 (1.82) | 4.37 (1.78) | 4.37 (1.69) | 0.157 |
| Alcohol intake in g/day | 5.24 (8.32) | 7.08 (10.25) | 7.28 (10.13) | 8.34 (14.2) | <0.001 |
| Years of university education | 4.92 (1.48) | 5.09 (1.50) | 5.13 (1.52) | 5.12 (1.58) | 0.055 |
| Sex (% male) | 41.2 | 40.1 | 38.4 | 45.3 | 0.001 |
| Snacking (%) | 35.8 | 29.1 | 32.5 | 35.4 | <0.001 |
| Special diet (%) | 6.24 | 6.60 | 7.25 | 9.73 | 0.007 |
| Hypertension (%) | 8.68 | 9.00 | 8.11 | 8.85 | 0.514 |
| Cholesterol >200 mg/dl (%) | 13.0 | 15.7 | 16.0 | 20.2 | <0.001 |
| Smoking (%) | | | | | |
|    Never | 59.5 | 49.5 | 41.8 | 28.8 | |
|    Current | 19.4 | 22.2 | 29.5 | 40.4 | |
|    Former | 18.3 | 25.9 | 26.4 | 27.1 | <0.001 |
| Marital status (%) | | | | | |
|    Single | 56.1 | 42.7 | 44.2 | 36.5 | |
|    Married | 41.6 | 54.7 | 53.1 | 60.7 | |
|    Other | 2.25 | 2.55 | 2.77 | 2.83 | <0.001 |
| Personality traits (range 0–10) | | | | | |
|    Psychological dependence | 3.69 (2.83) | 3.49 (2.87) | 3.53 (2.81) | 3.59 (2.97) | 0.236 |
|    Competitiveness | 6.99 (1.73) | 6.96 (1.77) | 6.96 (1.70) | 7.09 (1.73) | 0.086 |
|    Anxiety | 5.82 (2.22) | 5.80 (2.20) | 5.91 (2.13) | 6.22 (2.19) | 0.017 |

Data are mean (standard deviation), unless otherwise stated.

Table 2 presents hazard ratios (HR) and their 95% confidence intervals (CI) for the risk of depression in the crude and multivariable adjusted models. In the comparison across extreme categories of coffee consumption, participants who consumed at least 4 cups of coffee per day showed a 63% (HR = 0.37, 95% CI 0.15–0.95) lower risk of depression than participants who drank less than

1 cup of coffee per day. However, overall, we did not observe a linear dose–response association between coffee consumption and the incidence of depression ($p$ for trend = 0.22).

No significant interaction was found between total coffee consumption and sex, age, or smoking status in their association with incident depression ($p > 0.05$ for all of them).

**Table 2.** Hazard ratios (HR; 95% confidence intervals) for incidence of depression according to baseline total coffee consumption.

| Cups/Day | Total Coffee Consumption | | | | *p* for Trend |
|---|---|---|---|---|---|
| | <1 | 1 | >1 and <4 | ≥4 | |
| Cups/day (median) | 0.07 | 1 | 2.5 | 5 | |
| N | 5253 | 2667 | 5928 | 565 | |
| Cases | 64 | 39 | 91 | 5 | |
| Person-years | 51,145 | 26,065 | 60,705 | 6115 | |
| Crude HR | 1 (ref.) | 1.14 (0.76–1.70) | 1.12 (0.81–1.55) | 0.60 (0.24–1.50) | 0.963 |
| Model 1 | 1 (ref.) | 1.12 (0.75–1.67) | 1.09 (0.79–1.51) | 0.58 (0.23–1.45) | 0.923 |
| Model 2 | 1 (ref.) | 1.05 (0.70–1.58) | 0.95 (0.68–1.33) | 0.37 (0.15–0.95) | 0.220 |

Results from Cox regression models. Age was the underlying time variable in all analyses. Model 1: adjusted for sex and stratified for age (decades) and recruitment period. Model 2: adjusted for sex, alcohol intake (linear and quadratic term), years of university education, marital status, smoking, body mass index, total energy intake, adherence to the Mediterranean diet, between-meal snacking and following special diets, leisure-time physical activity (METS-h/week), hours of TV watching, hypertension at baseline, baseline high blood cholesterol, self-perception of competitiveness, anxiety, and psychological dependence, and use of anxiolytics, and stratified for age (decades) and recruitment period.

In further analyses, we specifically studied regular and decaffeinated coffee consumption (Table 3). The HR for the risk of depression associated with ≥4 cups per day of regular coffee compared to <1 cup per day was 0.44 (95% CI: 0.18–1.11; $p$ for trend = 0.141), in a model adjusted for the consumption of decaffeinated coffee consumption. On the other hand, decaffeinated coffee consumption was not associated with the risk of depression in the fully adjusted model.

**Table 3.** Subgroup analysis. Hazard ratios (95% confidence intervals) for incidence of depression according to baseline regular and decaffeinated coffee consumption.

| Cups/Day | Regular Coffee Consumption | | | | *p* for Trend |
|---|---|---|---|---|---|
| | <1 | 1 | >1 and <4 | ≥4 | |
| Cups/day (median) | 0 | 1 | 2.5 | 5 | |
| N | 6315 | 3433 | 4193 | 472 | |
| Cases | 84 | 49 | 61 | 5 | |
| Person-years | 61,621 | 34,065 | 43,130 | 5212 | |
| Crude HR | 1 (ref.) | 1.01 (0.71–1.44) | 0.97 (0.69–1.35) | 0.65 (0.26–1.60) | 0.569 |
| Model 1 | 1 (ref.) | 1.00 (0.70–1.42) | 0.96 (0.69–1.34) | 0.64 (0.26–1.59) | 0.533 |
| Model 2 | 1 (ref.) | 0.96 (0.67–1.37) | 0.84 (0.59–1.18) | 0.43 (0.17–1.07) | 0.095 |
| Additionally adjusted for decaffeinated coffee consumption | 1 (ref.) | 0.97 (0.68–1.39) | 0.87 (0.61–1.23) | 0.44 (0.18–1.11) | 0.141 |

| Cups/day | Decaffeinated Coffee Consumption | | | *p* for Trend |
|---|---|---|---|---|
| | <1 | 1 | >1 | |
| Cups/day (median) | 0 | 1 | 2.5 | |
| N | 12,700 | 1268 | 445 | |
| Cases | 167 | 21 | 11 | |
| Person-years | 127,007 | 12,674 | 4348 | |
| Crude HR | 1 (ref.) | 1.25 (0.79–1.96) | 1.90 (1.03–3.51) | 0.033 |
| Model 1 | 1 (ref.) | 1.20 (0.76–1.89) | 1.77 (0.96–3.26) | 0.065 |
| Model 2 | 1 (ref.) | 1.20 (0.76–1.89) | 1.54 (0.82–2.87) | 0.142 |
| Additionally adjusted for regular coffee consumption | 1 (ref.) | 1.15 (0.72–1.82) | 1.46 (0.78–2.76) | 0.218 |

Results from Cox regression models. Age was the underlying time variable in all analyses. Model 1: adjusted for sex and stratified for age (decades) and recruitment period. Model 2: adjusted for age, sex, alcohol intake (linear and quadratic term), years of university education, marital status, smoking, body mass index, total energy intake, adherence to the Mediterranean diet, between-meal snacking and following special diets, leisure-time physical activity (METS-h/week), hours of TV watching, hypertension at baseline, baseline high blood cholesterol, self-perception of competitiveness, anxiety, and psychological dependence, and use of anxiolytics.

## 4. Discussion

In this study, we found that participants who consumed at least four cups of coffee per day showed a lower risk of depression than participants who drank less than one cup of coffee per day. Nevertheless, we found no significant dose–response relationship between coffee consumption and the risk of depression.

The observed inverse association between extreme categories of coffee consumption is consistent with previous literature. On the one hand, several cross-sectional studies have assessed this association [18–22]. Some [18–20], but not all [21,22], found a significant inverse association between coffee consumption and the risk of depression. However, due to the cross-sectional design of these studies, reverse causality cannot be ruled out. On the other hand, there are three prospective studies that have longitudinally assessed the association between coffee consumption and the risk of depression [23–25]. These three prospective studies have been pooled in two independent meta-analyses [8,9]. The combined results suggested an inverse association between coffee consumption and the risk of depression. As far as the setting for the three prospective cohort studies was concerned, two of them had been conducted in the U.S. [23,24], and another one, including a smaller number (2232) of participants, in Finland [25]. Individually, the three studies described an inverse association between coffee consumption and the risk of depression. The strength of the association was highest in the Finnish cohort, which included only men [25]. In that study, a 75% reduction in the risk of depression was observed when heavy coffee drinkers were compared with non-coffee drinkers [25]. Nevertheless, the analyses were based on 73 events. Risk reductions in the other two cohorts were milder [23,24]. It is worth mentioning that our study had some differential characteristics with previous prospective studies. First, the mean age of participants in our cohort was 36 years, whereas the mean age was 53 years in the study by Ruusunen et al. [25], 62 years in the study with data from the NIH-AARP study [24], and 63 years in the Nurses' Health Study [23]. Also, in the Finnish study, the outcome was given by a discharge diagnosis of depressive disorder [25] and our outcome—consistent with the other two studies [22,23]—was ascertained through self-reported information.

When we separately assessed the association between regular and decaffeinated coffee consumption and the risk of depression, we found no significant association for decaffeinated coffee consumption. Out of the three longitudinal studies which have assessed the association between coffee consumption and depression risk, decaffeinated coffee consumption was associated with a lower risk of depression (HR ≥ 4 cups/day vs. none = 0.88 (95% CI 0.78–1.00), *p* for trend = 0.003) in the NIH-AARP study [24], but no significant association for decaffeinated coffee consumption was observed in the Nurses' Health Study [23]. In the Kuopio Ischaemic Heart Disease Risk Factor Study, decaffeinated coffee consumption was not specifically assessed [25].

In a dose–response meta-analysis on coffee consumption and depression risk, Grosso et al. observed a nonlinear J-shaped dose–response association with a peak (the lowest observed risk) for the inverse association at 400 mL/day, which was stable toward a slight increase for higher coffee consumption [9]. It is worth mentioning that the studies that contributed most—i.e., had a higher weight—in this meta-analysis had been conducted in the U.S. [23,24], where the typical serving size is bigger than in Spain. The greatest risk reduction in our study was observed for participants who consumed at least four cups of coffee per day compared to those who consumed less than one cup per day, but we were not able to draw conclusions for participants with heavier coffee consumptions.

There are two main hypotheses which could explain the association between the higher coffee intake and a possible reduction in the risk of depression. First, coffee is the main dietary source of caffeine. Caffeine is an alkaloid exerting a stimulant effect on the central nervous system and modulating the dopaminergic activity by nonspecific antagonism against A1/A2 adenosine receptors. A moderate amount of caffeine has a beneficial effect, improving psychomotor activity, vigilance level, and increasing the perception of feeling more energetic [26]. Second, coffee has a high concentration of polyphenols, such as chlorogenic acid and trigonelline, which have anti-inflammatory potential [7].

Thus, coffee consumption could protect against low-grade inflammation, which seems to be involved in the pathogenesis of depression [27]. In fact, coffee is the main dietary source of polyphenols [28,29] in some populations such as the U.S. or Northern Europe, where the other prospective studies on coffee and depression had been conducted [21–23]. Contrarily, in our cohort (data not shown), as well as in other Spanish cohorts, fruits—and not coffee—are the primary source of polyphenols [30]. Interestingly, in the prospective studies conducted so far [23–25], the analyses were not adjusted for an overall dietary pattern. Only one analysis [24] was adjusted for daily intake of folate and polyunsaturated fatty acids. Therefore, it was unknown if coffee had the same beneficial effect on participants beyond an overall healthy dietary pattern. To our knowledge, our study is the first one evaluating the association between coffee consumption and depression in which adherence to an overall healthy dietary pattern has been accounted for.

Some limitations of our study should be acknowledged. First, the SUN cohort is not a representative sample of the general population in the pure statistical sense. However, lack of representativeness does not preclude from establishing associations [31,32]. These associations can be generalized to other groups as long as no biological mechanism suggests that the association no longer holds for other populations. Second, dietary information was self-reported. Therefore, we cannot exclude some degree of nondifferential misclassification which could have biased our results more probably towards the null. However, the FFQ has been previously validated [12–14]. Third, due to the strict criteria used for the adjudication of the outcome together with some particular characteristics of our study participants—such as their high educational level and their high levels of health-consciousness related to voluntarily participating in a cohort—the incidence of depression in our cohort may seem relatively low in comparison with other studies. However, when we included as incident cases all participants with a medical diagnosis of depression, those who were using antidepressant medication, and the cases that occurred in the two first years, the overall incidence of depression in the cohort during follow-up was 6% (data not shown). In any case, this does not necessarily mean a bias in the sample, as Rothman and other methodologists have repeatedly considered regarding the nonrepresentative nature of most cohorts in the statistical sense of "representativeness" [32]. Fourth, although all the results were adjusted for potential confounders, we cannot exclude the presence of some residual confounding factors that could partly explain our results. Nevertheless, with subsequent adjustment of our models with a wide array of potential confounders, the association became stronger for total coffee consumption and for regular coffee consumption. Therefore, we believe it is unlikely that unmeasured confounders could explained the observed association. Fifth, coffee consumption was assessed only in the baseline questionnaire, assuming it was maintained over time. Nevertheless, previous studies have suggested that coffee consumption remains relatively stable over time [33]. Sixth, tea consumption was not very common in Spain by the time the FFQ was developed, and this item was thus not included in the FFQ. Therefore, we could not assess the specific association between tea consumption and incident depression.

Several strengths of this study deserve to be mentioned. The prospective longitudinal design of the study with an extended follow-up period, the relatively large sample size, the validated assessment of coffee consumption, the validated self-reported medical diagnosis of depression, the ability to control for a good number of potential confounding factors, and the high retention rate (91%) are strengths of our study. Additionally, the high educational level of our participants could contribute to increase the quality of the self-reported information and, thus, reduce the potential for misclassification bias. Furthermore, the exclusion of participants with a depression diagnosis or use of antidepressant medication at baseline or before the first two years of follow-up reduced the possibility of reverse causation bias due to subclinical cases of depression present at baseline. Baseline coffee consumption of participants with baseline depression or antidepressant use might be a consequence of their condition, rather than vice versa. Also, participants who were diagnosed during the first two years of follow-up might have already had some symptoms at the study inception, which might have conditioned their coffee consumption. Therefore, we excluded participants with self-reported depression or antidepressant use during the first two years of follow-up in order to ensure temporal sequence. Finally, the incident cases were defined as self-reported physician-diagnosed depression together with commencement of regular antidepressant medication. Self-reported medical diagnosis of depression showed an acceptable validity in a previous validation study [15]. In the present paper, we increased the specificity of our outcome by including as an additional criterion the commencement of regular antidepressant use. This definition is consistent with previous literature in this area [23] and is stricter. Eventually, this definition might have led to the underestimation of true cases and to a lower sensitivity, but to a higher specificity. Supposedly, with perfect specificity, the nondifferential sensitivity of disease detection would not bias the estimate for the relative risk [34].

## 5. Conclusions

In conclusion, higher coffee consumption was inversely associated with the incidence of depression in a Mediterranean cohort, although the linear dose–response association was not significant. Future studies with longitudinal design and intervention studies would be needed to investigate potential health benefits of coffee consumption.

**Author Contributions:** Conceptualization, M.Á.M.-G. and E.T.; methodology, M.Á.M-G., A.S.-V., and E.T.; software, A.M.N., D.A., and E.T.; validation, M.Á.M.-G. and A.S.-V.; formal analysis, A.M.N., D.A., and E.T.; investigation, E.T. and M.Á.M.-G.; resources, M.Á.M-G and A.S.-V.; data curation, M.Á.M.-G.; writing—original draft preparation, A.M.N. and D.A.; writing—review and editing, all authors; visualization, E.T.; supervision, E.T.; project administration, M.Á.M.-G.; funding acquisition, M.A.Á.-G. and A.S.-V.

**Funding:** This work was supported by the Spanish Government-Instituto de Salud Carlos III, the European Regional Development Fund (FEDER) (RD 06/0045, CIBER-OBN, Grants PI10/02658, PI10/02293, PI13/00615, PI14/01668, PI14/01798, PI14/01764, PI17/01795, and G03/140), the Navarra Regional Government (45/2011, 122/2014), and the University of Navarra.

**Acknowledgments:** The authors are indebted to the participants of the SUN study for their continued cooperation and participation. We are also grateful to the members of the Department of Nutrition of the Harvard School of Public Health (Willett W.C., Hu F.B., and Ascherio A.) who helped us to design the SUN study. We also thank the other members of the SUN Group: Alonso A., Barrio López M.T., Basterra-Gortari F.J., Benito Corchón S., Bes-Rastrollo M., Beunza J.J., Carlos S., Carmona L., Cervantes S., de Irala J., de la Fuente-Arrillaga C., de la Rosa P.A., Delgado-Rodríguez M., Donat-Vargas C., Donázar M., Eguaras S., Fernández-Montero A., Galbete C., García-López M., Gea A., Goñi Ochandorena E., Guillén Grima F., Hernández-Hernández A., Lahortiga F., Llorca J., López del Burgo C., Marí Sanchís A., Martí del Moral A., Martínez J.A., Núñez-Córdoba J.M., Pimenta A.M., Ramallal R., Rico A., Ruiz-Zambrana A., Ruiz-Canela M., Sánchez Adán D., Sayón-Orea C., Vázquez Ruiz Z., and Zazpe García I.

**Conflicts of Interest:** The authors declare no conflict of interest.

## References

1. Bromet, E.; Andrade, L.H.; Hwang, I.; Sasmpson, N.A.; Alonso, J.; de Girolamo, G.; de Graaf, R.; Demyttenaere, K.; Hu, C.; Iwata, N.; et al. Cross-national epidemiology of DSM-IV major depressive episode. *BMC Med.* **2011**, *9*, 90. [CrossRef] [PubMed]
2. World Health Organization. Depression. Global Health Observatory (GHO) Data. Available online: http://www.who.int/news-room/fact-sheets/detail/depression (accessed on 21 May 2018).
3. Chishti, P.; Stone, D.H.; Corcoran, P.; Williamson, E.; Petridou, E. EIROSAVE Working Group. Suicide mortality in the European Union. *Eur. J. Public Health* **2003**, *13*, 108–114. [PubMed]
4. Perez-Cornago, A.; de la Iglesia, R.; Lopez-Legarrea, P.; Abete, I.; Navas-Carretro, S.; Lacunza, C.I.; Lahortiga, F.; Martinez-Gonzalez, M.A.; Martinez, A.; Zulet, M.A. A decline in inflammation is associated wigh less depressive symptoms after a dietary intervention in metabolic syndrome patients: A longitudinal study. *Nutr. J.* **2014**, *13*, 36. [CrossRef] [PubMed]
5. Daly, M. The Relationship of C-Reactive Protein to Obesity-Related Depressive Symptoms: A Longitudinal Study. *Obesity* **2013**, *21*, 248–250. [CrossRef] [PubMed]
6. Morris, A.A.; Ahmed, Y.; Stoyanova, N.; Hooper, W.C.; De STaerke, C.; Gibbons, G.; Din-Dzietham, R.; Quyyumi, A.; Vaccarin, V. The Association between Depression and Leptin is Mediated by Adiposity. *Psychosom. Med.* **2012**, *74*, 483–488. [CrossRef] [PubMed]

7.  Godos, J.; Pluchinotta, F.R.; Marventano, S.; Buscemi, S.; Volti, G.L.; Galvano, F.; Grosso, G. Coffee components and cardiovascular risk: Beneficial and detrimental effects. *Int. J. Food Sci. Nutr.* **2014**, *65*, 925–936. [CrossRef] [PubMed]

8.  Wang, L.; Shen, X.; Wo, Y.; Zhang, D. Coffee and caffeine consumption and depression: A meta-analysis of observational studies. *Aust. N. Z. J. Psychiatry* **2016**, *50*, 228–242. [CrossRef] [PubMed]

9.  Grosso, G.; Micek, A.; Castellano, S.; Pajak, A.; Galvano, F. Coffee, tea, caffeine and risk of depression: A systematic review and dose-response meta-analysis of observational studies. *Mol. Nutr. Food Res.* **2016**, *60*, 223–234. [CrossRef] [PubMed]

10. Sánchez-Villegas, A.; Delgado-Rodríguez, M.; Alonso, A.; Schlatter, J.; Lahortiga, F.; Serra Majem, L.; Martínez-González, M.A. Association of the Mediterranean dietary pattern with the incidence of depression: The Seguimiento Universidad de Navarra /University of Navarra Follow-up (SUN) Cohort. *Arch. Gen. Psychiatry* **2009**, *66*, 1090–1098. [CrossRef] [PubMed]

11. Carlos, S.; De La Fuente-Arrillaga, C.; Bes-Rastrollo, M.; Razquin, C.; Rico-Campà, A.; Martínez-González, M.A.; Ruiz-Canela, M. Mediterranean Diet and Health Outcomes in the SUN Cohort. *Nutrients* **2018**, *10*. [CrossRef] [PubMed]

12. Martin-Moreno, J.M.; Boyle, P.; Gorgojo, L.; Maisonneuve, P.; Fernandez-Rodriguez, J.C.; Salvini, S.; Willett, W.C. Development and validation of a food frequency questionnaire in Spain. *Int. J. Epidemiol.* **1993**, *22*, 512–519. [CrossRef] [PubMed]

13. Fernández-Ballart, J.D.; Piñol, J.L.; Zazpe, I.; Corella, D.; Carrasco, P.; Toledo, E.; Perez-Bauer, M.; Martínez-González, M.A.; Salas-Salvadó, J.; Martín-Moreno, J.M. Relative validity of a semi-quantitative food-frequency questionnaire in an elderly Mediterranean population of Spain. *Br. J. Nutr.* **2010**, *103*, 1808–1816. [CrossRef] [PubMed]

14. De la Fuente-Arrillaga, C.; Ruiz, Z.V.; Bes-Rastrollo, M.; Sampson, L.; Martínez-González, M.A. Reproducibility of an FFQ validated in Spain. *Public Health Nutr.* **2010**, *13*, 1364–1372. [CrossRef] [PubMed]

15. Sanchez-Villegas, A.; Schlatter, J.; Ortuno, F.; Lahortiga, F.; Pla, J.; Benito, S.; Martinez-Gonzalez, M.A. Validity of a self-reported diagnosis of depression among participants in a cohort study using the Structured Clinical Interview for DSM-IV (SCID-I.). *BMC Psychiatry* **2008**, *8*, 43. [CrossRef] [PubMed]

16. Trichopoulou, A.; Costacou, T.; Bamia, C.; Trichopoulos, D. Adherence to a Mediterranean Diet and Survival in a Greek Population. *Engl. J. Med.* **2003**, *348*, 2599–2608. [CrossRef] [PubMed]

17. Lahortiga-Ramos, F.; Unzueta, C.R.; Zazpe, I.; Santiago, S.; Molero, P.; Sánchez-Villegas, A.; Martínez-González, M.Á. Self-perceived level of competitiveness, tension and dependency and depression risk in the SUN cohort. *BMC Psychiatry* **2018**, *18*, 241. [CrossRef] [PubMed]

18. Pham, N.M.; Nanri, A.; Kurotani, K.; Kuwahara, K.; Kume, A.; Sato, M.; Hayabuchi, H.; Mizoue, T. Green tea and coffee consumption is inversely associated with sepressive symptoms in a Japanese working population. *Public Health Nurt.* **2013**, *17*, 625–633. [CrossRef]

19. Park, R.J.; Moon, J.D. Coffee and depression in Korea: The fifth Koreal National Health and Nutrition Examination Survey. *Eur. J. Clin. Nutr.* **2014**, *69*, 501–504. [CrossRef] [PubMed]

20. Omagari, K.; Sakaki, M.; Tsujimoto, Y.; Shiogama, Y.; Iwanaga, A.; Ishimoto, M.; Yamaguchi, A.; Masuzumi, M.; Kawase, M.; Ichimura, M.; et al. Coffee consumption is inversely associated with depressive status in Japanes Patients with tipe 2 diabetes. *J. Clin. Biochem. Nutr.* **2014**, *55*, 134–142. [CrossRef] [PubMed]

21. Hintikka, J.; Tolmunen, T.; Honkalampi, K.; Haatainen, K.; Koivumaa-Honkanen, H.; Tanskanen, A.; Viinamäki, H. Daily tea drinking is associated with a low level of depressive symptoms in the Finnish general population. *Eur. J. Epidemiol.* **2005**, *20*, 359–363. [CrossRef] [PubMed]

22. Niu, K.; Hozawa, A.; Kuriyama, S.; Ebihara, S.; Guo, H.; Nakaya, N.; Ohmori-Matsuda, K.; Takahashi, H.; Masamune, Y.; Asada, M.; et al. Green tea consumption is associated with depressive symptoms in the elderly. *Am. J. Clin. Nutr.* **2009**, *90*, 1615–1622. [CrossRef] [PubMed]

23. Lucas, M.; Mirzaei, F.; Pan, A.; Okereke, O.; Willett, W.; O'Reilly, E.J.; Koenen, K.; Ascherio, A. Coffee, Caffeine, and Risk of Depression among Women. *Arch. Intern. Med.* **2011**, *171*, 1571–1578. [CrossRef] [PubMed]

24. Guo, X.; Park, Y.; Freedman, N.D.; Sinha, R.; Hollenbech, A.R.; Blair, A.; Chen, H. Sweetened Beverages, Coffee and Tea and Depression Risk among Older US Adults. *PLoS ONE* **2014**, *9*. [CrossRef] [PubMed]

25. Ruusunen, A.; Lehto, S.M.; Tolmunen, T.; Mursu, J.; Kaplan, G.A.; Voutilainen, S. Coffee, tea and caffeine intake and the risk of severe depression in middle-aged Finnish men: The Kuopio Ischaemic Heart Disease Risk Factor Study. *Public Health Nutr.* **2010**, *13*, 1215–1220. [CrossRef] [PubMed]

26. Adan, A.; Prat, G.; Fabbri, M.; Sanchez-Turet, M. Early effects of caffeinated and decaffeinated coffee on subjective state and gender differences. *Prog. Neuropsychopharmacol. Biol. Psychiatry* **2008**, *32*, 1698–1703. [CrossRef] [PubMed]

27. Sanchez-Villegas, A.; Martinez-González, M.A. Diet, a new target to prevent depression? *BMC Med.* **2013**, *11*, 3. [CrossRef] [PubMed]

28. Ovaskainen, M.L.; Törrönen, R.; Koponen, J.M.; Sinkko, H.; Hellström, J.; Reinivuo, H.; Mattila, P. Dietary intake and major food sources of polyphenols in Finnish adults. *J. Nutr.* **2008**, *138*, 562–566. [CrossRef] [PubMed]

29. Burkholder-Cooley, N.; Rajaram, S.; Haddad, E.; Fraser, G.E.; Jaceldo-Siegl, K. Comparison of polyphenol intakes according to distinct dietary patterns and food sources in the Adventist Health Study-2 cohort. *Br. J. Nutr.* **2016**, *115*, 2162–2169. [CrossRef] [PubMed]

30. Tresserra-Rimbau, A.; Medina-Remón, A.; Pérez-Jiménez, J.; Martínez-Gonzalez, M.A.; Covas, M.I.; Corella, D.; Salas-Salvado, J.; Gomez-Gracia, E.; Lapetra, J.; Aros, F.; et al. Dietary intake and major food sources of polyphenols in a Spanish population at high cardiovascular risk: The PREDIMED study. *Nutr. Metab. Cardiovasc. Dis.* **2013**, *23*, 953–959. [CrossRef] [PubMed]

31. Rothman, K.J.; Gallacher, J.E.; Hatch, E.E. Why representativeness should be avoided. *Int. J. Epidemiol.* **2013**, *42*, 1012–1014. [CrossRef] [PubMed]

32. Rothman, K.J. *Epidemiology. An Introduction*, 2nd ed.; Oxford University Press: New York, NY, USA, 2012.

33. Winkelmayer, W.; Stampfer, M.J.; Willett, W.C.; Curhan, G.C. Habitual Caffeine Intake and the Risk of Hypertension in Women. *JAMA* **2005**, *294*, 2330–2335. [CrossRef] [PubMed]

34. Greenland, S.; Lash, T.L. Bias analysis. In *Modern Epidemiology*, 3rd ed.; Rothman, K.J., Greenland, S., Lash, T.L., Eds.; Lippincott Williams and Wilkins: Philadelphia, PA, USA, 2008; p. 359.

*nutrients*

MDPI

*Article*

# Maternal and Paternal Caffeine Intake and ART Outcomes in Couples Referring to an Italian Fertility Clinic: A Prospective Cohort

Elena Ricci [1,*], Stefania Noli [2], Sonia Cipriani [1], Irene La Vecchia [2], Francesca Chiaffarino [1], Stefania Ferrari [1], Paola Agnese Mauri [1,2], Marco Reschini , Luigi Fedele [1,2] and Fabio Parazzini [1,2]

1   Dipartimento Madre-Bambino-Neonato, Fondazione IRCCS Ca' Granda Ospedale Maggiore Policlinico, 20122 Milan, Italy; son.cipriani@gmail.com (S.C.); francesca.chiaffarino@gmail.com (F.C.); stefania.ferrari@policlinico.mi.it (S.F.); paola.mauri@unimi.it (P.A.M.); luigi.fedele@unimi.it (L.F.); fabio.parazzini@unimi.it (F.P.)
2   Department of Clinical Sciences and Community Health, Università di Milano, Fondazione IRCCS Ca' Granda Ospedale Maggiore Policlinico, 20122 Milan, Italy; stefi.noli@gmail.com (S.N.); irene.lavecchia@unimi.it (I.L.V.)
3   Infertility Unit, Fondazione IRCCS Ca' Granda Ospedale Maggiore Policlinico, 20122 Milan, Italy; m.reschini@gmail.com
*   Correspondence: ed.ricci@libero.it; Tel.: +39-02-55032318; Fax: +39-02-550320252

Received: 15 July 2018; Accepted: 15 August 2018; Published: 17 August 2018

**Abstract:** Caffeine intake, a frequent lifestyle exposure, has a number of biological effects. We designed a cohort study to investigate the relation between lifestyle and assisted reproduction technique (ART) outcomes. From September 2014 to December 2016, 339 subfertile couples referring to an Italian fertility clinic and eligible for ART procedures were enrolled in our study. Sociodemographic characteristics, smoking, and usual alcohol and caffeine consumption in the year prior to ART were recorded. The mean age of participants was $36.6 \pm 3.6$ years in women and $39.4 \pm 5.2$ years in men. After oocytes retrieval, 293 (86.4%) underwent implantation, 110 (32.4%) achieved clinical pregnancy, and 82 (24.2%) live birth. Maternal age was the main determinant of ART outcome. In a model including women's age and college degree, smoking habits, calorie and alcohol intake for both partners, previous ART cycles, and partner's caffeine intake, we did not observe any association between caffeine intake and ART outcome. Using the first tertile of caffeine intake by women as a reference, the adjusted rate ratio (ARR) for live birth was 1.09 (95% confidence interval (CI) 0.79–1.50) in the second and 0.99 (95% CI 0.71–1.40) in the third tertiles. In conclusion, a moderate caffeine intake by women and men in the year prior to the ART procedure was not associated with negative ART outcomes.

**Keywords:** caffeine intake; assisted reproduction techniques; risk factors; implantation; clinical pregnancy; live birth

## 1. Introduction

*Caffeine Intake is Among the Most Common Lifestyle Exposure in Women and Men Alike*

Caffeine (1,3,7-trimethylxanthine) is found in coffee, tea, soft drinks (particularly cola-containing beverages and energy drinks), and chocolate. It easily crosses biologic membranes, is rapidly distributed throughout the body, and has been found in saliva, breast milk, the embryo, and the neonate [1]. The caffeine molecule is easily absorbed by humans, having approximately 100% bioavailability when taken by oral route and reaching a peak in the blood within 15–45 min after its consumption [2]. Caffeine has a number of biologic effects, including central nervous system stimulation, increased secretion of catecholamine, relaxation of smooth muscles, and stimulation of heart rate. Caffeine can also reach the follicular fluid, suggesting that it might exert a harmful role on the female reproductive process [3].

During the last decades, the relation between lifestyle factors and spontaneous fertility has been investigated in several observational studies, but, with regard to caffeine intake, few studies have analyzed the association between caffeine intake and in vitro fertilization (IVF) outcomes, showing inconsistent results. One study observed a negative association with live birth, when comparing women consuming >2–50 and >50 versus <2 mg/day of caffeine in the year prior to IVF [4], while other studies found no association between caffeine intake consumed just before or during IVF treatment and IVF outcomes [5,6]. In a study conducted in Boston [7], the adjusted percentage of cycles resulting in live birth for women in increasing categories of caffeine intake was 46% for <50 mg/day, 44%, 42%, 40% in intermediate intake categories, and 40% for >300 mg/day. On the other hand, Karmon et al. [8] recently found that caffeine intake was associated with a lower probability of achieving live birth after assisted reproduction techniques (ART), although this inverse association was limited to intracytoplasmic sperm injection (ICSI) cycles.

Thus, limited and conflicting data are available on the relation between caffeine intake and ART outcomes. In this paper, we analyzed the role of male and female caffeine consumption in ART outcomes, using data from a cohort study conducted in an Italian fertility center.

## 2. Methods

From September 2014 to December 2016, on randomly selected days, subfertile couples presenting for evaluation to the Fertility Unit of Fondazione IRCCS Ca' Granda, Ospedale Maggiore, Policlinico, Milan, and eligible for assisted reproduction technologies (ART), were invited to participate in an ongoing prospective cohort study on the role of lifestyle habits and diet on ART outcome. The study protocol was approved by the Ethical Review Board of Fondazione IRCCS Ca' Granda, Ospedale Maggiore, Policlinico (Milan, Italy). All procedures were conducted in accordance with the Helsinki Declaration and all participants provided written informed consent.

Study participation was proposed during the diagnostic phase. Couples were interviewed on the day of oocyte retrieval. On the same day, a semen sample was also collected and analyzed to proceed with in vitro fertilization (IVF) or intracytoplasmic sperm injection (ICSI). The time interval between the proposal of the study and the interview was generally less than one month. In the early period only women were interviewed, whereas partners' information collection started at a later stage.

Both partners of couples who agreed to participate were interviewed by centrally trained personnel, using a standard questionnaire to obtain information on general sociodemographic characteristics, personal and health history and habits (including smoking, physical activity, alcohol intake, and methylxanthine-containing beverages consumption). Couples who did not speak fluent Italian were excluded.

The overall participation rate was close to 95%. This high participation rate was mainly due to the fact that couples were interviewed during the period spent waiting for the different diagnostic stages, before the actual ART procedure. Considering this down time and the not sensitive character questions, couples did not usually refuse to answer the questionnaire.

The questionnaire included information on sociodemographic characteristics, anthropometric variables, and lifestyle factors—including tobacco smoking, alcohol and caffeine intake, and diet habits—as well as a problem-oriented personal medical history and reproductive history.

Information on diet was based on a reproducible and validated food frequency questionnaire (FFQ), including 78 foods, food groups (such as the major sources of animal fats (i.e., red meat, milk, cheese, ham, salami), folates, vitamins (vegetables and fruit), pasta and bread consumption, cake, sweets and chocolate, fish), and the most common Italian recipes [9–11]. Patients were asked to report their usual weekly food consumption in the last year. The FFQ includes the average weekly consumption of 78 food items or food groups. Energy and mineral, macro-, and micronutrient intakes were estimated using the most recent update of an Italian food consumption database [12].

The weekly numbers of drinks for several alcoholic beverages were elicited from the subjects. Taking into account the different ethanol concentrations, one drink corresponded to approximately 125 mL of wine, 330 mL of beer, and 30 mL of hard liquor (i.e., about 12.5 g of ethanol). Total alcohol intake, expressed in grams of ethanol per day (g/day), was computed as the sum of all reported alcoholic beverages. "Never drinkers" were patients who abstained from drinking lifelong; "ex-drinkers" were individuals who had abstained from drinking for at least 12 months at the time of interview. For the purpose of this study, we considered these subjects in the category "abstainers".

Further, questions included information on coffee and other methylxanthine-containing beverages (tea, cocoa, and decaffeinated coffee), and the average number of cups per day. Caffeine intake from coffee (60 mg per cup), cappuccino (75 mg per cup), tea (45 mg per cup), decaffeinated coffee (4 mg per cup), and chocolate (6 mg/10 g) was calculated [13].

A subject was considered a smoker if she had smoked more than one cigarette/day for at least one year; an ex-smoker if she had smoked more than one cigarette/day for at least one year, but had stopped more than one year before the interview, and a non-smoker if she had never smoked more than one cigarette/day.

Satisfactory reproducibility of questions on self-reported smoking and drinking habits in our study populations has been previously reported [14].

Patients were managed according to a standardized clinical protocol, as reported in detail elsewhere [15]. Couples underwent ART with conventional IVF or ICSI as clinically indicated.

Serum hCG assessment to detect pregnancy was performed 14 or 16 days after ovulation triggering or luteinizing hormone (LH) surge. Women with positive human corionic gonadotropin (hCG) values underwent a transvaginal sonography three weeks later. Clinical pregnancy was defined as the presence of at least one intrauterine gestational sac.

All clinical information (including infertility diagnoses) was collected from medical records.

*Statistical Analysis*

Multiple outcomes were considered in this analysis: (1) number of retrieved high quality oocytes; (2) undergoing embryo transfer (implantation); (3) clinical pregnancy; (4) live birth.

Categorical variables were described as frequency (N) and percentage (%) and compared using the Pearson or Mantel-Haenzsel (MH) chi-square, as appropriate. Continuous variables were described as means with standard deviation (SD) if normally distributed, or medians and interquartile ranges (IQR) if not normally distributed. Univariate analyses used were analysis of variance and Kruskal-Wallis test. The correlation between male and female caffeine consumption was evaluated by means of Spearman correlation rho, because caffeine consumption was not normally distributed.

In the multivariable models, we included as potential confounders variables associated with caffeine intake or ART outcomes at the univariate analysis. Thus, we accounted for women's age, education, tobacco smoking, alcohol intake, total energy intake, and previous ART cycles.

As regards the oocyte number, it was square-root transformed and included in a general linear equation with the aforementioned variables. We calculated the adjusted means in tertiles of women's caffeine intake, and according to its 95% confidence intervals (CIs). Then, these figures were back-transformed to medians and 95% CIs.

Using unconditional multiple logistic regression, we estimated rate ratios (RR) of each outcome and corresponding 95% CIs in categories of caffeine intake (approximate tertiles). In the logistic regression equation, we included woman's age, education, tobacco smoking, alcohol intake, total energy intake, and previous ART cycles. As regards men's variables, we included alcohol and calorie intake. Furthermore, we mutually adjusted men's and women's caffeine intake. In a second model, we combined categories of intake under and over the median (for each sex) and, using the lowest category (both the woman's and partner's intake under the median) as the reference, we calculated the RRs for ART failure in the other three groups.

Statistical significance was set at $p < 0.05$. All analyses were performed using SAS software, version 9.4 (SAS Institute, Inc., Cary, NC, USA).

## 3. Results

From September 2014 to December 2016, 501 women and 347 men were interviewed; since eight men did not provide complete information, the couples were excluded from the analysis. The final analysis included 339 couples, who provided complete information about their lifestyle and coffee/caffeine intake and underwent an ART cycle.

As regards women, the mean age was 36.6 years (standard deviation, SD, 3.6, range 27–45) and the mean body mass index (BMI) was 22.2 kg/m$^2$ (SD 3.7, range 17.0–41.0); 18 (5.4%) women were obese (BMI > 30.0 kg/m$^2$). As regards men, the mean age was 39.4 years (SD 5.2, range 27–60) and the mean BMI was 25.3 kg/m$^2$ (SD 3.0); 29 (8.8%) men were obese.

Table 1 shows the characteristics of women and men according to caffeine intake. There was no difference in terms of age, education, BMI, or cause for infertility in tertiles of caffeine consumption. A relationship was observed with smoking habits and alcohol intake both in men (chi-square $p = 0.001$ and 0.02, respectively) and women (chi-square $p = 0.002$ and 0.01, respectively). Women who had undergone previous ART cycles were more frequently in the lowest caffeine intake tertile ($p = 0.002$). Both men's and women's total energy intake increased by tertiles of caffeine consumption ($p < 0.0001$).

The correlation between male and female caffeine intake was statistically significant ($p = 0.0002$) but not very high (Spearman rho = 0.20).

After oocytes retrieval, 293 (86.4%) underwent embryo-transfer, 110 (32.4%) achieved clinical pregnancy, and 82 (24.2%) experienced a live birth, including eight twin births. Out of 28 interrupted clinical pregnancies, 27 were miscarriages and one was an induced abortion.

ART outcomes were not associated with any men's characteristics, whereas women's education was significantly related to implantation (RR for college degree 1.78, 95% CI 1.00–3.18) and age at clinical pregnancy (for women aged: 35–40 years, RR 1.11, 95% CI 0.80–1.54; ≥40 years, RR 1.80, 95% CI 1.09–2.98; chi-square for trend 5.05, $p = 0.025$) and live birth (for women aged: 35–39 years, RR 1.30, 95% CI 0.88–1.93; ≥40, RR 2.43, 95% CI 1.28–4.63; chi-square for trend 7.93, $p = 0.005$). In both outcomes, older women were at a higher risk of failure. At univariate analysis, no association was observed with men's or women's caffeine intake.

Mean gestational week at delivery was 39.2 (SD 1.9, range 34–42); this was not associated with maternal caffeine intake, either as a continuous variable or in tertiles (Spearman rho = 0.19, $p = 0.09$). Mean gestational age in tertile of maternal intake was 39.0 (SD 1.7), 39.4 (SD 2.0), and 39.3 (SD 2.1) in the first, second, and third tertiles, respectively. Excluding twins, mean birth weight was 3140 (SD 428), with no significant differences across groups of maternal caffeine intake.

Table 2 shows the adjusted number of retrieved oocytes, adjusted for women's age, education, smoking habits, and calorie and alcohol intake. Adjusted median number of oocytes was higher in the third tertile of caffeine intake, but this difference was not statistically significant.

**Table 1.** Demographic characteristics of 339 couples, according to caffeine intake.

| | Women | | | | | | Men | | | | | |
|---|---|---|---|---|---|---|---|---|---|---|---|---|
| | **First Tertile** | | **Second Tertile** | | **Third Tertile** | | **First Tertile** | | **Second Tertile** | | **Third Tertile** | |
| | 0–86 mg/day | 33.6% | 87–180 mg/day | 34.2% | 181–480 mg/day | 32.1% | 0–124 mg/day | 33.1% | 125–209 mg/day | 33.3% | 210–560 mg/day | 33.6% |
| | n = 114 | | n = 116 | | n = 109 | | n = 112 | | n = 113 | | n = 114 | |
| **Daily Caffeine Intake (mg/day), Median (IQR)** | 37 | 14–60 | 128 | 111–147 | 215 | 188–255 | 62 | 17–100 | 180 | 154–189 | 258 | 231–310 |
| **Age (years)** | | | | | | | | | | | | |
| <35 | 31 | 27.2 | 33 | 28.4 | 36 | 33.0 | 19 | 17.0 | 25 | 22.1 | 25 | 21.9 |
| 35–39 | 60 | 52.6 | 51 | 44.0 | 52 | 47.7 | 39 | 34.8 | 43 | 38.0 | 42 | 36.8 |
| ≥40 | 23 | 20.2 | 32 | 27.6 | 21 | 19.3 | 54 | 48.2 | 45 | 39.8 | 47 | 41.2 |
| **College Degree** | 62 | 54.4 | 67 | 57.8 | 53 | 48.6 | 42 | 37.5 | 42 | 37.2 | 51 | 44.7 |
| **Cause of Infertility** | | | | | | | | | | | | |
| Unexplained | 23 | 20.2 | 31 | 26.7 | 16 | 14.7 | 25 | 22.3 | 23 | 20.4 | 22 | 19.3 |
| Female factor only | 48 | 42.1 | 42 | 36.2 | 41 | 37.6 | 42 | 37.5 | 45 | 39.8 | 44 | 38.6 |
| Male and female factor | 43 | 37.7 | 43 | 37.1 | 52 | 47.7 | 45 | 40.2 | 45 | 39.8 | 48 | 42.1 |
| **BMI** | | | | | | | | | | | | |
| <18.5 | 12 | 10.5 | 4 | 3.5 | 12 | 11.0 | – | – | – | – | – | – |
| 18.5–24.9 | 82 | 71.9 | 86 | 76.1 | 78 | 71.6 | 46 | 45.1 | 55 | 48.7 | 45 | 39.5 |
| 25.0–29.9 | 17 | 14.9 | 13 | 11.5 | 14 | 12.8 | 48 | 47.1 | 48 | 42.5 | 58 | 50.9 |
| ≥30.0 | 3 | 2.6 | 10 | 8.8 | 5 | 4.6 | 8 | 7.8 | 10 | 8.8 | 11 | 9.6 |
| **Smoking Habits** | | | | | | | | | | | | |
| Never | **74** | **64.9** | **72** | **62.1** | **49** | **45.0** | **55** | **53.4** | **46** | **40.7** | **28** | **24.8** |
| Current | **15** | **13.2** | **15** | **12.9** | **34** | **31.2** | **24** | **23.3** | **33** | **29.2** | **50** | **44.2** |
| Former | **25** | **21.9** | **29** | **25.0** | **26** | **23.8** | **24** | **23.3** | **34** | **30.1** | **35** | **31.0** |
| **Alcohol Intake** | | | | | | | | | | | | |
| Abstainer | **40** | **35.1** | **33** | **28.4** | **23** | **21.1** | **14** | **13.6** | **9** | **8.0** | **8** | **7.0** |
| <1 unit/day | **72** | **63.2** | **75** | **64.7** | **73** | **67.0** | **57** | **55.3** | **66** | **58.4** | **53** | **46.5** |
| ≥1 unit/day | **2** | **1.7** | **8** | **6.9** | **13** | **11.9** | **32** | **31.1** | **38** | **33.6** | **53** | **46.5** |
| **Leisure Physical Activity** | | | | | | | | | | | | |
| <2 h/week | 57 | 50.0 | 66 | 57.4 | 69 | 63.3 | 36 | 35.3 | 48 | 42.9 | 49 | 43.8 |
| 2–4 | 44 | 38.6 | 43 | 37.4 | 32 | 29.4 | 37 | 36.3 | 44 | 39.3 | 38 | 33.9 |
| >4 | 13 | 11.4 | 6 | 5.2 | 8 | 7.3 | 29 | 28.4 | 20 | 17.9 | 25 | 22.3 |
| **Previous ART Cycle** | 81 | 71.0 | 60 | 51.7 | 54 | 49.5 | 67 | 59.8 | 63 | 58.4 | 62 | 54.4 |
| **Daily Calories Intake (Kcal/day), Median (IQR)** | 1589 | 1367–1924 | 1748 | 1497–2124 | 1871 | 1602–2179 | 1781 | 1480–2060 | 1960 | 1649–2262 | 2122 | 1753–2407 |

Sometimes the sums do not add up to the totals because of missing values. Bold: $p < 0.05$, IQR: median and interquartile range; BMI: body mass index.

Table 2. Rate ratios for failure of assisted reproduction technique (ART) outcomes.

| Number of High-Quality Oocytes (Adjusted * Median, 95% CI) | Implantation Failure n=46 (13.6%) n | % | Success n=293 (86.4%) n | % | ARR (95% CI) | Clinical Pregnancy Failure n=229 (67.6%) n | % | Success n=110 (32.4%) n | % | ARR (95% CI) | Live Birth Failure n=257 (75.8%) n | % | Success n=82 (24.2%) n | % | ARR (95% CI) |
|---|---|---|---|---|---|---|---|---|---|---|---|---|---|---|---|
| **Women** | | | | | | | | | | | | | | | |
| **Caffeine intake** | | | | | | | | | | | | | | | |
| First tertile 4.8 (4.0–5.7) | 14 | 30.4 | 100 | 34.1 | 1 | 74 | 32.3 | 40 | 36.4 | 1 | 84 | 32.7 | 30 | 36.6 | 1 |
| Second tertile 4.6 (3.9–5.4) | 20 | 43.5 | 96 | 32.8 | 1.34 (0.64–2.79) | 84 | 36.7 | 32 | 29.1 | 1.07 (0.76–1.50) | 95 | 37.0 | 21 | 25.6 | 1.09 (0.79–1.50) |
| Third tertile 5.3 (4.5–6.1) | 12 | 26.1 | 97 | 33.1 | 0.90 (0.38–2.10) | 71 | 31.0 | 38 | 34.6 | 1.00 (0.70–1.43) | 78 | 30.3 | 31 | 37.8 | 0.99 (0.71–1.40) |
| >90 percentile § 5.1 (3.8–6.7) | 5 | 10.9 | 29 | 9.9 | 0.58 (0.16–2.04) | 23 | 10.0 | 11 | 10.0 | 0.96 (0.51–1.80) | 24 | 9.3 | 10 | 12.2 | 0.99 (0.54–1.84) |
| **Men** | | | | | | | | | | | | | | | |
| **Caffeine intake** | | | | | | | | | | | | | | | |
| First tertile - | 12 | 26.1 | 100 | 34.1 | 1 | 71 | 31.0 | 41 | 37.3 | 1 | 83 | 32.3 | 29 | 35.4 | 1 |
| Second tertile - | 22 | 47.8 | 91 | 31.1 | 1.64 (0.78–3.44) | 76 | 33.2 | 37 | 33.6 | 1.01 (0.72–1.42) | 88 | 34.2 | 25 | 30.5 | 1.02 (0.75–1.41) |
| Third tertile - | 12 | 34.8 | 102 | 34.8 | 0.78 (0.32–1.87) | 82 | 35.8 | 32 | 29.1 | 1.07 (0.76–1.52) | 86 | 33.5 | 28 | 34.1 | 1.00 (0.72–1.40) |
| >90 percentile § - | 4 | 8.7 | 35 | 12 | 0.73 (0.17–3.16) | 27 | 11.8 | 12 | 10.9 | 1.12 (0.59–2.14) | 29 | 11.3 | 10 | 12.2 | 0.98 (0.54–1.83) |
| **Combined intake (W-M)** | | | | | | | | | | | | | | | |
| Low-low - | 14 | 30.4 | 81 | 27.6 | 1 | 64 | 28.0 | 31 | 28.2 | 1 | 71 | 27.6 | 24 | 29.3 | 1 |
| Low-high - | 10 | 21.7 | 65 | 22.2 | 0.79 (0.34–1.85) | 56 | 24.4 | 19 | 17.3 | 1.03 (0.71–1.50) | 62 | 24.1 | 13 | 15.8 | 1.09 (0.76–1.55) |
| High-low - | 12 | 26.1 | 61 | 20.8 | 1.22 (0.53–2.78) | 48 | 21.0 | 25 | 22.7 | 0.95 (0.64–1.42) | 57 | 22.2 | 16 | 19.5 | 1.06 (0.73–1.54) |
| High-high - | 10 | 21.7 | 86 | 29.4 | 0.60 (0.24–1.48) | 61 | 26.6 | 35 | 31.8 | 0.89 (0.61–1.30) | 67 | 26.1 | 29 | 35.4 | 0.93 (0.65–1.33) |

The final model included women's age class and college degree, smoking habits, calorie and alcohol intake for both men and women, previous ART cycles, and partner's caffeine intake. ARR: adjusted rate ratio; CI: confidence interval; * for women's age class and education, smoking habits, calorie and alcohol intake; § reference category: <10 percentile; W-M: women-men.

RRs for caffeine intake, after adjusting for variables that were associated with caffeine intake (smoking habits and alcohol intake, daily calories) or ART outcomes (women's age and education), was consistently higher in the intermediate class of women's intake, but this findings were not significant. Men's caffeine consumption was also not statistically significant; no dose-effect was suggested by the observed estimates.

We built a second model with four categories for combined couple's caffeine intake (lower and equal/higher than the median for their sex): using the group of lowest combined caffeine intake, we did not find significant associations between the outcomes and different couples' caffeine intake, nor did we observe trends suggesting a relationship.

Performing the analysis in groups of procedure (IVF or ICSI), we did not find any marked difference in the relationship between caffeine intake and ART outcomes (data not shown).

## 4. Discussion

This prospective study of couples undergoing IVF or ICSI found that caffeine intake by women, men, and the couple was not associated with implantation, clinical pregnancy, and live birth, adjusting for women's age class and college degree, smoking habits, calorie and alcohol intake for both men and women, and previous ART cycles. Similarly, caffeine consumption by women was not related to the number of oocytes retrieved.

Potential limitations of this study should be considered. All information on lifestyle habits was self-reported by the patients, so some underestimates may have occurred. However, in Italy, recommendations to avoid caffeine in pregnancy have not received widespread attention and are not routinely advocated by gynecologists before IVF or ICSI, and misreporting of this variable should be unlikely.

Other sources of bias, including selection or confounding factors, are also unlikely to have produced marked effects, especially considering that all subjects were interviewed in the same institution and that participation was practically complete.

With regard to other biases, we analyzed information on nutritional status, and their inclusion into the model did not change the estimated OR. Further, the questionnaire was satisfactorily reproducible: correlation coefficients were >0.65 for most frequently eaten food, and between 0.50 and 0.65 for others [16]. However, the exact amount of caffeine in caffeinated beverages is difficult to quantify. Although patients reported the number of cups of caffeinated beverages that they drink, the exact amount of milligrams of caffeine in a cup depends on the mix of the brew, how it is prepared, and the size of the cup. The questionnaire also asked questions about soda, but did not discriminate between caffeinated and non-caffeinated varieties. Although these factors are likely to underestimate the caffeine intake, a differential bias is unlikely.

Another potential limitation is study power. For example, with our data we can identify an RR of pregnancy loss for the third tertile of caffeine intake of about 1.8. Thus, our results cannot rule out modest effect sizes, which we were underpowered and thus difficult to detect.

The strengths of our study include its prospective design with complete follow-up and our ability to adjust for a wide range of potential confounders. We also obtained information on male partner diet, alcohol intake, and smoking habits, which previous studies have not included.

We did not found any statistically significant association between caffeine intake and ART success.

Our findings are not consistent with those of Klonoff-Cohen et al. [4], who observed an association between usual coffee intake and lower ART success rate in 221 couples undergoing IVF or gamete intra-Fallopian transfer. This relationship was significant even in women reporting an intake of 20–50 mg caffeine/day, less than the equivalent of one cup of coffee per day. However, no association was observed with intake in the week before or during the procedure, a fact that the authors ascribed to the possibility that during the ART procedure women refrained from or decreased coffee drinking.

In the study by Al-Saleh and colleagues [3], no relationship emerged between coffee consumption and pregnancy outcomes, yet the authors observed a decrease in the number of eggs retrieved and an increase in miscarriage frequency as caffeine intake increased.

On the contrary, a recent cohort study, including 300 women and 493 ART cycles, provided reassurance that low to moderate intakes of caffeine (e.g., <200 mg/day) in the year prior to infertility treatment initiation do not have an adverse effect on intermediate or clinical outcomes of ART [7]. Another recent study by Machtinger et al. [6] enrolled 340 women undergoing IVF from 2014 through 2016 and did not retrieve any association between coffee and caffeine consumption and ART outcomes, whereas a threat to reproductive success was attributable to sugared beverages, independent of their caffeine content.

Considering men's caffeine intake as well, Choi et al. [5] found no relationship with implantation, fertilization, or live birth in a cohort including 2474 couples and 4716 IVF cycles. Although higher caffeine intake by women was associated with a significantly lower peak estradiol level, it was not related to the number of oocytes retrieved, implantation, fertilization, or live birth rate.

In a survey conducted in Italy in 2005–2006 [17], 1245 women had a median caffeine intake of 116 mg/day (95th percentile 355 mg/day) and 1068 men had a median caffeine intake of 112 mg/day (95th percentile 330 mg/day). In our sample, levels of consumption were similar in men, with a higher median (180 mg/day) but a similar 95th percentile, but not in women, who showed a similar median (126 mg/day) but a lower 95th percentile intake (272 mg/day). About 80% of women in our group consumed less than 200 mg of caffeine, which is the limit that, according to the European Food Safety Agency (EFSA), does not give rise to safety concerns for the fetus [18].

## 5. Conclusions

Our study does not show an effect of moderate coffee intake by women, men, or the couple on oocyte quality and success rate after ART procedures. Considering that our sample represented a moderate consumption of caffeine, as well as alcohol and tobacco, we cannot evaluate the effect of higher intakes on IVF outcomes. Thus, conservatively, all women seeking pregnancy should be advised to maintain caffeine intakes within limits suggested by the EFSA.

**Author Contributions:** Conceptualization, F.P., I.L.V., and L.F.; Methodology, F.P. and E.R.; Validation, S.F., M.R., and P.A.M.; Formal Analysis, E.R. and S.C.; Investigation, S.N., M.R., and S.F.; Data Curation, M.R., S.F., and S.N.; Writing—Original Draft Preparation, E.R., F.C., and F.P.; Writing—Review and Editing, F.P., P.A.M., L.F., and S.N.

**Funding:** The authors have no funding to report.

**Acknowledgments:** We are indebted to Marta Castiglioni, Benedetta Gallotti, and Maria Cavadini for their valuable contribution to data collection and patients' counseling, as well as to Francesca Bravi for her support to the data analysis.

**Conflicts of Interest:** The authors declare no conflict of interest.

## References

1. Monteiro, J.P.; Alves, M.G.; Oliveira, P.F.; Silva, B.M. Structure-Bioactivity Relationships of Methylxanthines: Trying to Make Sense of All the Promises and the Drawbacks. *Molecules* **2016**, *21*, 974. [CrossRef] [PubMed]
2. Sepkowitz, K.A. Energy drinks and caffeine-related adverse effects. *JAMA* **2013**, *309*, 243–244. [CrossRef] [PubMed]
3. Al-Saleh, I.; El-Doush, I.; Grisellhi, B.; Coskun, S. The effect of caffeine consumption on the success rate of pregnancy as well various performance parameters of in-vitro fertilization treatment. *Med. Sci. Monit.* **2010**, *16*, CR598–CR605. [PubMed]
4. Klonoff-Cohen, H.; Bleha, J.; Lam-Kruglick, P. A prospective study of the effects of female and male caffeine consumption on the reproductive endpoints of IVF and gamete intra-fallopian transfer. *Hum. Reprod.* **2002**, *17*, 1746–1754. [CrossRef] [PubMed]
5. Choi, J.H.; Ryan, L.M.; Cramer, D.W.; Hornstein, M.D.; Missmer, S.A. Effects of caffeine consumption by women and men on the outcome of in vitro fertilization. *J. Caffeine Res.* **2011**, *1*, 29–34. [CrossRef] [PubMed]

6.  Machtinger, R.; Gaskins, A.J.; Mansur, A.; Adir, M.; Racowsky, C.; Baccarelli, A.A.; Hauser, R.; Chavarro, J.E. Association between preconception maternal beverage intake and in vitro fertilization outcomes. *Fertil. Steril.* **2017**, *108*, 1026–1033. [CrossRef] [PubMed]

7.  Abadia, L.; Chiu, Y.H.; Williams, P.L.; Toth, T.L.; Souter, I.; Hauser, R.; Chavarro, J.E.; Gaskins, A.J. The association between pre-treatment maternal alcohol and caffeine intake and outcomes of assisted reproduction in a prospectively followed cohort. *Hum. Reprod.* **2017**, *32*, 1846–1854. [CrossRef] [PubMed]

8.  Karmon, A.E.; Toth, T.L.; Chiu, Y.H.; Gaskins, A.J.; Tanrikut, C.; Wright, D.L.; Hauser, R.; Chavarro, J.E. Earth Study Team. Male caffeine and alcohol intake in relation to semen parameters and in vitro fertilization outcomes among fertility patients. *Andrology* **2017**, *5*, 354–361. [CrossRef] [PubMed]

9.  Franceschi, S.; Negri, E.; Salvini, S.; Decarli, A.; Ferraroni, M.; Filiberti, R.; Giacosa, A.; Gnagnarella, P.; Nanni, O.; Salvini, S. Reproducibility of an Italian food frequency questionnaire for cancer studies: results for specific food items. *Eur. J. Cancer* **1993**, *29*, 2298–2305. [CrossRef]

10. Franceschi, S.; Barbone, F.; Negri, E.; Decarli, A.; Ferraroni, M.; Filiberti, R.; Giacosa, A.; Talamini, R.; Nanni, O.; Panarello, G. Reproducibility of an Italian food frequency questionnaire for cancer studies. Results for specific nutrients. *Ann. Epidemiol.* **1995**, *5*, 69–75. [CrossRef]

11. Decarli, A.; Franceschi, S.; Ferraroni, M.; Gnagnarella, P.; Parpinel, MT.; La Vecchia, C.; Negri, E.; Salvini, S.; Falcini, F.; Giacosa, A. Validation of a food-frequency questionnaire to assess dietary intakes in cancer studies in Italy Results for specific nutrients. *Ann. Epidemiol.* **1996**, *6*, 110–118. [CrossRef]

12. Gnagnarella, P.; Parpinel, M.; Salvini, S.; Franceschi, S.; Palli, D.; Boyle, P. The update of the Italian food composition database. *J. Food Comp. Anal.* **2004**, *17*, 509–522. [CrossRef]

13. Tavani, A. Coffee and Health. 2013. Available online: http://www.coffeegroup.it/assets/img/curiosita/caffe-e-salute (accessed on 14 July 2018).

14. Ferraroni, M.; Decarli, A.; Franceschi, S.; Vecchia, C.L.; Enard, L.; Negri, E.; Parpinel, M.; Salvini, S. Validity and reproducibility of alcohol consumption in Italy. *Int. J. Epidemiol.* **1996**, *25*, 775–782. [CrossRef] [PubMed]

15. Benaglia, L.; Bermejo, A.; Somigliana, E.; Faulisi, S.; Ragni, G.; Fedele, L.; Garcia-Velasco, L.A. In vitro fertilization outcome in women with unoperated bilateral endometriomas. *Fertil. Steril.* **2013**, *99*, 1714–1719. [CrossRef] [PubMed]

16. D'Avanzo, B.; Vecchia, C.L.; Katsouyanni, K.; Negri, E.; Trichopoulos, D. Reliability of information on cigarette smoking and beverage consumption provided by hospital controls. *Epidemiology* **1996**, *7*, 312–315. [CrossRef] [PubMed]

17. INRAN-SCAI 2005–2006. (National Survey on Food Consumption in Italy) L'indagine Nazionale sui Consume Alimentary in Italia. Available online: http://nut.entecra.it/files/download/INRAN-SCAI/B2/appendice_14b2_acqua_e_bevande_analcoliche (accessed on 1 August 2018).

18. European Food Safety Agency. EFSA Panel on Dietetic Products, Nutrition and Allergies (NDA). Scientific Opinion on the Safety of Caffeine. 2015. Available online: https://efsa.onlinelibrary.wiley.com/doi/epdf/10.2903/j.efsa.2015.4102 (accessed on 14 July 2018).

**MDPI**

*Article*

# Coffee Consumption and Whole-Blood Gene Expression in the Norwegian Women and Cancer Post-Genome Cohort

Runa B. Barnung [1],*, Therese H. Nøst [1], Stine M. Ulven [2], Guri Skeie [1] and Karina S. Olsen [1]

[1] Department of Community Medicine, Faculty of Health Sciences, University of Tromsø-The Arctic University of Norway, 9037 Tromsø, Norway; therese.h.nost@uit.no (T.H.N.); guri.skeie@uit.no (G.S.); karina.standahl.olsen@uit.no (K.S.O.)

[2] Department of Nutrition, Institute for Basic Medical Sciences, University of Oslo, P.O. Box 1046 Blindern, 0317 Oslo, Norway; smulven@medisin.uio.no

* Correspondence: runa.b.barnung@uit.no; Tel.: +47-77-644-827

Received: 2 July 2018; Accepted: 7 August 2018; Published: 9 August 2018

**Abstract:** Norwegians are the second highest consumers of coffee in the world. Lately, several studies have suggested that beneficial health effects are associated with coffee consumption. By analyzing whole-blood derived, microarray based mRNA gene expression data from 958 cancer-free women from the Norwegian Women and Cancer Post-Genome Cohort, we assessed the potential associations between coffee consumption and gene expression profiles and elucidated functional interpretation. Of the 958 women included, 132 were considered low coffee consumers (<1 cup of coffee/day), 422 moderate coffee consumers (1–3 cups of coffee/day), and 404 were high coffee consumers (>3 cups of coffee/day). At a false discovery rate <0.05, 139 genes were differentially expressed between high and low consumers of coffee. A subgroup of 298 nonsmoking, low tea consumers was established to isolate the effects of coffee from smoking and potential caffeine containing tea consumption. In this subgroup, 297 genes were found to be differentially expressed between high and low coffee consumers. Results indicate differentially expressed genes between high and low consumers of coffee with functional interpretations pointing towards a possible influence on metabolic pathways and inflammation.

**Keywords:** whole-blood; mRNA; transcriptomics; gene expression; coffee; the Norwegian Women and Cancer Cohort (NOWAC)

## 1. Introduction

Coffee is consumed worldwide, and consumption rates in Norway (9.7 kg per capita) are surpassed only by Finland (12.3 kg per capita) [1]. On average, Norwegian women consume 454 grams of brewed coffee per day [2].

There has been a growing interest in studying the associations between coffee consumption and health in the recent decades. Some studies have indicated that coffee is beneficial to health, and it has been linked with a decreased risk of Alzheimer's, Parkinson, and type 2 diabetes [3–7]. Studies have also indicated that coffee has either has a neutral or a beneficial effect on the risk of cancer, specifically associations with a probable decreased risk of liver, and endometrial cancer [8].

Other studies have revealed detrimental health effects such as increased total cholesterol and triglycerides in blood, as well as certain negative pregnancy outcomes [9–13].

These diverse health effects may be attributed to different constituents of coffee, some of the most bioactive being caffeine, cafestol, kahweol, polyphenols, trigonellin, and polycyclic aromatic hydrocarbons [14,15].

Linking the different coffee constituents to health outcomes is challenging because of the individual variation in metabolism and physiological response to coffee. As an example, the metabolism of caffeine can vary up to 12-fold between individuals, mostly due to the variability of hepatic cytochrome p450 *(CYP)1A2* activity, which metabolizes over 95% of caffeine [16].

Genes associated with either coffee or caffeine intake have been identified in genome-wide association studies of single nucleotide polymorphism (SNPs). Some of the most well established SNPs are located in *CYP1A1* and *CYP1A2* (caffeine metabolism), and *AHR* (regulation of *CYP1A2*) [17,18]. SNPs in these genes were also confirmed as being associated with coffee consumption in a large meta-analysis of over 120,000 individuals together with SNPs in six other genes (*GCKR*, *ABCG2*, *MLXIPL*, *POR*, *BDNF*, and *EFCAB5*) [19]. Still, the knowledge from functional genomics studies using mRNA is limited, and especially gene expression studies in peripheral blood are scarce.

The health effects of coffee consumption can also be difficult to disentangle from other diet and lifestyle factors, as many of the constituents of coffee are also present in other dietary sources. For example, tea and certain soft drinks contain caffeine, while smoking can influence the same metabolic pathways as coffee.

The Norwegian Women and Cancer Cohort (NOWAC) started its questionnaire data collection in 1991, with the aim of being a national representative, population-based cohort study [20]. Collection of whole-blood samples viable for microarray gene expression started in 2003 [21].

By using dietary data and whole-blood derived, microarray based mRNA gene expression data from NOWAC, we assessed whether high versus low consumers of coffee had differentially expressed genes that could elucidate the possible relevant biological processes associated with coffee consumption.

## 2. Materials and Methods

### 2.1. Study Population

The NOWAC study consists of more than 170,000 women aged 30–70 years at recruitment. These women were randomly chosen from the Norwegian central person registry, and received an invitational letter and an eight-page lifestyle and food frequency questionnaire (FFQ). Approximately 50,000 of these women also later gave blood samples eligible for gene expression analysis (the Norwegian Women and Cancer Post-Genome Cohort), and answered a two-page questionnaire about current lifestyle at the time of blood sampling. Detailed information on NOWAC is available from Lund et al. [20], and on the NOWAC Post-Genome Cohort from Dumeaux et al. [21]. The present paper describes results from a subset of the NOWAC Post-Genome Cohort, where cancer-free women ($n$ = 977) originally enrolled as controls in one prediagnostic- and one postdiagnostic breast cancer case-control study were included. These controls were randomly drawn, but matched by age and time of inclusion in the NOWAC cohort. Women who either (1) did not answer the food frequency part of the questionnaire; (2) or did not answer the questions regarding tea and coffee consumption or (3) consumed less than 2500 KJ were excluded. Further details about dietary assessment are given below. From the 977 women in total, 958 women were left in the group "all women" after exclusion criteria were applied (Figure 1). As smoking and tea consumption are highly confounding variables to coffee consumption, we performed a subgroup analysis of 298 nonsmokers who drank less than an average of half a cup of tea per day to isolate the effects of coffee from smoking and tea consumption.

The women gave written informed consent to donate blood samples for gene expression analysis. The NOWAC study was conducted in accordance with the Declaration of Helsinki, and approved by the Norwegian Data Inspectorate and the Regional Ethical Committee of North Norway (reference: REK NORD 2010/2075). Collection and storage of human biological material was approved by the REK in accordance with the Norwegian Biobank Act (reference: P REK NORD 141/2008 Biobanken kvinner og kreft ref. 200804332-3).

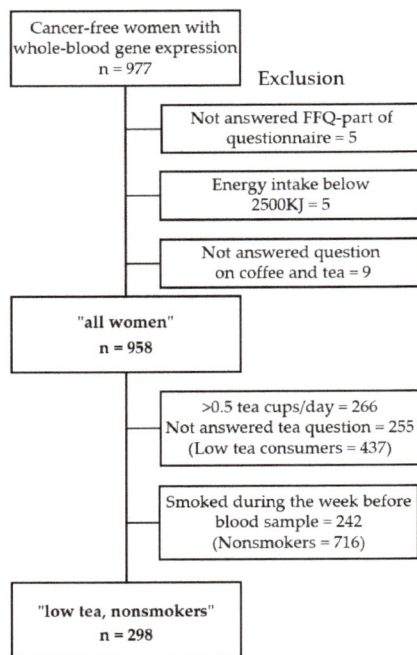

**Figure 1.** Flowchart showing the exclusion criteria in the study group and number of cancer-free women included in the study.

*2.2. Determination of Gene Expression Levels*

Non-fasting blood samples were collected using the PAXgene™ Blood RNA System (PreAnalytiX GmbH, CH–8634 Hombrechtikon, Switzerland), with buffers specially designed for the conservation of mRNA. The samples were mailed overnight to the Department of Community Medicine at the University of Tromsø-The Arctic University of Norway, and immediately frozen at −80 °C. The samples were sent to the Genomics Core Facility at the Norwegian University of Science and Technology, and processed according to the PAXgene Blood RNA Kit protocol. Total RNA was extracted and purified using the PAXgene Blood miRNA isolation Kit. RNA purity was assessed by NanoDrop ND 8000 spectrophotometer (ThermoFisher Scientific, Wilmington, DE, USA), and RNA integrity by Bioanalyzer capillary electrophoresis (Agilent Technologies, Palo Alto, CA, USA). Complementary RNA (cRNA) was prepared using the Illumina TotalPrepT-96 RNA Amplification Kit (Ambion Inc., Austin, TX, USA), and hybridized to Illumina HumanHT-12 Expression BeadChip microarrays (Illumina, Inc. San Diego, CA, USA). The raw microarray images were processed in Illumina GenomeStudio.

The preprocessing of the dataset was performed by the Norwegian Computing Center, and the methods are further described in Günter et al. [22]. In short, the preprocessing involved (1) removal of case-control pairs where either case or control was an outlier (determined by density plot, principal component analysis, or inspection of laboratory quality measures). (2) Background correction was performed using negative control probes (R package *limma*: Function nec), and finally (3) filtering out probes that either were reported to have poor quality in Illumina, were detectable in <1% of samples, or that were not annotated before mapping probes to genes. The dataset was then normalized on original scale by quantile normalization (R *lumi*: LumiN) and log2 transformed (R *lumi*:

LumiT). The packages R *lumi*: nuID2RefSeqID and R *illuminaHumanv4.db* were used to annotate the preprocessed dataset. The final dataset included 7741 probes and 977 individuals.

## 2.3. Dietary Assessment and Descriptive Variables

The FFQ contains questions on quantity and frequency of the most commonly consumed food items. From these, grams per day (g/d) of the food items and total energy intake (kJ/d) were estimated. Standard portion sizes and weights were taken from the official Norwegian Weight and Measures for Foods [23], and intake of energy, alcohol and nutrients from the Norwegian Food Composition table [24]. The FFQ has been validated by test-retest reproducibility and by comparison with repeated 24-h dietary recalls [25,26]. The test-retest study concluded that the FFQ performed within the reported range for similar instruments, and the comparison with 24-h dietary recalls found that the FFQ gave a good ranking especially for foods consumed frequently. Coffee was found to have the best Spearman's rank correlation coefficient (0.82) when the FFQ was compared to the 24-h dietary recalls [26].

Coffee consumption was self-reported based on the question: "How many cups of coffee do you normally drink of each brewing method?" with the different brewing methods being boiled, filtered, and instant. The frequency of consumption was divided into seven categories: Never/seldom, 1–6 cups per week, 1 cup per day, 2–3 cups per day, 4–5 cups per day, 6–7 cups per day, and 8+ cups per day. Interval midpoints of the frequencies were used to add the different brewing methods together. Average total coffee consumption was divided into the categories: Low (<1 cup/day), moderate ($\geq$1–$\leq$3 cups/day), and high (>3 cups/day). This categorization of coffee cups is similar to previously conducted studies on coffee consumption in the NOWAC cohort, but due to a lower sample population in the current study only one high consumption category was used [27,28].

A second version of the FFQ also included espresso (received by 205 of 977 women), only 9 of the 78 women who answered the question on espresso consumption replied something else than never/seldom. One cup of espresso was considered equal to one cup of coffee in the analyses.

One question on green tea and one question on black tea were combined for total tea consumption. For group characteristics, the variable g/d was used. However, the sum of the midpoints of the tea consumption frequency intervals was used for further establishing a subgroup "low tea, nonsmokers" that consisted of nonsmoking women who on average consumed less than half a cup of tea per day. This was done to isolate the effects of coffee from smoking and potential caffeine-containing tea consumption.

The women reported their physical activity level (both activity at home and at work) in the FFQ on a scale from 1 to 10, with 1 being very low and 10 being very high. Education was reported as years in school, including lower education. Both information on smoking status and BMI from self-reported height and weight were taken from the two-page questionnaire filled in at time of blood sampling. The smoking question asked if the women had smoked in the week prior to the blood sampling (yes/no).

## 2.4. Statistical Analysis

Potential confounders were investigated by comparing the categories of coffee consumption as described above using a Kruskal-Wallis test, robust ANOVAs, and a Chi-square test with $p < 0.05$ as the significance threshold; subsequent post hoc methods were then used to establish a significance between coffee consumption categories. Both Kruskal Wallis and robust ANOVA showed similar results, but since no variables were normally distributed except for "red and processed meat," Kruskal-Wallis with Dunn's post hoc rank sum test is presented in the tables. Based on these initial analyses, further analyses of differential gene expression between coffee consumption categories were performed on a subgroup of "low tea, nonsmoking" consumers (298 women), in addition to "all women." In the "low tea, nonsmoking" group, the differences in age, education, and meat and dairy consumption found in the "all women" group were no longer significant, and were therefore not adjusted for.

All analyses were performed using R v3.4.0 [29] and packages from R and the Bioconductor project. The R package *limma* [30] was used to find differentially expressed genes (false discovery rate (FDR) < 0.05 was used) between the three categories of coffee consumption. The lists of differentially expressed genes from *limma* were then used in *clusterProfiler* [31] to perform over-representation analysis (R *clusterProfiler*: EnrichGO) and to compare the enriched functional categories of each gene cluster between "all women" and "low tea, nonsmokers" (R *clusterProfiler*: CompareCluster) for biological processes within Gene Ontology (GO) terms. To ensure balanced comparisons between the gene lists of each group, the top 100 genes in each list were used to compare the groups.

## 3. Results

### 3.1. Descriptors

The group "all women" consisted of 958 women with a median coffee consumption of 525 grams of brewed coffee per day. Of these 958 women, 132 (13.8%) had a low coffee consumption (<1 cup of coffee/day), 422 (44.1%) were moderate coffee consumers (≥1–≤3 cups of coffee/day), and 404 (42.2%) were high coffee consumers (>3 cups of coffee/day) (Table 1). Filtered coffee was reported as the brewing method by 783 women, followed by instant coffee (205 women), boiled coffee (121 women), and espresso (nine women), with some women consuming more than one type of brewing method.

There was a higher percentage of women who smoked in the week before the blood sample was taken in the high coffee consumption group (36.8%) compared to both the low (14.4%) and moderate coffee consumption groups (17.3%). The high coffee consumption group also had the lowest median tea intake (0 g/d) of the three groups. The moderate group had higher median tea consumption (135 g/d) than the high coffee consumers, but the low coffee consumption group had a substantially higher intake than both moderate and high coffee consumers with a median of 405 g/d.

Further, a low education level was more frequent in the high coffee consumption group than in the two other groups. There was a higher median intake of dairy products in the high (179 g/d) and moderate (175 g/d) coffee consumption groups compared to the low consumption group (128 g/d). Median consumption of red and processed meat was slightly higher in the high coffee consumption group (93 g/d), compared to the moderate (86 g/d) and low (86 g/d) consumption groups. However, for red and processed meat, the actual difference in grams was small, and this is therefore unlikely to be of clinical relevance.

Table 2 describes the characteristics of the subgroup of women who did not smoke in the week before blood sample donation, and that drank less than 1–6 cups of tea per week (average of half a cup per day). This "low tea, nonsmoking" group consisted of 298 women with a median coffee consumption of 630 grams brewed coffee per day, of which 25 (8.4%) had a low coffee consumption, 139 (46.6%) were moderate coffee consumers, and 134 (45.0%) were in the high coffee consumption category.

In the "low tea, nonsmoking" group there was a difference in median energy intake among the coffee consumption categories, with a borderline significant difference ($p = 0.054$) between the high (7188 kJ/day) and low consumption group (6450 kJ/day), and a significant difference between the moderate (6625 kJ/day) and high group ($p = 0.034$).

**Table 1.** Descriptive statistics of "all women," low coffee consumers, moderate coffee consumers, and high coffee consumers.

| | All Women (n = 958) | Low Coffee Consumers (n = 132, 13.8%) | Moderate Coffee Consumers (n = 422, 44.1%) | High Coffee Consumers (n = 404, 42.2%) |
|---|---|---|---|---|
| Age, years **,a | 55 (48–61, 958) | 53 (47–60, 132) | 55 (48–61, 422) | 54 (48–60, 404) |
| BMI | 25.1 (20.2–33.4, 941) | 25.0 (19.8–33.7, 130) | 25.0 (20.3–32.8, 418) | 25.3 (20.2–33.8, 393) |
| Education (years in school) **,a,b,c | 12 (8–18, 916) | 13 (8–19, 128) | 12 (8–19, 398) | 11 (8–18, 390) |
| Smoking week before blood sample **,a,c | | | | |
|   Yes % (n) | 25.1% (240) | 14.4% (19) | 17.3% (73) | 36.8% (148) |
|   No % (n) | 74.9% (716) | 85.6% (113) | 82.7% (349) | 63.2% (254) |
| Physical activity level | 6 (3–9, 894) | 6 (3–9, 122) | 6 (3–9, 394) | 6 (2–8, 378) |
| Energy intake (kJ/d) | 7106 (4274–10409, 958) | 7052 (4087–10705, 132) | 7057 (4504–10094, 422) | 7172 (4050–10817, 404) |
| Alcohol consumption (g/d) | 2 (0–12, 958) | 2 (0–13, 132) | 2 (0–12, 422) | 2 (0–12, 404) |
| Dairy intake (g/d) **,a,b | 169 (15–635, 958) | 128 (6–614, 132) | 175 (20–634, 422) | 179 (16–641, 404) |
| Black and green tea consumption (g/d) **,a,b,c | 135 (0–945, 703) | 405 (0–1512, 99) | 135 (0–675, 309) | 0 (0–675, 295) |
| Red and processed meat intake (g/d) **,c | 89 (15–176, 958) | 86 (10–174, 132) | 86 (18–161, 422) | 93 (23–185, 404) |

Values presented as median (5 and 95 percentiles, number of women (n)) or % (n). Low coffee consumption (<1 cup/day), moderate coffee consumption (≥1–≤3 cups/day), and high coffee consumption (>3 cups/day). Categorical variables analyzed with $\chi^2$, continuous variables with Kruskal Wallis test, **: $p \leq 0.01$, a: $p \leq 0.05$ between low and moderate coffee consumption categories, b: $p \leq 0.05$ between moderate and high coffee consumption categories, c: $p \leq 0.05$ between low and high coffee consumption categories.

**Table 2.** Descriptive statistics of "low tea, nonsmoking women," low coffee consumers, moderate coffee consumers, and high coffee consumers.

| | Low Tea, Nonsmoking Women (n = 298) | Low Coffee Consumers (n = 25, 8.4%) | Moderate Coffee Consumers (n = 139, 46.6%) | High Coffee Consumers (n = 134, 45.0%) |
|---|---|---|---|---|
| Age, years | 55 (47–61, 298) | 55 (47–61, 25) | 55 (47–61, 139) | 55 (47–61, 134) |
| BMI | 25.7 (20.4–33.0, 293) | 25.8 (19.5–34.4, 24) | 25.4 (20.4–33.0, 138) | 26.2 (20.7–32.7, 131) |
| Education (years in school) | 12 (8–18, 288) | 14 (8–19, 24) | 12 (8–18, 133) | 12 (8–18, 131) |
| Physical activity level | 6 (3–9, 282) | 7 (2–9, 24) | 6 (3–9, 131) | 6 (3–8, 127) |
| Energy intake (kJ/d) **,b,c | 6911 (4251–10455, 298) | 6450 (3742–9486, 25) | 6625 (4440–9458, 139) | 7188 (4309–10995, 134) |
| Alcohol consumption (g/d) | 2 (0–12, 298) | 3 (0–13, 25) | 2 (0–13, 139) | 2 (0–9, 134) |
| Dairy intake (g/d) | 188 (24–617, 298) | 110 (18–491, 25) | 191 (24–618, 139) | 199 (28–623, 134) |
| Red and processed meat intake (g/d) | 90 (12–169, 298) | 80 (2–170, 25) | 87 (14–148, 139) | 95 (14–186, 134) |

Values presented as median (5 and 95 percentiles, number of women (n)) or % (n). Low coffee consumption (<1 cup/day), moderate coffee consumption (≥1–≤3 cups/day), and high coffee consumption (>3 cups/day). Continuous variables with Kruskal Wallis test **: $p \leq 0.01$, b: $p \leq 0.05$ between low and high coffee consumption categories; c: $p \leq 0.05$ between moderate and high coffee consumption categories.

## 3.2. Differential Gene Expression

When comparing high versus low coffee consumers in "all women," there were 139 significantly differentially expressed genes (FDR < 0.05) (Figure 2a, Table S1). The gene most differentially expressed (*LRRN3*) when comparing high versus low coffee consumers was also the only differentially expressed gene when comparing high versus moderate coffee consumption groups. When studying only those who did not smoke the week before blood sampling, 414 genes were significantly differentially expressed between high and low consumers (results not presented). In the group that consisted of the 298 women who neither smoked in the week before blood sampling nor drank more than an average of half a cup of tea per day ("low tea, nonsmoking"), 297 genes were significantly differentially expressed when comparing high versus low coffee consumers (Figure 2b, Table S2). Table 3 shows the top 20 significantly differentially expressed genes when comparing high versus low coffee consumers in the "low tea, nonsmoking" group. There were 36 genes in common between all the significantly differentially expressed genes in "all women" and "low tea, nonsmoking" groups, but there was only one gene in common between the top 50 genes for both groups.

**Table 3.** Top 20 significantly differentially expressed genes (false discovery rate < 0.05) between high and low coffee consumers in the "low tea, nonsmoking" group.

| Gene Symbol | Gene Name | Log Fold Change | Average Expression | T | *p*-Value |
|---|---|---|---|---|---|
| TLE3 | Transducin like enhancer of split 3 | −0.318 | 6.907 | −4.682 | 0.000004 |
| HLX | H2.0 like homeobox | −0.299 | 7.276 | −4.577 | 0.000007 |
| DDX18 | DEAD-box helicase 18 | 0.282 | 8.330 | 4.536 | 0.000008 |
| YRDC | YrdC N6-threonylcarbamoyltransferase domain containing | 0.159 | 7.034 | 4.415 | 0.000014 |
| KDM6B | Lysine demethylase 6B | −0.294 | 7.003 | −4.401 | 0.000015 |
| CANT1 | Calcium activated nucleotidase 1 | −0.264 | 8.524 | −4.397 | 0.000015 |
| WDR61 | WD repeat domain 61 | 0.213 | 7.608 | 4.387 | 0.000016 |
| MTSS1 | MTSS1, I-BAR domain containing | 0.260 | 7.383 | 4.375 | 0.000017 |
| MACF1 | Microtubule-actin crosslinking factor 1 | 0.248 | 7.815 | 4.336 | 0.000020 |
| PPP3CC | Protein phosphatase 3 catalytic subunit gamma | 0.241 | 7.482 | 4.284 | 0.000025 |
| FAM36A | Cytochrome c oxidase assembly factor | 0.190 | 6.781 | 4.274 | 0.000026 |
| TXK | TXK tyrosine kinase | 0.265 | 7.023 | 4.261 | 0.000027 |
| TFE3 | Transcription factor binding to IGHM enhancer 3 | −0.198 | 7.043 | −4.230 | 0.000031 |
| SPATA2L | Spermatogenesis associated 2 like | −0.220 | 7.134 | −4.204 | 0.000035 |
| DYSF | Dysferlin | −0.529 | 9.697 | −4.197 | 0.000036 |
| TTC13 | Tetratricopeptide repeat domain 13 | 0.219 | 7.541 | 4.193 | 0.000036 |
| LOC642684 | - | −0.136 | 6.339 | −4.188 | 0.000037 |
| CDK5RAP1 | CDK5 regulatory subunit associated protein 1 | 0.166 | 7.422 | 4.178 | 0.000039 |
| LOC441124 | - | −0.261 | 7.219 | −4.175 | 0.000039 |
| PHOSPHO1 | Phosphoethanolamine/phosphocholine phosphatase | −0.400 | 7.532 | −4.169 | 0.000040 |

Log Fold change: Log2 fold change between high and low coffee consumption; Average expression: Average log2-expression level for that gene; *t*: Moderated *t*-statistic.

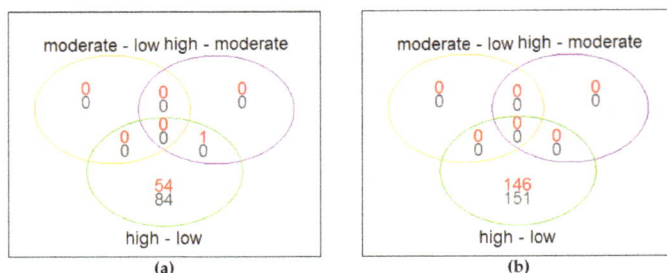

**Figure 2.** (**a**) Significantly up-(red) and down-(grey) regulated genes between coffee consumption categories for "all women." (**b**) Significantly up-(red) and down-(grey) regulated genes between coffee consumption categories for "low tea, nonsmokers."

## 3.3. Over-Representation Analysis

Over-representation analysis for the gene lists with significantly differentially expressed genes found no over-representation at FDR < 0.05. In the over-representation analysis for "all women" at *p*-value < 0.01 (*n* = 139 genes, Figure 3a), the top over-represented categories were involved in regulation and assembly of different tissues and cell constituents. In the "low tea, nonsmoking" group, processes related to immunological responses were indicated (*n* = 297 genes, Figure 3b). When separating the differentially expressed genes from the "low tea, nonsmoking" group into upregulated (146 genes) and downregulated (151 genes), the immunological responses were only apparent in the downregulated genes (Figure S1).

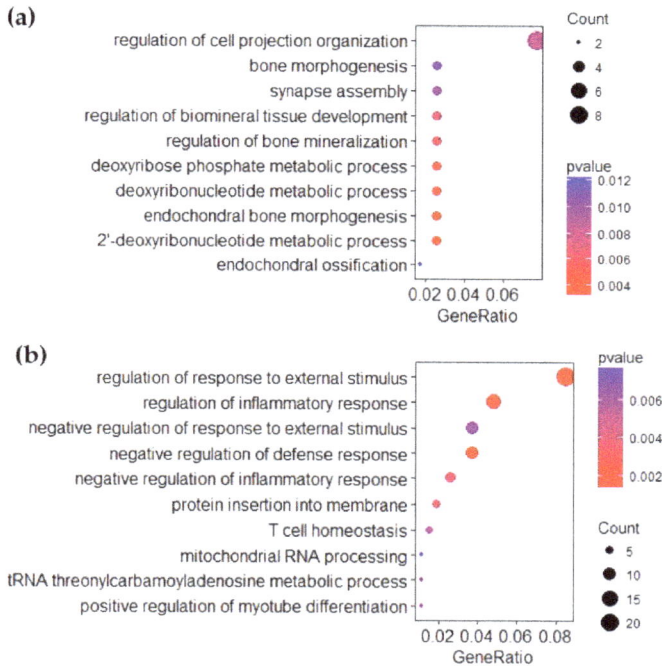

**Figure 3.** Over-representation analysis of Gene Ontology biological process categories. In the figure, the color of the dots indicates the *p*-value, the size of the dots indicates gene count, and the GeneRatio indicate the "number of genes in common between gene list and GO-category/number of genes in gene list." (**a**) Over-representation analysis for "all women," using the 139 significantly differentially expressed genes between high and low coffee consumers. (**b**) Over-representation analysis for "low tea, nonsmokers," using the 297 significantly differentially expressed genes between high and low coffee consumers.

Genes related to metabolic processes were indicated in ontology categories in a group comparison of high and low coffee consumers between "all women" and "low tea, nonsmokers" when using the top 100 significantly differentially expressed genes for both groups (Figure 4).

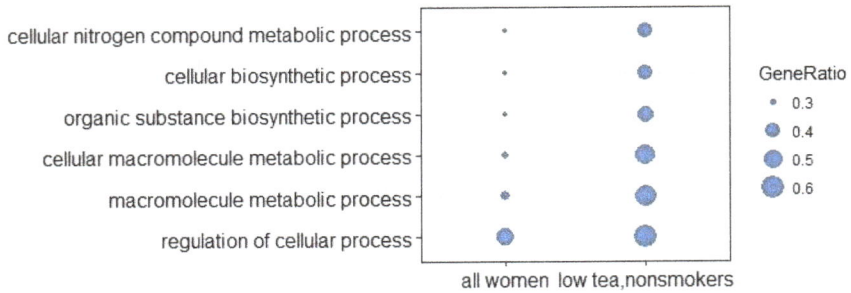

**Figure 4.** Group comparison of Gene Ontology biological process categories using a gene list of the top 100 significantly differentially expressed genes between high and low consumers in the "all women" group versus the "low tea, nonsmoking" group. GeneRatio indicates "number of genes in common between gene list and GO-category/number of genes in gene list."

## 4. Discussion

In this study of Norwegian women, 139 differentially expressed genes were found in whole-blood between self-reported high and low coffee consumers. Subgroup analyses with nonsmoking, low tea consumers yielded a separate set with 297 differentially expressed genes, but comparisons of the top 100 differentially expressed genes in both groups show similar tendencies towards gene ontologies involved in general metabolic processes. An over-representation analysis of GO biological process categories for the differentially expressed genes from the "low tea, nonsmoking" group pointed towards involvement in inflammation related processes. Both the "all women" and "low tea, nonsmoking" groups demonstrated modest fold changes, and the changes were both upregulation and downregulation of expression. This indicates effects from coffee consumption on whole-blood gene expression.

The median intakes of coffee consumption found in the current study were in accordance with the average consumption (560 g/d) among Norwegian women in the age group 50–59 [2]. Energy intake in the "low tea, nonsmoking" group was highest among high consumers of coffee. Few studies have investigated the influence of coffee consumption on energy intake. The studies that exist somewhat contradict our finding, with coffee consumption either having no effect on single meal energy intake or leading to a small daily decrease in energy intake [32].

Genes indicated from the gene expression profiles in this study have not previously been associated with coffee consumption. However, we were not able to distinguish the findings from coffee consumption in the full study group due to confounding from especially smoking. Smoking is strongly associated with coffee consumption, with smokers consuming more coffee than nonsmokers do, possibly due to an increased caffeine metabolism [33–35]. The two top differentially expressed genes (*LRRN3* and *PID1*) identified between current smokers and never smokers in a meta-analysis by Huan et al., [36] were the same two top differentially expressed genes between low and high consumers in the group "all women." *LRRN3* was also the only gene differentially expressed between the moderate and high coffee consumers in the same group. The observation of *LRRN3* and *PID1* indicate a strong influence of smoking on the gene expression profiles for "all women." However, *LRRN3* and *PID1* were not differentially expressed between high and low consumers of coffee in the "low tea, nonsmoking" group.

SNPs linked to several genes have previously been associated with coffee consumption [17–19], of these only *POR* was found to be significantly differentially expressed in the current study, and only in the group "low tea, nonsmokers." *POR* encodes P450 oxidoreductase that transfers electrons to microsomal *CYP 450* enzymes, which are needed for the metabolism of caffeine [19].

Notably, some of the most prominent candidate genes (*CYP1A1*, *CYP1A2* and *AHR*) involved in caffeine metabolism were filtered out from our expression data due to low detection rates, and we were therefore not able to assess the association between these and coffee consumption in the NOWAC cohort. Still, the fact that these genes had low detection rates indicates low expression of these genes in whole-blood. *CYP1A2* is mainly expressed in the liver, and only low levels of *CYP1A1* can usually be found in lymphocytes [37,38]. The association found between *POR* and coffee consumption might indicate that the *CYP1* genes are affected in other ways than by transcriptional regulation in whole-blood. In general, genetic background must also be considered, especially sex and ethnicity can impact the expression of *CYP1A2* [39,40].

Among the top 20 differentially expressed genes from the "low tea, nonsmoking" group, there were especially five genes, *TXK*, *HLX*, *KDM6B*, *SPATA2L*, and *CDK5RAP1*, that are of interest for further research concerning coffee consumption and gene expression. *TXK* and *HLX* are involved in development of T-helper 1 cells, which are necessary for human immune defense [41,42]. *KDM6B*, also known as *JMJD3*, takes part in inflammatory responses by participating in differentiation of macrophages [43], while *SPATA2L* is involved in processes related to inflammatory signaling [44]. *CDK5RAP1* is a repressor of *CDK5*, which is a cyclin-dependent protein known to be involved in neurodegenerative diseases like Parkinson's and Alzheimer's [45,46]. However, among these five genes, only *TXK* was in the GO biological processes involving inflammatory responses found in the over-representation analysis.

Inflammatory response processes were indicated in the over-representation analysis on "low tea, nonsmokers." It should be taken into consideration that monocytes and lymphocytes in whole-blood are immune cells, so an expression of immune-related processes should be expected, and is often found in studies concerning diet and gene expression [47,48]. Epidemiological studies have previously discovered that coffee consumption is associated with reduced risk of death attributed to inflammatory diseases, and that coffee consumption is negatively associated with inflammatory processes [49,50]. Another study found increased concentrations of inflammatory markers among both men and females that consumed >200 mL coffee per day compared to non-consumers [51]. Other indicated effects of coffee consumption have been. e.g., increased serum cholesterol [10], reduced risk of Parkinson's disease [4–6], and reduced risk of type 2 diabetes [7], which are all health endpoints caused by inflammation. Thus, associations between high coffee consumption and inflammatory indicators in peripheral blood could indicate markers of related pathways. In the healthy Norwegian population over 60% of the antioxidant intake is estimated to originate from coffee [52]. The increased intake of antioxidants among coffee consumers is a plausible source for the positive influence of coffee on inflammatory processes. Negative influences on inflammatory processes might be related to cafestol and kahweol, two coffee lipids mainly found in unfiltered coffee. In particular, cafestol is associated with increased serum cholesterol, which is a known underlying factor of atherosclerosis [53,54].

Five GO categories of different biosynthetic and metabolic processes were found to be the top categories in the comparison between genes identified for "all women" and the "low tea, nonsmoking" group. The "low tea, nonsmoking" group had a higher proportion of the top 100 differentially expressed genes involved in the metabolic processes than the group "all women". The metabolic processes evident in this comparison indicate that at least certain genes found to be associated with coffee consumption are involved in the metabolism of constituents of coffee. However, when looking at the over-representation analyses, these metabolic processes were not evident.

Some strengths and limitations of this study should be considered. Gene expression profiles represent a snapshot of the mRNA transcripts available in the whole-blood at the time of blood sampling, while the FFQ represent long term dietary intake. The indicated effects are therefore likely impacted by the discrepancy between this reported long term intake and short term mRNA snapshot. The blood samples were not collected in a fasting state, and we have no data on time since coffee consumption. Caffeine has a half-life of approximately 5.5 h, but other coffee metabolites have a half-life below one hour [55]. Therefore, both in the high and low consumers there could be participants whose gene expression profiles underestimates their differential expression compared to their FFQ reported intake.

This study used a relatively large number of women compared to many other whole-blood nutrigenomic studies, the higher sample size mitigates some of the concerns of limited impact and reliability found in other studies [48].

The FFQs used in this study were comprehensive and contained most of the commonly consumed food and beverages in Norway. However, there was no question designed to capture caffeine-containing beverages other than tea and coffee, and this might lead to some residual confounding in our analyses. Coffee consumption and other dietary exposures were assessed based on self-reported data. Thus, some misclassification could have occurred in the dietary exposures, although it was likely non-differential. The participants reported cups of coffee, but was not given an estimate of an average cup size, which would have allowed more detailed assessment of consumption. However, coffee showed good validation with a Spearman's rank correlation coefficient of 0.82 when the FFQ was compared to the 24-h dietary recalls where the women reported coffee consumption either in exact amount or based on cup sizes from a picture booklet [26]. Coffee differs in chemical constituents depending on variables such as bean type, roasting of bean, grinding of bean, and soaking time of coffee grinds. This information was not available from the FFQ. Taken together, we cannot rule out the possibility of coffee category misclassification, and for some women the classification might differ between reported coffee consumption and some of its constituents due to difference in cup size, brewing strength, and other factors. In this paper, we focus on coffee per se, rather than its constituents, as this is what people consume. No biomarker assessed in blood was used to affirm the estimates of coffee.

The gene expression data was not adjusted for age, education, or consumption of red and processed meat or dairy, even though high coffee consumers reported lower education level and a higher intake of red meat and dairy. Smoking and tea consumption are two known confounders for coffee consumption, and were also associated with coffee consumption in this study. For that reason, subgroup analyses targeting women with low tea consumption (<0.5 cup/day) and no smoking in the week before blood sampling were performed. Subsequently, the associations found between coffee consumption and dairy, red and processed meat, age, and education disappeared, indicating that smoking might be driving the differences observed in the full study group, and not coffee consumption per se.

Whole-blood samples were used in the NOWAC post-genome cohort since these are relatively non-invasive and practical for cohort studies. The PAX gene Blood RNA System made it possible to ship blood samples by mail overnight without having to freeze them first, while at the same time conserving the mRNA over time. Whole-blood has been considered as a surrogate biopsy material for other tissues, due to its transporting role where it both interacts with all tissues and organs and is exposed to bioactive molecules such as nutrients, metabolites, pollutants, and waste products [56]. This makes whole-blood a viable candidate for capturing gene expression profiles associated with dietary exposure [56]. The most transcriptionally active blood cells are the leukocytes, which are important in immune responses. The gene expression microarrays were performed on whole-blood samples lacking information regarding disease status and immune cell subtypes. Gene expression profiles vary depending on differences in cellular components of the whole-blood [57], and infections or autoimmune diseases can introduce differences in these cellular components. By quantification of the blood composition, genes specific to immune cells could have been better elucidated.

## 5. Conclusions

In this exploratory cross-sectional study, we show that coffee consumption is significantly associated to differentially expressed genes in whole-blood. To the best of our knowledge this is the first study using mRNA gene expression data to elucidate how coffee consumption influences gene expression in whole-blood. Our results indicate that the differentially expressed genes between high and low coffee consumers were associated with both metabolic and inflammatory processes. Some of the top genes found to be differentially expressed are especially interesting in relation to the effect on inflammatory processes associated with coffee consumption, and warrant further investigation. However, since this is an exploratory cross-sectional study based on self-reported coffee consumptions, the results presented herein must be interpreted with care.

**Supplementary Materials:** The following are available online at http://www.mdpi.com/2072-6643/10/8/1047/s1, Table S1: Significantly differentially expressed genes (FDR < 0.05) "all women," Table S2: Significantly differentially expressed genes (FDR < 0.05) "low tea-nonsmoking" group. Figure S1: Over-representation analyses of Gene Ontology biological process categories for up and down regulated genes in the "low tea, nonsmoking" group.

**Author Contributions:** Data curation, T.H.N. and K.S.O.; Formal analysis, R.B.B.; Methodology, R.B.B., T.H.N. and K.S.O.; Writing—original draft, R.B.B.; Writing—review and editing, R.B.B., T.H.N., S.M.U., G.S. and K.S.O.

**Funding:** This research received no external funding.

**Acknowledgments:** We gratefully thank all the Norwegian women participating in the NOWAC study. Bente Augdal and Merete Albertsen were responsible for the administration of the data collection and the biobanks. Thanks to Christopher Loe Olsen for critical reading and editing services. The publication charges for this article have been funded by a grant from the publication fund of UiT—The Arctic University of Norway.

**Conflicts of Interest:** The authors declare no conflict of interest.

## References

1. International Coffee Council. *Trends in Coffee Consumption in Selected Importing Countries*; International Coffee Council: London, UK, 2012.

2. The Norwegian Directorate of Health. *Norkost 3. En Landsomfattende Kostholdsundersøkelse Blant Menn og Kvinner i Norge i Alderen 18-70 år, 2010-11*; The Norwegian Directorate of Health: Oslo, Norway, 2012. (In Norwegian)

3. Liu, Q.P.; Wu, Y.F.; Cheng, H.Y.; Xia, T.; Ding, H.; Wang, H.; Wang, Z.M.; Xu, Y. Habitual coffee consumption and risk of cognitive decline/dementia: A systematic review and meta-analysis of prospective cohort studies. *Nutrition* **2016**, *32*, 628–636. [CrossRef] [PubMed]

4. Qi, H.; Li, S. Dose-response meta-analysis on coffee, tea and caffeine consumption with risk of parkinson's disease. *Geriatr. Gerontol. Int.* **2014**, *14*, 430–439. [CrossRef] [PubMed]

5. Hernan, M.A.; Takkouche, B.; Caamano-Isorna, F.; Gestal-Otero, J.J. A meta-analysis of coffee drinking, cigarette smoking, and the risk of parkinson's disease. *Ann. Neurol.* **2002**, *52*, 276–284. [CrossRef] [PubMed]

6. Costa, J.; Lunet, N.; Santos, C.; Santos, J.; Vaz-Carneiro, A. Caffeine exposure and the risk of parkinson's disease: A systematic review and meta-analysis of observational studies. *J. Alzheimers Dis.* **2010**, *20*, 221–238. [CrossRef] [PubMed]

7. Ding, M.; Bhupathiraju, S.N.; Chen, M.; van Dam, R.M.; Hu, F.B. Caffeinated and decaffeinated coffee consumption and risk of type 2 diabetes: A systematic review and a dose-response meta-analysis. *Diabetes Care* **2014**, *37*, 569–586. [CrossRef] [PubMed]

8. World Cancer Research Fund. American Institute for Cancer Research. In *Continous Update Project Expert Report 2018. Non-Alcoholic Drinks and the Risk of Cancer.*; World Cancer Research Fund: London, UK, 2018; pp. 34–39.

9. Jee, S.H.; He, J.; Appel, L.J.; Whelton, P.K.; Suh, I.; Klag, M.J. Coffee consumption and serum lipids: A meta-analysis of randomized controlled clinical trials. *Am. J. Epidemiol.* **2001**, *153*, 353–362. [CrossRef] [PubMed]

10. Cai, L.; Ma, D.; Zhang, Y.; Liu, Z.; Wang, P. The effect of coffee consumption on serum lipids: A meta-analysis of randomized controlled trials. *Eur. J. Clin. Nutr.* **2012**, *66*, 872–877. [CrossRef] [PubMed]

11. Grosso, L.M.; Bracken, M.B. Caffeine metabolism, genetics, and perinatal outcomes: A review of exposure assessment considerations during pregnancy. *Ann. Epidemiol.* **2005**, *15*, 460–466. [CrossRef] [PubMed]

12. Greenwood, D.C.; Thatcher, N.J.; Ye, J.; Garrard, L.; Keogh, G.; King, L.G.; Cade, J.E. Caffeine intake during pregnancy and adverse birth outcomes: A systematic review and dose-response meta-analysis. *Eur. J. Epidemiol.* **2014**, *29*, 725–734. [CrossRef] [PubMed]

13. Wikoff, D.; Welsh, B.T.; Henderson, R.; Brorby, G.P.; Britt, J.; Myers, E.; Goldberger, J.; Lieberman, H.R.; O'Brien, C.; Peck, J.; et al. Systematic review of the potential adverse effects of caffeine consumption in healthy adults, pregnant women, adolescents, and children. *Food Chem. Toxicol.* **2017**, *109*, 585–648. [CrossRef] [PubMed]

14. Ludwig, I.A.; Clifford, M.N.; Lean, M.E.; Ashihara, H.; Crozier, A. Coffee: Biochemistry and potential impact on health. *Food Funct.* **2014**, *5*, 1695–1717. [CrossRef] [PubMed]

15. Cano-Marquina, A.; Tarin, J.J.; Cano, A. The impact of coffee on health. *Maturitas* **2013**, *75*, 7–21. [CrossRef] [PubMed]

16. Cornelis, M.C. Toward systems epidemiology of coffee and health. *Curr. Opin. Lipidol.* **2015**, *26*, 20–29. [CrossRef] [PubMed]

17. Cornelis, M.C.; Monda, K.L.; Yu, K.; Paynter, N.; Azzato, E.M.; Bennett, S.N. Genome-wide meta-analysis identifies regions on 7p21 (ahr) and 15q24 (cyp1a2) as determinants of habitual caffeine consumption. *PLoS Genet.* **2011**, *7*, e1002033. [CrossRef] [PubMed]

18. Sulem, P.; Gudbjartsson, D.F.; Geller, F.; Prokopenko, I.; Feenstra, B.; Aben, K.K. Sequence variants at cyp1a1-cyp1a2 and ahr associate with coffee consumption. *Hum. Mol. Genet.* **2011**, *20*, 2071–2077. [CrossRef] [PubMed]

19. The Coffee and Caffeine Genetics Consortium; Cornelis, M.C.; Byrne, E.M.; Esko, T.; Nalls, M.A.; Ganna, A.; Paynter, N.; Monda, K.L.; Amin, N.; Fischer, K.; et al. Genome-wide meta-analysis identifies six novel loci associated with habitual coffee consumption. *Mol. Psychiatry* **2015**, *20*, 647–656. [CrossRef] [PubMed]

20. Lund, E.; Dumeaux, V.; Braaten, T.; Hjartaker, A.; Engeset, D.; Skeie, G.; Kumle, M. Cohort profile: The norwegian women and cancer study–nowac–kvinner og kreft. *Int. J. Epidemiol.* **2008**, *37*, 36–41. [CrossRef] [PubMed]

21. Dumeaux, V.; Borresen-Dale, A.L.; Frantzen, J.O.; Kumle, M.; Kristensen, V.N.; Lund, E. Gene expression analyses in breast cancer epidemiology: The norwegian women and cancer postgenome cohort study. *Breast Cancer Res.* **2008**, *10*, R13. [CrossRef] [PubMed]

22. Günter, C.; Holden, M.; Holden, L. *Preprocessing of Gene-Expression Data Related to Breast Cancer Diagnosis: Samba/35/14*; Norsk Regnesentral: Oslo, Norway, 2014.

23. Blaker, B.; Aarsland, M. *Mål og vekt for Matvarer*; Landsforeningen for Kosthold og Helse: Oslo, Norway, 1989. (In Norwegian)

24. Norwegian Food Safety Authority; The Norwegian Directorate of Health; University of Oslo. Norwegian Food Composition Database. Available online: www.matvaretabellen.no (accessed on 1 July 2018).

25. Parr, C.L.; Veierod, M.B.; Laake, P.; Lund, E.; Hjartaker, A. Test-retest reproducibility of a food frequency questionnaire (ffq) and estimated effects on disease risk in the norwegian women and cancer study (nowac). *Nutr. J.* **2006**, *5*, 4. [CrossRef] [PubMed]

26. Hjartaker, A.; Andersen, L.F.; Lund, E. Comparison of diet measures from a food-frequency questionnaire with measures from repeated 24-hour dietary recalls. The norwegian women and cancer study. *Public Health Nutr.* **2007**, *10*, 1094–1103. [CrossRef] [PubMed]

27. Lukic, M.; Licaj, I.; Lund, E.; Skeie, G.; Weiderpass, E.; Braaten, T. Coffee consumption and the risk of cancer in the norwegian women and cancer (nowac) study. *Eur. J. Epidemiol.* **2016**, *31*, 905–916. [CrossRef] [PubMed]

28. Gavrilyuk, O.; Braaten, T.; Skeie, G.; Weiderpass, E.; Dumeaux, V.; Lund, E. High coffee consumption and different brewing methods in relation to postmenopausal endometrial cancer risk in the norwegian women and cancer study: A population-based prospective study. *BMC Womens Health* **2014**, *14*, 48. [CrossRef] [PubMed]

29. R Core Team. *R: A Language and Environment for Statistical Computing*; R Foundation for Statistical computing: Vienna, Austria, 2017.

30. Ritchie, M.E.; Phipson, B.; Wu, D.; Hu, Y.; Law, C.W.; Shi, W.; Smyth, G.K. Limma powers differential expression analyses for rna-sequencing and microarray studies. *Nucleic Acids Res.* **2015**, *43*, e47. [CrossRef] [PubMed]

31. Yu, G.; Wang, L.G.; Han, Y.; He, Q.Y. Clusterprofiler: An r package for comparing biological themes among gene clusters. *OMICS* **2012**, *16*, 284–287. [CrossRef] [PubMed]

32. Schubert, M.M.; Irwin, C.; Seay, R.F.; Clarke, H.E.; Allegro, D.; Desbrow, B. Caffeine, coffee, and appetite control: A review. *Int. J. Food Sci. Nutr.* **2017**, *68*, 901–912. [CrossRef] [PubMed]

33. Bjorngaard, J.H.; Nordestgaard, A.T.; Taylor, A.E.; Treur, J.L.; Gabrielsen, M.E.; Munafo, M.R.; Nordestgaard, B.G.; Asvold, B.O.; Romundstad, P.; Davey Smith, G. Heavier smoking increases coffee consumption: Findings from a mendelian randomization analysis. *Int. J. Epidemiol.* **2017**, *46*, 1958–1967. [CrossRef] [PubMed]

34. Swanson, J.A.; Lee, J.W.; Hopp, J.W. Caffeine and nicotine—A review of their joint use and possible interactive effects in tobacco withdrawal. *Addict. Behav.* **1994**, *19*, 229–256. [CrossRef]

35. Grela, A.; Kulza, M.; Piekoszewski, W.; Senczuk-Przybylowska, M.; Gomolka, E.; Florek, E. The effects of tobacco smoke exposure on caffeine metabolism. *Ital. J. Food Sci.* **2013**, *25*, 76–82.

36. Huan, T.; Joehanes, R.; Schurmann, C.; Schramm, K.; Pilling, L.C.; Peters, M.J.; Magi, R.; DeMeo, D.; O'Connor, G.T.; Ferrucci, L.; et al. A whole-blood transcriptome meta-analysis identifies gene expression signatures of cigarette smoking. *Hum. Mol. Genet.* **2016**, *25*, 4611–4623. [CrossRef] [PubMed]

37. Yi, B.; Yang, J.Y.; Yang, M. Past and future applications of cyp450-genetic polymorphisms for biomonitoring of environmental toxicants. *J. Environ. Sci. Health C Environ. Carcinog. Ecotoxicol. Rev.* **2007**, *25*, 353–377. [CrossRef] [PubMed]

38. Spatzenegger, M.; Horsmans, Y.; Verbeeck, R.K. Cyp1a1 but not cyp1a2 proteins are expressed in human lymphocytes. *Pharmacol. Toxicol.* **2000**, *86*, 242–244. [CrossRef] [PubMed]

39. Denden, S.; Bouden, B.; Haj Khelil, A.; Ben Chibani, J.; Hamdaoui, M.H. Gender and ethnicity modify the association between the cyp1a2 rs762551 polymorphism and habitual coffee intake: Evidence from a meta-analysis. *Genet. Mol. Res.* **2016**, *15*, 1–11. [CrossRef] [PubMed]

40. McGraw, J.; Waller, D. Cytochrome p450 variations in different ethnic populations. *Expert Opin. Drug Metab. Toxicol.* **2012**, *8*, 371–382. [CrossRef] [PubMed]

41. Kashiwakura, J.; Suzuki, N.; Nagafuchi, H.; Takeno, M.; Takeba, Y.; Shimoyama, Y.; Sakane, T. Txk, a nonreceptor tyrosine kinase of the tec family, is expressed in t helper type 1 cells and regulates interferon gamma production in human t lymphocytes. *J. Exp. Med.* **1999**, *190*, 1147–1154. [CrossRef] [PubMed]

42. Mullen, A.C.; Hutchins, A.S.; High, F.A.; Lee, H.W.; Sykes, K.J.; Chodosh, L.A.; Reiner, S.L. Hlx is induced by and genetically interacts with t-bet to promote heritable t(h)1 gene induction. *Nat. Immunol.* **2002**, *3*, 652–658. [CrossRef] [PubMed]

43. De Santa, F.; Totaro, M.G.; Prosperini, E.; Notarbartolo, S.; Testa, G.; Natoli, G. The histone h3 lysine-27 demethylase jmjd3 links inflammation to inhibition of polycomb-mediated gene silencing. *Cell* **2007**, *130*, 1083–1094. [CrossRef] [PubMed]

44. Schlicher, L.; Brauns-Schubert, P.; Schubert, F.; Maurer, U. Spata2: More than a missing link. *Cell Death Differ.* **2017**, *24*, 1142–1147. [CrossRef] [PubMed]

45. Dhavan, R.; Tsai, L.H. A decade of cdk5. *Nat. Rev. Mol. Cell Biol.* **2001**, *2*, 749–759. [CrossRef] [PubMed]

46. Reiter, V.; Matschkal, D.M.S.; Wagner, M.; Globisch, D.; Kneuttinger, A.C.; Muller, M.; Carell, T. The cdk5 repressor cdk5rap1 is a methylthiotransferase acting on nuclear and mitochondrial rna. *Nucleic Acids Res.* **2012**, *40*, 6235–6240. [CrossRef] [PubMed]

47. De Mello, V.D.F.; Kolehmanien, M.; Schwab, U.; Pulkkinen, L.; Uusitupa, M. Gene expression of peripheral blood mononuclear cells as a tool in dietary intervention studies: What do we know so far? *Mol. Nutr. Food Res.* **2012**, *56*, 1160–1172. [CrossRef] [PubMed]

48. Olsen, K.S.; Skeie, G.; Lund, E. Whole-blood gene expression profiles in large-scale epidemiological studies: What do they tell? *Curr. Nutr. Rep.* **2015**, *4*, 377–386. [CrossRef] [PubMed]

49. Andersen, L.F.; Jacobs, D.R.; Carlsen, M.H.; Blomhoff, R. Consumption of coffee is associated with reduced risk of death attributed to inflammatory and cardiovascular diseases in the iowa women's health study. *Am. J. Clin. Nutr.* **2006**, *83*, 1039–1046. [CrossRef] [PubMed]

50. Schulze, M.B.; Hoffmann, K.; Manson, J.E.; Willett, W.C.; Meigs, J.B.; Weikert, C.; Heidemann, C.; Colditz, G.A.; Hu, F.B. Dietary pattern, inflammation, and incidence of type 2 diabetes in women. *Am. J. Clin. Nutr.* **2005**, *82*, 675–684. [CrossRef] [PubMed]

51. Zampelas, A.; Panagiotakos, D.B.; Pitsavos, C.; Chrysohoou, C.; Stefanadis, C. Associations between coffee consumption and inflammatory markers in healthy persons: The attica study. *Am. J. Clin. Nutr.* **2004**, *80*, 862–867. [CrossRef] [PubMed]

52. Svilaas, A.; Sakhi, A.K.; Andersen, L.F.; Svilaas, T.; Strom, E.C.; Jacobs, D.R.J.; Ose, L.; Blomhoff, R. Intakes of antioxidants in coffee, wine, and vegetables are correlated with plasma carotenoids in humans. *J. Nutr.* **2004**, *134*, 562–567. [CrossRef] [PubMed]

53. Halvorsen, B.; Ranheim, T.; Nenseter, M.S.; Huggett, A.C.; Drevon, C.A. Effect of a coffee lipid (cafestol) on cholesterol metabolism in human skin fibroblasts. *J. Lipid Res.* **1998**, *39*, 901–912. [PubMed]

54. Hurtubise, J.; McLellan, K.; Durr, K.; Onasanya, O.; Nwabuko, D.; Ndisang, J.F. The different facets of dyslipidemia and hypertension in atherosclerosis. *Curr. Atheroscler Rep.* **2016**, *18*, 82. [CrossRef] [PubMed]

55. Lang, R.; Dieminger, N.; Beusch, A.; Lee, Y.M.; Dunkel, A.; Suess, B.; Skurk, T.; Wahl, A.; Hauner, H.; Hofmann, T. Bioappearance and pharmacokinetics of bioactives upon coffee consumption. *Anal. Bioanal. Chem.* **2013**, *405*, 8487–8503. [CrossRef] [PubMed]

56. Mohr, S.; Liew, C.C. The peripheral-blood transcriptome: New insights into disease and risk assessment. *Trends Mol. Med.* **2007**, *13*, 422–432. [CrossRef] [PubMed]

57. Whitney, A.R.; Diehn, M.; Popper, S.J.; Alizadeh, A.A.; Boldrick, J.C.; Relman, D.A.; Brown, P.O. Individuality and variation in gene expression patterns in human blood. *Proc. Natl. Acad. Sci. USA* **2003**, *100*, 1896–1901. [CrossRef] [PubMed]

*nutrients*

MDPI

*Article*

# Coffee and Tea Consumption and the Contribution of Their Added Ingredients to Total Energy and Nutrient Intakes in 10 European Countries: Benchmark Data from the Late 1990s

Edwige Landais [1], Aurélie Moskal [2], Amy Mullee [2,3], Geneviève Nicolas [2], Marc J. Gunter [2],
Inge Huybrechts [2], Kim Overvad [4], Nina Roswall [5], Aurélie Affret [6], Guy Fagherazzi [6],
Yahya Mahamat-Saleh [6], Verena Katzke [7], Tilman Kühn [7], Carlo La Vecchia [8,9],
Antonia Trichopoulou [8], Elissavet Valanou [8], Calogero Saieva [10], Maria Santucci de Magistris [11],
Sabina Sieri [12], Tonje Braaten [13], Guri Skeie [13], Elisabete Weiderpass [14,15,16,17], Eva Ardanaz [18,19],
Maria-Dolores Chirlaque [19,20,21], Jose Ramon Garcia [22], Paula Jakszyn [23],
Miguel Rodríguez-Barranco [19,24,25], Louise Brunkwall [26], Ena Huseinovic [27], Lena Nilsson [28],
Peter Wallström [26], Bas Bueno-de-Mesquita [29,30], Petra H. Peeters [31], Dagfinn Aune [29], Tim Key [32],
Marleen Lentjes [33], Elio Riboli [29], Nadia Slimani [2] and Heinz Freisling [2,*]

1   UMR Nutripass, IRD-UM-Sup'Agro, 34394 Montpellier, France; edwige.landais@ird.fr
2   Nutrition and Metabolism Section, International Agency for Research on Cancer, 69372 Lyon, France;
    MoskalA@iarc.fr (A.M.); amy.mullee@ucd.ie (A.M.); nicolasg@iarc.fr (G.N.); gunterM@iarc.fr (M.J.G.);
    huybrechtsi@iarc.fr (I.H.); n.popovic@orange.fr (N.S.)
3   School of Public Health, Physiotherapy and Sports Science, Woodview House, University College Dublin,
    Belfield, Dublin 4, Ireland
4   Department of Public Health, Section for Epidemiology, Aarhus University, Bartholins Alle 2, room 2.26,
    DK-8000 Aarhus, Denmark; ko@ph.au.dk
5   Danish Cancer Society Research Center, Diet, Genes and Environment, Strandboulevarden 49,
    DK-2100 Copenhagen, Denmark; roswall@cancer.dk
6   Inserm CESP U1018, Gustave Roussy, Université Paris-Sud, Paris-Saclay, 94800 Villejuif, France;
    AURELIE.AFFRET@gustaveroussy.fr (A.A.); Guy.FAGHERAZZI@gustaveroussy.fr (G.F.);
    Yahya.MAHAMAT-SALEH@gustaveroussy.fr (Y.M.-S.)
7   German Cancer Research Center (DKFZ), Division of Cancer Epidemiology, 69120 Heidelberg, Germany;
    V.Katzke@Dkfz-Heidelberg.de (V.K.); t.kuehn@Dkfz-Heidelberg.de (T.K.)
8   Hellenic Health Foundation, 115 27 Athens, Greece; carlo.lavecchia@unimi.it (C.L.V.);
    atrichopoulou@hhf-greece.gr (A.T.); valanou@hhf-greece.gr (E.V.)
9   Department of Clinical Sciences and Community Health, Università degli Studi di Milano,
    20122 Milano, Italy
10  Molecular and Nutritional Epidemiology Unit, ISPO Cancer Prevention and Research Institute,
    50139 Florence, Italy; c.saieva@ispo.toscana.it
11  A.O.U. FEDERICO II, 80131 Naples, Italy; masantuc@unina.it
12  Epidemiology and Prevention Unit Fondazione IRCCS Istituto Nazionale dei Tumori, 20133 Milan, Italy;
    Sabina.sieri@istitutotumori.mi.it
13  Department of Community Medicine UiT, The Arctic University of Norway, 9037 Tromsø, Norway;
    tonje.braaten@uit.no (T.B.); guri.skeie@uit.no (G.S.)
14  Department of Community Medicine, Faculty of Health Sciences, University of Tromsø,
    The Arctic University of Norway, 9037 Tromsø, Norway; Elisabete.Weiderpass@kreftregisteret.no
15  Department of Research, Cancer Registry of Norway, Institute of Population-Based Cancer Research,
    NO-0304 Oslo, Norway
16  Department of Medical Epidemiology and Biostatistics, Karolinska Institutet, SE-171 77 Stockholm, Sweden
17  Genetic Epidemiology Group, Folkhälsan Research Center and Faculty of Medicine, University of Helsinki,
    00014 Helsinkiv, Finland
18  Navarra Public Health Institute, Pamplona, Spain IdiSNA, Navarra Institute for Health Research,
    31003 Pamplona, Spain; me.ardanaz.aicua@cfnavarra.es

[19]   CIBER Epidemiology and Public Health CIBERESP, 28029 Madrid, Spain;
       mdolores.chirlaque@carm.es (M.-D.C.); miguel.rodriguez.barranco.easp@juntadeandalucia.es (M.R.-B.)
[20]   Department of Epidemiology, Regional Health Council, IMIB-Arrixaca, 30008 Murcia, Spain
[21]   Department of Health and Social Sciences, Universidad de Murcia, 30008 Murcia, Spain
[22]   EPIC Asturias, Public Health Directorate, Asturias, 33006 Oviedo, Spain;
       joseramon.quirosgarcia@asturias.org
[23]   Unit of Nutrition, Environment and Cancer, Catalan Institute of Oncology, 08908 Barcelona, Spain;
       paujak.ico@gmail.com
[24]   Escuela Andaluza de Salud Pública. Instituto de Investigación Biosanitaria ibs, 18011 Granada, Spain
[25]   Hospitales Universitarios de Granada, Universidad de Granada, 18014 Granada, Spain
[26]   Clinical Science, Lund University, SE-221 00 Lund, Sweden; louise.brunkwall@med.lu.se (L.B.);
       peter.wallstrom@med.lu.se (P.W.)
[27]   Department of Internal Medicine and Clinical Nutrition, The Sahlgrenska Academy,
       University of Gothenburg, SE-405 30 Gothenburg, Sweden; ena.huseinovic@gu.se
[28]   Public Health and Clinical Medicine, Nutritional Research, Umeå University, and Arctic Research Centre at
       Umeå University, SE-901 85 Umeå, Sweden; lena.nilsson@umu.se
[29]   Department of Epidemiology and Biostatistics, The School of Public Health, Imperial College London,
       London W2 1PG, UK; basbuenodemesquita@gmail.com (B.B.-d.-M.); d.aune@imperial.ac.uk (D.A.);
       e.riboli@imperial.ac.uk (E.R.)
[30]   Department of Social & Preventive Medicine, Faculty of Medicine, University of Malaya,
       Kuala Lumpur 50603, Malaysia
[31]   University Medical Center Utrecht, 3584 CX Utrecht, The Netherlands; P.H.M.Peeters@umcutrecht.nl
[32]   Cancer Epidemiology Unit, Nuffield Department of Population Health, University of Oxford,
       Oxford OX3 7LF, UK; tim.key@ceu.ox.ac.uk
[33]   Strangeways Research Laboratories, Department of Public Health & Primary Care, University of Cambridge,
       Cambridge CB1 8RN, UK; marleen@srl.cam.ac.uk
*      Correspondence: freislingh@iarc.fr; Tel.: +33-472-738-664

Received: 19 April 2018; Accepted: 1 June 2018; Published: 5 June 2018

**Abstract:** Background: Coffee and tea are among the most commonly consumed nonalcoholic beverages worldwide, but methodological differences in assessing intake often hamper comparisons across populations. We aimed to (i) describe coffee and tea intakes and (ii) assess their contribution to intakes of selected nutrients in adults across 10 European countries. Method: Between 1995 and 2000, a standardized 24-h dietary recall was conducted among 36,018 men and women from 27 European Prospective Investigation into Cancer and Nutrition (EPIC) study centres. Adjusted arithmetic means of intakes were estimated in grams (=volume) per day by sex and centre. Means of intake across centres were compared by sociodemographic characteristics and lifestyle factors. Results: In women, the mean daily intake of coffee ranged from 94 g/day (~0.6 cups) in Greece to 781 g/day (~4.4 cups) in Aarhus (Denmark), and tea from 14 g/day (~0.1 cups) in Navarra (Spain) to 788 g/day (~4.3 cups) in the UK general population. Similar geographical patterns for mean daily intakes of both coffee and tea were observed in men. Current smokers as compared with those who reported never smoking tended to drink on average up to 500 g/day more coffee and tea combined, but with substantial variation across centres. Other individuals' characteristics such as educational attainment or age were less predictive. In all centres, coffee and tea contributed to less than 10% of the energy intake. The greatest contribution to total sugar intakes was observed in Southern European centres (up to ~20%). Conclusion: Coffee and tea intake and their contribution to energy and sugar intake differed greatly among European adults. Variation in consumption was mostly driven by geographical region.

**Keywords:** coffee; tea; European Prospective Investigation into Cancer and Nutrition; 24-h dietary recall

## 1. Introduction

Coffee and tea are the most widely consumed nonalcoholic beverages across the world [1,2]. Both beverages contain various antioxidants and phenolic compounds such as flavonoids or caffeine, some of which have been shown to have anticancer properties in laboratory conditions [3–6].

According to the third expert report of the World Cancer Research Fund (WCRF) and the Continuous Update Project (CUP), the evidence on the associations between cancer and the intakes of tea, and for many cancer sites, of coffee, were too limited in amount, consistency, and/or quality to draw conclusions, except for a probable decreased risk for cancers of the liver and endometrium among regular coffee drinkers [3,7].

Several systematic reviews and meta-analyses conducted subsequently also reported inconsistent results for the potential association of coffee or tea on certain types of cancers such as prostate, lung, colorectal, oesophageal, renal, or breast cancers. Indeed, whilst some of the studies reported inverse associations for tea or coffee (e.g., coffee and liver or prostate cancers, tea and lung cancer) [8–13], others did not observe any significant adverse or potential protective effects of such beverages [14–19].

A monograph conducted by the International Agency for Research on Cancer (IARC) in 2016 evaluating the carcinogenicity of drinking coffee to humans concluded that it was unclassifiable as to its carcinogenicity to humans [20].

Differences in tea- and coffee-drinking habits (e.g., green tea, black tea, with caffeine, decaffeinated) as well as the preparation processes, amount consumed, and additions such as sugar/milk are likely to vary by population and countries and could contribute to the inconsistencies found between studies comparing tea and coffee consumption and the risk of chronic diseases. Furthermore, the use of different assessment methods, such as distinct food frequency questionnaires, different variable definitions (e.g., food classification, serving sizes), or levels of detail to describe foods, may impede comparisons between studies [21].

Our main objective was to describe coffee and tea intake in men and women across 27 centres in the European Prospective Investigation into Cancer and Nutrition (EPIC) study using standardized 24-h dietary recall (24-HDR) data. We also estimated variation in intake levels according to selected sociodemographic, lifestyle, and anthropometric characteristics of study participants, and assessed the relative contribution of coffee and tea to intakes of total energy and selected nutrients (total sugars, calcium, magnesium, phosphorus).

## 2. Materials and Methods

### 2.1. Setting and Subjects

EPIC is a multicentre prospective cohort study investigating the association between diet and cancer and other chronic diseases in 23 centres in ten countries: Denmark, France, Germany, Greece, Italy, the Netherlands, Norway, Spain, Sweden, and the UK [22,23]. EPIC participants were mostly recruited from the general population between 1992 and 1998 and included 521,330 men and women aged 35–70 years; exceptions were France (health insurance members), Utrecht (The Netherlands) and Florence (Italy) (participants of breast cancer screening), and some centres in Spain and Italy (mostly blood donors). In the UK, a cohort consisting predominantly of vegetarians ('health-conscious' in Oxford) was considered separately from a 'general population' group recruited by general practitioners in Cambridge and Oxford. Most centres recruited both men and women, except Norway, France, Utrecht, and Naples, where only women were recruited. Details of the methods of recruitment and study design have been published previously [22,24,25]. All participants provided written informed consent, and the project was approved by ethical review boards of the IARC and local participating centres. In the present study, the initial 23 EPIC centres were redefined into 27 regions according to a geographical south–north gradient and relevant to analyses of dietary consumption and patterns [23].

The calibration substudy nested within the EPIC cohort was undertaken between 1995 and 2000 with the aim to partially correct for attenuation in diet–disease associations due to measurement errors. This has been obtained by rescaling the country-specific individual dietary intakes against the same reference dietary measurement obtained using a highly standardized 24-h dietary recall (24-HDR) [26]. The calibration population sample consisted of 36,994 participants, representing a random sample (~8%) of the total EPIC cohort, stratified by age, sex, and centre. Details of the population characteristics of the calibration study have been published previously [23,27–29]. In brief, each participant completed a single 24-HDR during a face-to-face interview, except in Norway, where it was conducted through a validated phone interview alternative [30]. A computer-based interview programme, named EPIC-Soft (recently renamed GloboDiet; IARC, Lyon, France), was developed to conduct standardized 24-HDR interviews [31,32] with the same structure and interview procedure across countries. The interviews were conducted over different seasons and days of the week. For logistical constraint reasons, interviews recalling diet on Saturday were conducted on Monday (instead of Sunday) in most countries, whereas for all other days of the week, the interviews were conducted the following day. Time and place of consumption were also collected.

*2.2. Dietary Variables*

The common food group classification used in the EPIC-Soft software, which has been described previously [23], was used to divide the overall coffee and tea group into four different subgroups as follows: coffee, split into three subgroups regarding caffeine content (with caffeine, partially decaffeinated, decaffeinated); tea, either black or green; herbal tea; and chicory and substitutes. Anything added to these beverages, e.g., milk or milk substitutes, sugar, and honey, was also taken into consideration, in order to evaluate the overall contribution of coffee and tea with their added ingredients to total energy and selected nutrients' intake (alcohol was a negligible ingredient to coffee in all cohorts). The beverages are expressed in grams per day as complete beverages (i.e., including the water for diluted beverages or reconstituted beverages from powder). The overall coffee and tea intake of individuals on the recall day was calculated by summing the amount of these four groups.

Places where coffee and tea could potentially be consumed were recorded as home, work, fast-food restaurant, bar, cafeteria, restaurant, friends' home, school, street, car/boat/plane, and other. These options were common across centres. After considering their distribution, some of these categories were merged as follows: work, school, and cafeteria into 'work'; other, street, and car/boat/ plane into 'other'; and fast-food restaurant with restaurant. The resulting places of consumption were: home, work, bar, restaurant, friends' place, and other place.

*2.3. Nutrient Databases*

Energy and nutrient intakes were estimated by means of standardized nutrient databases developed through the EPIC Nutrient DataBase (ENDB) project. Only relevant nutrients (sugar, calcium, magnesium, phosphorus) with regards to coffee and tea and their related added ingredients are reported. The rationale and procedures used to improve between-country comparability of the 26 nutrients included in this database are described elsewhere [33].

*2.4. Nondietary Variables*

Data on other lifestyle factors, including education (none or primary, secondary/technical, and university degree; completeness >98%), total physical activity (inactive, moderately inactive, moderately active, and active; completeness >86%) [34], and smoking status (never, former, current; completeness >98%), were collected at baseline through standardized questionnaires and clinical examinations and have been described elsewhere [22,23,35]. In most centres, age as well as body weight and height were self-reported by the participants during the 24-HDR interview. Individuals were classified according to age categories (35–44, 45–54, 55–64, 65–74 years) and body mass index (BMI; based on self-reported data) categories (BMI < 25 kg/m$^2$, BMI 25 to <30 kg/m$^2$, BMI $\geq$ 30 kg/m$^2$; no missing data). The time interval between the baseline questionnaires and the 24-HDR interview varied by country, ranging from one day to three years [23].

*2.5. Statistical Methods*

Centre-specific arithmetic means of coffee and tea intakes and standard errors of the mean (SEM) were calculated using generalized linear models, stratified by EPIC centre and sex. Fully adjusted models were adjusted for age, total energy intake, height, and weight (except for analyses stratified on BMI) and were weighted by season and day of recall to control for different distributions of 24-HDR interviews across seasons and days of the week. Means were also calculated for each type of coffee and tea as well as for decaffeinated versus caffeinated (including partially decaffeinated) coffee. If fewer than 20 persons were represented in a cross-classification (for example, centre, sex, and age group), the least-square mean was not reported in the table.

In order to compare means of coffee and tea across centres by categories of age, education, BMI, physical activity, and smoking status, we fitted regression models that included an interaction term between centre and each of the potentially associated factors at a time, to test whether the association of coffee and tea consumption with these factors differed across centres. These analyses were adjusted for age, total energy intake, height, and weight and weighted by season and day of recall, separately for men and women. Participants with missing data were omitted. Type III statistics of the GENMOD procedure in SAS were used to examine the partial effect of each variable; that is, the significance of a variable with all the other variables in the model. Tests for trends were computed across categories by using a score variable (from 1 up to the number of categories of a given variable).

The relative contribution of coffee and tea intake (overall and by type) to total energy and selected nutrient intakes (sugar, calcium, magnesium, phosphorus) were calculated by centre as the mean percentage of intake, stratified by centre; adjusted for sex, height, and weight; and weighted by season and weekday.

All the analyses were performed using SAS (version 9.4, SAS Institute, Cary, NC, USA).

### 3. Results

A total of 36,018 subjects with 24-HDR data were included in this analysis, after exclusion of 958 subjects aged under 35 or over 74 years because of low participation in these age categories and of 18 subjects without lifestyle variable data.

*3.1. Coffee and Tea Intakes*

The adjusted mean daily intake of coffee and tea varied widely across centres, ranging from 174 g/day and 170 g/day for men and women, respectively, in Greece to 1468 g/day and 1321 g/day in the UK general population (Table 1 for men and Table 2 for women). Overall, Northern European countries tended to drink more coffee and tea compared to Southern European countries (see Supplemental Materials, Table S1).

When describing consumption for the four different coffee and tea groups, the adjusted mean daily intake of coffee ranged from 107 g/day in Greek men (which corresponded to 0.9 cups) to 1016 g/day for men living in Aarhus (Denmark) (which corresponded to 5.5 cups) (Table 1) and from 94 g/day for Greek women (which corresponded to 0.6 cups) to 781 g/day for women from Aarhus (Denmark) (which corresponded to 4.4 cups) (Table 2). Among men, tea intake ranged from 18 g/day in San Sebastian (Spain) (which corresponded to 0.1 cups) to 928 g/day in the UK general population (which corresponded to 4.9 cups), and among women from 14 g/day in Navarra (Spain) (which corresponded to 0.1 cups) to 788 g/day in the UK general population (which corresponded to 4.3 cups). Across centres, the lowest consumption of herbal tea was observed in Umeå (Sweden) (0 g/day and 7 g/day for men and women, respectively) and the highest one in Germany (128 g/day for men in Potsdam and 202 g/day for women in Heidelberg). For both men and women, the lowest consumption of chicory and substitutes was reported in Sweden and Denmark, and the highest in UK health-conscious individuals (Tables 1 and 2).

Overall, in all centres but those in the UK, the amount of coffee consumed was higher than the amount of tea for both sexes.

**Table 1.** Mean daily intake of coffee and tea (g/day) by type in the EPIC calibration study population based on 24-H Dietary Recall among men across EPIC centres ordered from south to north.

| Country and Centre | n | Total Coffee and Tea * | | Coffee | | Tea * | | Herbal Tea | | Chicory and Substitutes | |
|---|---|---|---|---|---|---|---|---|---|---|---|
| | | Fully Adjusted Mean[1] | SEM[2] | Fully Adjusted Mean[1] | SEM[2] | Fully Adjusted Mean[1] | SEM[2] | Fully Adjusted Mean[1] | SEM[2] | Fully Adjusted Mean[1] | SEM[2] |
| Greece | 1324 | 173.5 | 13.3 | 106.7 | 12.2 | 47.9 | 9.2 | 18.3 | 4.4 | 0.6 | 1.8 |
| Spain | | | | | | | | | | | |
| Granada | 214 | 387.3 | 31.9 | 316.1 | 29.4 | 27.3 | 22.1 | 31.7 | 10.6 | 12.2 | 4.3 |
| Murcia | 243 | 302.0 | 30.0 | 202.9 | 27.7 | 25.1 | 20.8 | 53.5 | 10.0 | 20.5 | 4.0 |
| Navarra | 444 | 309.2 | 22.3 | 267.2 | 20.6 | 18.7 | 15.4 | 14.4 | 7.4 | 9.0 | 3.0 |
| San Sebastian | 490 | 270.2 | 21.4 | 192.9 | 19.7 | 17.7 | 14.8 | 28.1 | 7.1 | 31.6 | 2.9 |
| Asturias | 386 | 379.5 | 23.8 | 295.0 | 22.0 | 23.0 | 16.5 | 29.1 | 7.9 | 32.4 | 3.2 |
| Italy | | | | | | | | | | | |
| Ragusa | 168 | 222.6 | 36.0 | 160.3 | 33.2 | 47.6 | 25.0 | 4.9 | 12.0 | 9.7 | 4.8 |
| Florence | 271 | 270.1 | 28.2 | 187.2 | 26.0 | 45.5 | 19.6 | 9.0 | 9.4 | 28.4 | 3.8 |
| Turin | 676 | 260.9 | 18.0 | 171.7 | 16.6 | 56.3 | 12.5 | 13.6 | 6.0 | 19.3 | 2.4 |
| Varese | 327 | 392.6 | 25.8 | 277.9 | 23.8 | 70.1 | 17.9 | 14.6 | 8.6 | 29.9 | 3.5 |
| Germany | | | | | | | | | | | |
| Heidelberg | 1034 | 897.1 | 14.6 | 597.7 | 13.4 | 164.6 | 10.1 | 125.6 | 4.9 | 9.3 | 2.0 |
| Potsdam | 1233 | 843.9 | 13.2 | 578.7 | 12.2 | 126.8 | 9.2 | 128.2 | 4.4 | 10.3 | 1.8 |
| The Netherlands | | | | | | | | | | | |
| Bilthoven | 1020 | 960.5 | 15.1 | 698.1 | 13.9 | 235.0 | 10.5 | 21.6 | 5.0 | 5.8 | 2.0 |
| United Kingdom | | | | | | | | | | | |
| General population | 405 | 1467.7 | 23.1 | 523.9 | 21.3 | 927.8 | 16.0 | 9.5 | 7.7 | 6.4 | 3.1 |
| Health-conscious | 113 | 1222.4 | 43.9 | 439.0 | 40.5 | 620.5 | 30.4 | 113.1 | 14.6 | 49.9 | 5.9 |
| Denmark | | | | | | | | | | | |
| Copenhagen | 1356 | 1152.0 | 12.7 | 896.7 | 11.8 | 229.9 | 8.8 | 25.7 | 4.2 | 0.0 | |
| Aarhus | 567 | 1220.8 | 19.6 | 1015.5 | 18.0 | 184.9 | 13.6 | 18.0 | 6.5 | 2.3 | 2.6 |
| Sweden | | | | | | | | | | | |
| Malmö | 1421 | 855.7 | 13.2 | 727.1 | 12.1 | 133.5 | 9.1 | 0.0 | | 0.0 | |
| Umeå | 1342 | 785.6 | 12.8 | 626.1 | 11.8 | 160.4 | 8.9 | 0.0 | | 0.0 | |

* Either green or black tea, herbal tea excluded. [1] Adjusted for age, total energy intake, weight, and height and weighted by season and day of recall. [2] SEM: standard error of the mean.

**Table 2.** Mean daily intake of coffee and tea (g/day) by type in the EPIC calibration study population based on 24-H Dietary Recall among women across EPIC centres ordered from South to North.

| Country and Centre | n | Total Coffee and Tea * | | Coffee | | Tea * | | Herbal Tea | | Chicory and Substitutes | |
|---|---|---|---|---|---|---|---|---|---|---|---|
| | | Fully Adjusted Mean¹ | SEM² | Fully Adjusted Mean¹ | SEM² | Fully Adjusted Mean² | SEM² | Fully Adjusted Mean² | SEM² | Fully Adjusted Mean¹ | SEM² |
| Greece | 1368 | 170.3 | 12.5 | 93.8 | 10.3 | 54.5 | 9.8 | 20.0 | 5.5 | 2.0 | 3.4 |
| **Spain** | | | | | | | | | | | |
| Granada | 300 | 425.9 | 25.8 | 299.8 | 21.3 | 24.4 | 20.2 | 78.4 | 11.4 | 23.3 | 7.1 |
| Murcia | 304 | 389.9 | 25.6 | 289.3 | 21.1 | 20.9 | 20.1 | 62.0 | 11.4 | 17.7 | 7.1 |
| Navarra | 271 | 491.2 | 27.0 | 433.4 | 22.3 | 14.4 | 21.2 | 24.5 | 12.0 | 18.9 | 7.4 |
| San Sebastian | 244 | 468.3 | 28.4 | 360.8 | 23.5 | 20.8 | 22.3 | 39.9 | 12.6 | 46.8 | 7.8 |
| Asturias | 324 | 532.8 | 24.7 | 454.9 | 20.4 | 18.3 | 19.4 | 37.6 | 11.0 | 22.0 | 6.8 |
| **Italy** | | | | | | | | | | | |
| Ragusa | 137 | 201.0 | 38.1 | 147.8 | 31.4 | 32.4 | 29.9 | 12.0 | 16.9 | 8.9 | 10.5 |
| Naples | 403 | 297.2 | 22.3 | 226.6 | 18.4 | 41.5 | 17.5 | 10.3 | 9.9 | 18.8 | 6.1 |
| Florence | 783 | 328.1 | 15.9 | 226.3 | 13.1 | 54.8 | 12.5 | 17.2 | 7.1 | 29.7 | 4.4 |
| Turin | 392 | 312.3 | 22.4 | 194.9 | 18.5 | 73.3 | 17.6 | 21.3 | 10.0 | 22.8 | 6.2 |
| Varese | 795 | 404.1 | 15.9 | 262.8 | 13.1 | 97.0 | 12.5 | 25.1 | 7.0 | 19.1 | 4.4 |
| **France** | | | | | | | | | | | |
| South coast | 620 | 567.0 | 17.9 | 282.8 | 14.8 | 147.1 | 14.1 | 63.7 | 8.0 | 73.4 | 4.9 |
| South | 1425 | 651.7 | 11.9 | 280.7 | 9.8 | 228.7 | 9.3 | 64.4 | 5.3 | 78.0 | 3.3 |
| Northeast | 2059 | 656.0 | 9.9 | 323.3 | 8.2 | 200.3 | 7.8 | 62.1 | 4.4 | 70.3 | 2.7 |
| Northwest | 631 | 722.9 | 17.8 | 365.3 | 14.7 | 245.2 | 13.9 | 50.4 | 7.9 | 62.1 | 4.9 |
| **Germany** | | | | | | | | | | | |
| Heidelberg | 1087 | 968.7 | 13.6 | 557.6 | 11.2 | 193.1 | 10.7 | 202.2 | 6.1 | 15.8 | 3.8 |
| Potsdam | 1060 | 815.8 | 13.7 | 510.3 | 11.3 | 113.2 | 10.8 | 178.7 | 6.1 | 13.6 | 3.8 |
| **The Netherlands** | | | | | | | | | | | |
| Bilthoven | 1076 | 949.0 | 13.8 | 591.0 | 11.4 | 303.3 | 10.8 | 42.4 | 6.1 | 12.3 | 3.8 |
| Utrecht | 1870 | 1050.1 | 10.4 | 570.1 | 8.6 | 431.9 | 8.2 | 39.7 | 4.6 | 8.4 | 2.9 |
| **United Kingdom** | | | | | | | | | | | |
| General population | 570 | 1321.3 | 18.6 | 491.2 | 15.3 | 788.4 | 14.6 | 34.3 | 8.2 | 7.4 | 5.1 |
| Health-conscious | 196 | 1139.0 | 31.7 | 328.1 | 26.1 | 601.4 | 24.9 | 116.1 | 14.1 | 93.4 | 8.7 |
| **Denmark** | | | | | | | | | | | |
| Copenhagen | 1484 | 1009.3 | 11.6 | 631.7 | 9.6 | 315.8 | 9.1 | 61.4 | 5.2 | 0.5 | 3.2 |
| Aarhus | 510 | 1109.5 | 19.7 | 781.0 | 16.2 | 230.2 | 15.4 | 95.0 | 8.7 | 3.3 | 5.4 |
| **Sweden** | | | | | | | | | | | |
| Malmö | 1711 | 805.7 | 11.0 | 646.3 | 9.1 | 147.6 | 8.7 | 11.9 | 4.9 | 0.0 | 3.1 |
| Umeå | 1567 | 704.0 | 11.2 | 527.8 | 9.3 | 168.9 | 8.8 | 7.1 | 5.0 | 0.2 | |
| **Norway** | | | | | | | | | | | |
| South and East | 1004 | 892.8 | 14.4 | 643.7 | 11.8 | 190.1 | 11.3 | 57.1 | 6.4 | 1.9 | 4.0 |
| North and West | 793 | 894.5 | 16.0 | 690.9 | 13.2 | 135.1 | 12.6 | 67.4 | 7.1 | 1.1 | 4.4 |

* Either green or black tea, herbal tea excluded. ¹ Adjusted for age, total energy intake, weight, and height and weighted by season and day of recall. ² SEM: standard error of the mean.

## 3.2. Proportion of Consumers

In comparison with all centres, Greece had the highest proportion of individuals not consuming coffee nor tea over the previous day (27% and 31% for men and women, respectively), and Aarhus (Denmark) for men and Utrecht (The Netherlands) for women had the lowest proportion of nonconsumers (0.9% and 0.4%, respectively) (see Supplemental Materials, Figures S1 and S2). The proportion of men drinking only tea the previous day was the lowest in Ragusa (Italy) (0.6%) and the highest in the UK general population (23%). Women from Naples (Italy) and Navarra (Spain) had the lowest proportion of tea-only drinkers the previous day (0.7% in both cases) and the UK health-conscious population had the highest proportion (30%). The proportion of men and women drinking coffee only over the previous day was the lowest in the UK general population (10% and 12%, respectively) and the highest for both Italian men and women (Ragusa 87% and Naples 86%, respectively). Apart from in the UK, most of the men were coffee drinkers only. The same pattern was found for women in the UK as well as in The Netherlands.

Among coffee consumers from both sexes, the large majority of coffee consumed was coffee with caffeine (see Supplemental Materials, Figures S3 and S4). Overall, the mean percentage of decaffeinated coffee consumers slightly differed between sexes, with women tending to drink more decaffeinated coffee than men (8.8% vs. 6.0%). No south–north gradient was observed for the consumption of decaffeinated coffee. In Granada (Spain), men and women were the highest consumers of decaffeinated coffee (33% and 38%, respectively). In Malmö (Sweden), both men and women were the lowest consumers of decaffeinated coffee (0.3% and 0.6%, respectively).

## 3.3. Place of Consumption

When investigating the place of consumption, the large majority of coffee or tea consumed was consumed at home by both women and men. The percentage ranged from over 60% for both sexes in Denmark to almost 90% of all coffee and tea consumed in Italy (for men, the percentage ranged from 68% in Copenhagen (Denmark) to 88% in Florence (Italy), and for women, from 68% in Aarhus (Denmark) to 88% in Ragusa (Italy) (Figures 1 and 2).

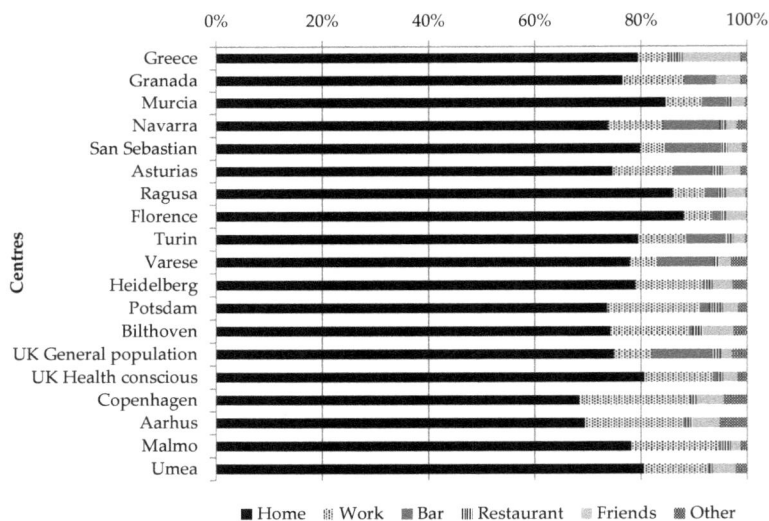

**Figure 1.** Proportion of coffee and tea consumption at different places of consumption, among men across EPIC centres; fully adjusted models among consumers only; "friends" refers to friends' place.

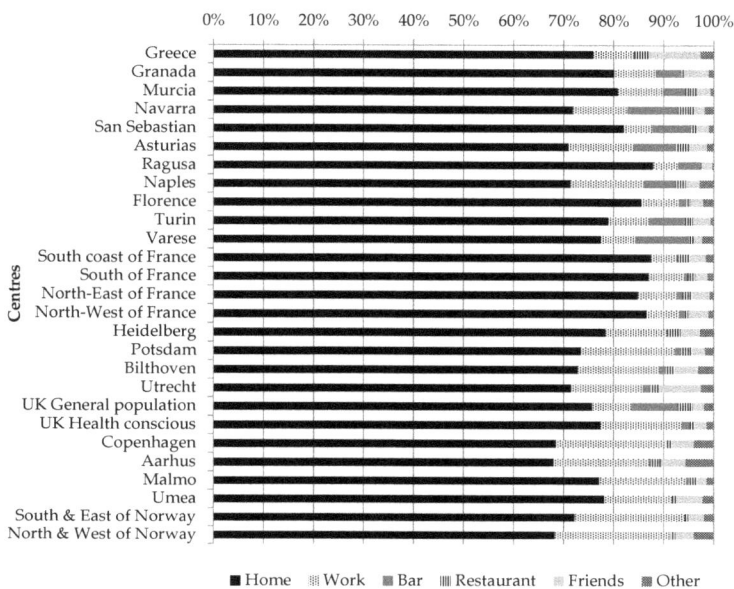

**Figure 2.** Proportion of coffee and tea consumption at different places of consumption, among women across EPIC centres; fully adjusted models among consumers only; "friends" refers to friends' place.

The second most important place of consumption was work, for which there was a south–north gradient as overall, for individuals living in the Northern part of Europe, coffee and tea were more frequently drunk at work compared to what was reported in the Southern part. The other important places of consumption were "bar" and "friends' place", for which a south–north gradient was observed. Indeed, for women living in the Northern part of Europe, coffee and tea were more frequently consumed at a friends' place rather that at a bar. The opposite pattern was observed for women living in South Europe, except for Greek women and women living in the South of France. A similar pattern was observed among men.

*3.4. Sociodemographic Factors*

When studying the age trends, overall, coffee and tea intake was significantly associated with age ($p < 0.0001$ in both sexes). Stratified by centre, a linear trend between coffee and tea consumption and age was only significant among four out of the 23 centres (Table 3), which could be related to lack of power due to stratification. In Greece and Florence (Italy), older men tended to drink significantly more coffee and tea compared to the younger ones. On the contrary, younger men from Malmö (Sweden), as well as younger women from Navarra (Spain), drank significantly more coffee and tea than their older counterparts on the day of the recall.

Education across all centres was significantly associated with coffee and tea consumption among both men and women ($p < 0.005$ and $p < 0.0001$, respectively). Overall, the amount of coffee and tea consumed was higher with higher education. Yet, when stratified by centre, the linear trend between coffee and tea intake and education was significant only in men from the UK general population (the less educated tended to drink more coffee and tea), as opposed to women from the South of France, Copenhagen (Denmark), and Umeå and Malmö (Sweden), where the more educated tended to drink more coffee and tea on the day of the recall compared to the less educated women (Table 4).

**Table 3.** Fully adjusted mean [1] daily intake of coffee and tea (g/day) by age group and sex in the EPIC calibration study population based on 24-H Dietary Recall across EPIC centres ordered from south to north.

| Country and Centre | Men n | Men All Mean[1] | Men All SEM[2] | Men 35–44 Mean[1] | Men 35–44 SEM[2] | Men 45–54 Mean[1] | Men 45–54 SEM[2] | Men 55–64 Mean[1] | Men 55–64 SEM[2] | Men 65–74 Mean[1] | Men 65–74 SEM[2] | Men p-Trend | Women n | Women All Mean[1] | Women All SEM[2] | Women 35–44 Mean[1] | Women 35–44 SEM[2] | Women 45–54 Mean[1] | Women 45–54 SEM[2] | Women 55–64 Mean[1] | Women 55–64 SEM[2] | Women 65–74 Mean[1] | Women 65–74 SEM[2] | Women p-Trend |
|---|---|---|---|---|---|---|---|---|---|---|---|---|---|---|---|---|---|---|---|---|---|---|---|---|
| **Greece** | 1324 | 173.5 | 13.3 | 116.4 | 37.9 | 137.9 | 26.9 | 184.5 | 23.1 | 191.6 | 22.0 | 0.034 | 1368 | 170.3 | 12.5 | 146.2 | 31.8 | 169.4 | 21.3 | 179.7 | 21.4 | 166.3 | 26.3 | 0.349 |
| **Spain** | | | | | | | | | | | | | | | | | | | | | | | | |
| Granada | 214 | 387.3 | 31.9 | 377.6 | 133.3 | 444.5 | 64.6 | 381.3 | 43.2 | 317.2 | 77.6 | 0.393 | 300 | 425.9 | 25.8 | 426.8 | 63.8 | 475.9 | 42.7 | 409.9 | 40.5 | 349.4 | 88.0 | 0.263 |
| Murcia | 243 | 302.0 | 30.0 | 319.8 | 85.3 | 342.1 | 54.6 | 277.5 | 41.4 | 343.7 | 113.6 | 0.970 | 304 | 389.9 | 25.6 | 408.0 | 49.7 | 463.4 | 43.3 | 351.2 | 42.0 | 268.1 | 144.8 | 0.177 |
| Navarra | 444 | 309.2 | 22.3 | 254.4 | 86.0 | 305.6 | 37.3 | 322.1 | 31.4 | 314.5 | 71.8 | 0.169 | 271 | 491.2 | 27.0 | 603.1 | 66.8 | 493.0 | 44.3 | 469.2 | 40.7 | 385.8 | 143.4 | 0.026 |
| San Sebastian | 490 | 270.2 | 21.4 | 304.0 | 46.5 | 282.9 | 29.4 | 264.3 | 41.0 | 298.8 | 119.3 | 0.754 | 244 | 468.3 | 28.4 | 421.4 | 59.5 | 482.5 | 45.2 | 513.7 | 47.5 | 454.6 | 164.6 | 0.571 |
| Asturias | 386 | 379.5 | 23.8 | 453.1 | 79.5 | 360.1 | 40.4 | 377.3 | 34.8 | 414.1 | 70.1 | 0.689 | 324 | 532.8 | 24.7 | 570.6 | 58.6 | 593.1 | 39.5 | 494.8 | 39.7 | 390.7 | 104.0 | 0.098 |
| **Italy** | | | | | | | | | | | | | | | | | | | | | | | | |
| Ragusa | 168 | 222.6 | 36.0 | 184.5 | 103.6 | 234.9 | 53.2 | 240.6 | 55.7 | 250.0 | 246.7 | 0.110 | 137 | 201.0 | 38.1 | 197.2 | 62.6 | 234.9 | 70.5 | 233.0 | 68.5 | 171.6 | 179.0 | 0.667 |
| Naples | | | | | | | | | | | | | 403 | 297.2 | 22.3 | 264.9 | 68.7 | 313.2 | 33.2 | 303.2 | 36.6 | 272.0 | 73.3 | 0.938 |
| Florence | 271 | 270.1 | 28.2 | 200.1 | 84.8 | 246.5 | 46.0 | 310.5 | 41.1 | 324.3 | 121.7 | 0.025 | 783 | 328.1 | 15.9 | 306.5 | 27.3 | 328.2 | 27.3 | 333.7 | 22.1 | 314.3 | 68.8 | 0.702 |
| Turin | 676 | 260.9 | 18.0 | 240.6 | 54.3 | 268.4 | 29.7 | 268.7 | 25.4 | 225.6 | 86.7 | 0.341 | 392 | 312.3 | 22.4 | 292.7 | 66.6 | 336.4 | 36.4 | 302.0 | 32.1 | 373.8 | 134.9 | 0.269 |
| Varese | 327 | 392.6 | 25.8 | 368.0 | 141.7 | 364.8 | 58.1 | 397.1 | 30.2 | 393.4 | 109.9 | 0.164 | 795 | 404.1 | 15.9 | 376.0 | 48.0 | 405.4 | 26.2 | 414.6 | 23.9 | 391.6 | 49.7 | 0.571 |
| **France** | | | | | | | | | | | | | | | | | | | | | | | | |
| South coast | | | | | | | | | | | | | 620 | 567.0 | 17.9 | 349.5 | 376.3 | 608.2 | 28.6 | 541.4 | 27.6 | 497.9 | 39.8 | 0.077 |
| South | | | | | | | | | | | | | 1425 | 651.7 | 11.9 | | | 662.4 | 17.9 | 638.8 | 18.7 | 577.0 | 28.2 | 0.404 |
| Northeast | | | | | | | | | | | | | 2059 | 656.0 | 9.9 | | | 663.0 | 15.0 | 648.4 | 15.4 | 619.2 | 23.8 | 0.121 |
| Northwest | | | | | | | | | | | | | 631 | 722.9 | 17.8 | | | 762.2 | 27.5 | 685.9 | 27.0 | 682.5 | 43.7 | 0.309 |
| **Germany** | | | | | | | | | | | | | | | | | | | | | | | | |
| Heidelberg | 1034 | 897.1 | 14.6 | 946.8 | 37.4 | 949.2 | 22.9 | 846.3 | 22.9 | 1496.3 | 241.5 | 0.324 | 1087 | 968.7 | 13.6 | 1005.3 | 22.9 | 999.2 | 24.1 | 951.6 | 22.8 | 538.2 | 154.5 | 0.267 |
| Potsdam | 1233 | 843.9 | 13.2 | 862.2 | 36.4 | 854.8 | 26.4 | 834.9 | 17.3 | 918.3 | 64.5 | 0.464 | 1060 | 815.8 | 13.7 | 816.5 | 26.5 | 879.4 | 26.6 | 806.9 | 19.8 | | | 0.227 |
| **The Netherlands** | | | | | | | | | | | | | | | | | | | | | | | | |
| Bilthoven | 1020 | 960.5 | 15.1 | 987.7 | 27.8 | 989.8 | 22.9 | 946.9 | 25.5 | 850.8 | 333.6 | 0.100 | 1076 | 949.0 | 13.8 | 926.1 | 23.9 | 1030.7 | 21.1 | 923.8 | 26.2 | 756.2 | 249.5 | 0.299 |
| Utrecht | | | | | | | | | | | | | 1870 | 1050.1 | 10.4 | | | 1040.7 | 17.2 | 1047.5 | 15.7 | 1024.7 | 22.0 | 0.522 |
| **United Kingdom** | | | | | | | | | | | | | | | | | | | | | | | | |
| General population | 405 | 1467.7 | 23.1 | 1311.7 | 73.8 | 1595.7 | 40.6 | 1418.3 | 42.2 | 1418.9 | 42.8 | 0.842 | 570 | 1321.3 | 18.6 | 1279.9 | 52.8 | 1385.3 | 30.0 | 1316.0 | 34.2 | 1227.8 | 40.9 | 0.560 |
| Health-conscious | 113 | 1222.4 | 43.9 | 1744.7 | 138.7 | 1141.8 | 68.3 | 1125.2 | 70.7 | 1411.1 | 127.9 | 0.548 | 196 | 1139.0 | 31.7 | 1180.6 | 95.9 | 1084.2 | 50.5 | 1256.0 | 50.5 | 910.2 | 88.2 | 0.447 |
| **Denmark** | | | | | | | | | | | | | | | | | | | | | | | | |
| Copenhagen | 1356 | 1152.0 | 12.7 | | | 1162.6 | 19.8 | 1143.5 | 16.6 | 1136.0 | 107.9 | 0.158 | 1484 | 1009.3 | 11.6 | | | 1092.5 | 18.6 | 949.4 | 14.8 | 954.9 | 91.0 | 0.355 |
| Aarhus | 567 | 1220.8 | 19.6 | | | 1263.9 | 26.7 | 1176.7 | 28.6 | 1030.8 | 227.0 | 0.088 | 510 | 1109.5 | 19.7 | | | 1153.2 | 26.4 | 1058.7 | 29.8 | 929.4 | 152.6 | 0.057 |
| **Sweden** | | | | | | | | | | | | | | | | | | | | | | | | |
| Malmö | 1421 | 855.7 | 13.2 | 804.0 | 41.4 | 1007.6 | 34.3 | 886.8 | 19.4 | 752.8 | 18.0 | 0.019 | 1711 | 805.7 | 11.0 | 792.5 | 26.1 | 865.8 | 18.6 | 781.8 | 17.6 | 744.8 | 17.2 | 0.141 |
| Umeå | 1342 | 785.6 | 12.8 | | | 824.2 | 23.0 | 755.7 | 17.1 | 782.9 | 52.9 | 0.420 | 1567 | 704.0 | 11.2 | | | 733.9 | 26.4 | 643.7 | 16.9 | 694.7 | 50.8 | 0.212 |
| **Norway** | | | | | | | | | | | | | | | | | | | | | | | | |
| South and East | | | | | | | | | | | | | 1004 | 892.8 | 14.4 | 853.6 | 30.9 | 909.2 | 17.4 | 1000.1 | 38.4 | | | 0.088 |
| North and West | | | | | | | | | | | | | 793 | 894.5 | 16.0 | 914.7 | 33.6 | 912.9 | 19.2 | 906.6 | 48.8 | | | 0.195 |

[1] Adjusted for total energy intake, weight, and height and weighted by season and day of recall. [2] SEM: standard error of the mean. If a group comprised fewer than 20 persons, mean intake is not presented.

Table 4. Fully adjusted mean [1] daily intake of coffee and tea (g/day) by education level and sex in the EPIC calibration study population based on 24-H Dietary Recall across EPIC centres ordered from south to north.

| Country and Centre | Men | | | | | | Women | | | | | |
|---|---|---|---|---|---|---|---|---|---|---|---|---|
| | n | All | None/Primary | Tech/Professional/Secondary | University | p-Trend | n | All | None/Primary | Tech/Professional/Secondary | University | p-Trend |
| Greece | 1319 | 171.8 (13.2) | 176.6 (19.1) | 145.4 (27.8) | 176.2 (23.2) | 0.993 | 1361 | 170.6 (12.5) | 162.2 (17.0) | 158.6 (25.1) | 181.7 (25.3) | 0.425 |
| Spain | | | | | | | | | | | | |
| Granada | 208 | 388.2 (32.1) | 375.0 (42.5) | 383.1 (79.8) | 409.8 (60.5) | 0.191 | 294 | 426.9 (26.0) | 434.8 (27.7) | 334.8 (93.1) | 362.5 (118.2) | 0.506 |
| Murcia | 243 | 300.3 (29.9) | 266.0 (35.2) | 385.3 (93.3) | 374.9 (68.6) | 0.384 | 304 | 390.6 (25.5) | 384.0 (29.6) | 360.5 (81.1) | 430.6 (62.9) | 0.547 |
| Navarra | 442 | 307.7 (22.2) | 301.1 (26.1) | 315.5 (47.7) | 322.7 (82.1) | 0.122 | 270 | 493.0 (27.0) | 486.3 (29.5) | 443.6 (89.7) | 601.2 (97.3) | 0.502 |
| San Sebastian | 488 | 269.8 (21.3) | 251.8 (26.9) | 281.3 (38.6) | 339.1 (68.8) | 0.118 | 242 | 464.7 (28.5) | 478.6 (33.4) | 454.3 (63.0) | 361.6 (103.0) | 0.207 |
| Asturias | 384 | 372.8 (23.8) | 372.7 (29.6) | 331.2 (49.7) | 427.2 (63.4) | 0.617 | 319 | 534.7 (24.9) | 533.6 (27.7) | 607.5 (72.6) | 422.8 (87.1) | 0.594 |
| Italy | | | | | | | | | | | | |
| Ragusa | 167 | 221.2 (35.9) | 200.2 (52.9) | 241.0 (57.1) | 221.6 (90.3) | 0.648 | 137 | 201.6 (38.0) | 224.3 (54.1) | 173.8 (60.0) | 192.0 (110.5) | 0.566 |
| Naples | | | | | | | 403 | 297.5 (22.3) | 287.9 (34.1) | 305.2 (34.1) | 283.0 (54.6) | 0.865 |
| Florence | 269 | 269.4 (28.2) | 264.6 (45.3) | 278.1 (42.0) | 256.0 (68.2) | 0.747 | 780 | 328.2 (15.9) | 314.1 (22.9) | 330.6 (26.3) | 357.6 (39.4) | 0.088 |
| Turin | 676 | 260.2 (17.9) | 242.8 (28.6) | 270.0 (24.6) | 270.6 (58.3) | 0.322 | 392 | 312.4 (22.4) | 299.7 (29.5) | 323.4 (39.5) | 338.4 (67.4) | 0.082 |
| Varese | 327 | 392.0 (25.6) | 422.2 (36.8) | 370.3 (37.0) | 279.4 (120.9) | 0.100 | 794 | 404.1 (15.9) | 408.9 (19.5) | 384.4 (30.3) | 407.6 (57.4) | 0.969 |
| France | | | | | | | | | | | | |
| South coast | | | | | | | 595 | 565.4 (18.2) | 521.6 (49.0) | 537.6 (25.1) | 624.1 (30.6) | 0.241 |
| South | | | | | | | 1358 | 649.5 (12.2) | 549.0 (36.4) | 626.7 (16.7) | 711.7 (19.7) | 0.016 |
| Northeast | | | | | | | 1984 | 658.6 (10.1) | 574.0 (28.0) | 652.0 (14.4) | 694.4 (15.9) | 0.108 |
| Northwest | | | | | | | 615 | 722.0 (17.9) | 616.7 (48.3) | 730.0 (23.6) | 755.5 (32.8) | 0.223 |
| Germany | | | | | | | | | | | | |
| Heidelberg | 1031 | 897.6 (14.5) | 854.8 (24.5) | 855.8 (24.2) | 995.5 (26.2) | 0.330 | 1085 | 970.1 (13.6) | 948.9 (26.1) | 996.4 (19.0) | 949.3 (28.1) | 0.995 |
| Potsdam | 1233 | 844.0 (13.2) | 829.3 (29.1) | 811.5 (23.4) | 871.3 (18.9) | 0.521 | 1060 | 816.5 (13.7) | 816.7 (26.8) | 835.4 (19.4) | 780.4 (27.3) | 0.550 |
| The Netherlands | | | | | | | | | | | | |
| Bilthoven | 1017 | 962.1 (15.0) | 1031.4 (39.7) | 928.9 (19.3) | 1002.6 (27.9) | 0.824 | 1071 | 951.3 (13.8) | 894.5 (35.8) | 930.9 (17.2) | 1058.2 (28.8) | 0.198 |
| Utrecht | | | | | | | 1869 | 1050.2 (10.4) | 1030.6 (20.9) | 1036.0 (13.2) | 1138.0 (26.6) | 0.305 |
| United Kingdom | | | | | | | | | | | | |
| General population | 335 | 1470.9 (25.0) | 1640.0 (63.0) | 1500.2 (33.3) | 1321.8 (47.0) | 0.045 | 448 | 1312.3 (20.7) | 1406.6 (43.8) | 1319.9 (27.5) | 1195.9 (44.2) | 0.065 |
| Health-conscious | 84 | 1299.3 (49.6) | | 1119.2 (85.9) | 1392.6 (60.2) | | 164 | 1186.3 (34.7) | | 1250.0 (53.2) | 1143.9 (45.4) | |
| Denmark | | | | | | | | | | | | |
| Copenhagen | 1355 | 1153.2 (12.7) | 1155.3 (23.2) | 1135.5 (20.0) | 1176.4 (22.5) | 0.656 | 1484 | 1009.5 (11.6) | 938.2 (22.3) | 1020.9 (14.8) | 1106.8 (32.3) | 0.007 |
| Aarhus | 567 | 1221.6 (19.5) | 1279.7 (34.3) | 1176.8 (29.5) | 1227.7 (38.5) | 0.663 | 510 | 1109.9 (19.6) | 1152.7 (36.0) | 1079.0 (24.9) | 1195.7 (65.8) | 0.763 |
| Sweden | | | | | | | | | | | | |
| Malmö | 1418 | 856.9 (13.1) | 844.5 (18.9) | 856.4 (22.0) | 886.3 (25.9) | 0.155 | 1708 | 805.8 (11.0) | 780.7 (17.3) | 809.5 (17.5) | 836.4 (22.8) | 0.012 |
| Umeå | 1338 | 787.5 (12.8) | 785.2 (21.0) | 761.3 (19.5) | 847.8 (27.3) | 0.506 | 1560 | 704.9 (11.2) | 656.8 (21.2) | 703.9 (16.7) | 756.5 (21.4) | 0.020 |
| Norway | | | | | | | | | | | | |
| South and East | | | | | | | 1004 | 893.7 (14.3) | 900.9 (34.3) | 890.0 (17.3) | 932.2 (34.8) | 0.494 |
| North and West | | | | | | | 793 | 895.5 (16.0) | 942.4 (33.8) | 886.9 (19.5) | 894.0 (44.6) | 0.408 |

[1] Adjusted for age, total energy intake, weight, and height and weighted by season and day of recall. If a group comprised fewer than 20 persons, mean intake is not presented.

## 3.5. Lifestyle Factors

Lifestyle factors such as smoking ($p < 0.001$ for men and women) and physical activity ($p < 0.01$ for men and $p = 0.03$ for women) were both associated with coffee and tea consumption. These two factors were still significant when considering coffee and tea separately in both men and women. Whilst there was a clear pattern for smoking, where current smokers drank more coffee and tea compared to "never" smokers, a similarly consistent pattern was not found for physical activity (see Supplemental Materials, Tables S2 and S3). Nevertheless, significant linear trends were found among men in Murcia (Spain, $p = 0.02$), Bilthoven (The Netherlands, $p = 0.04$), and Copenhagen (Denmark, $p = 0.04$), where active men tended to drink ~100 g/day less coffee and tea combined compared to inactive men. The opposite was observed for men from the UK general population ($p < 0.05$). Similar patterns were observed in women in these very same centres, although respective linear trends were statistically nonsignificant (all $p > 0.13$).

The overall association between BMI and coffee and tea consumption was not significant among women ($p = 0.06$), but was significant among men ($p < 0.001$), although with no clear pattern except for men from Potsdam (Germany), where normal-weight men tended to drink significantly more coffee and tea compared to obese men (Table S4).

## 3.6. Contribution to Energy and Micronutrients

The contribution of coffee and tea along with their added ingredients (i.e., milk, sugar, honey, etc.) to energy, sugar, calcium, magnesium, and phosphorus intakes was the lowest in Norway. The contribution of coffee and tea to energy intake ranged from 1.2% in the south and east of Norway to 8.2% in Asturias (Spain) (Table 5). The contribution to sugar intake ranged from 2.5% in the north and west of Norway to 23% in Varese (Italy). Coffee and tea contributed to more than one-fifth of sugar intake in five centres, all of them belonging to the southern centres (Granada, Navarra, Asturias, Naples, and Varese). The contribution of coffee and tea to calcium intake ranged from 3.3% in the north and west of Norway to 33% in Asturias (Spain). As for sugar, in Spain and in most Italian centres, coffee and tea contributed to more than one-fifth of calcium intake, reaching one-fourth and even one-third in some centres. The contribution of coffee and tea to magnesium intake ranged from 8.7% in Greece to 35% in France. Compared to other countries, in France, this contribution was higher and around 30%. The contribution of coffee and tea to phosphorus intake ranged from 1.6% in Norway to 19% in Murcia (Spain).

**Table 5.** Total mean intake of energy and selected nutrients, amount of energy and selected nutrients from coffee and tea, and percentage contribution of coffee and tea to the total mean intake of energy and selected nutrients in the EPIC calibration study population based on 24-H Dietary Recall, by center ordered from south to north.

| Country and Centre | Total Energy Intake (kcal)[1] | | | Sugar (g)[1] | | | Calcium (mg)[1] | | | Magnesium (mg)[1] | | | Phosphorus (mg)[1] | | |
|---|---|---|---|---|---|---|---|---|---|---|---|---|---|---|---|
| | Total Mean Intake (s.e[3]) | From CT[2] (s.e[3]) | % | Total Mean Intake (s.e[3]) | From CT[2] (s.e[3]) | % | Total Mean Intake (s.e[3]) | From CT[2] (s.e[3]) | % | Total Mean Intake (s.e[3]) | From CT[2] (s.e[3]) | % | Total Mean Intake (s.e[3]) | From CT[2] (s.e[3]) | % |
| Greece | 1939.2 (14.0) | 59.6 (1.7) | 3.3 | 79.0 (1.0) | 8.4 (0.3) | 11.8 | 986.1 (9.0) | 80.2 (2.7) | 11.4 | 318.1 (2.3) | 23.5 (0.9) | 8.7 | 1789.4 (10.9) | 68.1 (2.1) | 5.3 |
| Spain | | | | | | | | | | | | | | | |
| Granada | 2142.0 (31.3) | 153.1 (3.8) | 7.9 | 102.6 (2.3) | 19.9 (0.7) | 21.5 | 1027.5 (20.3) | 250.1 (6.1) | 28.1 | 369.6 (5.2) | 39.7 (1.9) | 11.7 | 1402.8 (24.4) | 191.0 (4.8) | 15.6 |
| Murcia | 2328.4 (30.3) | 133.3 (3.7) | 6.5 | 117.0 (2.2) | 18.7 (0.6) | 18.4 | 1011.4 (19.6) | 238.1 (5.9) | 27.8 | 403.9 (5.1) | 40.5 (1.9) | 11.3 | 1456.4 (23.7) | 183.0 (4.6) | 18.9 |
| Navarra | 2294.0 (26.6) | 140.0 (3.3) | 6.7 | 96.2 (1.9) | 18.4 (0.6) | 20.9 | 908.3 (17.2) | 254.2 (5.2) | 33.0 | 360.3 (4.5) | 41.9 (1.6) | 13.0 | 1447.9 (20.7) | 193.4 (4.1) | 14.6 |
| San Sebastian | 2456.0 (26.3) | 138.0 (3.2) | 6.1 | 110.3 (1.9) | 18.8 (0.6) | 18.4 | 976.8 (17.0) | 214.9 (5.1) | 25.3 | 411.1 (4.4) | 45.6 (1.6) | 12.6 | 1707.1 (20.5) | 179.2 (4.0) | 11.6 |
| Asturias | 2292.6 (26.6) | 170.3 (3.3) | 8.2 | 114.1 (1.9) | 22.3 (0.6) | 21.6 | 1040.4 (17.2) | 294.3 (5.2) | 33.3 | 393.6 (4.5) | 47.0 (1.6) | 13.0 | 1659.2 (20.8) | 224.8 (4.1) | 15.1 |
| Italy | | | | | | | | | | | | | | | |
| Ragusa | 2284.6 (40.6) | 81.1 (5.0) | 3.7 | 93.8 (3.0) | 15.1 (0.9) | 16.2 | 752.0 (26.3) | 76.5 (7.9) | 12.2 | 370.1 (6.8) | 30.0 (2.5) | 9.2 | 1358.7 (31.7) | 75.1 (6.2) | 6.2 |
| Naples | 2214.5 (35.6) | 99.1 (4.4) | 5 | 91.6 (2.6) | 17.7 (0.8) | 22.0 | 852.0 (23.0) | 137.1 (6.9) | 23.8 | 316.1 (6.0) | 43.0 (2.2) | 16.2 | 1394.2 (27.7) | 122.9 (5.4) | 12.1 |
| Florence | 2183.1 (21.9) | 88.9 (2.7) | 4.4 | 89.6 (1.6) | 12.8 (0.5) | 15.6 | 798.1 (14.1) | 134.2 (4.3) | 23.8 | 328.2 (3.7) | 38.5 (1.4) | 13.4 | 1374.1 (17.0) | 122.0 (3.3) | 10.8 |
| Turin | 2202.3 (21.7) | 99.5 (2.7) | 4.7 | 103.0 (1.6) | 17.7 (0.5) | 17.4 | 866.6 (14.0) | 95.7 (4.2) | 15.4 | 335.0 (3.6) | 34.7 (1.3) | 11.6 | 1349.4 (16.9) | 92.3 (3.3) | 7.9 |
| Varese | 2274.7 (21.2) | 138.6 (2.6) | 6.6 | 104.3 (1.5) | 22.3 (0.4) | 23.2 | 877.6 (13.7) | 152.2 (4.1) | 24.3 | 322.7 (3.6) | 44.0 (1.3) | 16.0 | 1413.2 (16.5) | 138.8 (3.2) | 11.8 |
| France | | | | | | | | | | | | | | | |
| South Coast | 2316.0 (28.8) | 78.7 (3.5) | 3.7 | 99.9 (2.1) | 10.3 (0.6) | 10.9 | 1037.1 (18.6) | 113.5 (5.6) | 13.7 | 405.6 (4.8) | 96.8 (1.8) | 28.6 | 1500.2 (22.4) | 92.7 (4.4) | 7.4 |
| South | 2271.3 (19.4) | 74.3 (2.4) | 3.3 | 103.1 (1.4) | 9.8 (0.4) | 9.6 | 956.5 (12.5) | 106.4 (3.8) | 13.9 | 395.5 (3.2) | 101.3 (1.2) | 30.3 | 1450.6 (15.1) | 88.2 (3.0) | 7 |
| Northeast | 2338.5 (16.3) | 70.3 (2.0) | 3.1 | 104.9 (1.2) | 9.0 (0.3) | 8.6 | 969.8 (10.5) | 97.8 (3.2) | 12.0 | 414.2 (2.7) | 112.9 (1.0) | 32.7 | 1470.8 (12.7) | 81.2 (2.5) | 6.1 |
| Northwest | 2297.5 (28.5) | 69.4 (3.5) | 3.0 | 100.8 (2.1) | 8.3 (0.6) | 7.9 | 917.4 (18.5) | 87.0 (5.6) | 11.2 | 439.9 (4.8) | 135.1 (1.8) | 35.4 | 1461.5 (22.3) | 76.1 (4.4) | 5.6 |
| Germany | | | | | | | | | | | | | | | |
| Heidelberg | 2154.1 (15.5) | 79.7 (1.9) | 3.9 | 102.2 (1.1) | 10.1 (0.3) | 10.1 | 1005.8 (10.1) | 104.9 (3.0) | 13.0 | 430.1 (2.6) | 48.9 (1.0) | 12.9 | 1332.6 (12.1) | 74.5 (2.4) | 6.4 |
| Potsdam | 2186.7 (14.8) | 54.4 (1.8) | 2.7 | 116.0 (1.1) | 6.1 (0.3) | 5.8 | 858.2 (9.6) | 47.2 (2.9) | 7.2 | 392.9 (2.5) | 40.0 (0.9) | 11.4 | 1275.3 (11.6) | 33.2 (2.3) | 3.1 |
| The Netherlands | | | | | | | | | | | | | | | |
| Bilthoven | 2224.9 (15.8) | 99.4 (1.9) | 4.4 | 119.2 (1.2) | 18.3 (0.3) | 13.6 | 968.0 (10.3) | 111.4 (3.1) | 15.0 | 353.0 (2.7) | 47.0 (1.0) | 15.3 | 1562.0 (12.4) | 71.0 (2.4) | 5.2 |
| Utrecht | 2254.6 (17.4) | 70.2 (2.1) | 3.1 | 120.4 (1.3) | 10.7 (0.4) | 8.8 | 1124.2 (11.3) | 123.8 (3.4) | 13.9 | 363.4 (2.9) | 44.2 (1.1) | 13.4 | 1644.3 (13.6) | 74.8 (2.7) | 4.8 |
| United Kingdom | | | | | | | | | | | | | | | |
| General population | 2039.6 (22.7) | 103.7 (2.8) | 5.3 | 113.4 (1.7) | 13.8 (0.5) | 12.1 | 987.7 (14.7) | 163.4 (4.4) | 19.7 | 321.0 (3.8) | 51.3 (1.4) | 17.9 | 1407.5 (17.7) | 157.2 (0.5) | 12.5 |
| Health-conscious | 2070.1 (40.3) | 68.8 (5.0) | 3.4 | 117.3 (2.9) | 7.2 (0.9) | 6.3 | 887.0 (26.1) | 104.0 (7.9) | 14.7 | 396.3 (6.8) | 39.1 (2.5) | 11.2 | 1314.8 (31.4) | 60.5 (6.2) | 5.3 |
| Denmark | | | | | | | | | | | | | | | |
| Copenhagen | 2235.4 (13.7) | 49.3 (1.7) | 2.3 | 99.5 (1.0) | 6.0 (0.3) | 5.0 | 960.0 (8.8) | 43.7 (2.7) | 6.1 | 365.1 (2.3) | 49.9 (0.8) | 15.3 | 1555.3 (10.6) | 38.2 (2.1) | 2.8 |
| Aarhus | 2383.2 (21.7) | 42.9 (2.7) | 2.0 | 105.5 (1.6) | 4.0 (0.5) | 3.6 | 1050.4 (14.0) | 38.3 (4.2) | 4.8 | 384.1 (3.6) | 56.7 (1.3) | 16.0 | 1632.9 (16.9) | 34.5 (3.3) | 2.4 |
| Sweden | | | | | | | | | | | | | | | |
| Malmö | 2039.6 (13.2) | 50.2 (1.6) | 2.6 | 96.0 (1.0) | 6.2 (0.3) | 6.4 | 869.4 (8.5) | 45.4 (2.6) | 7.4 | 304.8 (2.2) | 39.3 (0.8) | 14.7 | 1300.2 (10.3) | 33.2 (2.0) | 2.9 |
| Umeå | 2131.0 (13.3) | 41.0 (1.6) | 2.0 | 102.3 (1.0) | 5.6 (0.3) | 5.5 | 989.9 (8.6) | 39.8 (2.6) | 5.3 | 323.8 (2.2) | 36.2 (0.8) | 12.5 | 1417.0 (10.4) | 30.0 (2.0) | 2.4 |
| Norway | | | | | | | | | | | | | | | |
| South and East | 2092.8 (23.2) | 29.2 (2.9) | 1.2 | 99.3 (1.7) | 2.9 (0.5) | 2.7 | 814.7 (15.0) | 33.9 (4.5) | 5.5 | 363.3 (3.9) | 40.6 (1.4) | 12.7 | 1482.3 (18.1) | 26.4 (3.5) | 1.6 |
| North and West | 2075.5 (25.8) | 29.0 (3.2) | 1.3 | 100.4 (1.9) | 2.8 (0.5) | 2.5 | 815.2 (16.7) | 29.4 (5.8) | 3.3 | 364.0 (4.3) | 41.3 (1.8) | 9.4 | 1487.4 (20.1) | 22.0 (4.5) | 1.6 |

[1] Adjusted for sex, age, height, and weight and weighted by season and day of recall. [2] Coffee and tea. [3] s.e.: standard error.

## 4. Discussion

This is one of the largest population-based studies comparing coffee and tea consumption using a common, detailed, and standardized 24-h dietary recall method across 10 European countries participating in the EPIC study.

The amount of coffee and tea consumed varied widely across countries/centres and according to the type of beverages consumed. Average tea consumption was highest in the UK and lowest in Greece and Spain, while coffee consumption was highest in Denmark and lowest in Greece.

Apart from Greece, the majority of coffee and tea intakes from the previous day was consumed at home. Most coffee drinkers consumed caffeinated coffee.

These results are consistent with studies that used long-term dietary assessment methods in the EPIC cohort [36–38]. For coffee, the observed geographical differences might be due to different consumption habits. For instance, in countries such as Denmark, people tend to drink more diluted coffee in larger amounts, whilst in other countries such as Greece or Italy, people tend to drink stronger coffee in smaller amounts (e.g., Turkish coffee or ristretto coffee). Indeed, in Italy, the mean cup of coffee weighed 55 g, whereas in Denmark, the mean cup of coffee weighed 182 g.

Coffee and tea consumption also varied to some extent by sex, age, and education, with the direction of the associations being different across centres. For example, coffee and tea consumption combined decreased with level of education in the UK general population by about 200–300 g/day, comparing the population subgroup with primary education to that with a university degree (Table 4); whereas an opposite trend was observed in the two centers in Sweden (Malmö, Umea) and in Copenhagen (Denmark). In the remaining countries/centers, differences across level of education were less pronounced, which suggests that coffee and tea consumption is driven by country-specific dietary habits rather than characteristics at the individual level. Other studies that have investigated relationships between sociodemographic factors and coffee consumption also reported mixed results. For instance, the National Health and Nutrition Examination Survey (NHANES) 2003–2012 in the US observed that the mean usual intakes of coffee were higher in men than in women, in older versus younger individuals, and in lower- versus higher-educated individuals [39]. Different results were reported from the Japan Collaborative Cohort Study for Evaluation of Cancer Risk (JACC study), in which both men and women with high coffee consumption were younger and better educated [40]. A cross-sectional population-based survey conducted in Poland reported that higher coffee consumers were more likely to be women, younger, and with a medium–higher education [41]. The same study also reported that higher tea consumers were more likely to be women. These mixed results emphasize the fact that coffee and tea consumption differs with the population under investigation and explain why no homogeneity was found across the different EPIC centres.

In the present study, current smokers compared to former/"never" smokers tended to drink more coffee and tea. Other studies conducted in the US, but focusing on coffee only, reported that lifestyle factors such as smoking were related to coffee consumption. Also, in the National Institute of Health-American Association of Retired Persons Diet and Health Study, coffee drinkers where more likely to smoke [2]. A more recent study, also conducted in the US but using the NHANES 2003–2012 data, reported that the mean intake of coffee was higher among smokers versus "never" smokers [39]. The same pattern was also observed in Japan [40], Singapore [42], and Brazil [43].

Overall, BMI was associated to coffee and tea consumption among men, but with no clear patterns, and was not associated with coffee and tea consumption among women. This result, albeit different from what is generally reported in the literature [37,41,44,45], was not unexpected, considering the cross-sectional design and the use of a single 24-h dietary recall, and that the development of overweightness or obesity is a life course process.

The contribution of coffee and tea to sugar and calcium intakes was higher in Italy and Spain compared to other countries, reflecting different consumption habits and suggesting that in Southern European countries, people tended to add (more) sugar and milk, which both contribute to total sugar intake, to their coffee and tea. In those two countries, coffee and tea, with their added ingredients, contributed to about 20% to the overall sugar intakes, whilst in Norway, coffee and tea contributed to less than 3%. Given these results, it is recommended to consider both coffee and tea as potential major sources of sugar intake (free/added sugars) in dietary monitoring and public health surveillance. There are health concerns regarding added/free sugar consumption, and compared to carbonated soft drinks, coffee and tea with their added ingredients receive less attention. In a more positive sense, this also applies to coffee and tea as a source of calcium, where the milk added is rarely considered as a source of calcium.

The present study was based on a single 24-HDR and therefore did not reflect usual intakes of individuals. Hence, the interpretation of nonconsumers should be performed with caution due to the day-to-day variability. Indeed, the prevalence of tea or coffee nonconsumers was higher compared to the same prevalence measured with the EPIC country-specific Food Frequency Questionnaire assessing food intakes over the past 12 months [46]. However, considering the large sample size, except in Ragusa, and the fact that individuals usually drink such beverages on a daily basis, the population mean consumption levels should have been reasonably well captured. Indeed, when comparing the results of the present calibration study to the EPIC long-term consumption data, similar patterns were found [36–38]. Moreover, the standard error of the mean should be interpreted with caution because it is most likely overestimated due the day-to-day variation in consumption levels.

Data for the current study were collected in the mid to late 1990s, and coffee and tea intakes may have changed over time. Compared to more recent surveys conducted between 2003 and 2011 in Germany, Denmark, Spain, the UK, Italy, The Netherlands, and Sweden, where a similar dietary assessment method was used, i.e., 24-h dietary recalls, coffee intake in our study was lower, whilst tea intake was higher [47]. Such comparisons indicate that our study may serve as a common benchmark to evaluate trends in coffee and tea consumption over time in these countries.

However, some caution is warranted because the EPIC study populations were not necessarily representative of the corresponding national populations, and in several countries, they tended to be more "health-conscious".

Although the information about coffee was detailed, as individuals were asked to specify whether coffee was with caffeine or decaffeinated, the EPIC Nutrient DataBase does not contain information on caffeine content. Hence, for instance, one cup of coffee in Italy—where a 60-mL cup of espresso contains approximatively 80 mg of caffeine [48]—cannot be strictly compared with one cup of coffee in Denmark, where a 200-mL cup of filter coffee contains approximately 90 mg of caffeine [48]. However, caffeine intake across Europe, as reported in the European Food Safety Authority's fact sheets on caffeine [48], roughly confirm our findings based on consumed quantity of the beverages. For example, the estimated caffeine intakes in Greece (~30 mg/day) and Spain (67 mg/day) were lower as compared to Denmark (~320 mg/day) or Germany (~238 mg/day) [48]. The same reasoning applies for tea, as the different types of tea (green, white, black) differ in caffeine content [49]. The assessment of caffeine intake is of importance and therefore there is a need for collecting more detailed data, to add caffeine content in food composition tables or to use biomarkers, such as the dimethylxanthines theophylline or paraxanthine, in order to enable a more objective assessment of caffeine intake [49]. Additionally, the brewing method might also be considered when collecting data because of the consequences on the content of diterpenes [50] that have an anticarcinogenic activity [6].

The health benefits of coffee and tea consumption are still controversial [15,17,19]. Therefore, the use of a standardized method such as the one used in the present study, but with repeated assessments, to collect comparable dietary data across countries is of interest as it might help to investigate better associations between coffee and tea consumption and health outcomes. Moreover, such a method provides data that is not only geographically comparable, but is also comparable over time.

## 5. Conclusions

Levels of coffee and tea intake, and their contribution to energy and sugar intake, differed greatly among European adults. Variation in consumption was mostly driven by geographical region and to a lesser extent by individuals' characteristics.

**Supplementary Materials:** The following are available online at http://www.mdpi.com/2072-6643/10/6/725/s1. Figure S1: Percentages of consumers and nonconsumers of coffee and tea among men the day of the 24-h recall by centre; Figure S2: Percentages of consumers and nonconsumers of coffee and tea among women the day of the 24-h recall by centre; Figure S3: Percentage of consumers of coffee with caffeine and decaffeinated coffee among men who consumed coffee the day of the 24-h recall by centre; Figure S4: Percentage of consumers of coffee with caffeine and decaffeinated coffee among women who consumed coffee the day of the 24-h recall by centre; Table S1: Mean daily intake of coffee and tea (g/day) by type and country in the EPIC calibration study population based on 24-HDR among men and women; Table S2: Fully adjusted mean daily intake of coffee and tea (g/day) by smoking status and sex in the EPIC calibration study population based on 24-HDR across EPIC centres ordered from south to north; Table S3: Fully adjusted mean daily intake of coffee and tea (g/day) by physical activity level and sex in the EPIC calibration study population based on 24-HDR across EPIC centres ordered from south to north; Table S4: Fully adjusted mean daily intake of coffee and tea (g/day) by BMI group and sex in the EPIC calibration study population based on 24-HDR across EPIC centres ordered from south to north.

**Author Contributions:** N.S., H.F., and I.H. initiated the study; N.S. and G.N. took responsibility for dietary intake data and their interpretation; A.M. (Aurélie Moskal) and E.L. performed the statistical analyses; H.F. supervised the statistical analyses; E.L., A.M. (Aurélie Moskal), A.M. (Amy Mullee), G.N., I.H., M.J.G., and H.F. interpreted the results; E.L. wrote the manuscript taking into account comments from all co-authors; H.F., A.M. (Aurélie Moskal), I.H., M.J.G. and A.M. (Amy Mullee) contributed to the drafting of the manuscript; H.F. had primary responsibility for the final content. All other coauthors were local EPIC investigators involved in the collection of dietary data and other data. All authors critically revised the manuscript for intellectual content and approved the final version of the manuscript.

**Acknowledgments:** The EPIC study was supported by grants from the 'Europe Against Cancer' programme of the European Commission (SANCO); Ligue contre le Cancer (France); Société 3M (France); Mutuelle Générale de l'Education Nationale Institut National de la Santé et de la Recherche Médicale (INSERM); German Cancer Aid; German Cancer Research Center; German Federal Ministry of Education and Research; Danish Cancer Society; Health Research Fund (FIS) of the Spanish Ministry of Health; the participating regional governments and institutions of Spain; Cancer Research UK; Medical Research Council, UK; the Stroke Association, UK; British Heart Foundation; Department of Health, UK; Food Standards Agency, UK; the Wellcome Trust, UK; Greek Ministry of Health; Greek Ministry of Education; a fellowship honouring Vasilios and Nafsika Tricha (Greece); the Hellenic Health Foundation; Italian Association for Research on Cancer; Dutch Ministry of Health, Welfare and Sports; Dutch Ministry of Health; Dutch Prevention Funds; LK Research Funds; Dutch ZON (Zorg Onderzoek Nederland); World Cancer Research Fund (WCRF); Swedish Cancer Society; Swedish Scientific Council; Regional Government of Skane, Sweden; Catalan Institute of Oncology, Barcelona, Spain; Public Health Institute, Navarra, Spain; Andalusian School of Public Health, Granada, Spain; Public Health Department of Gipuzkoa, Health Department of the Basque Country, Donostia-San Sebastian, Spain; Murcia Health Council, Murcia, Spain; Health and Health Services Council, Principality of Asturias, Spain. This study was also supported by contracts from the US NCI (N02-PC-25023) and the EC (Contract No. SPC 2002332 for the EPIC and EuroFIR NoE Contract No. 513944). In addition, we wish to thank all study participants for their cooperation and all interviewers who participated in the fieldwork studies in each EPIC centre. The contribution of A.M. (Amy Mullee), to the work reported in this paper was undertaken during the tenure of an IARC-Ireland Postdoctoral Fellowship from the International Agency for Research on Cancer, funded by the Irish Cancer Society. We also acknowledge the Northern Sweden Diet Database and the funds supporting it, including the Swedish Research Council (VR), the Swedish Research Council for Health, Working Life and Welfare (FORTE), and the Västerbotten County Council.

**Conflicts of Interest:** The authors declare no conflict of interest.

## References

1. Graham, H.N. Green tea composition, consumption, and polyphenol chemistry. *Prev. Med.* **1992**, *21*, 334–350. [CrossRef]
2. Freedman, N.D.; Park, Y.; Abnet, C.C.; Hollenbeck, A.R.; Sinha, R. Association of Coffee Drinking with Total and Cause-Specific Mortality. *New Engl. J. Med.* **2012**, *366*, 1891–1904. [CrossRef] [PubMed]
3. World Cancer Research Fund/American Institute for Cancer Research. Diet, Nutrition, Physical Activity, and Cancer: A Global Perspective. Continuous Update Project Expert Report 2018. Available online: Dietandcancerreport.org (accessed on 29 May 2018).

4.  Lin, J.K. Mechanisms of cancer chemoprevention by tea and tea polyphenols. In *Tea and Tea Products Chemistry and Health-Promoting Properties*; Ho, C.T., Lin, J.K., Shahidi, F., Eds.; CRC Press: Boca Raton, FL, USA, 2009; pp. 161–176.

5.  Yang, C.S.; Wang, H.; Li, G.X.; Yang, Z.; Guan, F.; Jin, H. Cancer prevention by tea: Evidence from laboratory studies. *Pharmacol. Res.* **2011**, *64*, 113–122. [CrossRef] [PubMed]

6.  Cavin, C.; Holzhaeuser, D.; Scharf, G.; Constable, A.; Huber, W.W.; Schilter, B. Cafestol and kahweol, two coffee specific diterpenes with anticarcinogenic activity. *Food Chem. Toxicol.* **2002**, *40*, 1155–1163. [CrossRef]

7.  Norat, T.; Aune, D.; Navarro, D.; Abar, L. *The Associations between Food, Nutrition and Physical Activity and the Risk of Liver Cancer*; World Cancer Research Fund International: London, UK, 2015.

8.  Cao, S.; Liu, L.; Yin, X.; Wang, Y.; Liu, J.; Lu, Z. Coffee consumption and risk of prostate cancer: A meta-analysis of prospective cohort studies. *Carcinogenesis* **2014**, *35*, 256–261. [CrossRef] [PubMed]

9.  Gunter, M.J.; Murphy, N.; Cross, A.J.; Dossus, L.; Dartois, L.; Fagherazzi, G.; Kaaks, R.; Kühn, T.; Boeing, H.; Aleksandrova, K.; et al. Coffee drinking and mortality in 10 european countries: A multinational cohort study. *Ann. Intern. Med.* **2017**, *167*, 236–247. [CrossRef] [PubMed]

10. Lu, Y.; Zhai, L.; Zeng, J.; Peng, Q.; Wang, J.; Deng, Y.; Xie, L.; Mo, C.; Yang, S.; Li, S.; et al. Coffee consumption and prostate cancer risk: An updated meta-analysis. *Cancer Causes Control* **2014**, *25*, 591–604. [CrossRef] [PubMed]

11. Sang, L.X.; Chang, B.; Li, X.H.; Jiang, M. Consumption of coffee associated with reduced risk of liver cancer: A meta-analysis. *BMC Gastroenterol.* **2013**, *13*, 34. [CrossRef] [PubMed]

12. Tang, J.; Zheng, J.-S.; Fang, L.; Jin, Y.; Cai, W.; Li, D. Tea consumption and mortality of all cancers, CVD and all causes: A meta-analysis of eighteen prospective cohort studies. *Br. J. Nutr.* **2015**, *114*, 673–683. [PubMed]

13. Wang, L.; Zhang, X.; Liu, J.; Shen, L.; Li, Z. Tea consumption and lung cancer risk: A meta-analysis of case-control and cohort studies. *Nutrition* **2014**, *30*, 1122–1127. [CrossRef] [PubMed]

14. Je, Y.; Liu, W.; Giovannucci, E. Coffee consumption and risk of colorectal cancer: A systematic review and meta-analysis of prospective cohort studies. *Int. J. Cancer* **2009**, *124*, 1662–1668. [CrossRef] [PubMed]

15. Zheng, P.; Zheng, H.M.; Deng, X.M.; Zhang, Y.D. Green tea consumption and risk of esophageal cancer: A meta-analysis of epidemiologic studies. *BMC Gastroenterol.* **2012**, *12*, 165. [CrossRef] [PubMed]

16. Hu, Z.H.; Lin, Y.W.; Xu, X.; Chen, H.; Mao, Y.Q.; Wu, J.; Xu, X.L.; Zhu, Y.; Li, S.Q.; Zheng, X.Y.; et al. No association between tea consumption and risk of renal cell carcinoma: A meta-analysis of epidemiological studies. *Asian Pac. J. Cancer Prev.* **2013**, *14*, 1691–1695. [CrossRef] [PubMed]

17. Li, X.J.; Ren, Z.J.; Qin, J.W.; Zhao, J.H.; Tang, J.H.; Ji, M.H.; Wu, J.Z. Coffee consumption and risk of breast cancer: An up-to-date meta-analysis. *PLoS ONE* **2013**, *8*, e52681. [CrossRef] [PubMed]

18. Sang, L.X.; Chang, B.; Li, X.H.; Jiang, M. Green tea consumption and risk of esophageal cancer: A meta-analysis of published epidemiological studies. *Nutr. Cancer* **2013**, *65*, 802–812. [CrossRef] [PubMed]

19. Crippa, A.; Discacciati, A.; Larsson, S.C.; Wolk, A.; Orsini, N. Coffee Consumption and Mortality from All Causes, Cardiovascular Disease, and Cancer: A Dose-Response Meta-Analysis. *Am. J. Epidemiol.* **2014**, *180*, 763–775. [CrossRef] [PubMed]

20. Loomis, D.; Guyton, K.Z.; Grosse, Y.; Lauby-Secretan, B.; El Ghissassi, F.; Bouvard, V.; Benbrahim-Tallaa, L.; Guha, N.; Mattock, H.; Straif, K.; et al. Carcinogenicity of drinking coffee, mate, and very hot beverages. *Lancet Oncol.* **2016**, *17*, 877–878. [CrossRef]

21. Bingham, S.A. Limitations of the various methods for collecting dietary intake data. *Ann. Nutr. Metab.* **1991**, *35*, 117–127. [CrossRef] [PubMed]

22. Riboli, E.; Hunt, K.J.; Slimani, N.; Ferrari, P.; Norat, T.; Fahey, M.; Charrondière, U.R.; Hémon, B.; Casagrande, C.; Vignat, J.; et al. European Prospective Investigation into Cancer and Nutrition (EPIC): Study populations and data collection. *Public Health Nutr.* **2002**, *5*, 1113–1124. [CrossRef] [PubMed]

23. Slimani, N.; Kaaks, R.; Ferrari, P.; Casagrande, C.; Clavel-Chapelon, F.; Lotze, G.; Kroke, A.; Trichopoulos, D.; Trichopoulou, A.; Lauria, C.; et al. European Prospective Investigation into Cancer and Nutrition (EPIC) calibration study: Rationale, design and population characteristics. *Public Health Nutr.* **2002**, *5*, 1125–1145. [CrossRef] [PubMed]

24. Riboli, E.; Kaaks, R. The EPIC Project: Rationale and study design. European Prospective Investigation into Cancer and Nutrition. *Int. J. Epidemiol.* **1997**, *26* (Suppl. 1), S6–S14. [CrossRef] [PubMed]

25. Bingham, S.; Riboli, E. Diet and cancer—the European Prospective Investigation into Cancer and Nutrition. *Nat. Rev. Cancer* **2004**, *4*, 206–215. [CrossRef] [PubMed]

26. Ferrari, P.; Kaaks, R.; Fahey, M.T.; Slimani, N.; Day, N.E.; Pera, G.; Boshuizen, H.C.; Roddam, A.; Boeing, H.; Nagel, G.; et al. Within- and between-cohort variation in measured macronutrient intakes, taking account of measurement errors, in the European Prospective Investigation into Cancer and Nutrition study. *Am. J. Epidemiol.* **2004**, *160*, 814–822. [CrossRef] [PubMed]

27. Ferrari, P.; Day, N.E.; Boshuizen, H.C.; Roddam, A.; Hoffmann, K.; Thiebaut, A.; Pera, G.; Overvad, K.; Lund, E.; Trichopoulou, A.; et al. The evaluation of the diet/disease relation in the EPIC study: Considerations for the calibration and the disease models. *Int. J. Epidemiol.* **2008**, *37*, 368–378. [CrossRef] [PubMed]

28. Kaaks, R.; Plummer, M.; Riboli, E.; Esteve, J.; van Staveren, W. Adjustment for bias due to errors in exposure assessments in multicenter cohort studies on diet and cancer: A calibration approach. *Am. J. Clin. Nutr.* **1994**, *59*, 245S–250S. [CrossRef] [PubMed]

29. Kaaks, R.; Riboli, E.; van Staveren, W. Calibration of dietary intake measurements in prospective cohort studies. *Am. J. Epidemiol.* **1995**, *142*, 548–556. [CrossRef] [PubMed]

30. Brustad, M.; Skeie, G.; Braaten, T.; Slimani, N.; Lund, E. Comparison of telephone vs face-to-face interviews in the assessment of dietary intake by the 24 h recall EPIC SOFT program–the Norwegian calibration study. *Eur. J. Clin. Nutr.* **2003**, *57*, 107–113. [CrossRef] [PubMed]

31. Slimani, N.; Deharveng, G.; Charrondiere, R.U.; van Kappel, A.L.; Ocke, M.C.; Welch, A.; Lagiou, A.; van Liere, M.; Agudo, A.; Pala, V.; et al. Structure of the standardized computerized 24-h diet recall interview used as reference method in the 22 centers participating in the EPIC project. European Prospective Investigation into Cancer and Nutrition. *Comput. Methods Program Biomed.* **1999**, *58*, 251–266. [CrossRef]

32. Slimani, N.; Ferrari, P.; Ocke, M.; Welch, A.; Boeing, H.; Liere, M.; Pala, V.; Amiano, P.; Lagiou, A.; Mattisson, I.; et al. Standardization of the 24-h diet recall calibration method used in the european prospective investigation into cancer and nutrition (EPIC): General concepts and preliminary results. *Eur. J. Clin. Nutr.* **2000**, *54*, 900–917. [CrossRef] [PubMed]

33. Slimani, N.; Deharveng, G.; Unwin, I.; Southgate, D.A.; Vignat, J.; Skeie, G.; Salvini, S.; Parpinel, M.; Møller, A.; Ireland, J.; et al. The EPIC nutrient database project (ENDB): A first attempt to standardize nutrient databases across the 10 European countries participating in the EPIC study. *Eur. J. Clin. Nutr.* **2007**, *61*, 1037–1056. [CrossRef] [PubMed]

34. Lahmann, P.H.; Friedenreich, C.; Schuit, A.J.; Salvini, S.; Allen, N.E.; Key, T.J.; Khaw, K.T.; Bingham, S.; Peeters, P.H.; Monninkhof, E.; et al. Physical activity and breast cancer risk: The European Prospective Investigation into Cancer and Nutrition. *Cancer Epidemiol. Biomark. Prev.* **2007**, *16*, 36–42. [CrossRef] [PubMed]

35. Haftenberger, M.; Schuit, A.J.; Tormo, M.J.; Boeing, H.; Wareham, N.; Bueno-de-Mesquita, H.B.; Kumle, M.; Hjartåker, A.; Chirlaque, M.D.; Ardanaz, E.; et al. Physical activity of subjects aged 50-64 years involved in the European Prospective Investigation into Cancer and Nutrition (EPIC). *Public Health Nutr.* **2002**, *5*, 1163–1176. [CrossRef] [PubMed]

36. Michaud, D.S.; Gallo, V.; Schlehofer, B.; Tjønneland, A.; Olsen, A.; Overvad, K.; Dahm, C.C.; Teucher, B.; Lukanova, A.; Boeing, H.; et al. Coffee and tea intake and risk of brain tumors in the European Prospective Investigation into Cancer and Nutrition (EPIC) cohort study. *Am. J. Clin. Nutr.* **2010**, *92*, 1145–1150. [CrossRef] [PubMed]

37. Zamora-Ros, R.; Luján-Barroso, L.; Bueno-de-Mesquita, H.B.; Dik, V.K.; Boeing, H.; Steffen, A.; Tjønneland, A.; Olsen, A.; Bech, B.H.; Overvad, K.; et al. Tea and coffee consumption and risk of esophageal cancer: The European prospective investigation into cancer and nutrition study. *Int. J. Cancer* **2014**, *135*, 1470–1479. [CrossRef] [PubMed]

38. Sanikini, H.; Dik, V.K.; Siersema, P.D.; Bhoo-Pathy, N.; Uiterwaal, C.S.P.M.; Peeters, P.H.M.; González, C.A.; Zamora-Ros, R.; Overvad, K.; Tjønneland, A.; et al. Total, caffeinated and decaffeinated coffee and tea intake and gastric cancer risk: Results from the EPIC cohort study. *Int. J. Cancer* **2015**, *136*, E720–E730. [CrossRef] [PubMed]

39. Loftfield, E.; Freedman, N.D.; Dodd, K.W.; Vogtmann, E.; Xiao, Q.; Sinha, R.; Graubard, B.I. Coffee Drinking Is Widespread in the United States, but Usual Intake Varies by Key Demographic and Lifestyle Factors. *J. Nutr.* **2016**, *146*, 1762–1768. [CrossRef] [PubMed]

40. Yamada, H.; Kawado, M.; Aoyama, N.; Hashimoto, S.; Suzuki, K.; Wakai, K.; Suzuki, S.; Watanabe, Y.; Tamakoshi, A.; The JACC Study Group. Coffee Consumption and Risk of Colorectal Cancer: The Japan Collaborative Cohort Study. *J. Epidemiol.* **2014**, *24*, 370–378. [CrossRef] [PubMed]

41. Grosso, G.; Stepaniak, U.; Micek, A.; Topor-Mądry, R.; Pikhart, H.; Szafraniec, K.; Pająk, A. Association of daily coffee and tea consumption and metabolic syndrome: Results from the Polish arm of the HAPIEE study. *Eur. J. Nutr.* **2015**, *54*, 1129–1137. [CrossRef] [PubMed]

42. Ainslie-Waldman, C.E.; Koh, W.-P.; Jin, A.; Yeoh, K.G.; Zhu, F.; Wang, R.; Yuan, J.M.; Butler, L.M. Coffee Intake and Gastric Cancer Risk: The Singapore Chinese Health Study. *Cancer Epidemiol. Biomarkers Prev.* **2014**, *23*, 638–647. [CrossRef] [PubMed]

43. De Oliveira, R.A.M.; Araújo, L.F.; de Figueiredo, R.C.; Goulart, A.C.; Schmidt, M.I.; Barreto, S.M.; Ribeiro, A.L.P. Coffee Consumption and Heart Rate Variability: The Brazilian Longitudinal Study of Adult Health (ELSA-Brasil) Cohort Study. *Nutrients* **2017**, *9*, 741. [CrossRef] [PubMed]

44. Kim, J.-H.; Park, Y.S. Light coffee consumption is protective against sarcopenia, but frequent coffee consumption is associated with obesity in Korean adults. *Nutr. Res.* **2017**, *41*, 97–102. [CrossRef] [PubMed]

45. Vernarelli, J.A.; Lambert, J.D. Tea consumption is inversely associated with weight status and other markers for Metabolic Syndrome in U.S. adults. *Eur. J. Nutr.* **2013**, *52*, 1039–1048. [CrossRef] [PubMed]

46. Caini, S.; Masala, G.; Saieva, C.; Kvaskoff, M.; Savoye, I.; Sacerdote, C.; Hemmingsson, O.; Hammer Bech, B.; Overvad, K.; Tjønneland, A.; et al. Coffee, tea and melanoma risk: Findings from the European Prospective Investigation into Cancer and Nutrition. *Int. J. Cancer* **2017**, *140*, 2246–2255. [CrossRef] [PubMed]

47. EFSA. The EFSA Comprehensive European Food Consumption Database 2011. Available online: http://www.efsa.europa.eu/en/food-consumption/comprehensive-database (accessed on 18 August 2017).

48. EFSA. Fact Sheets on Caffeine. 2015. Available online: http://www.efsa.europa.eu/en/corporate/pub/efsaexplainscaffeine150527 (accessed on 30 May 2018).

49. Lang, R.; Dieminger, N.; Beusch, A.; Lee, Y.-M.; Dunkel, A.; Suess, B.; Skurk, T.; Wahl, A.; Hauner, H.; Hofmann, T. Bioappearance and pharmacokinetics of bioactives upon coffee consumption. *Anal. Bioanal. Chem.* **2013**, *405*, 8487–8503. [CrossRef] [PubMed]

50. Urgert, R.; Katan, M.B. The cholesterol-raising factor from coffee beans. *J. R. Soc. Med.* **1996**, *89*, 618–623. [CrossRef] [PubMed]

*nutrients*

*Article*

# Impact of Alcohol and Coffee Intake on the Risk of Advanced Liver Fibrosis: A Longitudinal Analysis in HIV-HCV Coinfected Patients (ANRS CO-13 HEPAVIH Cohort)

Issifou Yaya [1,2,*], Fabienne Marcellin [1,2], Marie Costa [1,2], Philippe Morlat [3,4], Camelia Protopopescu [1,2], Gilles Pialoux [5], Melina Erica Santos [6,7], Linda Wittkop [4,8], Laure Esterle [7], Anne Gervais [9], Philippe Sogni [10], Dominique Salmon-Ceron [11], Maria Patrizia Carrieri [1,2] and the ANRS CO13-HEPAVIH Cohort Study Group [†]

[1]    Aix Marseille Université, INSERM, IRD, SESSTIM, Sciences Economiques & Sociales de la Santé & Traitement de l'Information Médicale, 27 Bd Jean Moulin, 13005 Marseille, France; fabienne.marcellin@inserm.fr (F.M.); marie.costa@inserm.fr (M.C.); camelia.protopopescu@inserm.fr (C.P.); maria-patrizia.carrieri@inserm.fr (M.P.C.)
[2]    ORS PACA, Observatoire Régional de la Santé Provence-Alpes-Côte d'Azur, 27 Bd Jean Moulin, 13005 Marseille, France
[3]    Service de Médecine Interne, Hôpital Saint André, Centre Hospitalier Universitaire de Bordeaux, Université de Bordeaux, 1 Rue Jean Burguet, 33000 Bordeaux, France; philippe.morlat@chu-bordeaux.fr
[4]    ISPED, Inserm, Bordeaux Population Health Research Center, Team MORPH3EUS, UMR 1219, CIC-EC 1401, Université Bordeaux, F-33000 Bordeaux, France; linda.wittkop@u-bordeaux.fr
[5]    Département des Maladies Infectieuses, Hôpital Tenon, 4, rue de la Chine, 75020 Paris, France; gilles.pialoux@aphp.fr
[6]    Ministério da Saúde, Secretaria de Vigilância em Saúde, Departamento de Vigilância, Prevenção e Controle das IST, do HIV/Aids e das Hepatites Virais, Brasília 70719-040, Brazil; melinabtu@gmail.com
[7]    Faculdade de Ciências da Saúde, Programa de Pós-Graduação em Saúde Coletiva, Universidade de Brasília, Brasília 70910-900, Brazil; laure.esterle@u-bordeaux.fr
[8]    CHU de Bordeaux, Pole de Sante Publique, Service d'Information Medicale, F-33000 Bordeaux, France
[9]    Service des Maladies Infectieuses et Tropicales, AP-HP, Hôpital Bichat Claude Bernard, 46 rue Henri Huchard, 75018 Paris, France; anne.gervais@aphp.fr
[10]   INSERM U-1223, Institut Pasteur, Service d'Hépatologie, Hôpital Cochin, Université Paris Descartes, 27 rue du Faubourg Saint-Jacques, 75014 Paris, France; philippe.sogni@aphp.fr
[11]   Service Maladies Infectieuses et Tropicales, AP-HP, Hôpital Cochin, Université Paris Descartes, 27 rue du Faubourg-Saint-Jacques, 75014 Paris, France; dominique.salmon@aphp.fr
*     Correspondence: issifou.yaya@inserm.fr; Tel.: +33-658-217-022
†     The ANRS CO13-HEPAVIH Study Group is provided in the Acknowledgments.

Received: 29 April 2018; Accepted: 29 May 2018; Published: 31 May 2018

**Abstract:** Background: Coffee intake has been shown to modulate both the effect of ethanol on serum GGT activities in some alcohol consumers and the risk of alcoholic cirrhosis in some patients with chronic diseases. This study aimed to analyze the impact of coffee intake and alcohol consumption on advanced liver fibrosis (ALF) in HIV-HCV co-infected patients. Methods: ANRS CO13-HEPAVIH is a French, nationwide, multicenter cohort of HIV-HCV-co-infected patients. Sociodemographic, behavioral, and clinical data including alcohol and coffee consumption were prospectively collected using annual self-administered questionnaires during five years of follow-up. Mixed logistic regression models were performed, relating coffee intake and alcohol consumption to ALF. Results: 1019 patients were included. At the last available visit, 5.8% reported high-risk alcohol consumption, 27.4% reported high coffee intake and 14.5% had ALF. Compared with patients with low coffee intake and high-risk alcohol consumption, patients with low coffee intake and low-risk alcohol consumption had a lower risk of ALF (aOR (95% CI) 0.24 (0.12–0.50)). In addition, patients with high coffee intake had a lower risk of ALF than the reference group (0.14 (0.03–0.64) in high-risk alcohol drinkers and 0.11 (0.05–0.25) in low-risk alcohol drinkers). Conclusions: High coffee intake was associated with a low risk of liver fibrosis even in HIV-HCV co-infected patients with high-risk alcohol consumption.

**Keywords:** HIV-HCV co-infection; liver fibrosis; coffee; alcohol consumption

---

## 1. Background

Chronic hepatitis C virus (HCV) infection in patients co-infected with HIV who receive antiretroviral treatment (ART) accelerates hepatic complications including chronic inflammatory lesions of the liver, steatosis, liver fibrosis progression, liver cirrhosis and hepatocellular carcinoma (HCC) [1,2]. In addition, excessive alcohol consumption, which is associated with reduced liver function and steatosis in the general population, can increase the severity of fibrosis in HIV-HCV co-infected individuals due to the strong dose-response relationship between alcohol and liver fibrosis progression [3–5]. Furthermore, chronic alcohol consumption increases the risk of developing HCC, through inflammation of hepatic cells and metabolic disorders [6].

The consumption of certain beverages, such as coffee and green tea, has been shown to have hepatoprotective effects [7,8]. Coffee is one of the most consumed drinks in the world, especially in high-resource settings [9]. Coffee contains large amounts of bioactive compounds, including caffeine, diterpenes, melanoidins, and antioxidants, such as chlorogenic acids [10]. Dietary intake of coffee has been shown to be associated with human health [11], in particular with lower risk of mortality [12], cancer [13] and cardiovascular disease (CVD) [14]. Epidemiological studies have found an association between high coffee intake ($\geq$3 cups per day) and lower levels of liver enzymes, including aspartate aminotransferase (AST), alanine aminotransferase (ALT) and gamma-glutamyl transferase (GGT), which are markers of liver function [15–17]. In recent years, coffee intake has also been shown to modulate both the effect of ethanol on serum GGT activities in alcohol consumers and the risk of alcoholic cirrhosis in patients with chronic diseases [18]. In the context of HIV-HCV co-infection, high coffee intake has been found to have important benefits in terms of better adherence to treatment, less perceived toxicity [19,20], reduced levels of liver enzymes and lower risk of insulin resistance [15,17]. Several meta-analyses have also shown that coffee consumption is associated with a significant delay in the progression of liver fibrosis [21] and a reduced risk of HCC [22].

To our knowledge, no longitudinal study has ever analyzed the concomitant effects of coffee intake and alcohol consumption on liver fibrosis severity in HIV-HCV co-infected patients. This study aimed to analyze the impact of the interaction between high coffee intake and alcohol consumption on advanced liver fibrosis (ALF) among co-infected patients.

## 2. Materials and Methods

### 2.1. Study Design

This study is based on longitudinal data collected in the prospective, multicenter, observational ANRS CO13 HEPAVIH cohort, which recruited 1293 adult HIV-HCV co-infected individuals from 21 hospital centers throughout France between January 2006 and June 2014 [23].

Inclusion criteria in the cohort were as follows: being aged 18 years or more, HIV-1 infection and chronic HCV co-infection. Patients who had already cleared HCV, i.e., those who had a sustained virological response (SVR) to previous HCV treatment and those who had spontaneously cleared HCV, could also be included if eligible.

The study population included participants in the cohort with at least one measurement of alcohol consumption and coffee intake during the five first years of cohort follow-up. Patients with a history of liver transplant or clinical signs of decompensated liver cirrhosis at enrolment were excluded.

*2.2. Data Collection*

Throughout the follow-up, clinical/biological and socio-behavioral data were collected from medical records (clinical visits were scheduled annually, or every six months for cirrhotic patients) and annual self-administered questionnaires, respectively.

2.2.1. Outcome: Advanced Liver Fibrosis (ALF)

For the evaluation of liver fibrosis, we used patient age and serum markers—including ALT, AST, and platelets count—to calculate the FIB-4 index [24]. ALF was defined at each visit as a FIB-4 index >3.25.

2.2.2. Covariates

Clinical Variables

Clinical variables considered in the analyses included HIV plasma viral load, CD4 cell count, CDC clinical HIV stage, and time since antiretroviral therapy (ART) initiation at each follow-up visit. A detectable HIV viral load was defined as having a plasma HIV RNA level higher than the given hospital laboratory assay's threshold. Information concerning diabetes and current ART status history was also available at each follow-up visit.

We used the body mass index (BMI) to classify patients as obese if the BMI was >30.

We also recorded information about HCV genotype, exposure to HCV treatment before enrolment and during follow-up, and post-treatment HCV clearance.

Variables in the Self-Administered Questionnaire

Data on patients' socio-demographic characteristics (age, gender, educational level, marital status, and employment), coffee and tea consumption as well as psychoactive drug use were collected at enrolment and yearly thereafter using a self-administered questionnaire.

Data concerning patients' tobacco use were recorded during face-to-face medical interviews with physicians. Patients were asked about their experience of smoking (non-smoker, former smoker, and current smoker).

The AUDIT-C questionnaire was used to assess alcohol consumption during the previous six months. The number of alcohol units (AU) consumed per day (a standard drink, defined as one AU, contains 11–14 g of alcohol, and corresponds to one small bottle of beer, one medium glass of wine, or a shot of distilled spirits) was calculated for patients who reported they were current consumers. Alcohol consumption was defined as "high-risk" if it was >4 AU/day for men and >3 AU/day for women, and "low-risk" if it was ≤4 AU/day for men and ≤3 AU/day for women [25]. Binge drinking was defined as reporting to have consumed six alcoholic drinks or more on one occasion.

Coffee intake was investigated using a question referring to the 6 months prior to the given follow-up visit. Five answers were possible: never, occasionally, 1 cup/day, 2 cups/day and 3 cups or more/day (1 cup corresponding to 150–200 mL). Patients were classified as having high coffee intake if they reported drinking 3 cups of coffee or more/day.

A four-category variable combining alcohol consumption and coffee intake was also created (low coffee intake and low-risk alcohol consumption, low coffee intake and high-risk alcohol consumption, high coffee intake and low-risk alcohol consumption and high coffee intake and high-risk alcohol consumption).

The self-administered questionnaire also collected information about psychoactive drug use consumption including use of cannabis and other drugs (cocaine, heroin, crack, ecstasy, street buprenorphine, amphetamines) in the month prior to the visit, as well as patients' previous history of drug use.

*2.3. Statistical Analysis*

Participants' characteristics at the last available follow-up visit with completed self-administered questionnaire were compared according to fibrosis status using a Chi-square test or Fisher's exact test for categorical variables and Student's *t* test for continuous variables. For continuous variables, means and standard deviations were calculated while for categorical variables we calculated proportions.

All the variables included in this statistical analyze were used as time-varying covariates, as these variables were collected at the baseline and at each follow-up visits, except for gender. We used mixed-effects logistic regression models in order to take into account the correlations between repeated measurements. This type of models enables testing of both fixed (e.g., gender) and time-varying covariates (e.g., consumption behaviors), In the univariate analysis, we identified explanatory variables correlated with fibrosis status. Those with a liberal $p$-value $\leq 0.25$ were selected to be candidates for the final multivariable model.

The final multivariable model was built using a backward selection procedure, which was based on the likelihood ratio test ($p < 0.05$). Results were reported as adjusted odds ratios (aOR) with 95% confidence intervals (CI). Interactions between independent variables were also tested for.

Statistical analyses were performed using SAS software, version 9.4 (SAS Institute Inc., Cary, NC, USA).

## 3. Results

### 3.1. Patients' Characteristics at the Last Available Follow-Up Visit with a Completed Self-Administered Questionnaire

A total of 1019 patients were included in the study with a median follow-up of 5.0 years (IQR: 4.1–6.0). Men accounted for 69.7%. At the last available follow-up visit with a completed self-administered questionnaire, one third of patients had at least a high-school certificate and almost half (48.2%) were employed. Patient age varied between 19 and 75 years with a mean (SD) age of 47.8 ($\pm$6.4) years. In addition, 15.2% had ALF. The majority of patients (95.0%) were receiving ART. Only 36.9% were receiving or had received anti-HCV treatment. Elevated coffee intake was reported by 27.4% of the study patients, and patients without ALF were more likely ($p = 0.0002$) to report elevated coffee intake (29.3%) than those with ALF (14.1%). Almost 6% of the patients reported high-risk alcohol consumption. Patients with ALF were more likely ($p = 0.0018$) to report high-risk alcohol consumption (11.3%) than those without ALF (4.7%) (Table 1).

**Table 1.** Characteristics of HIV-HCV co-infected patients in the study population (*N* = 1019) at the last available follow-up visit with a completed self-administered questionnaire in the ANRS CO13 HEPAVIH cohort.

| | N (%) | Advanced Liver Fibrosis [1] | | |
| --- | --- | --- | --- | --- |
| | | No, 822 (84.8%) | Yes, 147 (15.2%) | *p*-Value (chi-2) |
| Age, years | | | | 0.0856 * |
| Mean (±SD) | 47.8 (±6.4) | 47.4 (±6.3) | 49.3 (±7.0) | |
| Median (IQR) | 48 (44–51) | | | |
| Gender | | | | 0.2880 |
| Male | 710 (69.7) | 568 (69.1) | 108 (73.5) | |
| Female | 309 (30.3) | 254 (30.9) | 39 (26.5) | |
| Living in a couple | | | | 0.8208 |
| No | 526 (51.9) | 418 (51.0) | 76 (52.1) | |
| Yes | 488 (48.1) | 401 (49.0) | 70 (47.9) | |
| High-school certificate | | | | 0.1331 |
| No | 576 (66.4) | 449 (65.2) | 97 (71.8) | |
| Yes | 291 (33.6) | 240 (34.8) | 38 (28.2) | |
| Employment | | | | 0.0100 |
| No | 525 (51.8) | 406 (49.7) | 90 (61.2) | |
| Yes | 489 (48.2) | 411 (50.3) | 57 (38.8) | |
| CDC clinical stage | | | | 0.5139 |
| Stage A | 446 (45.1) | 369 (45.0) | 63 (42.9) | |
| Stage B | 260 (26.3) | 221 (26.9) | 36 (24.5) | |
| Stage C | 283 (28.6) | 230 (28.1) | 48 (32.6) | |
| CD4 count, cells/mm$^3$ | | | | <0.0001 * |
| Mean (±SD) | 564 (±309) | 594 (±311) | 394 (±355) | |
| Median (IQR) | 516 (341–733) | | | |
| Body mass index [2] | | | | 0.2155 |
| Underweight or Normal | 749 (78.4) | 633 (79.6) | 103 (74.6) | |
| Overweight | 162 (17.0) | 129 (16.2) | 25 (18.1) | |
| Obese | 44 (4.6) | 33 (4.2) | 10 (7.3) | |
| Diabetes | | | | 0.1888 |
| No | 1008 (98.9) | 815 (99.1) | 144 (98.0) | |
| Yes | 11 (1.1) | 7 (0.9) | 3 (2.0) | |

Table 1. *Cont.*

| | N (%) | Advanced Liver Fibrosis [1] | | |
| --- | --- | --- | --- | --- |
| | | No, 822 (84.8%) | Yes, 147 (15.2%) | *p*-Value (chi-2) |
| **Receiving ART** | | | | 0.5111 |
| No | 50 (5.0) | 40 (4.9) | 9 (6.3) | |
| Yes | 951 (95.0) | 770 (95.1) | 135 (93.7) | |
| **HCV treatment status** | | | | <0.0001 |
| Not yet treated | 643 (63.1) | 518 (63.0) | 98 (66.7) | |
| On treatment | 62 (6.1) | 36 (4.4) | 19 (12.9) | |
| Treated, chronic HCV | 108 (10.6) | 79 (9.6) | 25 (17.0) | |
| Treated, HCV-cured | 206 (20.2) | 189 (23.0) | 5 (3.4) | |
| **Alcohol consumption** [3] | | | | 0.0018 |
| Low-risk | 916 (94.2) | 749 (95.3) | 125 (88.7) | |
| High-risk | 56 (5.8) | 37 (4.7) | 16 (11.3) | |
| **Binge drinking** [4] | | | | 0.0699 |
| No | 725 (72.4) | 594 (73.5) | 96 (66.2) | |
| Yes | 276 (27.6) | 214 (26.5) | 47 (33.8) | |
| **Coffee intake** | | | | 0.0002 |
| Low | 713 (72.6) | 561 (70.7) | 122 (85.9) | |
| High | 269 (27.4) | 232 (29.3) | 20 (14.1) | |
| **Cannabis consumption** | | | | 0.5252 |
| No | 620 (71.8) | 497 (71.4) | 92 (74.2) | |
| Yes | 244 (28.2) | 199 (28.6) | 32 (25.8) | |
| **Tobacco consumption** | | | | 0.4535 |
| Never | 116 (12.6) | 102 (13.3) | 13 (9.6) | |
| Past | 147 (15.9) | 123 (16.0) | 21 (15.4) | |
| Current | 659 (71.5) | 543 (70.7) | 102 (75.0) | |
| **Coffee intake-alcohol consumption** | | | | <0.0001 |
| Low coffee intake and high-risk alcohol consumption | 39 (4.0) | 25 (3.2) | 12 (8.5) | |
| Low coffee intake and low-risk alcohol consumption | 666 (68.9) | 529 (67.8) | 109 (77.3) | |
| High coffee intake and high-risk alcohol consumption | 15 (1.6) | 10 (1.3) | 4 (2.8) | |
| High coffee intake and low-risk alcohol consumption | 246 (25.5) | 216 (27.7) | 16 (11.4) | |

* *t*-test; [1]: FIB-4 $\geq$ 3.25; [2]: Underweight or Normal (BMI < 25), Overweight (BMI between 25 and 30), Obese (BMI > 30); [3]: High-risk alcohol consumption if >4 AU/day for men and >3 AU/day for women; and low-risk if $\leq$4 AU/day for men and $\leq$3 AU/day for women. [4]: defined as reporting to have consumed six alcohol drinks (units) or more on one occasion during the previous 6 months.

## 3.2. Factors Associated with ALF

In the univariate analyses, the following variables were significantly associated with higher odds of having an ALF ($p < 0.05$): older age, unemployment, lower $CD4^+$ cell count, obesity, not currently receiving ART, currently receiving HCV treatment, and high-risk alcohol consumption (Table 2). By contrast, high coffee intake and being HCV cured were significantly associated with lower odds of having ALF.

The multivariable analysis (Table 2) confirmed these results, except for the association with unemployment, which was no longer significant after multiple adjustment. Moreover, obesity increased the odds of having advanced fibrosis.

After multiple adjustment, compared with patients with low coffee intake and high-risk alcohol drinking who had a higher risk of advanced fibrosis (reference group), patients with low coffee intake and low-risk alcohol drinking had a 76% lower risk of ALF aOR (95% CI): 0.24 (0.12–0.50)). Among those with high coffee intake, high-risk alcohol consumption seemed to have no effect on liver fibrosis, with these drinkers having at least an 86% lower risk of advanced fibrosis than the reference group (0.14 (0.03–0.64) in high-risk alcohol drinkers and 0.11 (0.05–0.25) in low-risk alcohol drinkers) (Table 2 and Figure 1).

**Table 2.** Factors associated with ALF in HIV-HCV co-infected patients, mixed logistic regression models, ANRS CO13 HEPAVIH cohort (*N* = 1019).

| | Mixed-Effects Logistic Regression Models | | | |
| | Univariate Analyses | | Multivariable Analysis | |
| | OR (95% CI) | *p*-Value | AOR (95% CI) | *p*-Value |
|---|---|---|---|---|
| Age, years | 1.08 (1.05–1.1) | <0.001 | 1.18 (1.13–1.22) | <0.0001 |
| Gender | | | | |
| Male | 1 | | | |
| Female | 0.67 (0.41–1.09) | 0.109 | | |
| Living in a couple | | | | |
| No | 1 | | | |
| Yes | 0.91 (0.65–1.27) | 0.559 | | |
| High-school certificate | | | | |
| No | 1 | | | |
| Yes | 0.66 (0.40–1.10) | 0.114 | | |
| Employment | | | | |
| No | 1 | | | |
| Yes | 0.54 (0.40–0.74) | 0.0002 | | |
| CDC clinical stage | | 0.463 | | |
| Stage A | 1 | | | |
| Stage B | 1.25 (0.70–2.23) | 0.417 | | |
| Stage C | 1.37 (0.77–2.44) | 0.252 | | |
| CD4 count [1], cells/mm$^3$ | 0.85 (0.82–0.88) | <0.0001 | 0.74 (0.71–0.78) | <0.0001 |
| Obesity [2] | | <0.0021 | | 0.0031 |
| No | 1 | | 1 | |
| Yes | 4.24 (1.78–10.10) | 0.0021 | 5.93 (1.95–18.07) | 0.0031 |
| Diabetes | | | | |
| No | 1 | | | |
| Yes | 1.55 (0.75–3.21) | 0.229 | | |
| Receiving ART | | | | |
| No | 1 | | | |
| Yes | 0.23 (0.13–0.39) | <0.0001 | 0.18 (0.09–0.34) | <0.0001 |

Table 2. Cont.

| | Mixed-Effects Logistic Regression Models | | | |
| | Univariate Analyses | | Multivariable Analysis | |
| | OR (95% CI) | p-Value | AOR (95% CI) | p-Value |
|---|---|---|---|---|
| HCV treatment status | | <0.0001 | | <0.0001 |
| Not yet treated | 1 | | 1 | |
| On treatment/Treated, chronic HCV | 2.50 (1.79–3.49) | <0.0001 | 1.96 (1.33–2.90) | 0.0008 |
| Treated, HCV-cured | 0.07 (0.04–0.13) | <0.0001 | 0.04 (0.02–0.08) | <0.0001 |
| Alcohol consumption [3] | | | | |
| Low-risk | 1 | | | |
| High-risk | 3.2 (1.8–5.7) | 0.0002 | | |
| Binge drinking [4] | | | | |
| No | 1 | | | |
| Yes | 0.85 (0.62–1.17) | 0.3065 | | |
| Coffee intake | | | | |
| Low | 1 | | | |
| High | 0.49 (0.35–0.69) | <0.0001 | | |
| Cannabis consumption | | | | |
| No | 1 | | | |
| Yes | 1.09 (0.72–1.65) | 0.687 | | |
| Tobacco consumption | | 0.6354 | | |
| Never | 1 | | | |
| Past | 1.28 (0.28–5.93) | 0.7402 | | |
| Current | 1.67 (0.46–6.08) | 0.4095 | | |
| Coffee intake-alcohol consumption | | <0.0001 | | <0.0001 |
| Low coffee intake and high-risk alcohol consumption | 1 | | 1 | |
| Low coffee intake and low-risk alcohol consumption | 0.24 (0.13–0.46) | <0.0001 | 0.24 (0.12–0.50) | 0.0002 |
| High coffee intake and high-risk alcohol consumption | 0.26 (0.09–0.74) | 0.0117 | 0.14 (0.03–0.64) | 0.0114 |
| High coffee intake and low-risk alcohol consumption | 0.14 (0.07–0.27) | <0.0001 | 0.11 (0.05–0.25) | <0.0001 |

[1]: Intervals of 50 cell/mm$^3$; [2]: Underweight or Normal (BMI <25), Overweight (BMI between 25 and 30), Obese (BMI > 30); [3]: High-risk alcohol consumption if >4 AU/day for men and >3 AU/day for women; and low-risk if ≤4 AU/day for men and ≤3 AU/day for women; [4]: defined as reporting to have consumed six alcohol drinks (units) or more on one occasion during the previous 6 months.

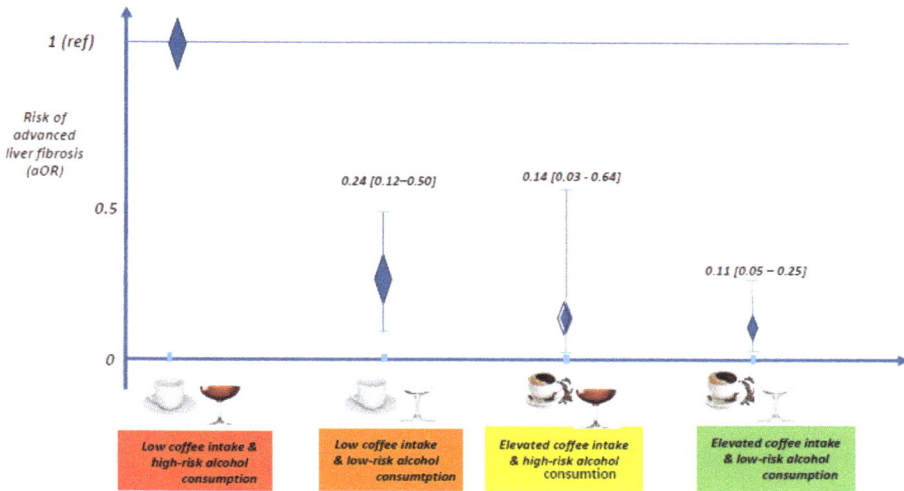

**Figure 1.** Risk of advanced liver fibrosis according to the pattern of coffee and alcohol consumption.

## 4. Discussion

In this longitudinal observational study of HIV-HCV co-infected patients from the French ANRS CO13 HEPAVIH cohort, after controlling for age, CD4, HCV clearance, ARV treatment and BMI, we found that there is an inverse relationship between alcohol intake and coffee consumption on the risk of ALF. This is a major result in a population where liver disease may persist even after HCV clearance, because of HIV-related risk factors [26]. This study also confirms that there is a strong inverse association between high coffee consumption and ALF. This important finding provides further evidence of the beneficial effect of coffee consumption on clinical issues in HIV-HCV co-infected patients.

Our findings are consistent with those of Stroffolini et al. [18] in a study conducted in Italy among patients who had either chronic hepatitis B or C. Their study showed that the association between high-risk alcohol consumption and the risk of cirrhosis decreased in individuals who consumed at least 3 cups of coffee/day. It has also been demonstrated that coffee minimizes the harmful effect of high-risk alcohol consumption on the functioning of the body and consequently on the health of the individual [27,28]. In Japan, Oze et al. conducted a case-control study to analyze the association between coffee and tea consumption and the risk of upper aerodigestive tract (UADT) cancer [27]. They demonstrated that drinking three cups of coffee or more per day was inversely associated with incidence of UADT cancer, but that this protective effect was observed only among people who had never smoked or drunk alcohol. In addition, in a study on mortality among 28,561 individuals in a cohort from three Eastern European countries [28], a mortality study stratified on alcohol consumption showed that drinking three cups of coffee/day or more was inversely associated with mortality irrespective of the level of alcohol consumption.

Other studies have shown that even in patients with chronic liver diseases, coffee consumption was associated with a decreased risk of alcoholic related cirrhosis [29,30]

In prior studies conducted in the ANRS HEPAVIH CO-13 cohort, we showed that high coffee intake had the following beneficial effects in HIV-HCV co-infected patients: reduced levels of liver enzymes, including aspartate aminotransferase (AST), alanine aminotransferase (ALT) and gamma-glutamyltransferase (GGT) [15], fewer self-reported side-effects during peg-IFN and ribavirin treatment [20], and a 50% reduction in mortality risk [25]. These findings are consistent with those of a meta-analysis of studies on patients with chronic hepatitis C which also showed an inverse relationship between coffee intake and the risk of liver fibrosis [7,31,32].

Coffee is a complex mixture of bioactive components including caffeine and polyphenols, like chlorogenic acids, although the precise chemical ingredient profile depends on the variety of coffee. These substances not only decrease the inflammation of liver cells (in the case of liver disease) by reducing the expression of inflammatory cytokines, but also demonstrate a well-documented anti-fibrotic effect [33,34]. It is also well documented that high alcohol consumption is strongly associated with several hepatic complications, including hepatic inflammation, steatosis and ALF in HIV-HCV co-infected patients [35,36]. In these patients, the two diseases are independently involved in fibrinogenesis—the inflammation of hepatic cells—and hepatocyte apoptosis in the liver [37–40]. This predisposes these patients to a higher risk of developing liver fibrosis and cirrhosis. These mechanisms may be accelerated by high-risk alcohol consumption. However, in our study, the latter effect seems to have been greatly diminished in HIV-HCV co-infected patients who drank at least 3 cups of coffee/day. Although the protective effect of coffee for several major health issues is becoming increasingly plausible, the mechanism by which coffee intake slows the progression of liver disease in HIV-HCV co-infected patients, and/or how it may inhibit the toxic effect of alcohol on the liver, is not understood. Some studies have reported antioxidant properties of certain components of coffee such as chlorogenic acids [41,42]. These properties help regulate the genes involved in the fibrogenesis process, and this could partially explain why patients in our study who consumed coffee were less likely to have ALF. Furthermore, just as has been reported for the effect of certain nutritional supplements—including L-cysteine, vitamin C and vitamin B1—on alcohol toxicity [43], coffee might also interact in the metabolization of blood alcohol into acetate, carbon dioxide and water, and thereby minimize the toxic effect. Another explanation, is that high coffee consumption may be associated with decreased alcohol consumption or the blocking of specific alcohol receptors in liver cells. Future studies are needed to better understand the interactions between consumption behaviors, including alcohol and coffee intake, and liver-related outcomes, such as liver fibrosis and liver stiffness, in HIV-HCV co-infected patients.

As reported elsewhere [44–46], HCV clearance in HIV-HCV co-infected patients with sustained virological response (SVR) was associated with a lower probability of having ALF, meaning that HCV clearance after antiviral therapy had a major impact on the natural course of the disease. A previous study in this cohort showed that in HIV-HCV co-infected patients, SVR after pegylated interferon-based treatment was significantly associated with improvement in liver stiffness [46]. In another study conducted among HIV-HCV co-infected patients in Spain [45], those who were treated for chronic HCV and cured with peg-IFN and ribavirin, experienced a significant reduction in liver fibrosis. Among chronic HCV patients in the United States, Fontana et al. [47] showed that serum levels in fibrosis markers decreased significantly in patients with SVR after peg-interferon- and ribavirin-based treatment for 24 to 48 weeks. In addition, Berenguer et al. [48] evaluated the clinical course of a cohort of HIV-HCV co-infected patients who were followed even after therapy with peg-interferon plus ribavirin, and showed that patients with SVR had significantly fewer liver-related events, including liver fibrosis, than those without SVR. The primary goal of the peg-Interferon and ribavirin treatment is the eradication of HCV, which may slow, stop or even reverse the progression of HCV infection events including liver fibrosis. In addition, successful HCV treatment leads to hepatic inflammation reduction and liver function improvements, even in patients with decompensated cirrhosis or transplant patients with chronic hepatitis C. These beneficial results are now being amplified by the generalized use of Direct Antiviral Analogues (DAA), which enable the treatment and cure of a large majority of HIV-HCV coinfected patients.

Importantly, our results revealed that co-infected patients on ARV treatment were less likely to have ALF. HIV infection is known to have a harmful effect on the natural history of HCV infection. In chronic hepatitis C patients, HIV co-infection was strongly associated with a rapid progression of hepatic complications including liver fibrosis and cirrhosis, due to immunosuppression [49,50].

Logically therefore, antiretroviral treatment in HIV-HCV co-infected patients should reduce the progression of liver complications [51]. However, several studies have shown persistent progression of liver fibrosis in HIV-HCV co-infected patients on antiretroviral treatment, which could be explained by a potential hepatotoxicity effect (including necroinflammation of hepatic cells) of certain categories of antiretroviral drugs [52,53]. In addition, closely related to the effectiveness of antiretroviral therapy, a greater CD4 count was associated with a decreased probability of having advanced fibrosis in the present work.

In our study, patients' body mass index was significantly associated with the risk of liver fibrosis, as obese HIV-HCV co-infected patients were six times more likely to have ALF. Several studies have shown a strong association between obesity and disease progression in chronic HCV patients [54–57]. In a study of an American cohort of chronic HCV patients with available liver biopsies, Younossi et al. [57] highlighted that overweight and obese patients were much more likely to have advanced fibrosis. In another study of chronic HCV patients [56], obese patients had a greater risk of advanced fibrosis. Similar results were found by El-Ray et al. In Egypt [55]. The harmful effects of obesity are caused by a state of chronic metabolic inflammation induced in the liver, which may predispose individuals to liver fibrosis and non-alcoholic fatty liver disease. This result, in our study, should be interpreted with caution, as the confidence interval (5.93 (1.95–18.07), $p = 0.0031$) is wide.

Finally, results from a meta-analysis [58,59] suggest that the effect of coffee does not depend on caffeine, as similar benefits on liver diseases including hepatocellular carcinoma were shown for caffeinated and decaffeinated coffee.

Our study had several limitations. First, it was observational in nature, meaning that the associations observed did not imply causality. Accordingly, more research (for example a randomized clinical trial) is needed to confirm these findings in this population. Second, although the sensitivity of FIB-4 was estimated to be only approximately 65% in a different study by Sterling et al. [19]. The FIB-4 index is nonetheless considered one of the most reliable non-invasive methods in the assessment of liver fibrosis in HIV-HCV co-infected patients. In this study, we did not use data from a DAA-based cohort but from a PEG-IFN-based one, and so treatment initiation rates and cure rates were much lower. However, the positive effect of coffee on ALF remained true both for patients cured and not. Finally, the behavioral data related to the consumption of alcohol and other substances were based on self-reports which could be affected by social desirability bias.

## 5. Conclusions

This observational study analyzed the combined effect of coffee intake and alcohol consumption on the risk of ALF. High coffee intake was associated with a significantly reduced risk of ALF in HIV-HCV co-infected patients, even in those with high-risk alcohol consumption. This finding confirms the need to systematically take into account coffee intake in the evaluation of liver fibrosis progression in this population. Further studies are needed not only to confirm our findings, but also to evaluate the dose-effect response of coffee consumption on liver fibrosis in HIV-HCV co-infected patients.

**Author Contributions:** Conceptualization, I.Y., F.M. and M.P.C.; Methodology, I.Y., C.P., M.P.C.; Validation, C.P. and M.P.C.; Formal Analysis, I.Y.; Investigation, P.M., G.P., A.G., P.S., D.S.-C.; Data Curation, F.M. and C.P.; Writing-Original Draft Preparation, I.Y., F.M. and M.P.C.; Writing-Review & Editing, I.Y., F.M., M.C., P.M., C.P., G.P., M.E.S., L.W., L.E., A.G., P.S., D.S.-C.; Supervision, M.P.C.; Project Administration, L.W., L.E.; Funding Acquisition, L.W., P.S., D.S.-C., M.P.C.

**Funding:** This work was supported by the French National Agency for Research on Aids and Viral Hepatitis (ANRS), with the participation of Abbott France; Glaxo-Smith-Kline; Roche; Schering-Plough; and INSERM's 'Programme Cohortes TGIR'.

**Acknowledgments:** We thank all the members of the ANRS CO13-HEPAVIH Study Group. Scientific Committee: D. Salmon (co-Principal investigator), L. Wittkop (co-Principal Investigator), P. Sogni (co-Principal Investigator), L. Esterle (project manager), P. Trimoulet, J. Izopet, L. Serfaty, V. Paradis, B. Spire, P. Carrieri, M.A. Valantin, G. Pialoux, J. Chas, I. Poizot-Martin, K. Barange, A. Naqvi, E. Rosenthal, A. Bicart-See, O. Bouchaud, A. Gervais, C. Lascoux-Combe, C. Goujard, K. Lacombe, C. Duvivier, D. Vittecoq, D. Neau, P. Morlat, F. Bani-Sadr, L. Meyer, F. Boufassa, S. Dominguez, B. Autran, A.M. Roque, C. Solas, H. Fontaine, D. Costagliola, L. Piroth, A. Simon,

D. Zucman, F. Boué, P. Miailhes, E. Billaud, H. Aumaître, D. Rey, G. Peytavin, V. Petrov-Sanchez, A. Pailhé. Clinical Centres (ward/participating physicians): APHP Cochin, Paris (Médecine Interne et Maladies Infectieuses: D. Salmon, R. Usubillaga; Hépato-gastro-entérologie: P. Sogni; Anatomo-pathologie: B. Terris; Virologie: P. Tremeaux); APHP Pitié-Salpêtrière, Paris (Maladies Infectieuses et Tropicales: C. Katlama, M.A. Valantin, H. Stitou; Hépato-gastro-entérologie: Y. Benhamou; Anatomo-pathologie: F. Charlotte; Virologie: S. Fourati); APHP Pitié-Salpêtrière, Paris (Médecine Interne: A. Simon, P. Cacoub, S. Nafissa); APHM Sainte-Marguerite, Marseille (Service d'Immuno-Hématologie Clinique: I. Poizot-Martin, O. Zaegel, H. Laroche; Virologie: C. Tamalet); APHP Tenon, Paris (Maladies Infectieuses et Tropicales: G. Pialoux, J. Chas; Anatomo-pathologie: P. Callard, F. Bendjaballah; Virologie: C. Le Pendeven); CHU Purpan, Toulouse (Maladies Infectieuses et Tropicales: B. Marchou; Hépato-gastro-entérologie: L. Alric, K. Barange, S. Metivier; Anatomo-pathologie: J. Selves ; Virologie: F. Larroquette); CHU Archet, Nice (Médecine Interne: E. Rosenthal; Infectiologie: A. Naqvi, V. Rio; Anatomo-pathologie: J. Haudebourg, M.C. Saint-Paul; Virologie: C. Partouche); APHP Avicenne, Bobigny (Médecine Interne – Unité VIH: O. Bouchaud; Anatomo-pathologie: M. Ziol; Virologie: Y. Baazia); Hôpital Joseph Ducuing, Toulouse (Médecine Interne: M. Uzan, A. Bicart-See, D. Garipuy, M.J. Ferro-Collados; Anatomo-pathologie: J. Selves; Virologie: F. Nicot); APHP Bichat-Claude-Bernard, Paris (Maladies Infectieuses:, A. Gervais, Y. Yazdanpanah; Anatomo-pathologie: H. Adle-Biassette; Virologie: G. Alexandre); APHP Saint-Louis, Paris (Maladies infectieuses: C. Lascoux-Combe, J.M. Molina; Anatomo-pathologie: P. Bertheau; Virologie:M.L. Chaix, C. Delaugerre, S. Maylin); APHP Saint-Antoine (Maladies Infectieuses et Tropicales:, K. Lacombe, J. Bottero; J. Krause P.M. Girard, Anatomo-pathologie: D. Wendum, P. Cervera, J. Adam; Virologie: C. Viala); APHP Bicêtre, Paris (Médecine Interne: C. Goujard, Y. Quertainmont, E. Teicher; Virologie: C. Pallier; Maladies Infectieuses: D. Vittecoq; APHP Necker, Paris (Maladies Infectieuses et Tropicales: O. Lortholary, C. Duvivier, C. Rouzaud, J. Lourenco, F. Touam, C. Louisin: Virologie: V. Avettand-Fenoel, A. Mélard); CHU Pellegrin, Bordeaux (Maladies Infectieuses et Tropicales: D. Neau, A. Ochoa, E. Blanchard, S. Castet-Lafarie, C. Cazanave, D. Malvy, M. Dupon, H. Dutronc, F. Dauchy, L. Lacaze-Buzy; Anatomo-pathologie: P. Bioulac-Sage; Virologie: P. Trimoulet, S. Reigadas); Hôpital Saint-André, Bordeaux (Médecine Interne et Maladies Infectieuses: Médecine Interne et Maladies Infectieuses: P. Morlat, D. Lacoste, F. Bonnet, N. Bernard, M. Hessamfar, J, F. Paccalin, C. Martell, M. C. Pertusa, M. Vandenhende, P. Merciéer, D. Malvy, T. Pistone, M.C. Receveur, M. Méchain, P. Duffau, C. Rivoisy, I. Faure, S. Caldato; Anatomo-pathologie: P. Bioulac-Sage; Virologie: P. Trimoulet, S. Reigadas); Hôpital du Haut-Levêque, Bordeaux (Médecine Interne: J.L. Pellegrin, J.F. Viallard, E. Lazzaro, C. Greib; Anatomo-pathologie: P. Bioulac-Sage; Virologie: P. Trimoulet, S. Reigadas); Hôpital FOCH, Suresnes (Médecine Interne: D. Zucman, C. Majerholc; Virologie: E. Farfour); APHP Antoine Béclère, Clamart (Médecine Interne: F. Boué, J. Polo Devoto, I. Kansau, V. Chambrin, C. Pignon, L. Berroukeche, R. Fior, V. Martinez ; Virologie : C. Deback); CHU Henri Mondor, Créteil (Immunologie Clinique: Y. Lévy, S. Dominguez, J.D. Lelièvre, A.S. Lascaux, G. Melica); CHU Hôtel Dieu, Nantes (Maladies Infectieuses et Tropicales: E. Billaud, F. Raffi, C. Allavena, V. Reliquet, D. Boutoille, C. Biron; Virologie: A. Rodallec, L. Le Guen) ; Hôpital de la Croix Rousse, Lyon (Maladies Infectieuses et Tropicales: P. Miailhes, D. Peyramond, C. Chidiac, F. Ader, F. Biron, A. Boibieux, L. Cotte, T. Ferry, T. Perpoint, J. Koffi, F. Zoulim, F. Bailly, P. Lack, M. Maynard, S. Radenne, M. Amiri; Virologie: C. Scholtes, T.T. Le-Thi); CHU Dijon, Dijon (Département d'infectiologie:, L. Piroth, P. Chavanet M. Duong Van Huyen, M. Buisson, A. Waldner-Combernoux, S. Mahy, R. Binois, A.L. Simonet-Lann, D. Croisier-Bertin); CH Perpignan, Perpignan (Maladies infectieuses et tropicales: H. Aumaître); CHU Robert Debré, Reims (Médecine interne, maladies infectieuses et immunologie clinique: F. Bani-Sadr, D. Lambert, Y Nguyen, J.L. Berger); CHRU Strasbourg (Le Trait d'Union: D Rey, M Partisani, ML Batard, C Cheneau, M Priester, C Bernard-Henry, E de Mautort, Virologie: P Gantner et S Fafi-Kremer), APHP Bichat-Claude Bernard (Pharmacologie: G. Peytavin).Data collection: F. Roustant, I. Kmiec, L. Traore, S. Lepuil, S. Parlier, V. Sicart-Payssan, E. Bedel, F. Touam, C. Louisin, M. Mole, C. Bolliot, M. Mebarki, A. Adda-Lievin, F.Z. Makhoukhi, O. Braik, R. Bayoud, M.P. Pietri, V. Le Baut, D. Bornarel, C. Chesnel, D. Beniken, M. Pauchard, S. Akel, S. Caldato, C. Lions, L. Chalal, Z. Julia, H. Hue, A. Soria, M. Cavellec, S. Breau, A. Joulie, P. Fisher, C. Ondo Eyene, S. Ogoudjobi, C. Brochier, V. Thoirain-Galvan. Management, statistical analyses: E. Boerg, P. Carrieri, V. Conte, L. Dequae-Merchadou, M. Desvallees, N. Douiri, L. Esterle, C. Gilbert, S. Gillet, R. Knight, F. Marcellin, L. Michel, M. Mora, S. Nordmann, C. Protopopescu, P. Roux, B. Spire, S. Tezkratt, A. Vilotitch, I. Yaya. T. Rojas, T. Barré. We especially thank all physicians and nurses who are involved in the follow-up of the cohort and all patients who took part in this study. We thank Jude Sweeney for the English revision and editing of our manuscript.

**Conflicts of Interest:** The authors declare no conflict of interest. The funding sources were not involved in the study design, data analysis, or in the writing and submission of the manuscript.

## References

1. Pawlotsky, J.-M. Pathophysiology of hepatitis C virus infection and related liver disease. *Trends Microbiol.* **2004**, *12*, 96–102. [CrossRef] [PubMed]

2. Castello, G.; Scala, S.; Palmieri, G.; Curley, S.A.; Izzo, F. HCV-related hepatocellular carcinoma: From chronic inflammation to cancer. *Clin. Immunol.* **2010**, *134*, 237–250. [CrossRef] [PubMed]

3. Moqueet, N.; Kanagaratham, C.; Gill, M.J.; Hull, M.; Walmsley, S.; Radzioch, D.; Saeed, S.; Platt, R.W.; Klein, M.B.; Canadian Co-infection Cohort Study (CTN 222). A prognostic model for development of significant liver fibrosis in HIV-hepatitis C co-infection. *PLoS ONE* **2017**, *12*, e0174205. [CrossRef] [PubMed]

4. Marcellin, P.; Asselah, T.; Boyer, N. Fibrosis and disease progression in hepatitis C. *Hepatology* **2002**, *36*, s47–s56. [PubMed]

5. Poynard, T.; Ratziu, V.; Charlotte, F.; Goodman, Z.; McHutchison, J.; Albrecht, J. Rates and risk factors of liver fibrosis progression in patients with chronic hepatitis C. *J. Hepatol.* **2001**, *34*, 730–739. [CrossRef]

6. Joshi, K.; Kohli, A.; Manch, R.; Gish, R. Alcoholic Liver Disease: High Risk or Low Risk for Developing Hepatocellular Carcinoma? *Clin. Liver Dis.* **2016**, *20*, 563–580. [CrossRef] [PubMed]

7. Liu, F.; Wang, X.; Wu, G.; Chen, L.; Hu, P.; Ren, H.; Hu, H. Coffee Consumption Decreases Risks for Hepatic Fibrosis and Cirrhosis: A Meta-Analysis. *PLoS ONE* **2015**, *10*, e0142457. [CrossRef] [PubMed]

8. Alferink, L.J.M.; Fittipaldi, J.; Kiefte-de Jong, J.C.; Taimr, P.; Hansen, B.E.; Metselaar, H.J.; Schoufour, J.D.; Ikram, M.A.; Janssen, H.L.A.; Franco, O.H.; et al. Coffee and herbal tea consumption is associated with lower liver stiffness in the general population: The Rotterdam study. *J. Hepatol.* **2017**, *67*, 339–348. [CrossRef] [PubMed]

9. Heckman, M.A.; Weil, J.; De Mejia, E.G. Caffeine (1, 3, 7-trimethylxanthine) in Foods: A Comprehensive Review on Consumption, Functionality, Safety, and Regulatory Matters. *J. Food Sci.* **2010**, *75*, R77–R87. [CrossRef] [PubMed]

10. Godos, J.; Pluchinotta, F.R.; Marventano, S.; Buscemi, S.; Li Volti, G.; Galvano, F.; Grosso, G. Coffee components and cardiovascular risk: Beneficial and detrimental effects. *Int. J. Food Sci. Nutr.* **2014**, *65*, 925–936. [CrossRef] [PubMed]

11. Grosso, G.; Godos, J.; Galvano, F.; Giovannucci, E.L. Coffee, Caffeine, and Health Outcomes: An Umbrella Review. *Annu. Rev. Nutr.* **2017**, *37*, 131–156. [CrossRef] [PubMed]

12. Grosso, G.; Micek, A.; Godos, J.; Sciacca, S.; Pajak, A.; Martínez-González, M.A.; Giovannucci, E.L.; Galvano, F. Coffee consumption and risk of all-cause, cardiovascular, and cancer mortality in smokers and non-smokers: A dose-response meta-analysis. *Eur. J. Epidemiol.* **2016**, *31*, 1191–1205. [CrossRef] [PubMed]

13. Alicandro, G.; Tavani, A.; La Vecchia, C. Coffee and cancer risk: A summary overview. *Eur. J. Cancer Prev.* **2017**, *26*, 424–432. [CrossRef] [PubMed]

14. Ding, M.; Bhupathiraju, S.N.; Satija, A.; van Dam, R.M.; Hu, F.B. Long-term coffee consumption and risk of cardiovascular disease: A systematic review and a dose-response meta-analysis of prospective cohort studies. *Circulation* **2014**, *129*, 643–659. [CrossRef] [PubMed]

15. Carrieri, M.P.; Lions, C.; Sogni, P.; Winnock, M.; Roux, P.; Mora, M.; Bonnard, P.; Salmon, D.; Dabis, F.; Spire, B.; ANRS CO13 HEPAVIH Study Group. Association between elevated coffee consumption and daily chocolate intake with normal liver enzymes in HIV-HCV infected individuals: Results from the ANRS CO13 HEPAVIH cohort study. *J. Hepatol.* **2014**, *60*, 46–53. [CrossRef] [PubMed]

16. Oh, M.G.; Han, M.A.; Kim, M.W.; Park, C.G.; Kim, Y.D.; Lee, J. Coffee consumption is associated with lower serum aminotransferases in the general Korean population and in those at high risk for hepatic disease. *Asia Pac. J. Clin. Nutr.* **2016**, *25*, 767–775. [PubMed]

17. Morisco, F.; Lembo, V.; Mazzone, G.; Camera, S.; Caporaso, N. Coffee and liver health. *J. Clin. Gastroenterol.* **2014**, *48* (Suppl. 1), S87–S90. [CrossRef] [PubMed]

18. Stroffolini, T.; Cotticelli, G.; Medda, E.; Niosi, M.; Del Vecchio-Blanco, C.; Addolorato, G.; Petrelli, E.; Salerno, M.T.; Picardi, A.; Bernardi, M.; et al. Interaction of alcohol intake and cofactors on the risk of cirrhosis. *Liver Int.* **2010**, *30*, 867–870. [CrossRef] [PubMed]

19. Freedman, N.D.; Everhart, J.E.; Lindsay, K.L.; Ghany, M.G.; Curto, T.M.; Shiffman, M.L.; Lee, W.M.; Lok, A.S.; Di Bisceglie, A.M.; Bonkovsky, H.L.; et al. Coffee Intake Is Associated with Lower Rates of Liver Disease Progression in Chronic Hepatitis C. *Hepatology* **2009**, *50*, 1360–1369. [CrossRef] [PubMed]

20. Carrieri, M.P.; Cohen, J.; Salmon-Ceron, D.; Winnock, M. Coffee consumption and reduced self-reported side effects in HIV-HCV co-infected patients during PEG-IFN and ribavirin treatment: Results from ANRS CO13 HEPAVIH. *J. Hepatol.* **2012**, *56*, 745–747. [CrossRef] [PubMed]

21. Shim, S.G.; Jun, D.W.; Kim, E.K.; Saeed, W.K.; Lee, K.N.; Lee, H.L.; Lee, O.Y.; Choi, H.S.; Yoon, B.C. Caffeine attenuates liver fibrosis via defective adhesion of hepatic stellate cells in cirrhotic model. *J. Gastroenterol. Hepatol.* **2013**, *28*, 1877–1884. [CrossRef] [PubMed]

22. Bai, K.; Cai, Q.; Jiang, Y.; Lv, L. Coffee consumption and risk of hepatocellular carcinoma: A meta-analysis of eleven epidemiological studies. *OncoTargets Ther.* **2016**, *9*, 4369–4375.

23. Loko, M.-A.; Salmon, D.; Carrieri, P.; Winnock, M.; Mora, M.; Merchadou, L.; Gillet, S.; Pambrun, E.; Delaune, J.; Valantin, M.A.; et al. The French national prospective cohort of patients co-infected with HIV

and HCV (ANRS CO13 HEPAVIH): Early findings, 2006–2010. *BMC Infect. Dis.* **2010**, *10*, 303. [CrossRef] [PubMed]

24. Sterling, R.K.; Lissen, E.; Clumeck, N.; Sola, R.; Correa, M.C.; Montaner, J.; Sulkowski, M.S.; Torriani, F.J.; Dieterich, D.T.; Thomas, D.L.; et al. Development of a simple noninvasive index to predict significant fibrosis in patients with HIV/HCV coinfection. *Hepatology* **2006**, *43*, 1317–1325. [CrossRef] [PubMed]

25. Carrieri, M.P.; Protopopescu, C.; Marcellin, F.; Rosellini, S.; Wittkop, L.; Esterle, L.; Zucman, D.; Raffi, F.; Rosenthal, E.; Poizot-Martin, I.; et al. Protective effect of coffee consumption on all-cause mortality of French HIV-HCV co-infected patients. *J. Hepatol.* **2017**. [CrossRef] [PubMed]

26. Labarga, P.; Fernandez-Montero, J.V.; de Mendoza, C.; Barreiro, P.; Pinilla, J.; Soriano, V. Liver fibrosis progression despite HCV cure with antiviral therapy in HIV-HCV-coinfected patients. *Antivir. Ther.* **2015**, *20*, 329–334. [CrossRef] [PubMed]

27. Oze, I.; Matsuo, K.; Kawakita, D.; Hosono, S.; Ito, H.; Watanabe, M.; Hatooka, S.; Hasegawa, Y.; Shinoda, M.; Tajima, K.; et al. Coffee and green tea consumption is associated with upper aerodigestive tract cancer in Japan. *Int. J. Cancer* **2014**, *135*, 391–400. [CrossRef] [PubMed]

28. Grosso, G.; Stepaniak, U.; Micek, A.; Stefler, D.; Bobak, M.; Pajak, A. Coffee consumption and mortality in three Eastern European countries: Results from the HAPIEE (Health, Alcohol and Psychosocial factors in Eastern Europe) study. *Public Health Nutr.* **2017**, *20*, 82–91. [CrossRef] [PubMed]

29. Kennedy, O.J.; Roderick, P.; Buchanan, R.; Fallowfield, J.A.; Hayes, P.C.; Parkes, J. Systematic review with meta-analysis: Coffee consumption and the risk of cirrhosis. *Aliment. Pharmacol. Ther.* **2016**, *43*, 562–574. [CrossRef] [PubMed]

30. Setiawan, V.W.; Porcel, J.; Wei, P.; Stram, D.O.; Noureddin, N.; Lu, S.C.; Le Marchand, L.; Noureddin, M. Coffee Drinking and Alcoholic and Nonalcoholic Fatty Liver Diseases and Viral Hepatitis in the Multiethnic Cohort. *Clin. Gastroenterol. Hepatol.* **2017**, *15*, 1305–1307. [CrossRef] [PubMed]

31. Jaruvongvanich, V.; Sanguankeo, A.; Klomjit, N.; Upala, S. Effects of caffeine consumption in patients with chronic hepatitis C: A systematic review and meta-analysis. *Clin. Res. Hepatol. Gastroenterol.* **2017**, *41*, 46–55. [CrossRef] [PubMed]

32. Larsson, S.C.; Wolk, A. Coffee Consumption and Risk of Liver Cancer: A Meta-Analysis. *Gastroenterology* **2007**, *132*, 1740–1745. [CrossRef] [PubMed]

33. Nieber, K. The Impact of Coffee on Health. *Planta Med.* **2017**, *83*, 1256–1263. [CrossRef] [PubMed]

34. Shi, H.; Dong, L.; Jiang, J.; Zhao, J.; Zhao, G.; Dang, X.; Lu, X.; Jia, M. Chlorogenic acid reduces liver inflammation and fibrosis through inhibition of toll-like receptor 4 signaling pathway. *Toxicology* **2013**, *303*, 107–114. [CrossRef] [PubMed]

35. Lim, J.K.; Tate, J.P.; Fultz, S.L.; Goulet, J.L.; Conigliaro, J.; Bryant, K.J.; Gordon, A.J.; Gibert, C.; Rimland, D.; Goetz, M.B. Relationship between alcohol use categories and noninvasive markers of advanced hepatic fibrosis in HIV-infected, chronic hepatitis C virus-infected, and uninfected patients. *Clin. Infect. Dis.* **2014**, *58*, 1449–1458. [CrossRef] [PubMed]

36. Marcellin, F.; Roux, P.; Loko, M.-A.; Lions, C.; Caumont-Prim, A.; Dabis, F.; Salmon-Ceron, D.; Spire, B.; Carrieri, M.P.; et al. High levels of alcohol consumption increase the risk of advanced hepatic fibrosis in HIV/hepatitis C virus-coinfected patients: A sex-based analysis using transient elastography at enrollment in the HEPAVIH ANRS CO13 cohort. *Clin. Infect. Dis.* **2014**, *59*, 1190–1192. [PubMed]

37. Balasubramanian, A.; Koziel, M.; Groopman, J.E.; Ganju, R.K. Molecular mechanism of hepatic injury in coinfection with hepatitis C virus and HIV. *Clin. Infect. Dis.* **2005**, *41* (Suppl. 1), S32–S37. [CrossRef] [PubMed]

38. Mascia, C.; Lichtner, M.; Zuccalà, P.; Vita, S.; Tieghi, T.; Marocco, R.; Savinelli, S.; Rossi, R.; Iannetta, M.; Campagna, M.; et al. Active HCV infection is associated with increased circulating levels of interferon-gamma (IFN-γ)-inducible protein-10 (IP-10), soluble CD163 and inflammatory monocytes regardless of liver fibrosis and HIV coinfection. *Clin. Res. Hepatol. Gastroenterol.* **2017**, *41*, 644–655. [CrossRef] [PubMed]

39. Negash, A.A.; Ramos, H.J.; Crochet, N.; Lau, D.T.; Doehle, B.; Papic, N.; Delker, D.A.; Jo, J.; Bertoletti, A.; Hagedorn, C.H.; et al. IL-1β Production through the NLRP3 Inflammasome by Hepatic Macrophages Links Hepatitis C Virus Infection with Liver Inflammation and Disease. *PLoS Pathog.* **2013**, *9*, e1003330. [CrossRef] [PubMed]

40. Czaja, A.J. Hepatic inflammation and progressive liver fibrosis in chronic liver disease. *World J. Gastroenterol.* **2014**, *20*, 2515–2532. [CrossRef] [PubMed]

41. Delgado-Andrade, C.; Rufián-Henares, J.A.; Morales, F.J. Assessing the antioxidant activity of melanoidins from coffee brews by different antioxidant methods. *J. Agric. Food Chem.* **2005**, *53*, 7832–7836. [CrossRef] [PubMed]

42. Gutiérrez-Grobe, Y.; Chávez-Tapia, N.; Sánchez-Valle, V.; Gavilanes-Espinar, J.G.; Ponciano-Rodríguez, G.; Uribe, M.; Méndez-Sánchez, N. High coffee intake is associated with lower grade nonalcoholic fatty liver disease: The role of peripheral antioxidant activity. *Ann. Hepatol.* **2012**, *11*, 350–355. [PubMed]

43. Lundberg, G.D. A Supplement That May Block. The Toxic Effects of Alcohol. Medscape. Available online: http://www.medscape.com/viewarticle/885865 (accessed on 6 October 2017).

44. Labarga, P.; Fernandez-Montero, J.V.; Barreiro, P.; Pinilla, J.; Vispo, E.; de Mendoza, C.; Plaza, Z.; Soriano, V. Changes in liver fibrosis in HIV/HCV-coinfected patients following different outcomes with peginterferon plus ribavirin therapy. *J. Viral Hepat.* **2014**, *21*, 475–479. [CrossRef] [PubMed]

45. Fernández-Montero, J.V.; Barreiro, P.; Vispo, E.; Labarga, P.; Sánchez-Parra, C.; de Mendoza, C.; Treviño, A.; Soriano, V. Liver fibrosis progression in HIV-HCV-coinfected patients treated with distinct antiretroviral drugs and impact of pegylated interferon/ribavirin therapy. *Antiviral Ther.* **2014**, *19*, 287–292. [CrossRef] [PubMed]

46. ANRS CO13 HEPAVIH Cohort. Regression of liver stiffness after sustained hepatitis C virus (HCV) virological responses among HIV/HCV-coinfected patients. *AIDS* **2015**, *29*, 1821–1830.

47. Fontana, R.J.; Bonkovsky, H.L.; Naishadham, D.; Dienstag, J.L.; Sterling, R.K.; Lok, A.S.; Su, G.L.; Halt-C Trial Group. Serum fibrosis marker levels decrease after successful antiviral treatment in chronic hepatitis C patients with advanced fibrosis. *Clin. Gastroenterol. Hepatol.* **2009**, *7*, 219–226. [CrossRef] [PubMed]

48. Berenguer, J.; Alvarez-Pellicer, J.; Martín, P.M.; López-Aldeguer, J.; Von-Wichmann, M.A.; Quereda, C.; Mallolas, J.; Sanz, J.; Tural, C.; Bellón, J.M.; et al. Sustained virological response to interferon plus ribavirin reduces liver-related complications and mortality in patients coinfected with human immunodeficiency virus and hepatitis C virus. *Hepatology* **2009**, *50*, 407–413. [CrossRef] [PubMed]

49. Benhamou, Y.; Bochet, M.; Di Martino, V.; Charlotte, F.; Azria, F.; Coutellier, A.; Vidaud, M.; Bricaire, F.; Opolon, P.; Katlama, C.; et al. Liver fibrosis progression in human immunodeficiency virus and hepatitis C virus coinfected patients. The Multivirc Group. *Hepatology* **1999**, *30*, 1054–1058. [CrossRef] [PubMed]

50. Graham, C.S.; Baden, L.R.; Yu, E.; Mrus, J.M.; Carnie, J.; Heeren, T.; Koziel, M.J. Influence of human immunodeficiency virus infection on the course of hepatitis C virus infection: A meta-analysis. *Clin. Infect. Dis.* **2001**, *33*, 562–569. [CrossRef] [PubMed]

51. Bräu, N.; Salvatore, M.; Ríos-Bedoya, C.F.; Fernández-Carbia, A.; Paronetto, F.; Rodríguez-Orengo, J.F.; Rodríguez-Torres, M. Slower fibrosis progression in HIV/HCV-coinfected patients with successful HIV suppression using antiretroviral therapy. *J. Hepatol.* **2006**, *44*, 47–55. [CrossRef] [PubMed]

52. Bani-Sadr, F.; Lapidus, N.; Bedossa, P.; De Boever, C.M.; Perronne, C.; Halfon, P.; Pol, S.; Carrat, F.; Cacoub, P.; French National Agency for Research on AIDS; et al. Progression of fibrosis in HIV and hepatitis C virus-coinfected patients treated with interferon plus ribavirin-based therapy: Analysis of risk factors. *Clin. Infect. Dis.* **2008**, *46*, 768–774. [CrossRef] [PubMed]

53. Suárez-Zarracina, T.; Valle-Garay, E.; Collazos, J.; Montes, A.H.; Cárcaba, V.; Carton, J.A.; Asensi, V. Didanosine (ddI) associates with increased liver fibrosis in adult HIV-HCV coinfected patients. *J. Viral Hepat.* **2012**, *19*, 685–693. [CrossRef] [PubMed]

54. Dyal, H.K.; Aguilar, M.; Bhuket, T.; Liu, B.; Holt, E.W.; Torres, S.; Cheung, R.; Wong, R.J. Concurrent Obesity, Diabetes, and Steatosis Increase Risk of Advanced Fibrosis among HCV Patients: A Systematic Review. *Dig. Dis. Sci.* **2015**, *60*, 2813–2824. [CrossRef] [PubMed]

55. El Ray, A.; Asselah, T.; Moucari, R.; El Ghannam, M.; Taha, A.A.; Saber, M.A.; Akl, M.; Atta, R.; Shemis, M.; Radwan, A.S.; et al. Insulin resistance: A major factor associated with significant liver fibrosis in Egyptian patients with genotype 4 chronic hepatitis C. *Eur. J. Gastroenterol. Hepatol.* **2013**, *25*, 421–427. [CrossRef] [PubMed]

56. Hu, S.X.; Kyulo, N.L.; Xia, V.W.; Hillebrand, D.J.; Hu, K.Q. Factors Associated with Hepatic Fibrosis in Patients with Chronic Hepatitis C: A Retrospective Study of a Large Cohort of U.S. Patients. *J. Clin. Gastroenterol.* **2009**, *43*, 758–764. [CrossRef] [PubMed]

57. Younossi, Z.M.; McCullough, A.J.; Ong, J.P.; Barnes, D.S.; Post, A.; Tavill, A.; Bringman, D.; Martin, L.M.; Assmann, J.; Gramlich, T.; et al. Obesity and non-alcoholic fatty liver disease in chronic hepatitis C. *J. Clin. Gastroenterol.* **2004**, *38*, 705–709. [CrossRef] [PubMed]

58. Kennedy, O.J.; Roderick, P.; Buchanan, R.; Fallowfield, J.A.; Hayes, P.C.; Parkes, J. Coffee, including caffeinated and decaffeinated coffee, and the risk of hepatocellular carcinoma: A systematic review and dose-response meta-analysis. *BMJ Open* **2017**, *7*, e013739. [CrossRef] [PubMed]
59. Godos, J.; Micek, A.; Marranzano, M.; Salomone, F.; Rio, D.D.; Ray, S. Coffee Consumption and Risk of Biliary Tract Cancers and Liver Cancer: A Dose-Response Meta-Analysis of Prospective Cohort Studies. *Nutrients* **2017**, *9*, 950. [CrossRef] [PubMed]

![nutrients logo] *nutrients*

MDPI

*Review*

# Caffeine in the Diet: Country-Level Consumption and Guidelines

**Celine Marie Reyes and Marilyn C. Cornelis \***

Department of Preventive Medicine, Northwestern University Feinberg School of Medicine, Chicago, IL 60611, USA; celine.reyes@northwestern.edu
\* Correspondence: Marilyn.cornelis@northwestern.edu; Tel.: +1-312-503-4548

Received: 26 October 2018; Accepted: 13 November 2018; Published: 15 November 2018

**Abstract:** Coffee, tea, caffeinated soda, and energy drinks are important sources of caffeine in the diet but each present with other unique nutritional properties. We review how our increased knowledge and concern with regard to caffeine in the diet and its impact on human health has been translated into food-based dietary guidelines (FBDG). Using the Food and Agriculture Organization list of 90 countries with FBDG as a starting point, we found reference to caffeine or caffeine-containing beverages (CCB) in 81 FBDG and CCB consumption data (volume sales) for 56 of these countries. Tea and soda are the leading CCB sold in African and Asian/Pacific countries while coffee and soda are preferred in Europe, North America, Latin America, and the Caribbean. Key themes observed across FBDG include (i) caffeine-intake upper limits to avoid risks, (ii) CCB as replacements for plain water, (iii) CCB as added-sugar sources, and (iv) health benefits of CCB consumption. In summary, FBDG provide an unfavorable view of CCB by noting their potential adverse/unknown effects on special populations and their high sugar content, as well as their diuretic, psycho-stimulating, and nutrient inhibitory properties. Few FBDG balanced these messages with recent data supporting potential benefits of specific beverage types.

**Keywords:** caffeine; coffee; tea; soda; energy drinks; mate; guidelines; country; consumption; population; public policy

---

## 1. Introduction

Caffeine is the most widely consumed psychostimulant in the world [1]. It occurs naturally in coffee beans, tea leaves, cocoa beans, and kola nuts, and is also added to foods and beverages. Important dietary sources include coffee, tea, yerba mate, caffeinated soda (cola-type), and energy drinks [2]. There is increasing public and scientific interest in the potential health consequences of habitual intake of these caffeine-containing beverages (CCB). Rigorous reviews of caffeine toxicity conclude that consumption of up to 400 mg caffeine/day in healthy adults is not associated with adverse effects [3–5]. Epidemiological studies support a beneficial role of moderate coffee intake in reducing risk of several chronic diseases, but heavy intake is likely harmful regarding pregnancy outcomes [6]. Health implications of regular tea, mate, and energy drink consumption are inconclusive and most concern for caffeinated soda intake currently pertains to its sugar content and relationship to obesity [7–12].

CCB also contribute a wealth of other compounds to the diet that have potential benefits or risks to health and thus it is imperative to consider the context (i.e., beverage type) in which caffeine is consumed [12–16]. Food-based dietary guidelines (FBDG) provide context-specific advice on healthy diets that are evidence-based and respond to a country's public health and nutrition priorities, sociocultural influences, and food production and consumption patterns, among other factors [17]. These factors change over time, and in turn, so do FBDG. Our knowledge and concern with regard to caffeine sources in the diet and their impact on human health has increased over the years. We therefore sought to review how such knowledge and concern has been translated into FBDGs and within the context of what each country actually consumes.

## 2. Material and Methods

### 2.1. Data Collection Strategy for Dietary Caffeine Guidelines

Figure S1 outlines our data collection strategy. We initially used the Food and Agriculture Organization (FAO) website, which provided general food-based dietary guidelines (FBDG) from each country (http://www.fao.org/nutrition/nutrition-education/food-dietary-guidelines/en/). Each country's page included the most recent publication date of the guidelines, intended audience, general FBDG messages, downloadable guidelines if available, and contact information of those governmental institutions that established the guidelines.

The general messages from FAO were the first resource for any guidelines pertaining to caffeine or CCB including coffee, tea, yerba mate, energy drinks, and carbonated soft drinks. Beverages were a focus of the current review because they are the primary contributors of caffeine in the diet [2,18]. Other caffeine sources, such as products containing cocoa and kola nut, contribute relatively small amounts to the diet [18]. We considered guidelines for the broader categories of soft drinks or sugar-sweetened beverages (SSB) since colas (typically containing caffeine) were rarely distinguished from these other beverages. Non-caffeine-containing teas were also considered because some FBDG provide different guideline for these teas and regular tea (i.e., black or green tea) that might provide additional insight into the underlying reason for the recommendations. We then accessed the downloadable materials if available to search for more caffeine-related messages. For materials published in foreign languages, we found translators using the Cochrane Task Exchange or personal contacts. Additionally, we used the contact information from the FAO page to inquire via email or web applications about any updated or additional caffeine-related guidelines that were not available via the FAO website. Finally, after these search efforts, for countries with limited or no information regarding dietary caffeine, we searched for publications related to national dietary guidelines and contacted the authors for further information.

Countries were classified according to the World Bank income classification [19]. We also used the non-comprehensive World Cancer Research Fund International NOURISHING database to identify actions in place by countries that attempt to regulate dietary caffeine consumption [20,21].

### 2.2. Data Resource for Dietary Caffeine Consumption

We adopted country-level volume sales of CCB as a proxy measure of CCB consumption and these were estimated using the Euromonitor Passport Global Market Information Database [22]. Euromonitor collects these data from trade associations, industry bodies, business press, company financial reports, and official government statistics. Specifically, we downloaded (bulk format) 2017 country-specific annual sales of (i) coffee, total brewed volume (liters); (ii) tea, total brewed volume (liters); (iii) "other hot drinks," total brewed volume (liters); (iv) carbonates, total volume (liters); (v) sports and energy drinks, total volume (liters); (vi) ready-to-drink (RTD) coffee, total volume (liters); and vii) RTD tea, total volume (liters). For each country, data for each beverage was presented as a proportion of total CCB volume sales. Total CCB volume sales were also expressed on a per capita basis using total population estimates for 2017 (also downloaded from Euromonitor). For each country, we additionally reviewed the 2017 detailed report to collect information on the most common type or category of each beverage sold. Additional details for "sports and energy drinks," RTD coffee and RTD tea were not systematically collected as they were not uniformly available across countries. Aside from including yerba mate, the "other hot drinks" category was deemed an unlikely key source of CCB and thus we only make reference to this category as appropriate.

## 3. Results

Using the FAO listing of the most recent FBDG from 90 countries as a starting point, we found any mentions of caffeine or CCB in 81 of these, which are summarized in Table 1. Sixty-six of these were published in the last ten years. The oldest guidelines were published by Venezuela (1991) and Greece (1999). Intended audiences for each FBDG are provided in Table S1. Most FBDG were intended for the general, healthy population over 2 years of age with several FBDG including specific guidelines for subgroups of the population such as children and pregnant/nursing mothers. Euromonitor annual volume sales of CCB in 2017 were available for 56 of the 90 FAO countries. Euromonitor data was not available for countries of the Near East (as defined by FAO). Figure 1 presents the percentage of caffeine-containing beverage volume sales per beverage per country. Subcategories of coffee, tea, and carbonates were assigned according to the most commonly, but not exclusively, consumed beverage type in that category. North America (defined by FAO as including Canada and USA) had the highest average country annual total CCB volume sales per capita (348 L/capita), followed by Europe (200 L/capita), Latin America and the Caribbean (153 L/capita), Asia and the Pacific (126 L/capita), and Africa (90 L/capita).

**Table 1.** Country Specific Guidelines for Dietary Caffeine.

| Country Publication Date | FAO key Messages and Published Guidelines [Other Resources and Personal Communication] |
|---|---|
| | **AFRICA** |
| Benin [LI] 2015 | *Drink carbonated drinks and sugar-sweetened beverages in moderation. These kinds of drinks only provide sugar and can promote obesity and diabetes.* |
| Nigeria [LMI] 2001 | Adults, elderly, pregnant, breast-feeding: Avoid coffee and tea during meals since these beverages hinder iron absorption. Diabetics: avoid coffee if hypertensive. |
| Seychelles [HI] 2006 | *Consume sugar, sugary foods, and sugary drinks in minimal amounts.* |
| Sierra Leone [LI] | *Use sugars and foods and drinks made with sugar in moderation.* |
| 2016 | Limit intake of beverages such as coffee and alcohol. Taken in excess, these will contribute to too much phosphorous, which depletes calcium levels. |
| South Africa [UMI] | *Use sugar and foods and drinks high in sugar sparingly.* |
| 2012 | Avoid giving tea, coffee, and sugary drinks and high-sugar, high-fat salty snacks to baby/child (6–36 months). |
| | Low nutrient-dense liquid and energy-dense sugar-sweetened, high-fat, and salty snacks exacerbate poor nutrient intake and displace nutrient-dense food in the diet. Tea and coffee have low nutrient content and polyphenols that inhibit iron bioavailability. |
| | Drink lots of clean, safe water including drinking water and other foods/beverages that contain water. Biological availability of water depends on the presence of various ingredients in foods and beverages (i.e., salt and carbohydrates accelerate water absorption; caffeine and alcohol have diuretic effects). |
| | Research has shown that the consumption of caffeinated beverages, such as tea and coffee, can add to the daily water balance in individuals who are used to ingesting these beverages. Acute increases in urine output only occur in individuals who are not accustomed to regular consumption of caffeinated beverages. |
| | **ASIA and the PACIFIC** |
| Afghanistan [LI] 2015 | *Reduce sugar intake and avoid sweet carbonated beverages.* |

**Table 1.** *Cont.*

| Country Publication Date | FAO key Messages and Published Guidelines [Other Resources and Personal Communication] |
|---|---|
| Australia [HI] 2013 | Drink plenty of water: Tea and coffee provide water, although they are not suitable for young children and large quantities can have unwanted stimulant effects in some people. |
| | Consumption of drinks with added sugars, such as soft drinks and cordials, fruit drinks, vitamin waters, and energy and sports drinks can increase the risk of excessive weight gain in both children and adults. |
| | Limit intake of foods and drinks containing added sugars such as confectionary, sugar-sweetened soft drinks and cordials, fruit drinks, vitamin waters, and energy and sports drinks. |
| Bangladesh [LMI] 2013 | *Consume less sugar, sweets, or sweetened drinks.* |
| | Drink coconut water and fresh fruit juices instead of carbonated drinks. |
| | Pregnancy and lactation: Iron supplements are poorly absorbed when they are taken with beverages such as coffee or tea or simultaneously with calcium supplements. They should be taken after a meal, preferably after breakfast or after lunch. |
| China [UMI] 2016 | For lactating women, overconsumption of coffee should be avoided. |
| | Consumption of sugar-sweetened beverages should be limited in all age groups. |
| Fiji [UMI] 2013 | Drinks like tea and coffee are diuretics, which make you produce more urine than usual. Therefore, if you drink lots of tea and coffee, you still need to drink plenty of water. Fizzy and sugary drinks are also not a good way to get your fluid. The high sugar content can cause unwanted weight gain or take the place of more nutritious foods in your diet. |
| India [LMI] 2011 | Lactating women: Excess intake of beverages containing caffeine like coffee and tea adversely affect fetal growth, and hence, should be avoided. Since drugs (antibiotics, caffeine, hormones, and alcohol) are secreted into the breast-milk and could prove harmful to the breast-fed infant, caution should be exercised by the lactating mother while taking medicines. |
| | Drink plenty of water and take beverages in moderation: Tea and coffee are popular beverages. They are known to relieve mental and muscular fatigue. This characteristic stimulating effect is due to their caffeine content. Caffeine stimulates the central nervous system and induces physiological dependence. Generally, low doses (20–200 mg) of caffeine produce mild positive effects like a feeling of well-being, alertness, and being energetic. Higher doses (>200 mg) can produce negative effects like nervousness and anxiety, especially in people who do not usually consume caffeine-containing beverages. Therefore, moderation in tea and coffee consumption is advised so that caffeine intake does not exceed the tolerable limits. Tannin is also present in tea and coffee and is known to interfere with iron absorption. Hence, tea and coffee should be avoided at least for one hour before and after meals. Excess consumption of coffee is known to increase blood pressure and cause abnormalities in heartbeat. In addition, an association between coffee consumption and elevated levels of total and low-density lipoprotein (LDL) cholesterol ("bad" cholesterol), triglycerides, and heart disease has been demonstrated. Therefore, individuals with heart disease need to restrict coffee consumption. Also, those who experience adverse effects from caffeine should stop drinking coffee. Besides caffeine, tea contains theobromine and theophylline. These are known to relax coronary arteries and thereby promote circulation. Tea also contains flavonoids and other antioxidant polyphenols, which are known to reduce the risk for coronary heart disease and stomach cancer. However, as a result of its caffeine content, excess tea consumption is deleterious to health. Decaffeinated coffee and tea are being marketed to obviate the adverse effects of caffeine. |

Table 1. *Cont.*

| Country<br>Publication<br>Date | FAO key Messages and Published Guidelines<br>[Other Resources and Personal Communication] |
|---|---|
| Indonesia<br>LMI<br>2014 | One should limit consumption of overripe fruits and sugary juice drinks to control blood glucose. It should be noted that other food containing simple carbohydrate (flour, bread, soy sauce), sweet fruits, juice, soda drinks, etc., also contain sugar. |
| | Caffeine, if consumed by pregnant women, has a diuretic and stimulant effect. Thus, pregnant women may experience increased urination which may lead to dehydration, and increase in blood pressure and heart rate if their coffee intake is not controlled. Other sources of caffeine are chocolate, tea, and energy supplement drinks. In addition to caffeine, coffee also contains a substance that may inhibit iron absorption. Caffeine consumption in pregnant women may affect fetus growth and development, since the metabolism is not fully developed. |
| | The National Institute of Health USA (1993) recommends safe caffeine consumption for pregnant women as 150–250 mg/day or two cups of coffee/day. Thus, it is recommended for pregnant women "to be wise in consuming coffee, limit your intake within safe range, i.e., 2 cups of coffee/day or avoid at all, as there is no nutrition content in the coffee." |
| | Children should not regularly drink sweet or soda drinks as they have high sugar contents. For daily liquid consumption, it is recommended for children to drink 1200–1500 mL water/day. |
| | Breastmilk may contain caffeine from the coffee consumed by nursing women. This may have unfavorable effects on babies since their metabolism are not fully developed to digest caffeine. Caffeine consumption in nursing women may be related with low production of breastmilk. Based on research in Harvard University, safe consumption of caffeine in nursing women is 300 mg/day or three cups of coffee/day. Based on research in Mayo Clinic Rochester Minnesota USA, consumption exceeding 300 mg/day will lower iron content in breastmilk by 30% compared to nursing women who do not take caffeine. |
| Japan<br>HI<br>2010 | *Drink plenty of water and tea. Moderate consumption of highly processed snacks, confectionary, and sugar-sweetened beverages.* |
| Malaysia<br>UMI<br>2010 | Consume foods and beverages low in sugar. |
| | Caffeine is naturally present in coffee, tea, and chocolate, and added to colas and other beverages. It has long been thought that consumption of caffeinated beverages, because of the diuretic effect of caffeine on reabsorption of water in the kidney, can lead to loss of body water. However, available data are inconsistent. Caffeine-containing beverages did not increase 24 h urine volume in healthy, free-living men when compared with other types of beverages for instance water, energy-containing beverages, or theobromine-containing beverages. In aggregate, available data suggest that higher doses of caffeine (above 180 mg per day) have been shown to increase urinary output, perhaps transiently, and that this diuretic effect occurs within a short time period. Hence, unless additional evidence becomes available indicating cumulative total water deficits in individuals with habitual intakes of significant amounts of caffeine, caffeinated beverages appear to contribute to the daily total water intake similar to that contributed by non-caffeinated beverages (Food and Nutrition Board, 2004). |
| Nepal<br>LI<br>2012 | *Consume less sugar, sweets, and sweetened drinks.* |

**Table 1.** *Cont.*

| Country Publication Date | FAO key Messages and Published Guidelines [Other Resources and Personal Communication] |
|---|---|
| New Zealand [H] 2008–2013 | **0–2 years** |
| | Coffee, tea, other caffeine-containing drinks, smart or energy drinks, herbal teas, and alcohol are unsuitable for infants and toddlers. |
| | Caffeine is a central nervous system stimulant, which can cause irritability and restlessness. |
| | Tea is not recommended because its tannin content inhibits the absorption of iron from the gut. This impact was confirmed by a New Zealand study that observed a significant association between tea drinking in infants and the presence of an iron deficiency. Herbal teas can have adverse effects on infants and toddlers, so are not recommended. |
| | **2–18 years** |
| | Children and young people should limit their intake of foods and drinks containing caffeine. Caffeine is a psychoactive stimulant drug that acts on the central nervous system, alters brain function, acts as a diuretic, and elevates blood pressure and metabolic rate. Acute adverse effects from caffeine that have been identified include anxiety, headaches, insomnia, irritation of the gastrointestinal tract, nausea, and depression. Long-term adverse effects from caffeine are not clear. Children may be more sensitive to adverse effects of caffeine than other groups in the population. An upper exposure of 2.5 mg/kg of body weight per day has been suggested as a cautious toxicological limit on which to base risk assessment for children, on the grounds of limited evidence. |
| | Energy drinks and energy shots contain large amounts of caffeine, a psychoactive stimulant drug, and often large amounts of sugar. They are not recommended for children and young people. Coffee and tea are not recommended for children less than 13 years of age. If drinking tea and coffee, it is recommended young people (13–18 years) limit their intake to one to two cups per day. They should avoid drinking tea at meal times, as this drink contains tannins and polyphenols, which can inhibit the absorption of nutrients, such as iron. |
| | **Pregnant and breast feeding** |
| | Caffeine is a mild central nervous system stimulant, present in chocolate and beverages such as coffee, tea, energy drinks, and cola. Caffeine readily crosses the placenta to the fetus and has also been found to stimulate metabolic rate. Many over-the-counter medications, such as cold and allergy tablets, headache medications, diuretics, and stimulants, also contain some caffeine. |
| | High doses of caffeine in pregnancy have been associated with increased risk of congenital abnormalities, pregnancy loss, low birth weight, and behavioral problems. Decaffeinated coffee appears to have no effect on birth weight. The effects of caffeine may be synergistic with those of smoking and alcohol. Caffeine is transferred into breast milk. The infant metabolizes and excretes caffeine slowly. High caffeine load in breast milk may lead to irritability and poor sleeping patterns and, occasionally, increased bowel activity. The benefits of breastfeeding outweigh any risks associated with the presence of caffeine in breast milk, however. Consuming caffeine-containing beverages immediately after the baby has fed will limit the amount of caffeine in the next feed. |
| | The UK Food Standards Agency advises pregnant women to limit their intake of caffeine to 300 mg per day. |
| | Breastfeeding women should consider their caffeine intake if the infant is irritable or wakeful. |

Table 1. Cont.

| Country Publication Date | FAO key Messages and Published Guidelines [Other Resources and Personal Communication] |
|---|---|
| | Older people |
| | Although there is some evidence that a high caffeine intake is a risk factor for fracture frequency or bone loss, there is also evidence to the contrary. High intakes of caffeine do increase urinary calcium excretion. Several large cohort studies have reported small but significant increases in either fracture frequency or bone loss associated with increased caffeine intake. Other studies have found no such association. Moderate caffeine intake is not associated with increased bone loss, and so a prudent recommendation would be for adequate dietary calcium intake together with moderate caffeine consumption in older adults. A "moderate" level of caffeine intake seems likely to be 300 mg or less of caffeine, which is equivalent to approximately: one large long black, three cappuccinos, four cups of plunger coffee, or six cups of tea. |
| | Adults |
| | Make plain water your first choice over other drinks; black tea and coffee are also popular and there is some evidence that both can provide benefits for health such as antioxidative properties. Tea and coffee both contain caffeine (a stimulant) and tea contains tannins, which lower the amount of iron that the gut absorbs. Therefore, the Ministry of Health recommends drinking only moderate amounts of tea and coffee. |
| Republic of Korea [HI] 2010 | Pregnant mothers should minimize coffee, black tea, soda, and chocolate. Young children should minimize the consumption of crackers/cookies, carbonated beverages, and fast food. Young adolescents should minimize instant and fast foods and carbonated beverages and should avoid junk food. |
| Sri Lanka [LMI] 2011 | *Consume less sugar, sweets, or sweetened drinks.* Tea without milk and sugar will have certain advantages as they have some antioxidants that will help to improve health. However, it is advisable to avoid tea or coffee closer to a main meal, as it will reduce iron absorption. |
| Thailand [UMI] 2008–2010 | Most food eaten on a daily basis as a main dish or dessert contain sugar. We also take additional sugar from soft drinks, candy, toffee, jelly, syrup, and sugar added to tea, coffee, and other beverages. Therefore, we often add excessive amounts of energy to our regular diet. Children who eat sugary foods often have a lower appetite and are prone to tooth decay. Thus, sugary foods should be limited in their diets. |
| Vietnam [LMI] 2013 | *Increase physical activity, maintain an appropriate weight, abstain from smoking, and limit your consumption of alcoholic/soft drinks and sweets.* |
| **NEAR EAST** | |
| Iran [UMI] 2015 | *Reduce your consumption of sugar, sweet foods and beverages, and soft drinks.* *During the day drink water and unsweetened beverages frequently.* |

**Table 1.** *Cont.*

| Country Publication Date | FAO key Messages and Published Guidelines [Other Resources and Personal Communication] |
|---|---|
| Lebanon [UMI] 2013 | *Limit intake of sugar, especially added sugar from sweetened foods and beverages.* |
| | Avoid drinking tea, coffee, and caffeine-containing carbonated beverages with meals, as food components in these beverages may decrease the absorption of iron in food. |
| | The consumption of the below food components can increase the urinary loss of calcium |
| | Salt (in excessive amounts) |
| | Caffeine (in amounts coming from more than three to four cups of coffee per day). |
| | Water is the preferred fluid to fulfill the body's daily fluid needs and is followed (in decreasing order) by: tea and coffee, low-fat and fat-free milk, non-sugar sweetened beverages that contain some nutritional benefits (such as fruit and vegetable juices), sugar-sweetened and nutrient-poor beverages (such as sweetened fruit juices and soft drinks). |
| | Although tea and coffee contain a good amount of water, these fluids contain caffeine, which increases urine excretion. As such, these fluids lead to a loss of water from the body if consumed in high amounts. |
| | Consumption of sugar-sweetened beverages has been linked to overweight, obesity, type 2 diabetes, and dental caries. Since fluid-derived energy is an important consideration in weight gain, drinking water should be the preferred beverage for hydration purposes. |
| Qatar [HI] 2015 | Limit sweetened food and beverages. Be aware of the amounts of sugar in hot and cold coffee beverages served in cafes. Avoid sweetened drinks such as soda, energy drinks, fruit drinks, vitamin waters, and sports drinks. Allow 30 min after meals before drinking tea to allow for absorption of iron from foods. |
| EUROPE | |
| Albania [UMI] 2008 | Infants in their first year of life |
| | Iron absorption is decreased with drinking tea and coffee |
| | Adults |
| | Consume 1–2 liters of beverages per day choosing those with no sugar added but potable water, mineral water, and fruit or plant teas. Take limited quantities of beverages containing caffeine (coffee, black or green tea, etc.). |
| | Women breastfeeding |
| | Avoid alcoholic beverages and products which contain caffeine. |

**Table 1.** *Cont.*

| Country Publication Date | FAO key Messages and Published Guidelines [Other Resources and Personal Communication] |
|---|---|
| | Women in menopause |
| | It is important to stay away from the alcoholic beverages, coffee, and spicy food because those influence in worsening the warmth. |
| | Children 2–3 years old |
| | Consume at least three glasses of sugarless juices (water, mineral water, original plant tea, fresh watered juicy fruits, no sugar added, etc.) per day. The beverages should not contain caffeine or alcohol. |
| | Children 4–6 years old |
| | Consume at least three to four glasses of sugarless juices per day. The beverages should not contain caffeine like teas or cola. |
| | Teenagers 13–18 years old |
| | Consume at least five to six glasses of sugarless juices per day. The beverages should not contain caffeine like teas or cola and alcohol. |
| Austria [HI] 2010 | *Non-alcoholic beverages: Drink at least 1.5 liters of fluid, preferably low-energy drinks in the form of water, mineral water, unsweetened fruit or herbal teas, or diluted fruit and vegetable juices. A daily moderate consumption of coffee, black tea (three to four cups) and other caffeinated beverages is acceptable.* |
| | *Processed foods high in fat, sugar, and salt: Some processed foods (such as sweets, pastries, fast food products, snacks, and soft drinks) are high in fat, sugar, and salt and are less desirable nutritionally. They should be consumed sparingly—a maximum of one small serving a day.* |
| | There is no objection about the daily moderate consumption of coffee, black tea (three to four cups) and other caffeinated drinks. |
| | Is coffee a liquid robber? In "normal" quantities (up to four cups/day) it is not. |
| Belgium [HI] 2005 | *Drink at least 1.5 liters of water every day (water, coffee, tea, etc.).* |
| | Pregnant women |
| | Avoid alcohol, tobacco, and large amounts of tea and coffee. |
| | 3 to 12 years |
| | Water is preferred over other beverages since the latter often contain a lot of sugar. |
| | 13 to 18 years |
| | Water, tea, or coffee are components of a "good breakfast." |

Table 1. *Cont.*

| Country Publication Date | FAO key Messages and Published Guidelines [Other Resources and Personal Communication] |
|---|---|
| | 60 and older |
| | Water, (weak) coffee, and tea are encouraged at each meal and throughout the day to maintain hydration. |
| | "Food for brain": tea and coffee are sources of antioxidants. |
| Bulgaria [UMI] 2006 | *Limit the consumption of sugar, sweets, and confectionery, and avoid sugar-containing soft drinks.* |
| | The high intake of sugar and sugar-containing foods and beverages leads to being overweight and obesity, which themselves increase the risk for hypertension, cardiovascular diseases, and type 2 diabetes. |
| | Try to drink tea or coffee without sugar. Prefer honey or brown sugar for sweetening. |
| Croatia [HI] 2002 | Black coffee can reduce calcium absorption. |
| Cyprus [HI] 2007 | *Reduce the consumption of sugar, preferring appropriate beverages and foods with reduced or no sugar.* |
| Denmark [HI] 2013 | Maximum amount of caffeine for an adult/day is 400 mg. |
| | Caffeine for children and youngsters is discouraged. |
| | Pregnant and breastfeeding women are encouraged to minimize their intake of caffeine. Furthermore, they are discouraged from drinking any energy drinks. |
| Estonia [HI] 2006 | *Limit the consumption of sweets and soft drinks.* |
| Finland [HI] 2014 | *Decrease consumption of soft drinks and sweet juices.* |
| | You should avoid having sugary drinks regularly, as they are associated with obesity and the risk of developing type 2 diabetes. A high intake of sugary drinks also affects your dental health. |

**Table 1.** *Cont.*

| Country Publication Date | FAO key Messages and Published Guidelines [Other Resources and Personal Communication] |
|---|---|
| France [HI] 2011 | *Limit the consumption of sugar and foods high in sugar (soft drinks, candies, chocolate, pastries, desserts, etc.)* |
| | Beverages: To quench your thirst, water is the only essential drink. You should drink at least a liter and a half a day, during and between meals, as it is or in the form of hot drinks (tea, herbal tea, coffee). If you are fond of soft drinks or sweet drinks, try to content yourself with one drink a day or only two or three at a party. The sugar contained in these drinks does not decrease your appetite and makes you easily gain weight. |
| | Consume sugary drinks with moderation. Instead, choose fruit juices without added sugar. If you drink a lot of soft drinks, and do not manage to replace with water, drink light versions to reduce your sugar intake. In tea or herbal teas, a little spoon of honey is always better than a cube of sugar. For children, it is best to regularly offer tap water, spring water, or mineral water (flat or carbonated), with plant extracts and sugar-free; plain or flavored milk with cocoa, vanilla, orange blossom, or syrup (fruit based concentrates); or pure fruit juice (no added sugar) or a squeezed fruit juice. |
| Georgia UMI 2005 | *Do not drink tea while eating plant meals rich in iron (e.g., vegetables, legumes), because tea limits the availability of non-heme iron.* |
| | *Sweet drinks can be replaced by sufficient amounts of unsweetened liquids, e.g., boiled water.* |
| Germany [HI] 2013 | Drink water with or without carbon dioxide and low-energy drinks. Drink sweetened drinks rarely. These are energetic and can increase the emergence of obesity. |
| Greece [HI] 1999 | Always prefer water over soft drinks. |
| | Non-alcoholic beverages, including sodas, have not been conclusively linked to health effects. |
| | Fruit juices are likely to share some of the benefits of fruits, whereas other beverages have been criticized for their high content in simple carbohydrates. |
| | Simple sugars are plentiful in deserts, and also exist, or are added, in beverages, like coffee, tea, fruit juices, soft drinks, and colas. Simple sugars have glycemic effects, mainly comparable to or less than those of starch from cooked foods. |
| Hungary [HI] 2004–2016 | *Avoid the frequent consumption of foods or drinks rich in added sugar. To quench your thirst, drink water or mineral water instead of sugary drinks.* |
| | Children: Consumption of energy drinks is not recommended because it has high caffeine and high sugar content. |
| Ireland [HI] 2012 | *Limit high fat, sugar, and salt foods from the top shelf of the pyramid to no more than once or twice a week (includes soft drinks and juice).* |
| | Fluids: Water is the best drink for hydration and the safest for teeth. At least 8–10 cups of fluid are needed every day, and this can come from fluids in the foods eaten as well as water, milk, tea, and coffee. |
| | Other sources of excessive sugar in the form of sweets, soft drinks, cakes, biscuits, and confectionery are not integrated with nutritious high-fiber foods. Limiting these foods will help to lower calories, fat, sugar, and salt. Although "diet" soft drinks (sugar free) can be used sometimes for variety, they are acidic and if taken too frequently, they can still harm teeth. |
| | Pregnant women: Caffeine in the mother's diet can reach the baby and may be harmful. High caffeine intakes during pregnancy are not advisable, and mothers should aim to keep their caffeine intake below 200 mg of caffeine per day (an accompanying table provides caffeine content of the most common caffeine-containing foods and drinks). Also limit consumption of caffeine while breastfeeding. |

**Table 1.** *Cont.*

| Country Publication Date | FAO key Messages and Published Guidelines [Other Resources and Personal Communication] |
|---|---|
| Israel [HI] 2008 | Eat/drink sweets, snacks, and sweetened drinks sparingly. |
| Italy [HI] 2003 | *Consume appropriate amounts of sugars, sweets, and sugar-sweetened beverages.*<br><br>The balance of water must be maintained by drinking essentially water, both tap and bottled, safe and controlled. Remember that certain drinks (such as orange juice, sodas, fruit juices, coffee, and tea), in addition to providing water, also bring other substances that contain calories (for example, simple sugars) or that are pharmacologically active (for example caffeine). These should be used in moderation. |
| Latvia [HI] 2003–2008 | *Limit consumption of salt and sugar and products containing them.*<br><br>Do not wait until you feel thirsty—use liquid regularly! The day should consist of eight glasses of liquid including water and herbal teas.<br><br>[The Centre for Disease Prevention and Control infographics stresses the importance of using non-sweetened and non-caffeinated drinks like pure water (2 liters a day) or herbal tea for hydration needs. Usage of these liquids instead of coffee or tea is also recommended for mental health and heart disease prevention.<br><br>It is recommended for pregnant women to drink water every day and limit caffeine consumption to 200 mg per day. Exclusion of energy drinks during this period is also recommended. As coffee and tea reduces iron absorption, reduced intake of these drinks is recommended whilst using iron supplements. Limited coffee and tea usage for vegetarians is also recommended, especially during meals as these drinks reduce calcium absorption] |
| Malta [HI] 2016 | *Avoid adding sugar to your tea or coffee. Avoid energy drinks. Avoid soft and sweetened drinks especially in children.*<br><br>Avoid energy drinks: Energy drinks generally have a high sugar and caffeine content as well as other stimulant substances. They can easily contribute to an excess energy intake and may cause a variety of problems such as headaches, restlessness, insomnia, stomach upset, fast heartbeat, tooth decay, and anxiety. |
| Netherlands [HI] 2016 | *Drink three cups of tea daily.*<br><br>*Replace unfiltered coffee by filtered coffee.*<br><br>*Minimize consumption of sugar-containing beverages.*<br><br>Tea: In the context of this advisory report, the term "tea" covers green tea and black tea. Green tea comes from the tea plant, but unlike black tea, it has not undergone oxidation. Herbal teas and, for example, rooibos are outside the scope of this report. The Committee concludes that it has been convincingly demonstrated that the consumption of tea reduces the risk of stroke. That conclusion is based on the fact that randomized controlled trials (RCTs) show that three cups of green tea or five cups of black tea a day reduce blood pressure, while the consumption of tea is associated with a lower risk of stroke in cohort studies. In addition, the consumption of black tea and the consumption of green tea are plausibly associated with a lower risk of diabetes. |

**Table 1.** *Cont.*

| Country Publication Date | FAO key Messages and Published Guidelines [Other Resources and Personal Communication] |
|---|---|
| | Coffee: The way that coffee is prepared, whether it is filtered or not, makes a difference to its influence on health. That is because filtering can remove the cholesterol-raising substances cafestol and kahweol from coffee. In the context of this advisory report, "filtered coffee" covers coffee made using a filter machine, coffee made using coffee pods, instant coffee, and vending-machine coffee made using liquid coffee concentrate. Unfiltered coffee includes boiled coffee, cafetière coffee, Greek coffee, and Turkish coffee. Espresso and coffee from vending machines that use fresh coffee may count either as filtered or as unfiltered, depending on the type of machine, the type and amount of coffee, and the type of filter used. The Committee concludes that it has been convincingly demonstrated in RCTs that unfiltered coffee increases LDL cholesterol, which is known to be a causal risk factor for coronary heart disease. Coffee consumption is associated with lower risks of coronary heart disease, stroke, and diabetes in cohort studies, which relate predominantly to filtered coffee consumption. |
| | Sugar-containing beverages: Examples include fruit juice drinks and "nectars," carbonated drinks ("pops" and "sodas"), ice tea, vitamin-fortified water, and sports drinks made by the addition of sugar. The Committee concludes that it has been convincingly demonstrated that the consumption of drinks with added sugar increases the risk of diabetes. That conclusion is based on the fact that RCTs have shown that drinks with added sugar increase body weight, while cohort studies indicate an association of the consumption of drinks with added sugar with a higher risk of diabetes. |
| Norway [HI] 2014 | *Find the right balance between how much energy you consume through food and drink and how much energy you use by being physically active.* |
| | *Limit your consumption of food and drink with a high sugar content.* |
| | People at risk of developing heart disease and diabetes should limit their consumption of non-filtered coffee. |
| | It is recommended that drinks with added sugar or caffeine not be offered in schools. |
| Poland [HI] 2010 | *Moderate your intake of sugar and sweets.* |
| | Drink tea and coffee in moderation. It is better to choose green tea, fruit tea, or herbal tea and rarely drink black teas and coffee. |
| | Exclude or limit sweetened beverages, especially sodas (carbonated drinks). Their excess in diet favors even more occurrence of overweight and obesity than consumption of great amounts of sugar and sweets. Carbon dioxide can suppress thirst. |
| | Tea is a good source of water, which you can drink but in smaller amounts. Fruit and herbal teas are best, rarely drink dark tea, which should not be too strong. Try not to sweeten tea, and if you do not like the bitter taste, decrease the amount of added sugar. |

Table 1. *Cont.*

| Country Publication Date | FAO key Messages and Published Guidelines [Other Resources and Personal Communication] |
|---|---|
| Portugal [HI] 2003 | *Prefer water to beverages containing added sugar, alcohol, and caffeine.*<br><br>The recommended minimum consumption of liquids per day is 1.5 to 3 liters, depending on the activity and state of health of the individual. Although water is the best drink to satisfy your thirst, you can also drink other drinks that do not contain added sugar, alcohol, or caffeine. Natural fruit juices and caffeine-free teas are examples of these beverages. Coffee and some teas and soft drinks contain caffeine, a stimulant substance whose intake should be limited to a maximum of 300 mg per day. In the case of children adolescents and their consumption is discouraged. Despite the designation, decaffeinated beverages are not completely exempt of this substance. In tea, the absorption of caffeine is slower than in coffee, which means the stimulating effect is lower but lasts longer. |
| Romania [UMI] 2006 | *Eat highly processed foods high in sugar sparingly.*<br><br>Caffeine, nicotine, and warm foods also increase the thermic effect of food.<br><br>Various seeds, unprocessed grains, sprouted wheat and wheat bran, nuts, vegetables, green vegetables, hard water, coffee, tea, and cocoa are good sources of magnesium.<br><br>Pregnant women<br><br>Iron supplements must be administered between meals, preferable with liquids containing ascorbic acid (increases the iron absorption), avoiding concurrent administration of tea, milk, or coffee.<br><br>Caffeine: There are controversies about an acceptable level of the caffeine intake during pregnancy. In USA, the Food and Drug Administration (FDA) recommends limiting the intake of caffeine during pregnancy and if possible complete avoidance, given the teratogenic effects seen in animal studies. In humans, large doses of caffeine (over 300 mg/day) are associated with low birth weight.<br><br>Caffeinated beverages should also be avoided while breastfeeding.<br><br>Elderly<br><br>Avoid alcohol abuse, excess intake of caffeine and unnecessary medications. Tea and coffee have no contribution to calorie intake unless milk, cream, or sugar is added. They contain caffeine and theobromine and their excess use may cause insomnia and irritability.<br><br>Tea is an important source of bioflavonoids with antioxidant properties that might protect against cardiovascular disease. Cocoa contains significant amounts of iron, proteins, fats, and carbohydrates; however, because of the quantities in which it is usually consumed, it loses its nutritional value. |
| Slovenia [HI] 2011 | *Consume enough fluids, preferably drinking water, mineral water, unsweetened fruit or herbal teas or diluted fruit and vegetable juices.* |
| Spain [HI] 2008 | Avoid excessive consumption of sugary soft drinks and juices with added sugar. The studies warn about the relationship between the excessive consumption of these soft drinks and the increase in childhood obesity. Do not consume them as a substitute for water. |

**Table 1.** *Cont.*

| Country Publication Date | FAO key Messages and Published Guidelines [Other Resources and Personal Communication] |
|---|---|
| Sweden [HI] 2015 | *Less sugar - Hold back on the sweets, pastries, ice creams, and other products containing lots of sugar. Cut back on sweet drinks in particular.* <br><br> Water is by far the best drink for quenching thirst—much better than fizzy drinks, juice, soft drinks, and sports drinks. Sweet drinks in particular increase the risk of obesity as they contain lots of calories but do not make you feel full. |
| Switzerland [HI] 2011 | Beverages: Consume 1–2 liters per day, preferably in the form of unsweetened drinks such as tap water/mineral water or fruit/herb tea. Caffeinated drinks, such as coffee, black tea, and green tea, can also count towards your fluid intake. |
| Turkey [UMI] 2006 | Reduce the consumption of sugary beverages and sweets and choose foods containing less sugar. Instead of sugar-added soft drinks, please prefer skim milk, ayran (watered yogurt), and kefir. Instead of drinking beverages containing sugar, water should be preferred. Drink sugar-free tea and herbal teas. |
| United Kingdom [HI] 2016 | *If consuming foods and drinks high in fat, salt, or sugar, have these less often and in small amounts.* <br><br> Aim to drink six to eight glasses of fluid every day. Water, lower fat milk, and sugar-free drinks including tea and coffee all count. Fruit juice and smoothies also count towards your fluid consumption, although they are a source of free sugars and so you should limit consumption to no more than a combined total of 150 mL per day. Swap sugary soft drinks for diet, sugar-free, or no added sugar varieties to reduce your sugar intake in a simple step. <br><br> The European Food Safety Authority (EFSA) opinion confirms the safety of daily caffeine intakes of up to 3 mg per kg of body weight for children and adolescents (3–18 years) and up to 400 mg for adults. |
| **LATIN AMERICA and the CARIBBEAN** | |
| Antigua and Barbuda [HI] 2013 | *Reduce the intake of food and drinks that are high in sugars and fats.* |
| Argentina [UMI] 2015 | *Limit the consumption of sugary drinks and foods high in fats, sugar, and salt.* |
| Bahamas [HI] 2002 | Eating too much high-seasoned and sweet foods increases your risk of developing heart disease, high blood pressure, and type II diabetes. Sodas, fruit-flavored drinks, and desserts like guava duff and tarts contain lots of sugar. |
| Barbados [HI] 2017 | *Choose food and beverages with less added sugar every day.* <br><br> Many foods and beverages with added sugar are high in calories and have little nutritional value. These extra calories can contribute to overweight and obesity. High sugar intake can also increase blood pressure and risk of death from cardiovascular disease. |

**Table 1.** *Cont.*

| Country<br>Publication<br>Date | FAO key Messages and Published Guidelines<br>[Other Resources and Personal Communication] |
|---|---|
| Belize [HI]<br>2012 | Replace sweet drinks with water. Add less sugar when preparing foods and drinks. Foods that contain a high amount of sugar include soft drinks, box drinks, cakes, ice cream, puddings, sweets, jams and jellies, and condensed milk. Benefits: better weight control, better control of blood sugar levels, and fewer problems with dental caries/tooth decay. |
| Bolivia [LMI]<br>2013 | *Avoid the over consumption of sugar, sweets, sodas, and alcoholic drinks.*<br><br>*Reduce the consumption of tea and coffee, replacing them with milk, fruit juices or "apis".*<br><br>Avoid the exaggerated consumption of sugar, sweets, soft drinks: Exaggerated consumption of these products cause tooth decay and deteriorate health.<br><br>Certain substances such as caffeine and alcohol have a diuretic effect and increase the urinary losses of water and electrolytes. Adequate water intake can be covered not only with water, but also with food or liquids that do not contain caffeine and alcohol. |
| Brazil [UMI]<br>2014 | *Avoid consumption of ultra-processed foods.*<br><br>*Because of their ingredients, ultra-processed foods, such as salty fatty packaged snacks, soft drinks, sweetened breakfast cereals, and instant noodles, are nutritionally unbalanced. As a result of their formulation and presentation, they tend to be consumed in excess, and displace natural or minimally processed foods. Their means of production, distribution, marketing, and consumption damage culture, social life, and the environment.*<br><br>Examples of natural/minimally processed foods: tea, herbal infusions, coffee, and tap, spring, and mineral water.<br><br>Breakfast examples: Fruits and coffee with milk are a constant part of the first meal of the day.<br><br>Water: Pure water (or, as preferred by some people, "seasoned" with lime slices or mint leaves) is the best option. Brazilians also consume water in the form of coffee and tea, in which case sugar should be reduced to a minimum or not added at all. |
| Chile [HI]<br>2013 | *If you want to maintain a healthy weight, avoid eating sugar, sweets, sugar-sweetened juices, and beverages.*<br><br>The [lactating] mother should be well fed, eat healthy, drink water, and not consume any alcoholic drinks, tobacco, drugs, or excessive tea or coffee.<br><br>Children 6–11 months<br><br>Artificial sweeteners (saccharin, aspartame, sucralose, stevia, or other) should not be used in children's foods under 2 years of age directly or in preparations or commercial products labeled "light" or "diet" or other similar ones. Powdered drinks, juices or nectars with sugar, soft drinks, and in general any sugary drink or with artificial sweeteners, are not recommended, nor necessary. |

Table 1. *Cont.*

| Country Publication Date | FAO key Messages and Published Guidelines [Other Resources and Personal Communication] |
|---|---|
| Colombia UMI 2014 | *To maintain a healthy weight, reduce the consumption of packaged products, fast foods, soft drinks, and sweetened drinks.* |
| | Pregnant women |
| | Avoid the consumption of products like soft drinks, energy drinks, and sugary drinks that favor the onset of diabetes, being overweight, and obesity. |
| | Lactating mothers |
| | Do not consume alcoholic beverages or energy drinks. |
| Costa Rica UMI 2010 | Decrease the amount of sugar that is used to sweeten your drinks. |
| | Choose juices or drinks that are 100% natural, without sugar. Avoid the consumption of pastries, cookies, condensed milk, dulce de leche, soft drinks, sweets, chocolates, ice creams, jellies, and jams. |
| | In addition to water, you can consume other liquids like juices, tea, broths, and soups. |
| Cuba UMI 2009 | Decrease the consumption of all types of sweets (homemade, industrial, candies, jams and others), as well as sweetened drinks. |
| Dominica UMI 2007 | *Choose less sweet foods and drinks.* |
| Dominican Republic UMI 2009 | Prefer medium and low energy foods to achieve energy balance. Foods high in energy include: fritters, desserts, oils, margarine, butter, lard, mayonnaise, picaderas, pork rinds, bacon, fast food, snacks, dressings, stuffed crackers, sausages, pastries, and carbonated and alcoholic beverages. |

Table 1. *Cont.*

| Country Publication Date | FAO key Messages and Published Guidelines [Other Resources and Personal Communication] |
| --- | --- |
| El Salvador LMI 2012 | *Avoid eating sugary foods and drinks, chips, sausages, sweets, highly processed foods, and canned foods.* |
| | These foods have delicious taste and appetizing appearance, but they provide large amounts of saturated fat and calories. Therefore, the consequences of frequent consumption are: becoming overweight and obesity, the appearance of diseases of the heart, diabetes mellitus, hypertension, tumors and cancers, and dental caries. |
| Grenada UMI 2016 | *Choose to use less sweet foods and drinks.* |
| Guatemala LMI 2012 | Avoid the consumption of carbonated water, drinks energizers, bottled drinks with artificial flavors, packaged juices, etc., because they contain excess sugar, preservatives, and dyes that are harmful to health. Coffee consumption is not recommended as a substitute for water. Coffee stimulates acidity or secretion of gastric acids and produces discomfort in cases of diseases of the digestive system. The effect of tea, especially black tea, is similar. |
| Honduras LMI 2013 | Do not substitute water for coffee or tea, as they can cause acidity and other problems in the digestive system. |
| | Avoid the consumption of soft drinks, energy drinks, bottled beverages, packaged juices, juices and natural soft drinks, etc., since they generally contain large amounts of sugar, dyes, and preservatives that are harmful to health. |
| Jamaica UMI 2015 | *Reduce intake of sugary foods and drinks.* |
| | These include honey, syrup, jam, sweetened carbonated beverages, condensed milk, sweet snacks, and desserts. Benefits: reduces risk of becoming overweight/obesity, hypertension, diabetes, heart diseases, and other chronic illness. |
| Mexico UMI 2015 | *Drink plenty of plain water. Drink plain aguas frescas or flavored water without added sugar instead of sweetened drinks such as soft drinks, juices, and aguas frescas.* |
| | Decrease the consumption of energy-containing beverages such as soft drinks, nectars, and sweetened drinks with fruit flavors. Mexico has one of the highest consumption of sugary drinks in the world for all age groups from 1 year old. The types of beverages that contribute the greatest to energy in the population are: soft drinks (carbonated and not carbonated); drinks made with fruit juices (with or without sugar), which are taken as natural juices; fresh waters; juices made with 100% fruit; and flavored milks. |
| | Beverage classes according to their energy content, nutritional value, and risks to health on a scale that classifies drinks from the most (level 1) to the least (level 5) healthy. |
| | Level 1: Drinking water; Jamaican water without sugar (six to eight glasses of water per day) |
| | Level 2: Low-fat milk (1%) or fat-free and sugar-free (max two glasses a day) |
| | Level 3: Coffee, tea, and fruit water without sugar (maximum four cups a day) |
| | Level 4: Drinks with high calorific value and limited health benefits (fruit juices, whole milk, fruit smoothies with sugar or honey, and sports drinks) (maximum 1/2 cup per day) |
| | Level 5: Drinks with sugar and low nutrient content |

**Table 1.** *Cont.*

| Country Publication Date | FAO key Messages and Published Guidelines [Other Resources and Personal Communication] |
|---|---|
| Panama [UMI] 2013 | *Avoid sodas, iced tea, and sugary drinks. Prefer natural juices without sugar.* |
| | The frequent or daily consumption of sodas or sugary drinks leads to the appearance of obesity, diabetes, hypertension, and high cholesterol, as well as cardiovascular and renal diseases. Avoid all colors and flavors of sodas, iced teas of any flavor, and sugary drinks with colorants, packaged, or with powder. |
| Paraguay [UMI] 2015 | *Drink less carbonated beverages and artificial juices because they damage your health.* |
| | Consume less sweets, sodas, and sweetened drinks to stay healthy. Foods high in simple sugars contribute empty calories without special nutrients, and so excessive consumption can harm your health when these foods rich in sugars are eaten in excess, they are accumulated in the body in the form of fat, which is considered a risk factor of obesity, cardiovascular disease (heart and arteries), diabetes, dental cavities, and others. |
| Saint Kitts and Nevis [HI] 2010 | *Limit the use of foods and drinks with added salt and sugar.* |
| Saint Lucia [UMI] 2007 | *Choose fewer beverages and foods preserved or prepared with added sugar.* |
| Saint Vincent and the Grenadines [UMI] 2006 | *Reduce the intake of sugar: use less sugar, sweet foods, and drinks.* |
| Uruguay [HI] 2016 | *Base your diet on natural foods, and avoid the regular consumption of ultra-processed products with excessive contents of fat, sugar, and salt.* |
| | *Prefer water to other beverages. Limit sodas, artificial juices, and flavored waters.* |
| Venezuela [UMI] 1991 | Soft drinks and other sugary drinks only provide calories. In contrast, natural fruit juices prepared at home provide an addition of calories, vitamins, and minerals |

**Table 1.** *Cont.*

| Country Publication Date | FAO key Messages and Published Guidelines [Other Resources and Personal Communication] |
|---|---|
| | **NORTH AMERICA** |
| Canada [H] 2007 | Limit foods and beverages high in calories, fat, sugar or salt (sodium) [many listed, i.e., soft drinks, sports and energy drinks, and sweetened hot or cold drinks, hot chocolate, specialty coffee]. |
| | Have a glass of low fat milk rather than pop or fruit drinks. Use low-fat evaporated milk instead of cream or coffee whitener in coffee or tea. |
| | Soft drinks, sports drinks, energy drinks, and alcoholic beverages can add a significant number of calories to the diet. These drinks may also contain caffeine or sodium. |
| | Health Canada: https://www.canada.ca/en/health-canada/services/food-nutrition/food-safety/food-additives/caffeine-foods/foods.html |
| | A review (Nawrot et al, Food Additives and Contaminants, 2003) undertaken by Health Canada scientists has considered the numerous studies dealing with caffeine and its potential health effects. It has re-confirmed that for the average adult, moderate daily caffeine intake at dose levels of 400 mg/day is not associated with any adverse effects. Data has shown, however, that women of childbearing age and children may be at greater risk from caffeine. |
| | Recommended max caffeine intake levels. |
| | 4–6 years: 45 mg/day |
| | 7–9 years: 62.5 mg/day |
| | 10–12 years: 85 mg/day |
| | Women planning to become pregnant, pregnant women, and breast feeding mothers: 300 mg/day |
| | Suggestions for adolescents 13+ years: 2.5 mg/kg body weight. |

Table 1. *Cont.*

| Country Publication Date | FAO key Messages and Published Guidelines [Other Resources and Personal Communication] |
|---|---|
| United States HI 2016 | *Limit calories from added sugars and saturated fats and reduce sodium intake.* |
| | When choosing beverages, both the calories and nutrients they may provide are important considerations. Beverages that are calorie-free, especially water, or that contribute beneficial nutrients, such as fat-free and low-fat milk and 100% juice, should be the primary beverages consumed. Milk and 100% fruit juice should be consumed within recommended food group amounts and calorie limits. Sugar-sweetened beverages, such as soft drinks, sports drinks, and fruit drinks that are less than 100% juice, can contribute excess calories while providing few or no key nutrients. If they are consumed, amounts should be within overall calorie limits and limits for calories from added sugars. |
| | Caffeine is not a nutrient; it is a dietary component that functions in the body as a stimulant. Caffeine occurs naturally in plants (e.g., coffee beans, tea leaves, cocoa beans, and kola nuts). It is also added to foods and beverages (e.g., caffeinated soda and energy drinks). |
| | Moderate coffee consumption (three to five 8-oz cups/day or providing up to 400 mg/day of caffeine) can be incorporated into healthy eating patterns. This guidance on coffee is informed by strong and consistent evidence showing that, in healthy adults, moderate coffee consumption is not associated with an increased risk of major chronic diseases (e.g., cancer) or premature death, especially from CVD. |
| | Individuals who do not consume caffeinated coffee or other caffeinated beverages are not encouraged to incorporate them into their eating pattern. |
| | Limited and mixed evidence is available from randomized controlled trials examining the relationship between those energy drinks which have high caffeine content and cardiovascular risk factors and other health outcomes. |
| | Caffeinated beverages, such as some sodas or energy drinks, may include calories from added sugars, and although coffee itself has minimal calories, coffee beverages often contain added calories from cream, whole or 2% milk, creamer, and added sugars, which should be limited. The same considerations apply to calories added to tea or other similar beverages. |
| | Those who choose to drink alcohol should be cautious about mixing caffeine and alcohol together or consuming them at the same time. |
| | Women who are capable of becoming pregnant or who are trying to, or who are pregnant, and those who are breastfeeding should consult their health care providers for advice concerning caffeine consumption. |

HI: high-income; LI: low-income; LMI: lower-middle-income; UMI: upper-middle-income; CVD: cardiovascular disease.

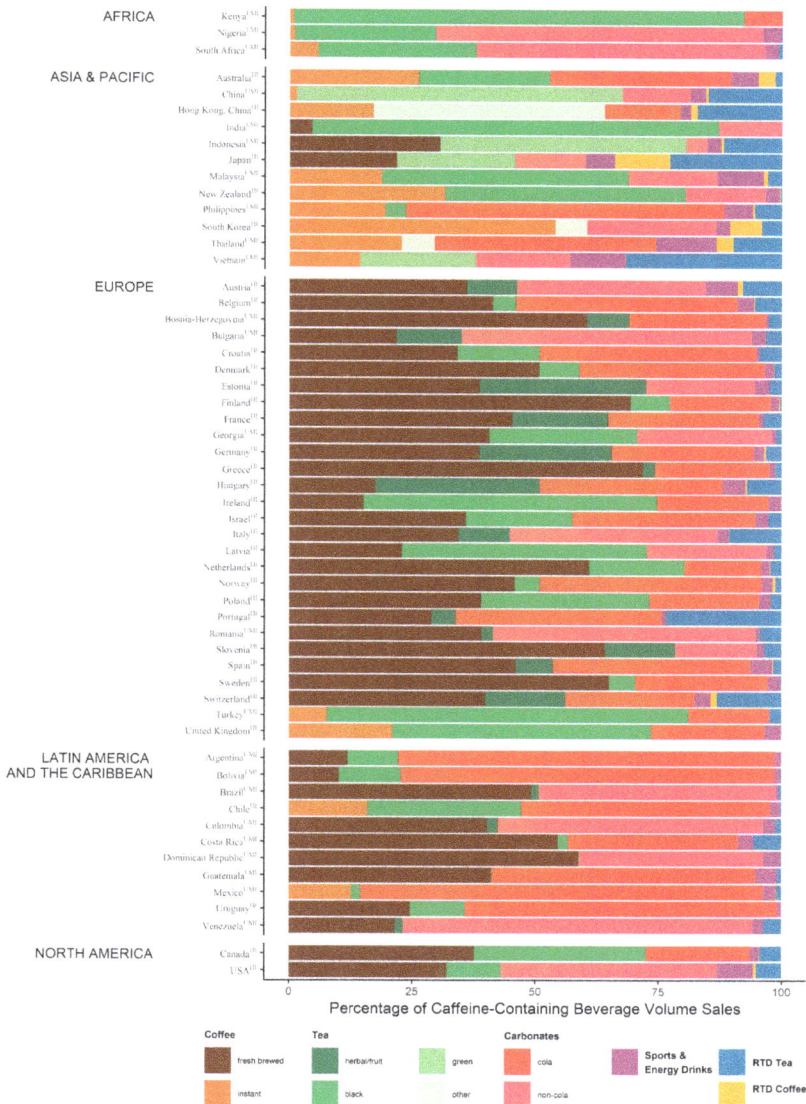

**Figure 1.** Percentage of caffeine-containing beverage volume sales per beverage (Euromonitor 2017). Subcategories of coffee, tea, and carbonates were assigned according to the most commonly, but not exclusively, consumed beverage type in that category. Countries were classified by income based on World Bank 2017. HI: high-income; LI: low-income; LMI: lower-middle-income; UMI: upper-middle-income. Data for RTD beverages were incomplete for Kenya, Nigeria, South Africa, India, Bosnia-Herzegovina, Croatia, Estonia, Georgia, Latvia, Slovenia, Bolivia, Costa Rica, Dominican Republic, Guatemala, Uruguay, and Venezuela.

## 3.1. Africa

The most commonly consumed CCB in African countries include tea and carbonated soda. Tea is typically of the black type while carbonated drinks are commonly non-cola-type (unlikely caffeinated including lime, ginger ale, tonic water, orange carbonates, and "other"). When coffee is consumed, it is usually of the regular (not decaffeinated, >95% of sales) and instant type. Data for RTD coffee/tea were not complete for these selected African countries.

Five African countries have published FBDG that consider dietary caffeine sources in some context. Nigeria, Sierra Leone, and South Africa discourage high coffee and tea intake because they inhibit iron bioavailability or increase phosphorous levels. South Africa's guidelines for ages 5+ nevertheless support the intake of these beverages as a means to attain adequate fluid intake, further noting that any diuretic effects of caffeine are only a concern for individuals unaccustomed to regular caffeine intake. Most FBDG discourage caffeinated soda, but only due to their high sugar content.

## 3.2. Asia and the Pacific

Tea and carbonated soda are the leading CCB sold across Asia and the Pacific countries. High carbonated soda consuming countries prefer the cola type, while low consuming countries prefer non-colas. Black, green, and "other" teas are major tea types consumed. RTD teas are also popular in Japan, Hong Kong, and Vietnam. Most coffee that is consumed is of the regular (>97% of sales) and instant type; instant mixes (coffee, sugar, and cream powder) are especially popular in South Korea, Hong Kong, Malaysia, Thailand, Philippines, Vietnam, and China.

Six of the fifteen included countries of Asia and the Pacific express caution concerning the iron inhibitory effects of coffee and tea, particularly when these beverages are consumed with meals. China, India, Indonesia, New Zealand, and Korea all advise pregnant and lactating women to minimize their intake of CCB. Indonesia and New Zealand further cite research supporting caffeine limits of 250–300 mg/day for these women. Potential diuretic effects of caffeine are discussed in guidelines for Fiji, Indonesia, and Malaysia. According to Fiji and Indonesia, heavy tea and coffee consumers may need to adjust their water intake, while Malaysian guidelines note little concern regarding the diuretics effects of CCB in amounts typically consumed. India's guidelines discuss the stimulant effects of caffeine present in coffee and tea and advise moderation when consuming these beverages. Excess consumption of coffee was viewed unfavorably for cardiovascular health, while any potential benefits noted for tea consumption were off-set by its caffeine content. The majority of FBDG discouraged caffeinated soda due to its high sugar content. New Zealand further referenced the caffeine content of these beverages, discouraging the intake of these and other caffeine-containing beverages among children and adolescents. With some concern of caffeine's impact on bone health, older people in New Zealand are advised to consume no more than 300 mg of caffeine per day. Moderate amounts of tea and coffee are also advised for adults; advice that aims to balance the beneficial and potentially adverse properties of these beverages attributable to polyphenol, caffeine, and tannin content. Sri Lanka also noted that tea without milk and sugar has some antioxidants that benefit health.

## 3.3. Near East

FBDG for Iran advise the general population to reduce soft drink consumption in the context of reducing overall sugar intake. Lebanon's guidelines advise individuals to avoid consuming coffee, tea, or caffeinated sodas with meals as they inhibit dietary iron absorption. Despite notes concerning caffeine's diuretic effects, tea and coffee are the preferred beverages (after water) for hydration. Sweetened beverage intake should be limited according to Qatar's guidelines and in this context, soda and energy drinks are discouraged and careful attention made to the amount of sugar added to coffee.

*3.4. Europe*

Overall, coffee and carbonated soft drinks are the top CCB sold in Europe. Netherlands consumes the largest volume of coffee per capita than any other country in the Euromonitor database, followed by Finland and Sweden. U.K. and Turkey prefer instant coffee while the rest of Europe prefers fresh-brewed coffee. Decaffeinated coffee accounts for ≈8% of coffee sold in Spain and U.K. and <5% for other parts of Europe. Most carbonated soft drinks sold are of the cola-type. Ireland, Turkey, U.K., and Latvia prefer tea over other CCB. Ireland consumes the more tea per capita than any other country in the Euromonitor database. Black and fruit/herbal teas are the most commonly consumed teas across Europe. Sports and energy drinks and RTD teas are consumed at varying amounts across Europe while RTD coffee consumption is uncommon.

Thirty European countries have published FBDG that consider dietary caffeine sources in some context. Albania, Georgia, Latvia, and Romania are the only set of guidelines noting the iron inhibitory effects of coffee and tea. Latvia and Croatia's guidelines stated that coffee and tea can reduce calcium absorption. Albania, Latvia, and Portugal were the only FBDG that discouraged the intake of CCB to meet daily water requirements. Most FBDG that specifically mention limiting soda or the broader SSB category do so in the context of limiting sugar intake. Energy drinks are discouraged in Malta guidelines due to their sugar as well as stimulant content. Albania, Belgium, Denmark, Ireland, Latvia, and Romania advise pregnant and lactating women to minimize consumption of coffee, tea, or other CCB (≤200–300 mg caffeine/day). Albania, Denmark, Hungary, and Portugal discourage caffeine intake among children. In Denmark and the U.K., adults are advised to limit caffeine intake to 400 mg/day. In Portugal, this limit is set to 300 mg/day. Netherlands' guidelines recommend the daily consumption of three cups of green or black tea on the basis of research showing it reduces risk of stroke, blood pressure, and possibly diabetes. Similar benefits are stated for coffee consumption, but the Dutch are only advised to replace unfiltered coffee with filtered coffee due to known cholesterol-raising substances present in the former. Romania notes that tea is an important source of bioflavonoids with antioxidant properties that might protect against cardiovascular disease (CVD) but does not provide recommendations for tea per se. In contrast, Latvia discourages the use of coffee or tea in place of water or herbal teas for hydration, in part, for mental health and heart disease prevention.

*3.5. Latin America and the Caribbean*

Carbonated soda (mostly cola-type) and coffee (mostly fresh-brewed) are the most commonly sold CCB in Latin America and the Caribbean. Argentina and Uruguay are also heavy consumers of yerba mate [22]. Uruguay has the highest per capita consumption of yerba mate in the world [22–24]. Other CCB are less commonly consumed across this region compared to other regions of the world.

Twenty-six countries of Latin America and the Caribbean have published FBDG with some mention of dietary caffeine. All countries advised limiting SSB (including soda). Only seven guidelines made specific reference to coffee, tea or caffeine. Yerba mate was not specifically mentioned in any FBDG. Pregnant and lactating mothers in Chile are advised to limit tea and coffee intake while Colombian guidelines advise they avoid energy drinks. In FBDG of Bolivia, Guatemala and Honduras, coffee, tea, and caffeine more generally, were discouraged as substitutes for water because they are diuretics, acidic and/or lead to digestive system problems. In Mexican guidelines, non-sweetened coffee and tea are limited to four cups/day. In Brazil, unsweetened coffee and tea were acceptable substitutes for water.

*3.6. North America*

In North America, fresh-brewed coffee and carbonated sodas are the most commonly sold CCB. Tea (mostly black) and other CCB are common as well. The USA consumed the most carbonated soda and sports and energy drinks per capita than any other country in the Euromonitor database.

Canadian and American guidelines for CCB were based on evidence compiled and reviewed, in part, for the purpose of setting national guidelines [4,25]. Canada's FBDG include caffeine upper limits ranging from 45 to 85 mg/day for ages 4 through 12 years, 2.5 mg/kg body weight for adolescents aged 13+, 400 mg/day for adults, and 300 mg/day for pregnant or breastfeeding women, as well as women planning to become pregnant. In the USA, three to five cups of coffee/day (providing up to ≈400 mg/day caffeine) is considered safe for adults, yet individuals who do not consume regular coffee or other caffeinated beverages are not encouraged to begin doing so. Pregnant and breastfeeding women are encouraged to consult their health care providers for advice concerning caffeine intake. Sodas and energy drinks are discouraged but more with regard to their sugar content. Caution is also advised when mixing caffeine and alcohol.

## 4. Discussion and Conclusions

The goal of the current review was to provide the first world summary of guidelines pertaining to dietary caffeine consumption. CCB, while major contributors to caffeine in the diet, also present with other unique nutritional properties. We therefore leveraged existing FBDG since they emphasize food-specific rather than nutrient-specific advice on healthy diets and are developed by interdisciplinary teams of experts with many sources of information reviewed in the process [17]. We begin our discussion with country differences in consumption habits that extend the macro-level consumption data we present in the current report. These are followed by key themes observed across country FBDG including (i) caffeine-intake upper limits to avoid potential health risks, (ii) CCB as replacements for plain water, (iii) CCB as added-sugar sources, and (iv) health benefits of caffeine-containing beverage consumption.

Consumption habits are greatly affected by factors such as geographical origin, culture, lifestyle, social behavior, and economic status. Although regular coffee dominates over decaffeinated coffee across countries, coffee brewing methods differ and these are only partly captured by Euromonitor data used in the current report. While drip filter coffee is the most popular brewing method worldwide, plunger coffee dominates in northern Europe. Turkish coffee is popular in the Middle East, Greece, Turkey, and Eastern Europe, and Espresso and Moka methods are the most common in Italy, Spain, and Portugal [26–37]. Tea habits also vary around the world [38–40]. For example, Western countries generally drink black tea, made by pouring boiling water over a teabag in a pot or mug and allowing it to steep before consuming (either with or without milk and/or sugar). In India, Pakistan, and some Middle Eastern countries, black tea is largely prepared by boiling the black leaves in a pan for several minutes prior to consumption (often together with water, milk, and sugar). In China and Japan, the drink is normally prepared from green tea by infusing it in hot (but not boiling) water and only the second and subsequent infusions are consumed [40]. Yerba mate is consumed in several South American countries, where it originated, but is less common to other parts of the world [12,22]. Grounded and dried yerba mate leaves and stems are widely consumed in the form of infusions, such as chimarrão and tererê, prepared with hot and cold water, respectively [15]. Differences in brewing methods as well as the type and processing of beans/leaves/stems used are relevant since all affect the sensorial quality and the amount and type of compounds in a "cup" of coffee, tea, or mate [12,15,41]. For example, one needs to consume about three Turkish and five Espresso coffee cups to acquire the same amount of caffeine in one American cup [41]; details to consider when comparing country guidelines. In contrast to the aforementioned natural sources of caffeine, there is little evidence, to our knowledge, in support of a true cultural component to consumption of caffeine-added beverages such as caffeinated soda and energy drinks. A global and concerning pattern is that caffeine-added beverages, which have potential health risks and no benefits, are the primary contributors to caffeine in the diet of children and adolescents [18,42–52].

FBDG respond to a country's public health and nutrition priorities, sociocultural influences and food production and consumption patterns, among other factors [17]. Historically, the FBDG have focused on undernutrition and included guidelines aimed at consuming a diverse diet to address energy and nutrient gaps [53]. With time, many FBDG have evolved to include guidelines to support healthy lifestyle and specific recommendations to target various age groups [53]. In general, FBDG of some countries such as Sri Lanka, Sierra Leone, and Bangladesh address nutrient inadequacies in the population [54–59], while those of other countries, such as India, Thailand, Iran, Lebanon, and Brazil, address the double burden of undernutrition and overnutrition [54,60–65]. FBDG of developed and high-income countries, much of Europe and North America, are largely intended for prevention of chronic disease, adverse symptoms, or side effects [25,66,67]. These nutritional priorities partly determined if and how dietary caffeine sources were incorporated into guidelines.

For infants, children and adolescents, CCB consumption is often simply discouraged in FBDG. Canada is an exception and provides quantitative upper limits for caffeine intake according to age. For some countries, such as Nigeria, South Africa, Greece, and Mexico, it is not uncommon to introduce tea to the diet of children <2 years of age [68–72] and thus caffeine guidelines targeting this age group are highly relevant. Only thirteen country FBDG, spanning Asia and the Pacific, Europe, Latin America, and North America, advise pregnant/nursing women to avoid CCB. Eight of these advised specific caffeine limits, which ranged from 200 to 300 mg/day. Small epidemiological studies report that over 60% of women drink caffeine-containing beverages during pregnancy, but total amounts are generally below advised limits [23,73–77]. For adults, Denmark, U.K., Portugal, Canada, and USA advise to limit caffeine intake to 300 or 400 mg/day. FBDG for Australia, Indonesia, New Zealand, Denmark, Hungary, Malta, Colombia, USA, and Canada state specific concerns for energy drinks, generally defined as any drink with >150 mg of caffeine/liter, but often contain other bioactive ingredients and sugar [16,78]. Some guidelines to avoid or limit caffeine intake were based on human or animal studies of pregnancy outcomes, fetal development, and acute caffeine effects (including diuresis, see below). Other guidelines were in place as a safety-precaution since the long-term adverse effects of caffeine are not clear.

Water is essential for life and thus a staple recommendation in all FBDG. Coffee, tea, and yerba mate are naturally non-caloric beverages and currently make important contributions to total fluid intake for many countries [79–82], but whether they are suitable substitutes for plain water varies by country guidelines and likely reflects the nutritional priorities of the country. For example, FBDG of African countries stress the importance of consuming "enough safe" water as opposed to listing adequate water-substitutes. They were nevertheless concerned about the iron absorption inhibitory effects of tannins present in coffee and tea [83–87], as were the FBDG of several countries of Asia and the Pacific. For some FBDG, whether coffee or tea were adequate water-substitutes was often dependent on whether the diuretic effects of caffeine were considered significant by the country. European and North American countries rarely noted these diuretic effects.

Caffeine-containing soda is a major contributor to sugar intake along with other SSB and implicated in obesity and other metabolic disease around the globe [88]. Guidelines concerning reductions in soda are thus geared towards reducing sugar intake as opposed to monitoring caffeine intake. Coffee and tea become significant sources of added sugar and energy in the diet for countries such as China, Korea, Malaysia, Spain, Italy, Brazil, and Uruguay that prefer to prepare coffee and tea with sugar and cream, or for countries where instant coffee mixes are highly consumed [37,43,44,82,89–93]. These habits are often overlooked and may off-set any benefits that coffee and tea might offer over other beverage types [89,92]. In our review of guidelines, Sri Lanka, Thailand, Qatar, Bulgaria, France, Greece, Italy, Malta, Poland, Turkey, Brazil, Mexico, and the USA advised careful attention to the amount of sugar added to coffee and tea.

While there is currently no evidence of health benefits for caffeine-added beverages, recent reviews concerning coffee, and perhaps tea, suggest some benefits with coffee and tea consumption [6,10]. Some FBDGs make reference to these benefits and a few of these also provide specific guidelines. In 2015, the USA dietary guidelines committee reviewed the literature concerning coffee, specifically, as well as total caffeine on health. Potential benefits of three to five cups of coffee/day were discussed in the committee's Scientific Report [25]. The favorable message, however, could not yet be applied to children or pregnant women or for an equivalent amount of total caffeine (from any source). The Netherlands also point to benefits of green and black tea consumption and *recommend* three cups/day. Interestingly, Poland *discourages* consumption of black tea in particular, and Latvia *discourages* the use of coffee or tea for hydration, in part, for mental health and heart disease prevention. India also provides an in-depth look at coffee, tea, and caffeine. Coffee and caffeine are viewed negatively and potential benefits with tea are off-set by its caffeine content. Mexican guidelines classify beverages from the most (level 1) to the least (level 5) healthy according to their energy content, nutritional value, and risks to health. Coffee and tea (without sugar) are level 3 beverages, *limited* to four cups/day. No African studies encourage coffee and tea consumption for health. Taken together, inconsistencies concerning health benefits (and risks) of coffee and tea consumption were observed across FBDG and this may be due to the breadth of research on the topic (function of FBDG development date and country-relevancy) or nutritional priorities of the country.

Most guidelines pertaining to dietary caffeine are evidence-based but there are some exceptions. Portugal's guidelines, for example, state "in tea, the absorption of caffeine is slower than in coffee, which means the stimulating effect is lower but lasts longer." Albania guidelines advise menopausal women to avoid coffee (among other foods/beverages) because it worsens "warming." Peer-reviewed literature supporting these statements were often not available. Missing from guidelines was information on known between-person variation in caffeine metabolism, resulting from lifestyle or genetic factors [94,95]. However, despite enthusiasm for "personalized-caffeine recommendations," further studies are warranted before they can be included in FBDG.

While FBDG help individuals optimize their caffeine habits, many countries regulate caffeine intake at the food manufacturing level by setting limits to the amount of caffeine added to foods [78,96]. Several countries have specifically enacted measures to regulate the labeling, distribution, and sale of energy drinks [2,8,97,98]. For example, Denmark, Turkey, Norway, Uruguay, Sweden, Lithuania, Latvia, and Iceland have banned or restricted sales to children or those <18 years of age, while Hungary and Mexico apply an additional tax to energy drinks [20,21,99,100]. The USA, Canada, and Mexico have further restrictions on the sale of caffeinated alcohol beverages [101,102]. In view of the health risks associated with the widespread consumption of SSB, many national governments have also taken action to reduce consumption of SSB [20,21,103,104]. While these actions are not targeting caffeine, per se, they are targeting a subset of SSB that contain caffeine which include colas and energy drinks. Unfortunately, all policies in place to regulate caffeine intake are challenged by the fact that major dietary sources of caffeine (i.e., coffee and tea) are exempt since they naturally contain caffeine [105,106].

Our data collection strategy for dietary caffeine guidelines was systematic and comprehensive but may be incomplete. We relied on FAO as a starting point which may have missed FBDG of certain countries or may not have been updated with the latest FBDG. Our approach offered several opportunities to address the latter. Our efforts to search and contact secondary resources was often met with limited success. As described elsewhere, the Euromonitor Passport is not a scholarly database and the data have similar limitations to official government trade and economic statistics [107]. Euromonitor data capture sales volume only, an imperfect measure of consumption because it does not capture products distributed through informal food systems or wastage [108]. Moreover, some beverage categories are not exclusively CCB. For example, sports drinks without caffeine are consumed in greater quantities than energy drinks and thus contributions to overall CCB by the "sports drinks and energy drinks" is likely overestimated. However, these data are abundant, less biased than survey data, and they have been consistently reported across countries and time using standardized measures [107]. Despite these limitations, the current review is a starting resource for country-level guidelines and consumption data pertaining to dietary caffeine.

In summary, FBDG provide an unfavorable view of caffeinated-beverages by noting their potential adverse/unknown effects on special populations as well as their diuretic, psycho-stimulating and nutrient inhibitory properties. Few FBDG balanced these messages with recent data supporting potential benefits of specific beverage-types. FBDG can serve to guide a wide range of food and nutrition policies and programs with the unique opportunity to favorably impact diets and the food system [17]. FBDG undergo review and revisions in keeping with changes in nutrition priorities of a country and advancements in nutrition research. We therefore anticipate modifications to guidelines pertaining to caffeine in future releases of FBDG.

**Supplementary Materials:** The following are available online at http://www.mdpi.com/2072-6643/10/11/1772/s1, Figure S1: Data Collection Strategy for Dietary Caffeine Guidelines; Table S1: Country Specific Guidelines Pertaining to Dietary Caffeine.

**Author Contributions:** C.M.R. and M.C.C. designed the data collection framework and collected the data. M.C.C. conceptualized the paper and wrote the first draft. All authors revised and approved the final the manuscript.

**Funding:** This work was funded by the National Institute on Aging (K01AG053477 to MCC).

**Acknowledgments:** The authors would like to thank the following individuals for providing supplemental data on dietary caffeine guidelines: Olu Adetokunbo, Lydia K. Browne, Antje Gahl, Neliya Mikushinska, Ul-Aziha bt. Muhammad, Zsuzsanna Nagy-Lőrincz, Rebone Ntsie, Hólmfríður Þorgeirsdóttir, Lāsma Pikele, Jessica Priem, Silke Restemeyer, Laura Rossi, Guro Smedshaug, Ilze Vamža, Joka van Dusseldorp-Dijkstra, Sara Upplysningen, Dagny Løvoll Warming, and Yurun Wu; as well as the following individuals for their assistance in translating FBDG: H.M. Abdulazeem, Corneliu C Antonescu, Eric Cifaldi, Marina Dujmović, Diane Gal, Janet Lyu, Pawel Rykowski, Morten Svan, and Noni Tobing. Finally, we thank FAO staff (Melissa Vargas) and Northwestern University's Galter Health Sciences Library research staff (Eileen Wafford, Annie Wescott) for support and guidance on data collection, and Alan Kuang for assistance on data presentation.

**Conflicts of Interest:** The authors declare no conflicts of interest.

## References

1. Fredholm, B.B.; Battig, K.; Holmen, J.; Nehlig, A.; Zvartau, E.E. Actions of caffeine in the brain with special reference to factors that contribute to its widespread use. *Pharmacol. Rev.* **1999**, *51*, 83–133. [PubMed]

2. Zucconi, S.; Volpato, C.; Adinolfi, F.; Gandini, E.; Gentile, E.; Loi, A.; Fioriti, L. Gathering consumption data on specific consumer groups of energy drinks. *EFSA Support. Publ.* **2013**, *10*, 394E. [CrossRef]

3. Wikoff, D.; Welsh, B.T.; Henderson, R.; Brorby, G.P.; Britt, J.; Myers, E.; Goldberger, J.; Lieberman, H.R.; O'Brien, C.; Peck, J. Systematic review of the potential adverse effects of caffeine consumption in healthy adults, pregnant women, adolescents, and children. *Food Chem. Toxicol.* **2017**, *109*, 585–648. [CrossRef] [PubMed]

4. Nawrot, P.; Jordan, S.; Eastwood, J.; Rostein, J.; Hugenholtz, A.; Feeley, M. Effects of caffeine on human health. *Food Addit. Contam.* **2003**, *20*, 1–30. [CrossRef] [PubMed]

5. EFSA Panel on Dietetic Products Nutrition and Allergies. Scientific opinion on the safety of caffeine. *EFSA J.* **2015**, *13*, 4102.

6. Poole, R.; Kennedy, O.J.; Roderick, P.; Fallowfield, J.A.; Hayes, P.C.; Parkes, J. Coffee consumption and health: Umbrella review of meta-analyses of multiple health outcomes. *BMJ* **2017**, *359*, j5024. [CrossRef] [PubMed]

7. Higgins, J.P.; Babu, K.; Deuster, P.A.; Shearer, J. Energy drinks: A contemporary issues paper. *Curr. Sports Med. Rep.* **2018**, *17*, 65–72. [CrossRef] [PubMed]

8. Breda, J.J.; Whiting, S.H.; Encarnação, R.; Norberg, S.; Jones, R.; Reinap, M.; Jewell, J. Energy drink consumption in Europe: A review of the risks, adverse health effects, and policy options to respond. *Front. Public Health* **2014**, *2*, 134. [CrossRef] [PubMed]

9. Vuong, Q.V. Epidemiological evidence linking tea consumption to human health: A review. *Crit. Rev. Food Sci. Nutr.* **2014**, *54*, 523–536. [CrossRef] [PubMed]

10. Hayat, K.; Iqbal, H.; Malik, U.; Bilal, U.; Mushtaq, S. Tea and its consumption: Benefits and risks. *Crit. Rev. Food Sci. Nutr.* **2015**, *55*, 939–954. [CrossRef] [PubMed]

11. Ruanpeng, D.; Thongprayoon, C.; Cheungpasitporn, W.; Harindhanavudhi, T. Sugar and artificially sweetened beverages linked to obesity: A systematic review and meta-analysis. *QJM Int. J. Med.* **2017**, *110*, 513–520. [CrossRef] [PubMed]

12. Heck, C.I.; de Mejia, E.G. Yerba mate tea (ilex paraguariensis): A comprehensive review on chemistry, health implications, and technological considerations. *J. Food Sci.* **2007**, *72*, R138–R151. [CrossRef] [PubMed]

13. Spiller, M.A. The chemical components of coffee. In *Caffeine*; Spiller, G.A., Ed.; CRC: Boca Raton, FL, USA, 1998; pp. 97–161.

14. Mitchell, D.C.; Hockenberry, J.; Teplansky, R.; Hartman, T.J. Assessing dietary exposure to caffeine from beverages in the US population using brand-specific versus category-specific caffeine values. *Food Chem. Toxicol.* **2015**, *80*, 247–252. [CrossRef] [PubMed]

15. Da Silveira, T.F.F.; Meinhart, A.D.; de Souza, T.C.L.; Cunha, E.C.E.; de Moraes, M.R.; Godoy, H.T. Chlorogenic acids and flavonoid extraction during the preparation of yerba mate based beverages. *Food Res. Int.* **2017**, *102*, 348–354. [CrossRef] [PubMed]

16. McLellan, T.M.; Lieberman, H.R. Do energy drinks contain active components other than caffeine? *Nutr. Rev.* **2012**, *70*, 730–744. [CrossRef] [PubMed]

17. Food and Agriculture Organization of the United Nations. Food-Based Dietary Guidelines. Available online: http://www.fao.org/nutrition/education/food-dietary-guidelines/background/en/ (accessed on 1 August 2018).

18. Mitchell, D.C.; Knight, C.A.; Hockenberry, J.; Teplansky, R.; Hartman, T.J. Beverage caffeine intakes in the US. *Food Chem. Toxicol.* **2013**, *63*, 136–142. [CrossRef] [PubMed]

19. World Bank. World Bank Country and Lending Groups: Country Classification 2017. Available online: https://datahelpdesk.worldbank.org/knowledgebase/articles/906519 (accessed on 1 September 2018).

20. Hawkes, C.; Jewell, J.; Allen, K. A food policy package for healthy diets and the prevention of obesity and diet-related non-communicable diseases: The nourishing framework. *Obes. Rev.* **2013**, *14*, 159–168. [CrossRef] [PubMed]

21. WCRF International Nourishing Framework. Available online: https://www.wcrf.org/int/policy/nourishing-database (accessed on 1 October 2018).

22. Euromonitor Euromonitor International. Available online: https://www.euromonitor.com/ (accessed on 1 September 2018).

23. Matijasevich, A.; Barros, F.C.; Santos, I.S.; Yemini, A. Maternal caffeine consumption and fetal death: A case–control study in Uruguay. *Paediatr. Perinat. Epidemiol.* **2006**, *20*, 100–109. [CrossRef] [PubMed]

24. Ronco, A.L.; Stefani, E.; Mendoza, B.; Deneo-Pellegrini, H.; Vazquez, A.; Abbona, E. Mate intake and risk of breast cancer in Uruguay: A case-control study. *Asian Pac. J. Cancer Prev.* **2016**, *17*, 1453–1461. [CrossRef] [PubMed]

25. Millen, B.; Lichtenstein, A.; Abrams, S. *Scientific Report of the 2015 Dietary Guidelines Advisory Committee*; US Department of Agriculture: Washington, DC, USA, 2015.

26. Petracco, M. Technology 4: Beverage preparation: Brewing trends for the new millennium. In *COFFEE Recent Developments*; Blackwell Science Ltd.: London, UK, 2001.

27. Panagiotakos, D.B.; Lionis, C.; Zeimbekis, A.; Makri, K.; Bountziouka, V.; Economou, M.; Vlachou, I.; Micheli, M.; Tsakountakis, N.; Metallinos, G.; et al. Long-term, moderate coffee consumption is associated with lower prevalence of diabetes mellitus among elderly non-tea drinkers from the Mediterranean islands (Medis study). *Rev. Diabet. Stud.* **2007**, *4*, 105–111. [CrossRef] [PubMed]

28. Lopez-Garcia, E.; Guallar-Castillon, P.; Leon-Munoz, L.; Graciani, A.; Rodriguez-Artalejo, F. Coffee consumption and health-related quality of life. *Clin. Nutr.* **2014**, *33*, 143–149. [CrossRef] [PubMed]

29. Sanchez, J.M. Methylxanthine content in commonly consumed foods in Spain and determination of its intake during consumption. *Foods* **2017**, *6*, 109. [CrossRef] [PubMed]

30. D'Amicis, A.; Scaccini, C.; Tomassi, G.; Anaclerio, M.; Stornelli, R.; Bernini, A. Italian style brewed coffee: Effect on serum cholesterol in young men. *Int. J. Epidemiol.* **1996**, *25*, 513–520. [CrossRef] [PubMed]

31. Ferraroni, M.; Tavani, A.; Decarli, A.; Franceschi, S.; Parpinel, M.; Negri, E.; La Vecchia, C. Reproducibility and validity of coffee and tea consumption in Italy. *Eur. J. Clin. Nutr.* **2004**, *58*, 674. [CrossRef] [PubMed]

32. Leurs, L.J.; Schouten, L.J.; Goldbohm, R.A.; van den Brandt, P.A. Total fluid and specific beverage intake and mortality due to IHD and stroke in The Netherlands Cohort study. *Br. J. Nutr.* **2010**, *104*, 1212–1221. [CrossRef] [PubMed]

33. Gavrilyuk, O.; Braaten, T.; Skeie, G.; Weiderpass, E.; Dumeaux, V.; Lund, E. High coffee consumption and different brewing methods in relation to postmenopausal endometrial cancer risk in the Norwegian women and cancer study: A population-based prospective study. *BMC Women's Health* **2014**, *14*, 48. [CrossRef] [PubMed]

34.  Lof, M.; Sandin, S.; Yin, L.; Adami, H.O.; Weiderpass, E. Prospective study of coffee consumption and all-cause, cancer, and cardiovascular mortality in Swedish women. *Eur. J. Epidemiol.* **2015**, *30*, 1027–1034. [CrossRef] [PubMed]
35.  Sousa, A.G.; da Costa, T.H. Usual coffee intake in Brazil: Results from the national dietary survey 2008–9. *Br. J. Nutr.* **2015**, *113*, 1615–1620. [CrossRef] [PubMed]
36.  Miranda, A.M.; Steluti, J.; Goulart, A.C.; Bensenor, I.M.; Lotufo, P.A.; Marchioni, D.M. Coffee consumption and coronary artery calcium score: Cross-sectional results of Elsa-Brasil (Brazilian longitudinal study of adult health). *J. Am. Heart Assoc.* **2018**, *7*, e007155. [CrossRef] [PubMed]
37.  Bezerra, I.N.; Goldman, J.; Rhodes, D.G.; Hoy, M.K.; Moura Souza, A.; Chester, D.N.; Martin, C.L.; Sebastian, R.S.; Ahuja, J.K.; Sichieri, R.; et al. Difference in adult food group intake by sex and age groups comparing Brazil and United States nationwide surveys. *Nutr. J.* **2014**, *13*, 74. [CrossRef] [PubMed]
38.  Beresniak, A.; Duru, G.; Berger, G.; Bremond-Gignac, D. Relationships between black tea consumption and key health indicators in the world: An ecological study. *BMJ Open* **2012**, *2*, e000648. [CrossRef] [PubMed]
39.  Grigg, D. The worlds of tea and coffee: Patterns of consumption. *GeoJournal* **2002**, *57*, 283–294. [CrossRef]
40.  Astill, C.; Birch, M.R.; Dacombe, C.; Humphrey, P.G.; Martin, P.T. Factors affecting the caffeine and polyphenol contents of black and green tea infusions. *J. Agric. Food Chem.* **2001**, *49*, 5340–5347. [CrossRef] [PubMed]
41.  Derossi, A.; Ricci, I.; Caporizzi, R.; Fiore, A.; Severini, C. How grinding level and brewing method (Espresso, American, Turkish) could affect the antioxidant activity and bioactive compounds in a coffee cup. *J. Sci. Food Agric.* **2018**, *98*, 3198–3207. [CrossRef] [PubMed]
42.  Paulsen, M.M.; Myhre, J.B.; Andersen, L.F. Beverage consumption patterns among Norwegian adults. *Nutrients* **2016**, *8*, 561. [CrossRef] [PubMed]
43.  Pereira, R.A.; Souza, A.M.; Duffey, K.J.; Sichieri, R.; Popkin, B.M. Beverage consumption in Brazil: Results from the first national dietary survey. *Public Health Nutr.* **2015**, *18*, 1164–1172. [CrossRef] [PubMed]
44.  Landais, E.; Moskal, A.; Mullee, A.; Nicolas, G.; Gunter, M.J.; Huybrechts, I.; Overvad, K.; Roswall, N.; Affret, A.; Fagherazzi, G. Coffee and tea consumption and the contribution of their added ingredients to total energy and nutrient intakes in 10 European countries: Benchmark data from the late 1990s. *Nutrients* **2018**, *10*, 725. [CrossRef] [PubMed]
45.  Yamada, H.; Kawado, M.; Aoyama, N.; Hashimoto, S.; Suzuki, K.; Wakai, K.; Suzuki, S.; Watanabe, Y.; Tamakoshi, A. Coffee consumption and risk of colorectal cancer: The Japan collaborative cohort study. *J. Epidemiol.* **2014**, *24*, 370–378. [CrossRef] [PubMed]
46.  Mesirow, M.S.; Welsh, J.A. Changing beverage consumption patterns have resulted in fewer liquid calories in the diets of US children: National health and nutrition examination survey 2001–2010. *J. Acad. Nutr. Diet.* **2015**, *115*, 559–566.e4. [CrossRef] [PubMed]
47.  Storey, M.L.; Forshee, R.A.; Anderson, P.A. Beverage consumption in the US population. *J. Am. Diet. Assoc.* **2006**, *106*, 1992–2000. [CrossRef] [PubMed]
48.  Loftfield, E.; Freedman, N.D.; Dodd, K.W.; Vogtmann, E.; Xiao, Q.; Sinha, R.; Graubard, B.I. Coffee drinking is widespread in the United States, but usual intake varies by key demographic and lifestyle factors-3. *J. Nutr.* **2016**, *146*, 1762–1768. [CrossRef] [PubMed]
49.  Rybak, M.E.; Sternberg, M.R.; Pao, C.-I.; Ahluwalia, N.; Pfeiffer, C.M. Urine excretion of caffeine and select caffeine metabolites is common in the US population and associated with caffeine intake-4. *J. Nutr.* **2015**, *145*, 766–774. [CrossRef] [PubMed]
50.  Nergiz-Unal, R.; Akal Yildiz, E.; Samur, G.; Besler, H.T.; Rakicioglu, N. Trends in fluid consumption and beverage choices among adults reveal preferences for Ayran and black tea in central Turkey. *Nutr. Diet.* **2017**, *74*, 74–81. [CrossRef] [PubMed]
51.  Pena, A.; Lino, C.; Silveira, M.I. Survey of caffeine levels in retail beverages in Portugal. *Food Addit. Contam.* **2005**, *22*, 91–96. [CrossRef] [PubMed]
52.  Martins, A.; Ferreira, C.; Sousa, D.; Costa, S. Consumption patterns of energy drinks in Portuguese adolescents from a city in northern Portugal. *Acta Med. Port.* **2018**, *31*, 207–212. [CrossRef] [PubMed]
53.  Andrade, J.; Andrade, J. Food-Based Dietary Guidelines: An Overview. Washington: Integrating Gender and Nutrition within Agricultural Extension Services and USAID, 2016. Available online: https://www.agrilinks.org/sites/default/files/resource/files/ING%20TN%20(2016_10%20)%20Food%20Based%20Dietary%20Guideline%20-%20Overview%20(Andrade,%20Andrade).pdf (accessed on 1 August 2018).

54. World Health Organization. *Regional Consultation on Food-Based Dietary Guidelines for Countries in the South-East Asia Region, 6–9 December 2010, New Delhi, India*; World Health Organization: New Delhi, India, 2010.

55. Bangladesh Institue of Research and Rehabilitation in Diabetes, Endocrine and Metabolic Disorders. *Dietary Guidelines for Bangladesh*; Bangladesh Institue of Research and Rehabilitation in Diabetes, Endocrine and Metabolic Disorders: Dhaka, Bangladesh, 2013.

56. Nutrition Division Federaal Ministry of Health; World Health Organization. *Food-Based Dietary Guideline for Nigeria*; Federal Ministry of Health: Abuja, Nigeria, 2006.

57. Vorster, H.H.; Badham, J.; Venter, C. An introduction to the revised food-based dietary guidelines for South Africa. *S. Afr. J. Clin. Nutr.* **2013**, *26*, S5–S12.

58. German Ministry of Food Agriculture and Consumer Protection; Food and Agriculture Organization of the United Nations. *Sierra Leone Food Based Dietary Guideline for Healthy Eating*. 2016. Available online: https://afro.who.int/publications/sierra-leone-food-based-dietary-guidelines-healthy-eating-2016 (accessed on 1 August 2018).

59. Harika, R.; Faber, M.; Samuel, F.; Kimiywe, J.; Mulugeta, A.; Eilander, A. Micronutrient status and dietary intake of iron, vitamin a, iodine, folate and zinc in women of reproductive age and pregnant women in Ethiopia, Kenya, Nigeria and South Africa: A systematic review of data from 2005 to 2015. *Nutrients* **2017**, *9*, 1096. [CrossRef] [PubMed]

60. National Institute of Nutrition. *Dietary Guidelines for Indians*; National Institute of Nutrition: Hyderabad, India, 2011.

61. Goshtaei, M.; Ravaghi, H.; Sari, A.A.; Abdollahi, Z. Nutrition policy process challenges in Iran. *Electron. Phys.* **2016**, *8*, 1865. [CrossRef] [PubMed]

62. Hwalla, N.; Nasreddine, L.; Farhat Jarrar, S. The Food-Based Dietary Guideline Manual for Promoting Healthy Eating in the Lebanese Adult Population. The Faculty of Agricultural and Food Sciences, American University of Beirut, Lebanese National Council for Scientific Research (CNRS), Ministry of Public Health, 2013. Available online: www.aub.edu.lb/fafs/nfsc/Documents/LR-e-FBDG-EN-III.pdf (accessed on 14 July 2015).

63. Fonseca, V.M. Aspects of the Brazilian nutritional situation. *Cienc. Saude Colet.* **2014**, *19*, 1328–1329.

64. Ministry of Health of Brazil. *Dietary Guidelines for the Brazilian Population*; Ministry of Health of Brazil: Brasília, Brazil, 2015.

65. Vorster, H.H.; Kruger, A.; Margetts, B.M. The nutrition transition in Africa: Can it be steered into a more positive direction? *Nutrients* **2011**, *3*, 429–441. [CrossRef] [PubMed]

66. National Health and Medical Research Council. *Australian Dietary Guidelines Summary*; National Health and Medical Research Council: Canberra, Australia, 2013.

67. Kromhout, D.; Spaaij, C.; De Goede, J.; Weggemans, R. The 2015 Dutch food-based dietary guidelines. *Eur. J. Clin. Nutr.* **2016**, *70*, 869. [CrossRef] [PubMed]

68. Nwankwo, B.O.; Brieger, W.R. Exclusive breastfeeding is undermined by use of other liquids in rural Southwestern Nigeria. *J. Trop. Pediatr.* **2002**, *48*, 109–112. [CrossRef] [PubMed]

69. Tympa-Psirropoulou, E.; Vagenas, C.; Psirropoulos, D.; Dafni, O.; Matala, A.; Skopouli, F. Nutritional risk factors for iron-deficiency Anaemia in children 12–24 months old in the area of Thessalia in Greece. *Int. J. Food Sci. Nutr.* **2005**, *56*, 1–12. [CrossRef] [PubMed]

70. Urkin, J.; Adam, D.; Weitzman, D.; Gazala, E.; Chamni, S.; Kapelushnik, J. Indices of iron deficiency and Anaemia in Bedouin and Jewish toddlers in southern Israel. *Acta Paediatr.* **2007**, *96*, 857–860. [CrossRef] [PubMed]

71. González-Castell, D.; González de Cosío, T.; Rodríguez-Ramírez, S.; Escobar-Zaragoza, L. Early consumption of liquids different to breast milk in Mexican infants under 1 year: Results of the probabilistic national health and nutrition survey 2012. *Nutr. Hosp.* **2016**, *33*, 9. [PubMed]

72. Mennella, J.A.; Turnbull, B.; Ziegler, P.J.; Martinez, H. Infant feeding practices and early flavor experiences in Mexican infants: An intra-cultural study. *J. Am. Diet. Assoc.* **2005**, *105*, 908–915. [CrossRef] [PubMed]

73. Rachidi, S.; Awada, S.; Al-Hajje, A.; Bawab, W.; Zein, S.; Saleh, N.; Salameh, P. Risky substance exposure during pregnancy: A pilot study from Lebanese mothers. *Drug Healthc. Patient Saf.* **2013**, *5*, 123. [PubMed]

74. Bailey, H.D.; Lacour, B.; Guerrini-Rousseau, L.; Bertozzi, A.-I.; Leblond, P.; Faure-Conter, C.; Pellier, I.; Freycon, C.; Doz, F.; Puget, S. Parental smoking, maternal alcohol, coffee and tea consumption and the risk of childhood brain tumours: The ESTELLE and ESCALE studies (SFCE, France). *Cancer Causes Control* **2017**, *28*, 719–732. [CrossRef] [PubMed]

75. Da Mota Santana, J.; Alves de Oliveira Queiroz, V.; Monteiro Brito, S.; Barbosa Dos Santos, D.; Marlucia Oliveira Assis, A. Food consumption patterns during pregnancy: A longitudinal study in a region of the north east of Brazil. *Nutr. Hosp.* **2015**, *32*, 130–138. [PubMed]

76. Tollånes, M.C.; Strandberg-Larsen, K.; Eichelberger, K.Y.; Moster, D.; Lie, R.T.; Brantsæter, A.L.; Meltzer, H.M.; Stoltenberg, C.; Wilcox, A.J. Intake of caffeinated soft drinks before and during pregnancy, but not total caffeine intake, is associated with increased cerebral palsy risk in the Norwegian mother and child cohort study–3. *J. Nutr.* **2016**, *146*, 1701–1706. [CrossRef] [PubMed]

77. Peacock, A.; Hutchinson, D.; Wilson, J.; McCormack, C.; Bruno, R.; Olsson, C.A.; Allsop, S.; Elliott, E.; Burns, L.; Mattick, R.P. Adherence to the caffeine intake guideline during pregnancy and birth outcomes: A prospective cohort study. *Nutrients* **2018**, *10*, 319. [CrossRef] [PubMed]

78. Rosenfeld, L.S.; Mihalov, J.J.; Carlson, S.J.; Mattia, A. Regulatory status of caffeine in the United States. *Nutr. Rev.* **2014**, *72*, 23–33. [CrossRef] [PubMed]

79. Guelinckx, I.; Ferreira-Pêgo, C.; Moreno, L.A.; Kavouras, S.A.; Gandy, J.; Martinez, H.; Bardosono, S.; Abdollahi, M.; Nasseri, E.; Jarosz, A. Intake of water and different beverages in adults across 13 countries. *Eur. J. Nutr.* **2015**, *54*, 45–55. [CrossRef] [PubMed]

80. Özen, A.; Bibiloni, M.D.M.; Pons, A.; Tur, J. Fluid intake from beverages across age groups: A systematic review. *J. Hum. Nutr. Diet.* **2015**, *28*, 417–442. [CrossRef] [PubMed]

81. Platania, A.; Castiglione, D.; Sinatra, D.; Urso, M.D.; Marranzano, M. Fluid intake and beverage consumption description and their association with dietary vitamins and antioxidant compounds in Italian adults from the Mediterranean healthy eating, aging and lifestyles (meal) study. *Antioxidants* **2018**, *7*, 56. [CrossRef] [PubMed]

82. De Stefani, E.; Boffetta, P.; Ronco, A.L.; Deneo-Pellegrini, H.; Acosta, G.; Mendilaharsu, M. Dietary patterns and risk of bladder cancer: A factor analysis in Uruguay. *Cancer Causes Control* **2008**, *19*, 1243. [CrossRef] [PubMed]

83. Ahmad Fuzi, S.F.; Koller, D.; Bruggraber, S.; Pereira, D.I.; Dainty, J.R.; Mushtaq, S. A 1-h time interval between a meal containing iron and consumption of tea attenuates the inhibitory effects on iron absorption: A controlled trial in a cohort of healthy UK women using a stable iron isotope. *Am. J. Clin. Nutr.* **2017**, *106*, 1413–1421. [CrossRef] [PubMed]

84. Hurrell, R.F.; Reddy, M.; Cook, J.D. Inhibition of non-haem iron absorption in man by polyphenolic-containing beverages. *Br. J. Nutr.* **1999**, *81*, 289–295. [PubMed]

85. Thankachan, P.; Walczyk, T.; Muthayya, S.; Kurpad, A.V.; Hurrell, R.F. Iron absorption in young Indian women: The interaction of iron status with the influence of tea and ascorbic acid. *Am. J. Clin. Nutr.* **2008**, *87*, 881–886. [CrossRef] [PubMed]

86. Nelson, M.; Poulter, J. Impact of tea drinking on iron status in the UK: A review. *J. Hum. Nutr. Diet.* **2004**, *17*, 43–54. [CrossRef] [PubMed]

87. Savolainen, H. Tannin content of tea and coffee. *J. Appl. Toxicol.* **1992**, *12*, 191–192. [CrossRef] [PubMed]

88. Hu, F.B.; Malik, V.S. Sugar-sweetened beverages and risk of obesity and type 2 diabetes: Epidemiologic evidence. *Physiol. Behav.* **2010**, *100*, 47–54. [CrossRef] [PubMed]

89. International Coffee Organization. *Coffee in China*; International Coffee Organization: Milan, Italy, 2014.

90. Amarra, M.S.V.; Khor, G.L.; Chan, P. Intake of added sugar in Malaysia: A review. *Asia Pac. J. Clin. Nutr.* **2016**, *25*, 227–240. [PubMed]

91. Saw, W.; Shanita, S.N.; Zahara, B.; Tuti, N.; Poh, B.K. Dietary intake assessment in adults and its association with weight status and dental caries. *Pak. J. Nutr.* **2012**, *11*, 1066.

92. Je, Y.; Jeong, S.; Park, T. Coffee consumption patterns in Korean adults: The Korean national health and nutrition examination survey (2001–2011). *Asia Pac. J. Clin. Nutr.* **2014**, *23*, 691–702. [PubMed]

93. Guallar-Castillon, P.; Munoz-Pareja, M.; Aguilera, M.T.; Leon-Munoz, L.M.; Rodriguez-Artalejo, F. Food sources of sodium, saturated fat and added sugar in the Spanish hypertensive and diabetic population. *Atherosclerosis* **2013**, *229*, 198–205. [CrossRef] [PubMed]

94. Gunes, A.; Dahl, M.L. Variation in cyp1a2 activity and its clinical implications: Influence of environmental factors and genetic polymorphisms. *Pharmacogenomics* **2008**, *9*, 625–637. [CrossRef] [PubMed]

95. Cornelis, M.C.; Kacprowski, T.; Menni, C.; Gustafsson, S.; Pivin, E.; Adamski, J.; Artati, A.; Eap, C.B.; Ehret, G.; Friedrich, N. Genome-wide association study of caffeine metabolites provides new insights to caffeine metabolism and dietary caffeine-consumption behavior. *Hum. Mol. Genet.* **2016**, *25*, 5472–5482. [CrossRef] [PubMed]

96. Food Standards Australia New Zealand. *Australia New Zealand Food Standards Code Standard 2.6.4 Formulated Caffeinated Beverages*; Food Standards Australia New Zealand: Kingston, Australia, 2014.

97. Thomson, B.; Schiess, S. Risk profile: Caffeine in Energy Drinks and Energy Shots. Institute of Environmental Science & Research Limited, April 2010. Available online: https://www.foodsafety.govt.nz/elibrary/industry/Risk_Profile_Caffeine-Science_Research.pdf (accessed on 1 September 2018).

98. Higgins, J.P.; Tuttle, T.D.; Higgins, C.L. Energy beverages: Content and safety. *Mayo Clin. Proc.* **2010**, *85*, 1033–1041. [CrossRef] [PubMed]

99. Hungarian National Institute for Health Development (NIHD). *Impact Assessment of the Public Health Product Tax*; NIHD: Budapest, Hungary, 2013.

100. Australian Food News. Lithuania Ban on Energy Drink Sales to under 18s Comes in with Broader Restrictions and Warnings. Available online: http://www.ausfoodnews.com.au/2014/11/05/lithuania-ban-on-energy-drink-sales-to-under-18s-comes-in-with-broader-restrictions-and-warnings.html (accessed on 1 September 2018).

101. Food and Drug Administration US Department of Health and Human Services. *Update on Caffeinated Alcoholic Beverages*; US Food and Drug Administration: Silver Spring, MD, USA, 2010.

102. Attwood, A.S. Caffeinated alcohol beverages: A public health concern. *Alcohol Alcohol.* **2012**, *47*, 370–371. [CrossRef] [PubMed]

103. Popkin, B.M.; Hawkes, C. Sweetening of the global diet, particularly beverages: Patterns, trends, and policy responses. *Lancet Diabetes Endocrinol.* **2016**, *4*, 174–186. [CrossRef]

104. Caro, J.C.; Smith-Taillie, L.; Ng, S.W.; Popkin, B. Designing a food tax to impact food-related non-communicable diseases: The case of Chile. *Food Policy* **2017**, *71*, 86–100. [CrossRef] [PubMed]

105. Lachenmeier, D.W.; Winkler, G. Caffeine content labeling: A prudent public health policy? *J. Caffeine Res.* **2013**, *3*, 154–155. [CrossRef]

106. Kole, J.; Barnhill, A. Caffeine content labeling: A missed opportunity for promoting personal and public health. *J. Caffeine Res.* **2013**, *3*, 108–113. [CrossRef] [PubMed]

107. Stuckler, D.; McKee, M.; Ebrahim, S.; Basu, S. Manufacturing epidemics: The role of global producers in increased consumption of unhealthy commodities including processed foods, alcohol, and tobacco. *PLoS Med.* **2012**, *9*, e1001235. [CrossRef] [PubMed]

108. Pomerleau, J.; Lock, K.; McKee, M. Discrepancies between ecological and individual data on fruit and vegetable consumption in fifteen countries. *Br. J. Nutr.* **2003**, *89*, 827–834. [CrossRef] [PubMed]

*nutrients*

MDPI

*Review*

# Key Findings and Implications of a Recent Systematic Review of the Potential Adverse Effects of Caffeine Consumption in Healthy Adults, Pregnant Women, Adolescents, and Children

Candace Doepker [1,*], Kara Franke [2], Esther Myers [3], Jeffrey J. Goldberger [4],
Harris R. Lieberman [5], Charles O'Brien [6], Jennifer Peck [7], Milton Tenenbein [8], Connie Weaver [9]
and Daniele Wikoff [2]

[1]    ToxStrategies, Cincinnati, OH 41075, USA
[2]    ToxStrategies, Asheville, NC 28804, USA; kfranke@toxstrategies.com (K.F.);
       dwikoff@toxstrategies.com (D.W.)
[3]    EF Myers Consulting, Trenton, IL 62293, USA; efmyers@efmyersconsulting.com
[4]    Cardiovascular Division, University of Miami Miller School of Medicine, 1120 NW 14th Street, Suite 1124,
       Miami, FL 33136, USA; j-goldberger@miami.edu
[5]    Military Nutrition Division, US Army Research Institute of Environmental Medicine, Kansas Street,
       Natick, MA 01760-5007, USA; harris.r.lieberman.civ@mail.mil
[6]    Department of Psychiatry, University of Pennsylvania, 3535 Market Street, 4th Floor,
       Philadelphia, PA 19104, USA; obrien@mail.med.upenn.edu
[7]    College of Public Health, University of Oklahoma Health Sciences Center, 801 NE 13th Street,
       Oklahoma City, OK 73104, USA; Jennifer-Peck@ouhsc.edu
[8]    Department of Pediatrics and Child Health, Department of Community Health Sciences,
       Children's Hospital, University of Manitoba, 840 Sherbrook Street, Room CE 208,
       Winnipeg, MB R3A 1S1, Canada; mtenenbein@exchange.hsc.mb.ca
[9]    Department of Nutrition Science, Purdue University, 700 W. State Street, West Lafayette, IN 47907, USA;
       weavercm@purdue.edu
*      Correspondence: cdoepker@toxstrategies.com; Tel.: +1-513-206-9929

Received: 23 August 2018; Accepted: 12 October 2018; Published: 18 October 2018

**Abstract:** In 2016–2017, we conducted and published a systematic review on caffeine safety that set out to determine whether conclusions that were presented in the heavily cited Health Canada assessment, remain supported by more recent data. To that end, we reviewed data from 380 studies published between June 2001 and June 2015, which were identified from an initial batch of over 5000 articles through a stringent search and evaluation process. In the current paper, we use plain language to summarize our process and findings, with the intent of sharing additional context for broader reach to the general public. We addressed whether caffeine doses previously determined not to be associated with adverse effects by Health Canada (400 mg/day for healthy adults, 300 mg/day for pregnant women, 2.5 mg/kg body weight/day for adolescents and children, and 10 g/day for acute effects) remain appropriate for five outcome areas (acute toxicity, cardiovascular toxicity, bone & calcium effects, behavior, and development and reproduction) in healthy adults, pregnant women, adolescents, and children. We used a weight-of-evidence approach to draw conclusions for each of the five outcomes, as well as more specific endpoints within those outcomes, which considered study quality, consistency, level of adversity, and magnitude of response. In general, updated evidence confirms the levels of intake that were put forth by Health Canada in 2003 as not being associated with any adverse health effects, and our results support a shift in caffeine research from healthy to sensitive populations.

**Keywords:** caffeine; coffee; systematic review; pregnancy; safety

## 1. Introduction

Consumption of caffeine remains a topic of popular interest, but it is also often a cause of confusion for medical professionals, nutritionists, and the public. The editors of this special issue of *Nutrients*, related to the impact of coffee and caffeine on human health, invited us to provide a summary of the recently published article, "Systematic Review of the Potential Adverse Effects of Caffeine Consumption in Healthy Adults, Pregnant Women, Adolescents and Children", for a broad audience. The large (64-page) systematic review was published in Food and Chemical Toxicology in April 2017, received much attention in the press, and was chosen "Best Paper of the Year" by the Editors of the journal [1]. The format of the paper followed a systematic review (SR) approach, which used an established and recognized framework that was specifically chosen to ensure transparency. Staying true to this framework required a large amount of documentation, which rendered the paper groundbreaking in terms of content but perhaps challenging to read and digest. At the same time, tracking statistics have demonstrated that the general public, in fact, has an interest in the SR findings with regard to caffeine. Scientific findings lose their value if they cannot be easily comprehended by diverse audiences. The Institute of Medicine (IOM) also recognizes this fact, and their guidance related to systematic reviews suggests that plain-language summaries can improve the work's usability for general audiences [2]. Thus, the aim of this paper is to provide a plain-language summary of this important review, and the reader is referred to the original work for full references [1]. We hope that this approach will allow the findings to be more understandable and help individuals make educated decisions regarding their (or their patients') consumption of caffeine.

Caffeine (1,3,7-trimethylxanthine) is a pharmacologically active component of many foods, beverages, dietary supplements, and drugs. Interestingly, it is also used to treat very ill, often premature, newborns afflicted with apnea (temporary cessation of breathing) [3]. Caffeine is probably best recognized for its use as a flavor in cola-type beverages, and for its natural occurrence in some seeds, such as coffee and cocoa. Coffee is one of the major contributors of caffeine to the diet [4] and it has been consumed safely for centuries, as have black and green tea. Energy drinks entered the market in the 1980s, introducing another popular source of caffeine. A number of other caffeine-added products have also attempted entry into the marketplace, such as maple syrup, beef jerky, donuts, and chewing gum. These products, with varying degrees of success, have attempted to provide novel sources of caffeine to the consumer.

The long history of caffeine use and the wide array of new products offered as sources suggest that consumers continue to desire caffeine's pharmacological effects. In the last decades, caffeine has received both favorable and unfavorable attention from various stakeholders, such as the scientific community, the press, and Non-Government Organizations. Any general internet search yields many consumer questions related to the health and safety of caffeine. Mixed messaging in the press related to benefits and potential adverse effects, combined with the possible difficulty of assessing one's own exposure to caffeine, can lead to a great deal of uncertainty for the consumer. To address this concern in the United States, health-care professionals made a public request in the form of a letter to the FDA to gather data related to overall caffeine safety [5]. As part of this request for more investigation, the IOM's Food and Nutrition Board and Board on Health Science Policy hosted a two-day workshop in August of 2013, entitled, "Caffeine in Food and Dietary Supplements: Examining Safety". This workshop provided a public forum for discussion and examination of the potential health hazards of caffeine, which were later summarized in a large (190-page) publication [6]. The bulk of the data presented at that time came from the Oak Ridge National Laboratory (ORNL) report that was commissioned by the FDA [7]. The IOM's public forum event to discuss caffeine safety was not unprecedented—in the past couple of decades, many other countries have initiated discussions about the use of caffeine in food and beverages, with the intent of better understanding the consumption practices and potential safety concerns (India [8]; Australia and New Zealand [9], Europe [10], and Canada [11]). The European Food Safety Authority (EFSA) has the most recent publication of such an effort [10].

Most of the authoritative reviews or discussions mentioned above allowed for some sort of public and stakeholder input, either via submission of public comments directly or participation in public forums for discussion, and three major themes or requests continually surfaced: (1) help the consumer understand how much caffeine is actually in food and beverages (exposure); (2) help the consumer understand what level of caffeine is safe (risk); and (3) better elucidate what sort of adverse effects are associated with particular doses (dose-effect). Throughout the discussions and various publications, another commonality was the repeated references to one particular publication—Nawrot et al. (2003) [11]—and subsequent references to the suggested "safe values" for ingestion of caffeine those authors put forward.

Nawrot et al. (2003) [11] is a peer-reviewed publication from Health Canada, which conducted a narrative, but not systematic, review of scientific literature. We believe that at least part of the reason this article has been so heavily cited is that it is easy to read and covers multiple areas of interest related to caffeine. In developing their conclusions, Nawrot et al. (2003) [11] reviewed many potential adverse-event areas; however, given the voluminous scope, they focused primarily on five outcomes (1) acute toxicity (defined herein as abuse, overdose, and potential death); (2) cardiovascular; (3) bone and calcium; (4) behavior; and (5) development and reproductive toxicity. The authors also touched on genotoxicity, mutagenicity, and carcinogenicity, but these have not been a focal point of concern for caffeine outside of reproductive toxicity. The authors concluded after conducting their qualitative review that the consumption of up to 300 mg/day for pregnant women and 2.5 mg/kg body weight/day for children is not associated with adverse effects. They went on to conclude that an intake dose of up to 400 mg caffeine/day is not associated with adverse effects in healthy adults [11]. Importantly, since Nawrot et al. [11] was published in 2003, more than 10,000 papers on caffeine-related topics have been published, and of those, more than 5000 address effects or exposure in humans. In addition, 800+ reviews related to various human health effects of caffeine have also been published (nearly all are specific to a particular adverse endpoint category).

With this as background and in light of the wealth of new data in the peer-reviewed literature, and because Health Canada's work is so commonly referenced in discussions and debates over caffeine safety, the goal of our systematic review was to investigate whether or not the Nawrot et al. (2003) [11] conclusions remain current as an acceptable level of protection to the healthy general public. We chose the same outcomes for evaluation, because these endpoints reflect importance, as documented in other comprehensive evaluations [6,10–13], and indicate stakeholder interest. Therefore, it is useful to determine whether the values that were put forth by Nawrot et al. (2003) [11] remain appropriate and as such can still serve as a basis to assure the typical healthy caffeine consumer of a reasonable certainty of no harm. This evaluation also allows scientists to move on from this question and focus more on sensitive subpopulations that may be at greater risk.

Thus, the need for our systematic review was established. Specifically, our objective was to determine whether the literature published since the 2003 Health Canada review supports the conclusion that caffeine consumption at amounts up to 400 mg/day for healthy adults, 300 mg/day for healthy pregnant women, and 2.5 mg/kg body weight/day for healthy children is not associated with adverse effects. We also evaluated the consumption of 2.5 mg/kg body weight/day in adolescents, although this was not specifically addressed by Nawrot et al. (2003) [11].

## 2. Materials and Methods

The Systematic Review (SR) was conducted using the IOM's *Finding What Works in Health Care—Standards for Systematic Reviews* as guidance [14]. The overall work flow of the systematic review is shown in Figure 1 and it included problem formulation; developing a protocol; conducting a systematic search (informed by a librarian) of three databases; screening of literature for inclusion/exclusion; critically appraising individual studies; conducing endpoint, outcome, and overall syntheses and weight-of-evidence analyses; and, reporting the systematic review.

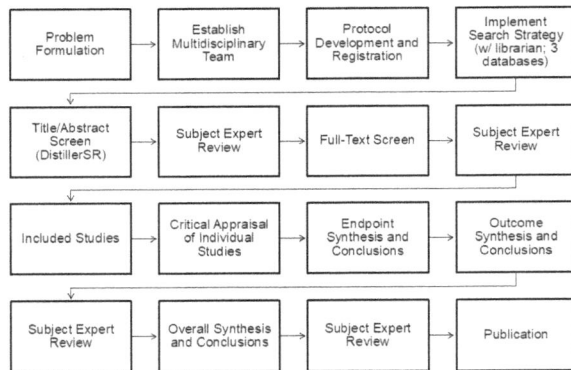

**Figure 1.** Work flow of the systematic review.

Consistent with IOM recommendations, the first step that is involved establishing a team with appropriate expertise and experience (Table 1). The project team was composed of eight scientists from ToxStrategies with a range of expertise, as well as a scientific advisory board (SAB), of which each member had expertise in an outcome (e.g., cardiovascular) evaluated in the review.

**Table 1.** Project team and roles for the systematic review.

| Entity | Description | Roles |
|---|---|---|
| Scientific Review Team: ToxStrategies | Scientists with a range of expertise (caffeine, toxicology, epidemiology, systematic review, literature searching, etc.) | Develop and perform the systematic review (SR) (consistency in application of SR process, independent assessment, documentation) |
| Scientific Advisory Board (SAB) | Multidisciplinary experts (systematic review, behavior, cardiovascular, bone & calcium, reproduction & development, acute, pharmacokinetics—PhDs and MDs from academic, private, and clinical practices) | Provide input, review, and approval; develop protocol, conclusions |
| Sponsor: ILSI North America | Members of the ILSI-North America Working Group (additional funding through two unrestricted grants from the American Beverage Association and the National Coffee Association) | Budgetary |

**Develop the Population Exposure Comparator Outcome (PECO).** As part of the IOM framework problem formulation, the specific research question or objective addressed in the systematic review was based on a "PECO" format (which is different from the PICO (population, intervention, comparator, and outcome) format that is often used in nutrition and clinical medicine). Specifically, the PECO was:

"For (population), is caffeine intake above (dose), compared to intakes (dose) or less, associated with adverse effects on (outcome)?" As an example, for healthy adults, the PECO would be, "For healthy adults, is caffeine intake above 400 mg/day, compared to 400 mg/day or less, associated with adverse cardiovascular effects?"

The SR focused on five outcomes (Figure 2): acute, cardiovascular, bone and calcium, behavior, and development and reproduction (further descriptions of the endpoints included within each of these outcomes can be found in the results section of each outcome. It should be noted and emphasized that, within each outcome (e.g., cardiovascular), there were many endpoints (e.g., morbidity, mortality, blood pressure, heart rate, etc.). A sixth outcome, pharmacokinetics (PK), was included as a contextual topic; the objective was to generally characterize the current understanding of caffeine kinetics and critically review any information that advances the science. Thus, this topic particularly pertained to the differences and similarities between our populations of interest, characterization of kinetics in children and adolescent populations of interest, and characterization of kinetic parameters (particularly fast/slow phenotypes) in the context of the outcomes of interest.

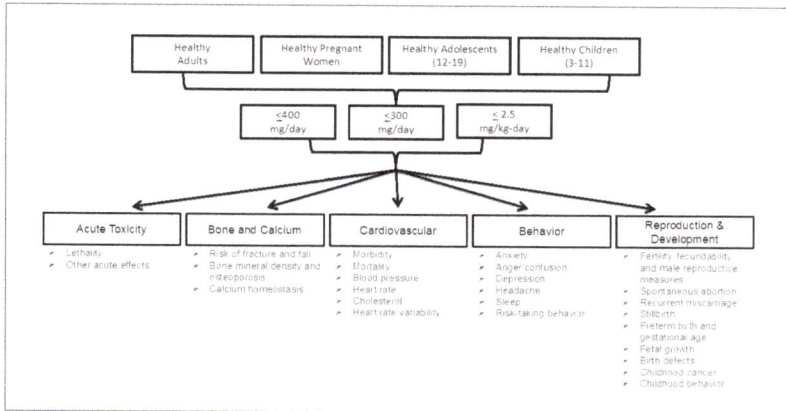

**Figure 2.** Populations, dose/intake levels, outcomes, and endpoints evaluated.

Four populations were evaluated: healthy adults, healthy pregnant women, healthy adolescents (aged $\geq$12–$\geq$19 years), and healthy children (aged $\geq$3–<12 years). For all outcomes, except acute, the daily intake (exposure) values that were evaluated were based on those established by Nawrot et al. (2003) as acceptable levels of daily intake. Thus, the exposure values (the "E" in the PECO) were 400 mg/day (10 g for acute), 300 mg/day, and 2.5 mg/kg body weight/day for adults, pregnant women, and adolescents and children, respectively. Similarly, comparators (the "C" in the PECO) were $\leq$400 mg/day for adults (10 g for acute), $\leq$300 mg/day for pregnant women, and $\leq$2.5 mg/kg body weight/day for adolescents and children. Thus, for example, we investigated whether the literature supports a finding that a daily exposure of 400 mg caffeine per day is safe for adults (the exposure), or rather, whether the literature supports the safety of daily exposures to less than 400 mg caffeine body weight per day for adults (the comparator).

**Protocol Registration.** Consistent with expectations for transparency as part of the framework, a protocol for each outcome was developed and registered on PROSPERO (PROSPERO protocol nos. CRD42015026704, CRD42015027413, CRD42015026673, CRD42015026609, and CRD42015026736; https://www.crd.york.ac.uk/PROSPERO/). Each protocol included: (1) context and rationale for the review; (2) study selection and screening criteria; (3) descriptions of outcome measures, time points, and comparison groups; (4) search strategy; (5) procedures for study selection; (6) data extraction strategy; (7) approach for critically appraising individual studies; and (8) method for evaluating the body of evidence. The objective of registering a protocol is to make the approach apparent a priori, as is consistent with the IOM guidelines and standard practice of systematic review.

**Literature Search.** A comprehensive search strategy was iteratively developed and employed with the assistance of a librarian who had expertise in the conduct of SRs. Three databases were searched: PubMed, EMBASE, and the Cochrane Database of Systematic Reviews. DistillerSR (a software tool that facilitates systematic review) was used for screening and selecting studies, as well as for documenting the extraction and evaluation of data. It is important to note that, to be included in the SR, studies had to provide a quantitative estimate or measurement of individual exposure to a caffeine source associated with an adverse effect. We included many forms of caffeine, such as coffee, tea, chocolate, cola-type beverages, energy drinks, supplements, medicines, and energy shots. For included studies, basic information that was reported by the author was extracted from each study (i.e., direct extraction of information from the text), along with other selected information needed to inform the PECO questions (e.g., dose/exposure calculations) that may have required interpretation by the analysts. For example, the exposure (dose) of caffeine was extracted directly from the studies when the authors of the studies evaluated caffeine directly or reported findings based on the amount of caffeine in given sources. In cases where this was not directly reported, the reviewers standardized the quantity of caffeine; this process was explained in supplementary materials to the original publication, and the interested reader can find more details there.

**Individual Study Evaluation.** During extraction of information from an individual study, the level of adversity (potential for harm) of the endpoints within the study was characterized [15]. That is, the reviewer noted whether the study evaluated a clinical (e.g., morbidity or mortality) or physiological endpoint (e.g., blood pressure changes), as well as the importance of the effect for decision making (e.g., mortality vs. blood pressure changes). Additionally, from each study and each eligible endpoint within a study, specific values were selected or determined in order to compare to the PECO (i.e., the conclusions of Nawrot et al., 2003 [11]). This involved identifying effect and no-effect levels. Specifically, we endeavored to establish a lowest-observed-effect level (LOEL), or, preferably, a no-observed-effect level (NOEL) (e.g., a daily exposure of X caffeine/day was without effects on Y endpoint in study Z), which could then be used for comparison to the PECO.

Following data extraction, individual studies were assessed for the risk of bias (internal validity) using the National Toxicology Program's Office of Health Assessment and Translation (OHAT) Risk of Bias Rating Tool for Human and Animal Studies [15]. Bias is differentiated from the broader concept of quality of the methodology and is aimed at assessing the systematic error—a measure of whether the design and conduct of a study compromised the credibility of the link between exposure and outcome [14–16]. This approach evaluated what are called "specific domains" based on study type (i.e., controlled trial vs. observational study). Specific domains related to bias included selection, confounding, performance, detection/measurement, attrition/missing data, reporting, and other types of bias. Each domain was rated from "definitely low risk of bias" to "definitely high risk of bias" per the OHAT tool. These ratings for individual studies were then considered in the weight-of-evidence assessment when developing conclusions for the endpoint, outcome, and overall (Figure 3).

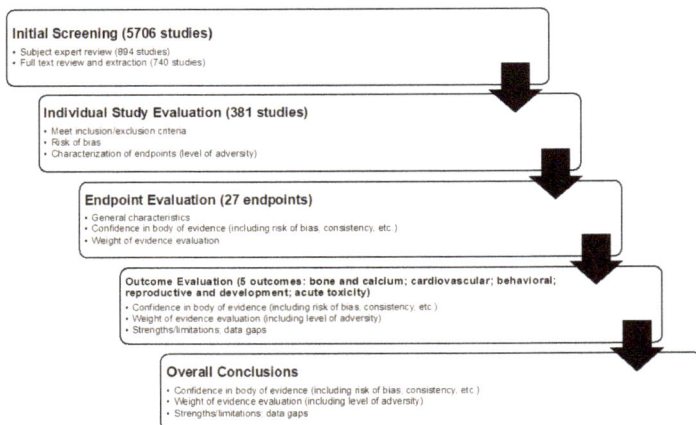

**Figure 3.** Review process, from initial evaluation to reaching overall conclusions.

**Determination of Weight of Evidence.** Following the appraisal of individual studies, the body of evidence was evaluated using a weight-of-evidence approach for each endpoint, each outcome, and overall (Figure 3). Similar to the approach and conclusions of Nawrot et al. (2003) [11], the objective in the weight-of-evidence assessment was not to find the most protective amount or the lowest amount associated with an effect, *per se*, but rather, to make a determination that is based on the body of evidence as a whole, which included considerations for positive and negative findings, quality of data, level of adversity, consistency, and magnitude of effect (for studies with effects below the comparator). The weight-of-evidence approach implemented was based on the framework established by the IOM [14] and it was complemented by guidance from the National Toxicology Program handbook on systematic reviews [17], given the specific application to toxicological assessments. We also relied on the GRADE (Grades of Recommendation, Assessment, Development and Evaluation) process in determining and implementing our weight-of-evidence approach [18,19].

In evaluating and conducting a qualitative synthesis of the body of evidence, data were described based on the volume of data above and below the comparator, as well as the types of effects and quality of evidence of data that are above and below the comparator. An initial level of confidence in the evidence was assigned based on key features of study design: controlled exposure, exposure prior to outcome, individual outcome data, and comparison group used [17]. Then, using expert judgement, a number of additional factors were considered for the overall body of evidence, which yielded increases or decreases in the confidence level. These factors included the following: overall risk of bias, indirectness (when the population, exposure, or outcome differ from those in which we were interested), magnitude of effect, confounding, and overall consistency [17–19]. Consideration of endpoint importance in terms of the endpoint's degree of adversity [18,19] was also important in reaching weight-of-evidence conclusions.

**Weight-of-evidence determinations** were made by endpoint, outcomes, and overall (Figure 4). Such determinations were also made by population, because the comparators were different for healthy adults, pregnant women, and children. Conclusions were developed by categorizing evidence relative to the comparator (an intake value not associated with adverse effects) as follows: comparator is acceptable (i.e., evidence supports the Nawrot et al., 2003 [11], conclusions regarding intake), comparator is too high (i.e., evidence suggests the comparator is too high for a given endpoint), or comparator is too low (i.e., evidence suggests the comparator could be higher for a given endpoint). Using a similar approach, conclusions were also developed for the outcome. When developing outcome conclusions, clinical endpoints with a high level of adversity were given the most weight. Several tools were used to facilitate and support the weight-of-evidence evaluation, including generation of evidence tables, risk-of-bias heat maps, summary plots of selected NOEL/LOEL data from individual studies, and a tabular summary of the confidence in the evidence for each outcome and endpoint. Conclusions were not developed for endpoints that contained fewer than five studies; in these instances, summary thoughts were provided, but data were determined to be insufficient to reach a conclusion.

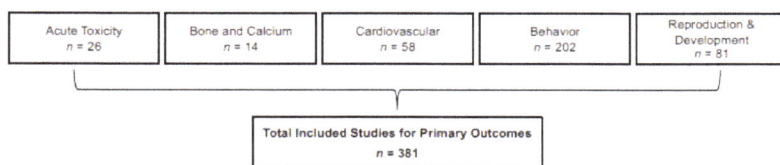

**Figure 4.** Number of studies that met the SR inclusion criteria and were reviewed for each endpoint.

Transparency in Reporting. All data from the systematic review were placed in a freely available Agency for Healthcare Research and Quality (AHRQ) Systematic Review Database Repository (SRDR).

## 3. Results

Throughout this section, the reader is reminded to refer to the original paper for extensive references [1]. This approach (not including full references here) was chosen to best fulfill the goal of simplifying the text so that this summary can accomplish its aim—i.e., to provide ease of reading and understanding for diverse audiences. Figure 5 below summarizes the key findings from each outcome, as well as perspective that is related to confidence in the value based on our analysis. The manner in which these conclusions were reached is discussed in each section for the respective outcomes below.

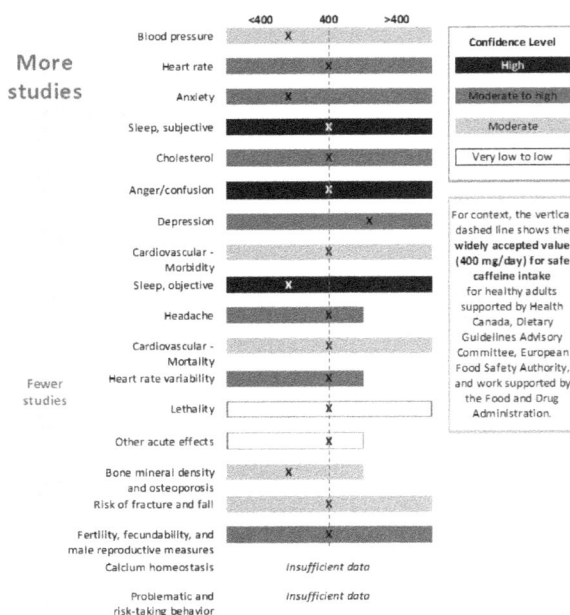

**Figure 5.** Summary of the spectrum of data and our endpoint conclusions for healthy adults, weighted for level of confidence in the body of evidence considering risk of bias, magnitude, consistency, and other factors. Shading indicates that data reported effects at the corresponding intake level (<400, 400, or >400 mg caffeine/day), and darker shading indicates increased confidence in the body of evidence (from very low to high). X indicates the SR weight-of-evidence conclusion for the level of intake not associated with significant health effects. Although effects were observed at exposures below 400 mg (e.g., blood pressure, bone mineral density and osteoporosis), these results did not affect the overall conclusion of the SR, due to considerable variability in individuals' sensitivity to caffeine and potential confounding, and the effects were limited to physiological effects following acute exposure, and subgroups of clinical endpoints, such as those with low calcium intake. Such effects were generally of low magnitude, and/or were of overall low or negligible consequence to downstream effects. Several studies also showed a lack of effects on clinical endpoints at exposures above 400 mg.

*3.1. Literature Searching*

All databases were searched on 8 June 8 2015. Following removal of duplicates, 5706 records of human studies were identified. Following committee reviews, internal quality-control efforts, and SAB review of title and abstract screening, 740 records were carried forward to full text review (Figure 3). The most common reasons for exclusion during title and abstract review were as follows: outcomes not included in the SR (e.g., cancer), unhealthy populations, co-exposures (e.g., alcohol), study was focused on benefit or therapy, and in vitro studies. Following a full text review, a total of 381 studies (plus 46 for contextual pharmacokinetic discussion) were included in this SR relevant to the five outcomes considered for healthy adults, children, adolescents, or pregnant women. Almost half of the studies (42%) specifically evaluated caffeine as a source; the majority of the remaining studies evaluated coffee (21%), tea (12%), and soda (9%) as sources of caffeine, whereas the other studies evaluated caffeine from energy drinks, chocolate, medicine, and other sources. In 77% of the studies, the exposure (dose) of caffeine did not need to be standardized (i.e., the author either evaluated caffeine directly or reported findings based on the amount of caffeine in the given sources). With respect to study type, more than half of the studies (63%) were controlled trials. The remaining were observational studies as follows: cohort studies (14%), case-control studies (9%), cross-sectional studies (5%), and meta-analyses (2%). Seven percent of the publications were case reports or case series, all of which were associated with the acute outcome (these were excluded for other outcomes). The majority of the literature (79%) identified and reviewed the involved adult populations. Literature characterizing the outcomes of interest in other populations was much more limited, including studies that involved pregnant women (14%), adolescents (aged 12–19 years) (4%), or children (aged 3–11 years) (2%). Data were extracted by the research team and rated for risk of bias and indirectness (internal and external validity). Selected no- and low-effect intakes were assessed relative to the population-specific comparator. See Figure 4 for the specific number of studies reviewed per outcome.

### 3.2. Endpoint Evaluation

Results by outcome are discussed below. Often, observational studies relied on food frequency questionnaires and thus used categorical exposure groups based on self-reported exposure (e.g., <1 cup/day, 1–3 cups/day). Thus, the studies that directly evaluated caffeine (i.e., low level of indirectness) were given more weight in the body-of-evidence assessment relative to those that evaluated caffeine via the consumption of coffee or other substances, such as soda, tea, and chocolate, which needed to be standardized by the reviewer. It should also be noted that the general lack of mention of pregnant women in each section, outside of the outcome for reproductive effects, is a result of the lack of studies investigating this subpopulation. Figure 5 is a graphical depiction of the key findings discussed below.

To deliver the key findings from the original work in an easy-to-follow format, we have chosen to omit the original references that are cited extensively in the SR. However, the reader will find that the summary format follows that of the original text, and full references can be found therein: Food and Chemical Toxicology 109 (2017) 585–648 [1].

### 3.2.1. Bone and Calcium

The potential for caffeine to adversely affect bone metabolism was raised in Nawrot et al. (2003) [11], and this was likely considered as an area of concern due to work that originated in the 1980s in the lab of Heaney and Recker [20]. This work examined the effect of caffeine on the calcium economy in the bone, and concerns regarding risk of osteoporosis followed soon after. Because this was an important outcome of interest raised by Nawrot, we specifically looked for literature that investigated the relationship between caffeine and risk of fracture and fall, bone mineral density (BMD) and osteoporosis, and metabolic impacts on calcium homeostasis. The majority of the studies reviewed evaluated associations between caffeine consumption and BMD or bone mineral content (BMC); in some studies, these data were also used to characterize osteopenia. Results were found to vary by bone site. Overall, there were 14 studies that met the inclusion criteria, because they permitted comparison to the conclusions of Nawrot et al. (2003) [11]. Most studies were observational (including large cohorts, such as the Nurses' Health Study), although randomized controlled trials were included as well, and the study populations were healthy adults (with the exception of one study that also included adolescents).

In reviewing studies for this outcome, we recognized that calcium intake was a potential confounding factor that was not accounted for equally in all studies. Effects of caffeine on bone are most often associated with increased urinary calcium excretion. Altered calcium balance through perturbing calcium excretion can influence bone mass. However, urinary calcium excretion is affected by calcium intake, so calcium intake needed to be considered in the analysis. This was reported by the aforementioned Heaney and Recker (1982) [19], the research group that first identified caffeine as a potential risk; however, they later concluded that individuals who ingest the recommended daily allowance of calcium are not at risk of effects from caffeine on calcium economy of the bone (Heaney, 2001) [21]. To this end, it is noted that almost 20% of the United States (US) adult population does not consume the estimated average requirement of calcium [22]. Other important common variables accounted for in studies included age, weight, body mass index (BMI), other nutrient intake, alcohol consumption, smoking habits, and physical activity level.

Exposures evaluated in the evidence base ranged from below 20 mg/day up to 760 mg/day. For risk of fracture and fall, the majority, but not all, of the data demonstrated a lack of effects at levels below and well above (up to 760 mg/day) the comparator of 400 mg/day, with a moderate level of confidence. It is worth noting that there was no significant concern for those with adequate calcium intake. For BMD and osteoporosis, the majority of studies reviewed support a finding that the comparator of 400 mg/day in healthy adults is not harmful, although more evidence is needed for effects of caffeine intake above the comparator, because only one study examined such exposure. Calcium homeostasis was also reviewed, but only two studies met the inclusion criteria, and thus, no conclusion was developed. No data for children, adolescents, or pregnant women were available.

Weight of Evidence for Outcome. Overall, the recent evidence is consistent with the conclusions reached by Nawrot et al. (2003) [11] for bone and calcium endpoints. Individual studies generally had a low risk of bias. When the weight of evidence was considered, 400 mg/day was found to be an acceptable intake that should not cause concern with regard to adverse effects on bone or calcium-related endpoints, particularly when individuals are consuming adequate amounts of calcium. When effects were observed at levels below 400 mg/day, they were physiological effects that followed an acute exposure, or they occurred in population subgroups; and they were generally of low health impact Limitations of the data included uncertainty in exposure estimates, ambiguity regarding calcium intake, and a high level of indirectness. Due to factors such as the consideration of only females and only one site (as opposed to fracture risk at all sites evaluated), as well as the use of different consumption groupings by study authors, the uncertainty associated with assessing caffeine exposure (particularly relative to calcium consumption), and the lack of consistently observed effects (above or below the comparator), a moderate to low level of confidence was placed on this conclusion.

### 3.2.2. Cardiovascular

Caffeine is a central nervous system stimulant, and its pharmacological activity involves non-specific antagonism of the adenosine receptor, which in terms of the cardiovascular system, produces various effects [23]. For that reason, extensive literature reports both caffeine's acute effects (e.g., blood pressure, heart rate) and its chronic effects (e.g., heart disease) on this system. With this background, we considered the effects of caffeine on mortality, morbidity, blood pressure, heart rate, cholesterol, and heart-rate variability. A key factor in evaluation of endpoints other than mortality and morbidity was the consideration of level of adversity, or how much a measured endpoint actually affects a person's overall state of health, in both the short and long term. For example, elevated heart rate, while considered an "adverse effect", is a temporary state, and occasional increases in heart rate do not affect one's overall health status.

Overall, there were 203 studies that, after full review, met the inclusion criteria of the SR, because they permitted comparison to the conclusions of Nawrot et al. (2003) [11]. A large majority of the included studies were randomized controlled trials (RCTs), and the remaining were observational studies, meta-analyses of observational studies, and one meta-analysis of RCTs. Exposure was well defined in the RCTs, with most studies administering pure caffeine in pill/capsule or liquid form in a single "acute" exposure or dose, which meant a high level of directness. Often in the clinical studies, participants had fasted or abstained from caffeine consumption for some number of hours or an entire day before exposure. Some study designs involved pre-treating individuals, followed by a challenge of caffeine. Most studies involved healthy adult populations, while only 11 involved children or adolescents; however, not enough evidence existed for children to reach an overall conclusion for that population. Most of the controlled trials evaluated few, if any, potential confounders, whereas the majority of the observational studies included analyses accounting for many common risk factors for cardiovascular disease (CVD) (e.g., age, sex, smoking, alcohol consumption, BMI).

Quantified exposures generally ranged from below 50 mg/day to more than 800 mg/day. About one-half of the data points were below the comparator of ≤2.5 mg/kg body weight in studies of children and/or adolescents. There was a moderate to high level of confidence, depending on the endpoint. The endpoint of cardiac mortality was reviewed, and the majority of evidence supports a conclusion that 400 mg caffeine/day in healthy adult populations is an acceptable intake that is not associated with significant concern. Even at higher intakes, up to ~822 mg/day, there are no consistently reported effects on mortality; further, several studies reported findings that suggest protective effects. Regarding cardiovascular morbidity, when all data were considered collectively, and considering the greater utility of meta-analyses, evidence supports that 400 mg caffeine/day in healthy adult populations is an acceptable intake that is not associated with significant effects for this endpoint. Some studies, including two meta-analyses, reported a lack of effects above the comparator (suggesting that the comparator is too low). In several cases, associations were observed only in specific genotypes, highlighting the potential role of kinetic influence on pharmacodynamics (PD; discussed below in the pharmacokinetics section). No data were available for pregnant women, adolescents, or children.

Blood pressure was a heavily studied endpoint, with more than 100 controlled trials using exposures ranging from 50 mg to 1 g/day and considering different aspects of blood pressure. It is important to note that chronically elevated blood pressure is a known risk factor for CVD [24], whereas intermittent blood pressure elevations, such as those that are associated with exercise, are not. Taken together, studies were relatively consistent in demonstrating that exposures to caffeine at intakes both below and above the comparator of 400 mg/day have the potential to minimally increase blood pressure (often only a few mmHg) in all the populations evaluated. The biological significance of this small magnitude of change is difficult to interpret relative to the determination of adversity, because such a determination is likely to be conditional. When the evidence is considered collectively, findings suggest that the comparator of 400 mg/day in healthy adults is too high if one is considering only the potential for caffeine to cause a physiological change in blood pressure (which may or may not be adverse). However, when considering the small magnitude of changes in this physiological parameter, as well as the lack of information demonstrating an association between chronic caffeine-mediated blood pressure increases relative to known cardiovascular risk factors, the comparator of 400 mg/day is likely acceptable with a moderate to high level of confidence. Regarding the comparator of 2.5 mg/kg body weight/day in children, findings were mixed with regard to changes in blood pressure (but as noted above, blood pressure changes may not necessarily be adverse). As in the healthy adult population, when considering the small magnitude of changes and the lack of association between chronic caffeine-mediated blood pressure increases and known cardiovascular risk factors, evidence shifts to support the comparator of 2.5 mg/kg body weight/day with a moderate to high level of confidence. Additionally, results indicate that it would be prudent to evaluate blood pressure in children and/or adolescents with significant caffeine intake and consider limiting such intake for those with significant caffeine-mediated blood pressure rise. There were no data for pregnant women.

Mainly controlled trials evaluated heart rate, with exposures ranging from <100 to 780 mg of caffeine/day, often evaluated during exercise. Collectively and with a moderate to high level of confidence, data supported that the comparator of 400 mg caffeine/day in healthy adults is acceptable in terms of not raising meaningful concern regarding the adverse effects on heart rate. Heart rate was often, but not always, significantly increased during or after exercise at a wide range of caffeine exposures, with the reported increase in these studies considered to be a beneficial (i.e., performance-enhancing) effect (heart-rate increase during exercise is a key mechanism to improve cardiac output). For children and adolescents, the data support a relationship between caffeine exposure and decreased heart rate; however, further characterization of exposures associated with such an effect were difficult, given that changes were observed in studies both below and above the

Nawrot et al. (2003) [11] comparator of 2.5 mg/kg. Thus, it was determined that the evidence base was insufficient to render a conclusion regarding appropriateness of the comparator for potential impacts of caffeine consumption on heart rate in children and adolescents. There were no data for pregnant women.

Caffeine effects on cholesterol were investigated in controlled trials, with exposures ranging from 180 to 475 mg caffeine/day; relatively consistent data showed a lack of effect of caffeine consumption on cholesterol at intakes below and above the comparator. This supports a conclusion that, for cholesterol, 400 mg/kg is an acceptable comparator in healthy adults, with a moderate to high level of confidence. No data were available for pregnant women, children, or adolescents.

Heart-rate variability (HRV) was the final endpoint evaluated in the category of cardiovascular effects, with a moderate to high level of confidence. Exposures ranging from 40 to 500 mg caffeine/day were investigated in controlled trials, and most subjects were habitual consumers of caffeine or coffee, whereas others were relatively caffeine naïve or not specified. Taken together, there was no consistent effect of caffeine on HRV at intakes below or above the comparator, thus supporting that 400 mg caffeine/day in healthy adults is an acceptable intake that is not associated with significant change in heart-rate variability.

Weight of Evidence for Outcome. Overall, the recent evidence is consistent with the conclusions of Nawrot et al. (2003) [11], and we maintain a moderate to high level of confidence in the evidence base. Most of the studies were clinical trials that were designed to specifically evaluate caffeine, so the level of indirectness was low. When the weight of evidence was considered, 400 mg/day was concluded to be an acceptable intake that is not associated with significant concern for adverse cardiovascular health effects in healthy adults. In general, evidence for clinical endpoints (mortality, morbidity) indicated that 400 mg/day is too conservative, and consuming higher amounts of caffeine would still be safe. While effects were seen for physiological endpoints (e.g., blood pressure, heart rate) at intakes below 400 mg/day, it remains unclear what amount of change would be considered adverse in a clinical or toxicological context. Data in children and adolescents were limited to 11 studies that evaluated physiological endpoints. Therefore, it was determined that the evidence base was insufficient to render a conclusion regarding the appropriateness of the comparator for assessing the potential impacts of caffeine consumption on cardiovascular outcomes in these populations. The available data for blood pressure and heart rate are inconsistent; several studies that report physiological changes are described below.

### 3.2.3. Behavioral

As discussed in the Pharmacokinetics/Pharmacodynamics section of this article, caffeine is probably best known for two of the behavioral effects it exerts on the body through antagonism of the adenosine receptor: increasing mental alertness and vigor. Although it may seem remiss to not include these effects here, because this systematic review was intended to look only at potential adverse effects, these mood states were not relevant to the inclusion criteria. Instead, the main categories that encompass potential caffeine-related adverse effects were mood, withdrawal, headache, and sleep, which were similar to those that were described in Nawrot et al. (2003) [11]. One newer category that was not covered in Nawrot et al. (2003) [11] was that of "risk-taking behavior", which has become a topic of heightened interest in adolescents and young adults with the rise in popularity of energy-drink consumption in these cohorts.

After full review, 80 studies met the inclusion criteria of the SR, because they permitted a comparison to Nawrot et al. (2003) [11] conclusions. The majority of these were RCTs with healthy adults. For sensitive populations, only five studies were found that met the requirement for quantitative information; these studies were conducted in children or adolescents, and no studies in pregnant women met the criteria. In the controlled trials, a large number administered pure caffeine, which led to a low level of indirectness.

As has been described elsewhere in this summary, confounding remains an important consideration. For the endpoint of behavior, confounders, such as smoking, age, and sex, and sometimes anxiety sensitivity or sleep behavior, were taken into consideration by the authors, depending on the endpoint objective. Most studies evaluated caffeine intake that fell at or below the comparator of 400 mg/day, with a quantified exposure range from 60 mg/day up to approximately 1.2 g/day. Overall confidence in this data set was moderate to high.

A number of endpoints represented potential behavioral effects, and for this reason, major categories were used for simpler designations, and subdivisions within each category were discussed. For example, the category of "mood" was subdivided to include anxiety and other general mood states. In studying this endpoint, the majority of studies were randomized controlled trials, and within the study design, questionnaires, such as the Profile of Mood States (POMS) or visual analogue scales (VAS), were frequently used to summarize perceptions by subjects. Using this form, subjects could use common terms such as vigor, depression, fatigue, anger, and confusion, as well as anxiety, to gauge their mood state. It is important to note that these dimensions represent nonclinical mood states, and changes to them do not necessarily indicate negative effects. In our review, we also wanted to note (as did Nawrot et al., 2003 [11]) that, when evaluating anxiety, some of the potential associated manifestations, such as "tension", "jitteriness", "nervousness", and "worry", must be also considered in light of caffeine's pharmacologic ability to increase alertness and arousal, and thus, these can be associated effects. Taken together, some but not all evidence, primarily from RCTs involving single/short-term caffeine exposure (range 70–1200 mg caffeine/day) and subjective measures of anxiety, suggests that the comparator of 400 mg/day can lead to increases, albeit small, in measures of anxiety in adults. There were no data for pregnant women.

Tolerance to the stimulant effects of caffeine occurs with repeated dosing over several days, and this explains why the effects on increased blood pressure are largely temporary and not usually clinically important in the long term. The opposite of tolerance is withdrawal, which is reported as sleepiness and fatigue if the usual dose of caffeine is omitted for a day. This has been clinically recognized by the diagnosis of "caffeine withdrawal" by the American Psychiatric Association (DSM-5, p. 506) [25].

"Anger" and "confusion" were other subdivisions of mood for which a number of RCTs used doses ranging from 70 to 1200 mg caffeine. Confusion included difficulty concentrating and bewilderment or muddled perception. Overall, the data suggest that the comparator of 400 mg/day is an acceptable daily intake that is not associated with significant concern regarding anger and confusion. There were mixed findings when doses were administered above the comparator—well-rested individuals manifested no effect, but at very high doses (1200 mg/day given as 400 mg 3×/day for seven days), there was a significant increase in POMS anger scores. There were no data for pregnant women.

Depression and related endpoints were investigated in mostly RCTs, but also a fair number of observational studies where exposures ranged from 80 to 1200 mg caffeine/day. Similar to Nawrot et al. (2003), the finding from our review indicated no effects of caffeine, even at very high exposures, on scores of depression. Taken together, the weight of evidence suggests with moderate to high confidence that the comparator of 400 mg/day of caffeine is an acceptable intake. A few studies indicated a decreased risk of depression effect that is associated with exposure to caffeine. There were no data in pregnant women.

Headache was another category of relevance and interest, due to both "acute" effects and potential "withdrawal" effects of caffeine. Ratings of headaches (pain or severity), which are often captured via customized questionnaires or a VAS, were not significantly increased in any of the controlled trials that evaluated the effect of acute caffeine ingestion doses below the comparator of 400 mg.

For adults, the weight of evidence supports, with a moderate to high level of confidence, that consumption of ≤400 mg caffeine is not associated with an increase in headaches. However, like the evidence presented in Nawrot et al. (2003), observational studies do indicate a potential link between caffeine use and headache prevalence in some individuals, although some of this effect is likely due to withdrawal-related symptoms. There were no data for pregnant women.

Sleep was a category divided by subjective and objective categories, because the types of endpoints evaluated by each metric vary (i.e., different endpoints of sleep). The subjective effects are those that looked at perceptions of "sleepiness"—mood states, such as fatigue, tiredness, drowsiness, or weariness that are often measured with POMS or VAS questionnaires. Objective measures included sleep latency, duration, and efficiency, all of which are quantitated for the night(s) following caffeine intake. Of the large number of controlled trials that were reviewed, the majority demonstrate that the comparator of 400 mg caffeine/day is acceptable as an intake that is generally not associated with concern regarding adverse effects on sleep. There were a few cases in which prolonged dosing was associated with increased fatigue, but the magnitude of these changes was difficult to assess. Caffeine's mode of action in the central nervous system (CNS) helps, in part, to explain why most caffeine doses tested in these studies may indeed provide some benefit on this endpoint by reducing perceived fatigue; however, higher doses might disrupt sleep and lead to an increase in fatigue when consumed over the course of several days.

Objective effects of sleep were evaluated in controlled studies and observational studies. With respect to the data obtained via objective measures of sleep in adults, results indicate that the comparator of 400 mg caffeine/day is likely too high as an intake, in that it would be expected to disrupt sleep when administered with the intention to do so. Specifically, ingestion of caffeine, even at doses below the comparator, can lead to delayed sleep onset and decreases in sleep quality and efficiency, but this is particularly the case when caffeine is consumed near bedtime. Overall, caffeine at doses both above and below the comparator might provide short-term benefits to improve perceived fatigue, but, depending on the dose and timing, may also disrupt sleep, leading to increased fatigue the following day. There were no data for pregnant women.

The available literature for children and adolescents included in this SR was scant, but the higher-quality studies suggest no major adverse effects on the observed endpoints at doses near or less than 2.5 mg/kg. Above this comparator for all mood endpoints (anger, confusion, anxiety, depression) measured in children and adolescents, it was determined that data were insufficient to develop refined conclusions regarding the potential effects of caffeine. However, the two studies identified that fit the criteria for inclusion suggested no effect of caffeine on mood parameters in adolescents. Regarding headache and sleep, like the other endpoints, it was concluded that there are insufficient quantitative data to evaluate with confidence the effect of caffeine dose on sleep in children and adolescent populations. Based on the limited data, and similar to adults, considerations, such as timing and duration of dose, are likely to be important for these populations. Regarding headache, for children and adolescent populations, there was not enough information, high quality or otherwise, to fully evaluate the appropriateness of the comparator. More targeted research is required to identify sensitive subpopulations in these younger groups, to better quantify the levels at which adverse behavioral effects are observed, and to better understand the link between caffeine consumption and adverse effects.

Regarding risk-taking behavior, there is sparse evidence that caffeine is associated with an increase in risk-taking behavior in adults. This latter effect is a research area that has seemingly attracted more attention since the work by Nawrot et al. (2003) was published, particularly for younger consumers. Unfortunately, the majority of these studies did not provide quantitative caffeine values for comparison to the comparator value of 400 mg/day.

Weight of Evidence for Outcome. When the weight of evidence was considered, the comparator, 400 mg caffeine/day, was found to be an acceptable intake that is not associated with significant concern for adverse behavioral effects in adults. However, intake below the comparator may affect some sensitive individuals who are prone to anxiety or sleep disruption. Often, observed effects below the comparator (e.g., anxiety) were limited to subgroups or the timing of dose (e.g., sleep), whereas others were complicated by consumer status (e.g., headache and fatigue). For some endpoints (depression, headache, sleep (subjective), and anger/confusion), there was largely a lack of effects reported, and in some cases, data suggested that intakes higher than the comparator were without effect. There is a moderate to high level of confidence in the body of evidence supporting this conclusion. Confidence was increased by the overall low risk of bias and low level of indirectness, although the variability that was introduced by sensitive subpopulations was a key limitation that precluded a higher level of confidence. It was determined that the evidence base was insufficient to render a conclusion regarding appropriateness of the comparator (2.5 mg caffeine/day) for the potential impacts of caffeine consumption on behavior outcomes in these populations. Overall, the body of literature reviewed for children and adolescents was generally of lower quality when compared to the data for adults.

### 3.2.4. Reproductive and Development

Caffeine as a reproductive and/or developmental potential hazard has been and continues to be a point of much discussion. General searching of the internet suggests that pregnant women want to know whether they can have caffeine or not. For this outcome, 58 studies were carried forward as meeting the inclusion criteria of the SR, because they permitted comparison to the Nawrot et al. (2003) [11] conclusions. All of these were focused on adults, with the majority studying pregnant women. As opposed to other outcome areas, a large majority of these were observational, relying on self-reports of caffeine consumption from coffee, soda, and tea in most cases; chocolate, caffeine-containing medications, and energy drinks were the source in a few of the reports. Many of these observational studies, such as the Danish Cohort and Birth Defect Registry, used data from very large, population-based cohorts, meaning that more than 50,000 pregnancies were examined per report. Figure 6 summarizes the key findings for this outcome.

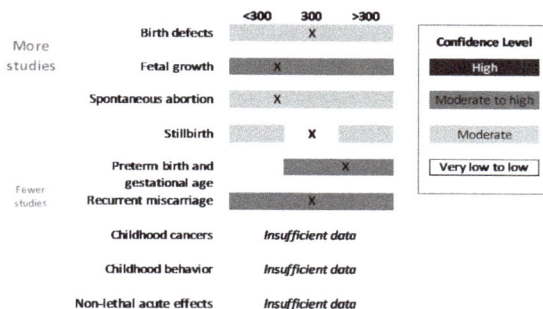

**Figure 6.** Summary of the spectrum of data and our endpoint conclusions specific to pregnant women, weighted for level of confidence in the body of evidence considering risk of bias, magnitude, consistency, etc. Shading indicates that data reported effects at the corresponding intake level (<300, 300, or >300 mg caffeine/day), and darker shading indicates increased confidence in the body of evidence (from very low to high). X indicates the weight-of-evidence conclusion. Although some effects were seen at intakes lower than 400 mg (e.g., fetal growth, spontaneous abortion), these results did not affect the overall conclusion of the SR due to considerable variability in findings and potential confounding.

Common variables accounted for in such analyses included maternal characteristics, such as race, age, weight, BMI, smoking (some using cotinine as a marker), and alcohol consumption. Other factors that were more specific to endpoints of concern were also considered, such as history of pregnancy or miscarriage, partner characteristics, family history of condition, gestational age at birth, and maternal nutrient and supplement intake. Some studies included changes in caffeine consumption during pregnancy as a variable, although most studies did not. Nausea was evaluated as a confounder in most studies, although the extent to which information was collected and incorporated varied. Although confounding factors need to be considered in all epidemiological studies and they were factored into the risk of bias for all endpoints, one unique factor affects reproductive studies in particular. This is a phenomenon known as the "pregnancy signal": nausea, aversion to smells or tastes, and vomiting are associated with a healthy pregnancy, which then leads to the avoidance of strong smells, including coffee. When not properly controlled for, such avoidance can lead to a misperception that the caffeine (e.g., coffee) is the cause of a pregnancy loss, when in fact, the pregnancy was already in jeopardy, as manifested by the lack of pregnancy signal (i.e., the mother felt no aversion to strong smells) and is correlated with low hormone levels [26,27]. Without specific analysis of coffee aversion, it is difficult to ascertain whether an increased incidence of spontaneous abortion in a study is due to higher caffeine consumption, or if reduced caffeine consumption is occurring in healthier pregnancies due to the pregnancy signal (i.e., reverse causation).

Many potential adverse-event outcomes were reviewed, and the confidence in the evaluation of the comparator varied. The comparator of 400 mg/day was considered acceptable, with a moderate to high level of confidence on the endpoints of fecundability (the ability to conceive during a given menstrual cycle), fertility, and male reproductive measures. However, due to significant limitations to fully accommodate for the pregnancy signal, the confidence was decreased to a moderate level for the comparator of ≤300 mg/d as an acceptable intake that would be associated with no significant concern for spontaneous abortion, recurrent miscarriage, and stillbirth. Preterm birth and gestational age were considered together, and because the data consistently showed a lack of effects, both above and below the comparator of 300 mg/day, the data suggest that the comparator could be higher. Fetal growth was an endpoint for which the body of evidence was difficult to assess despite there being a large number of studies. The biological significance of the birth-weight changes is evaluated more robustly in studies that assess small for gestational age (SGA) or intrauterine growth restriction (IUGR). These types of studies, as a whole, did not support effects occurring below the comparator of 300 mg/d. However, the low magnitude of effect (measures of association between 1.0 and 2.0 for studies below the comparator)—as well as the observation that, in many cases, the effects were limited to single measures and/or subgroups or were not clinically relevant changes—reduced overall confidence in the data, suggesting that the comparator may be too high. Many types of birth defects have been studied for associations with caffeine exposure: cardiovascular malformations, choanal atresia, cleft lip (with or without cleft palate), cleft palate only, persistent cryptorchidism, and various other individual birth defects, including anotia/microtia, esophageal atresia, diaphragmatic hernia, omphalocele, or gastroschisis. For all of these birth defects, there was no association with maternal caffeine consumption at or above the comparator of 300 mg/day. Additionally, some weak to moderate but inconsistent associations were reported for anorectal atresia, limb defects, and neural tube defects. Thus, although the evidence base is broad with respect to the type of birth defects and underlying etiologies, data were relatively consistent in demonstrating a lack of effects following consumption of caffeine at intakes up to 300 mg/day in healthy pregnant women. Based on the underlying study types (observational), low risk of bias, and consistency in findings, there was a moderate level of confidence in this conclusion.

Mixed findings for childhood cancers (CNS tumor and childhood leukemia) and their association with maternal consumption of caffeine were attributed to problems with design related to improper control for recall bias (i.e., the phenomenon of individuals experiencing adverse outcomes tending to report more exposure than other individuals, even when no difference may exist). That is, it is generally recognized by epidemiologists that, when asking mothers to recall what they may have ingested during pregnancy after giving birth to a child with a birth defect or disease, they will try to find a cause. For this reason, an alternative study design is for both the case and control populations to have adverse conditions manifested; otherwise, there is a high likelihood of recall bias [28]. Another stronger study design option would be a nested case-control design with prospective assessment of exposure. This topic of confounding was acknowledged by both the authors and observers at the International Agency for Cancer Research (IARC) in the recent review of the potential carcinogenesis of coffee, in which IARC concluded that, overall, coffee drinking was unclassifiable as to its carcinogenicity to humans [12]. The limited number of studies, combined with the significant impact of potential recall bias, precluded the development of a conclusion for this SR but highlights the need for additional research that accommodates this significant bias in the future.

Another area of much interest in public forums has been prenatal exposure to caffeine and subsequent changes in childhood behavior. Only a few studies were included that related to this endpoint. Because data were limited, and all pertained to different behavioral changes, no conclusion was developed; however, the lack of effects observed in all studies suggests that this is not an endpoint of concern. A number of studies that were included in the review (meeting criteria) fell into the category designated as "other reproductive endpoints", because only one study was identified per endpoint. These included pregnancy-induced hypertension and/or preeclampsia, and median age at menopause, as well as maternal stress.

Weight of Evidence for Outcome. The current body of evidence characterizing this endpoint is generally consistent with what was reported by Nawrot et al. (2003); the majority of studies included in the SR do not report reproductive or developmental effects at levels below the relevant comparator. Although effects below 300 mg/day (or 400 mg/day, in the case of males and nonpregnant females) cannot be ruled out with the currently available data, the effects seen at these levels were primarily limited to isolated reports of congenital malformations [29,30] or childhood cancers [31,32], and the findings were of relatively low magnitude.

### 3.2.5. Acute Toxicity

Acute effects that are associated with caffeine consumption can include a wide spectrum of symptoms, with headache, nausea, vomiting, fever, tremors, hyperventilation, dizziness, anxiety, tinnitus, and agitation at the milder end of the spectrum [33]. More severe effects resulting from caffeine intoxication can include abdominal pain, altered consciousness, rigidity, and seizures, as well as abnormal heart rhythms and reduced blood flow to the heart [34]. Many of these changes would be expected at very high doses, considering caffeine's ability to stimulate the central nervous system, among other physiological effects [35].

In the SR, we investigated studies addressing death or non-lethal effects following an acute exposure [1]. Acute toxicity as an outcome of interest for the systematic review was defined as abuse, overdose, and potential death due to caffeine. Forty-six full-text papers were reviewed, and 26 were found to meet the criteria, because they permitted comparison to the conclusions of Nawrot et al. (2003) [11]. All 26 were case reports or case series, most of which were associated with emergency department (ED) visits and/or suicide-related events. This was the only endpoint in the systematic review for which case reports were allowed; while the SR authors recognize that these types of reports are not generalizable (because they investigate one incident and not trends within a population), more robust types of data were not identified for this endpoint.

Of the 26 included, the majority of reports were in adults, with four covering adolescents and two evaluating pregnant women. All of the reports involved very high doses of caffeine (up to 50 g) being delivered over a very short time frame, and in most reports, the authors delivered only brief discussions of the amount of caffeine ingested. In about one-half of the reports, caffeine was consumed as a powder or tablet (sleep aid), and the remaining reports involved energy drinks, with a few involving cola. Coffee and green tea received mentions, but they were not the major sources of caffeine in these intoxications. However, confidence in exposure characterization was low, due to mainly self-reporting with corroboration of friends/relatives as the source. Because Nawrot et al. presented 10 g as the acute lethal dose, 10 g/person was the comparator [11].

Key Findings Described in the Body-of-Evidence Characterization: Overall, the current body of evidence related to acute toxicity of caffeine is generally consistent with what was reported by Nawrot et al. (2003) [11], which suggests the potential for death following acute exposures of approximately 10 g of caffeine. The review of the data also supports a lack of nonlethal acute effects at or below exposures of 400 mg/day. However, there is very low to low confidence associated with this conclusion because of the reliance on case reports, ambiguity of exposure levels, and high risk of bias (e.g., case reports are not published when there is no effect). It is notable that each case appeared to have a unique spectrum of adverse events, although vasospasm, seizure, mania, hypokalemia, and muscle weakness were commonly reported. Nearly all of the case reports describing fatalities involved caffeine powder and tablets, whereas the case reports that were associated with other acute (non-lethal) effects generally involved rapid consumption of caffeinated beverages over a short time.

### 3.2.6. Caffeine Pharmacokinetics (PK) and Pharmacodynamics (PD)

Simply put, PK refers to the rates of absorption, metabolism, and excretion of caffeine, and PD refers to the effects of caffeine upon the body. In general, the PK/PD of caffeine is well understood; however, we were particularly interested in any new science with respect to differences and similarities between populations of interest, in the context of the five main outcome areas. The review found that most recent research has been in the area of caffeine metabolism focused on how one's own genetic makeup leads to interindividual differences in how caffeine is handled by the body.

The most common PK/PD topic reviewed was in relation to how small nucleotide polymorphisms (SNPs) have been characterized, further helping to elucidate individual differences in caffeine metabolism and even consumption practices. This type of work evaluates changes at the allele level in genes and the resultant changes in how one's body handles exposure to caffeine. As an example, caffeine is a known antagonist of the adenosine receptor, and research has shown that the ADORA2A gene encodes specifically the adenosine $A_{2A}$ receptor; polymorphisms in this gene can affect individual sensitivity to caffeine. Effects can include different sensitivities in feelings of anxiousness following decreased caffeine intake. A fair amount of pharmacogenomic research pertains to two other alleles that are commonly studied: CYP1A2*1F (variant rs762551, genotype AA) and the CYP1A2*1K alleles. These alleles are of interest, because they are associated with increased and decreased caffeine metabolism, respectively. Our findings suggest that epigenetic trends or effects, including further characterizations of SNPs believed to be associated with consumption practices (e.g., self-regulation), as well as specific effects, including several behavioral endpoints (i.e., mood, tolerance, withdrawal), can be important when interpreting overall findings, as well as future endeavors, to characterize sensitive effects or sensitive populations.

### 4. Discussion

The article, "Systematic Review of the Potential Adverse Effects of Caffeine Consumption in Healthy Adults, Pregnant Women, Adolescents and Children [1]", summarized herein, provides a comprehensive assessment of evidence in the peer-review literature regarding caffeine safety. Results demonstrated that the conclusions from Health Canada established in 2003 [11] still hold true today. That is, moderate caffeine consumption—up to 400 mg/day in healthy adults, 300 mg/day in healthy pregnant women, or 2.5 mg/kg body weight/day in children and adolescents—is unlikely to be associated with adverse effects. The Special Issue of Nutrients afforded us the opportunity to provide a plain-language summary of the systematic review, thus improving the usability of the SR for health-care professionals and consumers of caffeine.

Serious considerations were given to the strengths and weaknesses of the systematic review. Key strengths included: (1) Use of the systematic review format based on IOM standards (IOM, 2011) [14]; this format imparts transparency and rigor to the review process (and subsequent confidence in the overall assessment); (2) Assessment of five health outcomes (reproductive and developmental toxicity, behavior, cardiovascular, bone and calcium homeostasis, and acute toxicity); (3) Assessment of four populations (healthy adults, healthy pregnant women, healthy adolescents, healthy children); (4) A large evidence base (>5000 studies considered for eligibility, >381 included across the five outcomes); (5) A multidisciplinary team consisting of subject-matter experts and systematic-review experts; (6) Full transparency in analysis and reporting via the registration of systematic review protocols on PROSPERO, use of the AHRQ Systematic Review Data Repository, and open access to both this summary and the systematic review publication in *Food and Chemical Toxicology*. Additionally, the review sponsor supports a website containing all relevant resources (http://ilsina.org/caffeine-systematic-review-2017).

Weaknesses of the systematic review included: (1) The large volume of information reviewed precluded the ability to discuss or present all aspects of each study (e.g., all findings, critical appraisal of individual study strengths and limitations); (2) The evidence base was complex and heterogeneous. Study design and reporting varied widely, both within an outcome or endpoint and between outcomes and endpoints; for example, different methods were used to assess caffeine intake, or different approaches were used to measure effects on sleep; (3) Limitations in the overall evidence base did not allow for an assessment of chronic exposures for all endpoints evaluated in the review; for example, data from studies that reported physiological endpoints (e.g., blood-pressure changes) were most often obtained from short-term (often single-exposure) controlled trials; (4) Not all study designs properly controlled for confounding; (5) Various sources of potential bias (pregnancy signal and recall bias) were discussed briefly here, but the reader is also referred to an article in this special issue devoted solely to this topic [27]; (6) Difficulties encountered in characterizing exposure (discussed in more detail below).

One of the largest areas of uncertainty in the underlying body of evidence assessed herein, and one of much interest to the consumer, is that of exposure. In the case of the SR, confidence in the characterization of exposure for each individual study was not high. Several of the caffeine sources that were included in the SR are complex mixtures with other potentially active compounds, and the amount of caffeine within each source can be highly variable. This is a problem for coffee in particular [4], which was the primary substance evaluated in >20% of studies assessed in this SR. To address this, we attempted to standardize this metric in the SR. It should be noted, however, that the evidence also contains a large number of controlled trials in which exposure was well characterized, although these studies were associated primarily with physiological endpoints. Providing consumers with information that is related to caffeine levels contained in specific products (e.g., better product labeling) will help them to make educated decisions regarding their personal exposure level.

From recent literature, one can see that other aspects of caffeine consumption are important to consider when determining caffeine safety; for example, the conditions under which various sources of caffeine are consumed and whether caffeine consumption is habitual or not. Our SR evaluated consumption of total caffeine amounts within a day; however, as consistent with the kinetic behavior of caffeine, effects may vary based on how the caffeine is consumed within a day. The most dramatic examples of this are the case studies that report lethality events that are associated with rapid and excessive consumption of capsules or powders (the comparator for lethality (10 g) is equivalent to ~100 cups of coffee). This concern is supported by recent FDA activity designating pure or highly concentrated caffeine in powder or liquid as unlawful (FDA guidance, 2018; https://www.fda.gov/newsevents/newsroom/pressannouncements/ucm604485.htm). Therefore, it is important for the consumer to understand such nuances of exposure. To that end, considering the wide array of caffeine-containing products in the marketplace, and hence, the potential for exposure to caffeine, the consumer's own perception of the effects of caffeine and self-limitation will remain an important area of research. A recent review by Nehlig (2018) [36] provides insight into consumer self-limiting based on objective (what caffeine does to the body that may not be recognized by the consumer) and subjective effects (the caffeine effects sought by the consumer) of caffeine. Further research will likely continue in the area of interindividual sensitivity and consumption practices, as related to genetic makeup [37].

Based on our findings, we would suggest that any discussion with consumers or patients should consider the magnitude and level of the adversity of effects. That is, the pharmacological effects of caffeine are anticipated to cause certain physiological changes and thus require some characterization of the level of significance to health (because not all physiological changes are adverse). An example is that caffeine intake is expected to result in increased alertness, which is often desirable; however, under some conditions (such as prior to bedtime), this is an adverse effect, leading to difficulty sleeping. Another good example is that, while data suggest that caffeine intake can result in changes to heart rate or blood pressure, it is less clear at what level these effects are clinically significant.

The findings of the SR support the safety of standard consumption practices in the United States, because both mean and upper-end estimated intakes (mean of 165 mg/day and 90th percentile of 395 mg/day, all ages) are below the comparator value evaluated herein. Findings of this assessment, however, also confirm that there is no "bright-line" safe exposure, because potential effects depend on many conditional factors; further, there is some limited evidence that self-regulation reduces consumption [38]. With regard to child and adolescent populations, limited data were identified; however, based on the available studies reviewed, there is no evidence to suggest a need for a change from the recommendation of 2.5 mg/kg body weight/day. Our review supports that additional research would be valuable in this area, as well as in other areas that were identified as having insufficient information—a finding similar to that of other investigators (e.g., Ruxton 2014 [39]). This includes more research on effects in sensitive populations and establishing better quantitative characterization of interindividual variability, as well as subpopulations (e.g., unhealthy populations, those with preexisting conditions), conditions (e.g., co-exposures), and outcomes (e.g., exacerbation of risk-taking behavior) that could render individuals at greater risk relative to healthy adults and pregnant women.

In addition to the area of self-regulation mentioned above, this work identified other suggested research areas, listed here per outcome area. Bone & calcium: more research in non-adult populations as well as a better understanding of caffeine's effects on physiology and the role of calcium would be valuable. Cardiovascular disease: a better understanding of dose-response relationships following chronic exposure for some endpoints (e.g., endothelial function and heart rate variability) would be useful. Additionally, for certain physiological effects, research should better characterize what, if any, magnitude of change may be considered harmful. Behavior: more research is necessary on children and adolescents; particularly with regards to caffeine's effects on sleep and risk-taking behavior. It would also be helpful if more consideration for/or a better understanding of the effects of caffeine withdrawal on these endpoints. The are no data available on pregnant women that fit the quantitative inclusion criteria, so studies designed to account for this would be beneficial. Finally, investigating a better understanding of the effects of caffeine on anxiety and sleep in sensitive subpopulations as well as in individuals with polymorphisms (e.g., ADORA2A) would be of use. Reproductive and developmental: more research is necessary to understand the effect of caffeine on childhood cancer and childhood behavior with properly designed/controlled studies. In addition, more consideration and accounting for the pregnancy signal would be beneficial. Overall, as noted for all outcomes, better exposure characterization in pregnant women to reduce measurement error, which continues to be a major challenge for observational study design, would be valuable. Acute: the main identified research need in this area is improved exposure characterization; testing of blood concentrations would prove valuable.

## 5. Conclusions

In conclusion, the results of the SR support the guidance values that were characterized over a decade ago by Health Canada [11] and reinforce integrative assessments from other authoritative groups (EFSA, 2015) [10]. Recognizing that individuals may differ in their own level of sensitivity to caffeine, our conclusions, as well as those of Health Canada, are intended to provide guidance on safe levels of consumption for healthy consumers.

**Author Contributions:** Conceptualization, C.D. and D.W.; Methodology, D.W.; Validation, C.D., D.W.; Formal Analysis, C.D., E.M., J.J.G., H.R.L., C.O., J.P., M.T., C.W., D.W.; Investigation, C.D., E.M., J.J.G., H.R.L., C.O., J.P., M.T., C.W., D.W.; Data Curation, X.X.; Writing—Original Draft Preparation, C.D., K.F., D.W.; Writing—Review &

Editing, C.D., K.F., E.M., J.J.G., H.R.L., C.O., J.P., M.T., C.W., D.W.; Visualization, K.F., D.W.; Supervision, D.W.; Project Administration, C.D., K.F.; Funding Acquisition, C.D.

**Funding:** The SR was sponsored by the North American Branch of the International Life Sciences Institute (ILSI) Caffeine Working Group. ILSI North America also received unrestricted grants from the American Beverage Association (ABA) and the National Coffee Association (NCA). The two unrestricted grantors received periodic progress reports but did not participate in any aspect of the systematic review. Neither ILSI North America nor the two grantors (ABA and NCA) had any influence over the grading, evaluation, or interpretation of the data collected in this review. ILSI North America is a public, nonprofit foundation that provides a forum for government, academia, and industry, to advance understanding of scientific issues related to the nutritional quality and safety of the food supply by sponsoring research programs, educational seminars, and workshops, as well as publications. ILSI North America receives support primarily from its industry membership. ABA is the national trade association that represents the US nonalcoholic beverage industry. NCA is the national trade association that represents the US coffee industry. ToxStrategies and some of the members of the SAB received funds or an honorarium from ILSI North America for conducting the work on the SR of caffeine. ToxStrategies is a private consulting firm providing services on toxicology and risk assessment issues to private and public organizations.

**Acknowledgments:** The opinions or assertions contained herein are the private views of the author(s) and are not to be construed as official or as reflecting the views of the Army or the Department of Defense. Citations of commercial organizations and trade names in this report do not constitute an official Department of the Army endorsement or approval of the products or services of these organizations.

**Conflicts of Interest:** The researchers' scientific conclusion and professional judgements were not subject to the funders' control; the contents of this manuscript reflect solely the view of the authors. We have no additional conflicts of interest to declare.

## References and Note

1. Wikoff, D.; Welsh, B.; Henderson, R.; Brorby, G.P.; Britt, J.; Myers, E.; Goldberger, J.; Lieberman, H.R.; O'Brien, C.; Peck, J.; et al. Systematic review of the potential adverse effects of caffeine consumption in healthy adults, pregnant women, adolescents, and children. *Food Chem. Toxicol.* **2017**, *109*, 585–648. [CrossRef] [PubMed]

2. Institute of Medicine (IOM). *Initial National Priorities for Comparative Effectiveness Research*; The National Academies Press: Washington, DC, USA, 2009.

3. Schmidt, B.; Roberts, R.S.; Davis, P.; Doyle, L.; Barrington, K.J.; Ohlsson, A.; Solimano, A.; Tin, W. Caffeine therapy for apnea of prematurity. *N. Engl. J. Med.* **2006**, *20*, 2112–2121. [CrossRef] [PubMed]

4. Mitchell, D.C.; Knight, C.A.; Hockenberry, J.; Teplansky, R.; Hartman, T.J. Beverage caffeine intake in the US. *Food Chem. Toxicol.* **2014**, *63*, 136–142. [CrossRef] [PubMed]

5. Arria, A.M.; O'Brien, M.C.; Griffiths, R.R.; Crawford, P.B. Letter to Dr. Hamburg re: The use of caffeine drinks, 19 March 2013.

6. Institute of Medicine (IOM). *Caffeine in Food and Dietary Supplements: Examining Safety—Workshop Summary*; The National Academies Press: Washington, DC, USA, 2014.

7. Milanez, S. *Adverse Health Effects of Caffeine: Review and Analysis of Recent Human and Animal Research*; Oak Ridge National Laboratory: Oak Ridge, TN, USA, 2011.

8. Food Safety and Standards Authority of India (FSSAI). *FSSAI Notifies Caffeine Level for Energy Drinks*; FSSAI: New Delhi, India, 2016.

9. Smith, P.F.; Smith, A.S.; Miners, J.; McNeil, J.; Proudfoot, A. *Report from the Expert Working Group on the Safety Aspects of Dietary Caffeine*; Australia New Zealand Food Authority: Wellington, New Zealand, 2000.

10. European Food Safety Authority (EFSA). EFSA Scientific Opinion on the safety of caffeine. *EFSA J.* **2015**, *13*, 4102–4120.

11. Nawrot, P.; Jordan, S.; Eastwood, J.; Rotstein, J.; Hugenholtz, A.; Feeley, M. Effects of caffeine on human health. *Food Addit. Contam.* **2003**, *20*, 1–30. [CrossRef] [PubMed]

12. IARC. Coffee, tea, matte, methylxanthines and methylglyoxal. In *IARC Monographs On the Evaluation of Carcinogenic Risks to Humans*; Word Health Organization: Geneva, Switzerland, 1991; p. 51.

13. Loomis, D.; Guyton, K.Z.; Grosse, Y.; Lauby-Secretan, B.; El Ghissassi, F.; Bouvard, V.; Benbrahim-Tallaa, L.; Guha, N.; Mattock, H.; Straif, K. Carcinogenicity of drinking coffee, mate, and very hot beverages. *Lancet Oncol.* **2016**, *17*, 877–878. [CrossRef]

14. Institute of Medicine (IOM). *Finding What Works in Health Care: Standards for Systematic Reviews*; The National Academies Press: Washington, DC, USA, 2011. [CrossRef]

15. *OHAT Risk of Bias Rating Tool for Human and Animal Studies*; National Toxicology Program, Office of Health Assessment and Translation (OHAT): Research Triangle Park, NC, USA, 2015.

16. Rooney, A.A.; Boyes, A.L.; Wolfe, M.S.; Bucher, J.R.; Thayer, K.A. Systematic review and evidence integration for literature-based environmental health science assessments. *Environ. Health Perspect.* **2014**, *122*, 711–718. [CrossRef] [PubMed]

17. OHAT. *Handbook for Conducting a Literature-Based Health Assessment Using OHAT Approach for Systematic Review and Evidence Integration*; National Toxicology Program: Research Triangle Park, NC, USA, 2015.

18. Guyatt, G.H.; Oxman, A.D.; Kunz, R.; Atkins, D.; Brozek, J.; Vist, G.; Alderson, P.; Glasziou, P.; Falck-Ytter, Y.; Schünemann, H.J. GRADE guidelines: 2. Framing the question and deciding on important outcomes. *J. Clin. Epidemiol.* **2011**, *64*, 395–400. [CrossRef] [PubMed]

19. Guyatt, G.; Oxman, A.D.; Akl, E.A.; Kunz, R.; Vist, G.; Brozek, J.; Norris, S.; Falck-Ytter, Y.; Glasziou, P.; deBeer, H.; et al. GRADE guidelines: 1. Introduction—GRADE evidence profiles and summary of findings tables. *J. Clin. Epidemiol.* **2011**, *64*, 383–394. [CrossRef] [PubMed]

20. Heaney, R.P.; Recker, R.R. Effects of nitrogen, phosphorus, and caffeine on calcium balance in women. *J. Lab. Clin. Med.* **1982**, *99*, 46–55. [PubMed]

21. Heaney, R.P.; Rafferty, K. Carbonated beverages and urinary calcium excretion. *Am. J. Clin. Nutr.* **2001**, *74*, 343–347. [CrossRef] [PubMed]

22. Blumberg, J.B.; Frei, B.; Fulgon, V.L.; Weaver, C.M.; Zeisel, S.H. Contribution of dietary supplements to nutritional adequacy in various adult age groups. *Nutrients* **2017**, *9*, 1325. [CrossRef] [PubMed]

23. Benowitz, N.L.; Jacob, P., 3rd; Mayan, H.; Denaro, C. Sympathomimetic effects of paraxanthine and caffeine in humans. *Clin. Pharmacol. Ther.* **1995**, *58*, 684–691. [CrossRef]

24. Mozaffarian, D.; Benjamin, E.J.; Go, A.S.; Arnett, D.K.; Blaha, M.J.; Cushman, M.; Das, S.R.; de Ferranti, S.; Després, J.P.; Fullerton, H.J.; et al. Heart disease and stroke statistics-2016 update: A report from the American Heart Association. *Circulation* **2016**, *133*, e38–e360. [CrossRef] [PubMed]

25. American Psychiatric Association. *Diagnostic and Statistical Manual of Mental Disorders*, 5th ed.; DSM-5; American Psychiatric Publishing: Washington, DC, USA, 2013.

26. Lawson, C.C.; LeMasters, G.K.; Wilson, K.A. Changes in caffeine consumption as a signal of pregnancy. *Reprod. Toxicol.* **2004**, *18*, 625–633. [CrossRef] [PubMed]

27. Stein, Z.; Susser, M. Miscarriage, caffeine, and the epiphenomena of pregnancy: The causal model. *Epidemiology* **1991**, *2*, 163–167. [PubMed]

28. Leviton, A. Biases inherent in studies of coffee consumption in early pregnancy and the risks of subsequent events. **2018**. [CrossRef]

29. Chen, L.; Bell, E.M.; Browne, M.L.; Druschel, C.M.; Romitti, P.A.; Schmidt, R.J.; Burns, T.L.; Moslehi, R.; Olney, R.S. Maternal caffeine consumption and risk of congenital limb deficiencies. *Birth Defects Res. A Clin. Mol. Teratol.* **2012**, *94*, 1033–1043. [CrossRef] [PubMed]

30. Miller, E.A.; Manning, S.E.; Rasmussen, S.A.; Reefhuis, J.; Honein, M.A. Maternal exposure to tobacco smoke, alcohol and caffeine, and risk of anorectal atresia: National Birth Defects Prevention Study 1997–2003. *Paediatr. Perinat. Epidemiol.* **2009**, *23*, 9–17. [CrossRef] [PubMed]

31. Bonaventure, A.; Rudant, J.; Goujon-Bellec, S.; Orsi, L.; Leverger, G.; Baruchel, A.; Bertrand, Y.; Nelken, B.; Pasquet, M.; Michel, G.; et al. Childhood acute leukemia, maternal beverage intake during pregnancy, and metabolic polymorphisms. *Cancer Causes Control* **2013**, *24*, 783–793. [CrossRef] [PubMed]

32. Plichart, M.; Menegaux, F.; Lacour, B.; Hartmann, O.; Frappaz, D.; Doz, F.; Bertozzi, A.I.; Defaschelles, A.S.; Pierre-kahn, A.; Icher, C.; et al. Parental smoking, maternal alcohol, coffee and tea consumption during pregnancy and childhood malignant central nervous system tumours: The ESCALE study (SFCE). *Eur. J. Cancer Prev.* **2008**, *17*, 376–383. [CrossRef] [PubMed]

33. Rudolph, T.; Knudsen, K. A case of fatal caffeine poisoning. *Acta Anaesthesiol. Scand.* **2010**, *54*, 521–523. [CrossRef] [PubMed]

34. Holmgren, P.; Nordén-Pettersson, L.; Ahlner, J. Caffeine fatalities: Four case reports. *Forensic Sci. Int.* **2004**, *139*, 71–73. [CrossRef] [PubMed]

35. Hering-Hanit, R.; Gadoth, N. Caffeine-induced headache in children and adolescents. *Cephalalgia* **2003**, *23*, 332–335. [CrossRef] [PubMed]

36. Nehlig, A. Interindividual differences in caffeine metabolism and factors driving caffeine consumption. *Pharmacol. Rev.* **2018**, *70*, 384–411. [CrossRef] [PubMed]

37. Cornelis, M.C.; Monda, K.L.; Yu, K.; Paynter, N.; Azzato, E.M.; Bennett, S.N.; Berndt, S.I.; Boerwinkle, E.; Chanock, S.; Chatterjee, N.; et al. Genome-wide meta-analysis identifies regions on 7p21 (AHR) and 15q24 (CYP1A2) as determinants of habitual caffeine consumption. *PLoS Genet.* **2011**, *7*, e1002033. [CrossRef] [PubMed]

38. Griffiths, R.R.; Bigelow, G.E.; Liebson, I.A.; O'Keeffe, M.; O'Leary, D.; Russ, N. Human coffee drinking: Manipulation of concentration and caffeine dose. *J. Exp. Anal. Behav.* **1986**, *45*, 133–148. [CrossRef] [PubMed]

39. Ruxton, C.H. The suitability of caffeinated drinks for children: A systematic review of randomized controlled trials, observational studies and expert panel guidelines. *J. Hum. Nutr. Diet* **2014**, *27*, 342–357. [CrossRef] [PubMed]

*nutrients*

MDPI

*Review*

# The Influence of Caffeine Expectancies on Sport, Exercise, and Cognitive Performance

**Akbar Shabir [1], Andy Hooton [1], Jason Tallis [2] and Matthew F. Higgins [1],***

[1]  Sport, Outdoor and Exercise Science, Kedleston Campus, University of Derby, Kedleston Road, Derby DE22 1GB, UK; a.shabir2@derby.ac.uk (A.S.); a.hooton@derby.ac.uk (A.H.)

[2]  Centre for Applied Biological and Exercise Sciences, Coventry University, Priory Street, Coventry CV1 5FB, UK; ab0289@coventry.ac.uk

*   Correspondence: m.higgins@derby.ac.uk

Received: 1 September 2018; Accepted: 15 October 2018; Published: 17 October 2018

**Abstract:** Caffeine (CAF) is widely consumed across sport and exercise for its reputed ergogenic properties, including central nervous stimulation and enhanced muscular force development. However, expectancy and the related psychological permutations that are associated with oral CAF ingestion are generally not considered in most experimental designs and these could be important in understanding if/how CAF elicits an ergogenic effect. The present paper reviews 17 intervention studies across sport, exercise, and cognitive performance. All explore CAF expectancies, in conjunction with/without CAF pharmacology. Thirteen out of 17 studies indicated expectancy effects of varying magnitudes across a range of exercise tasks and cognitive skills inclusive off but not limited to; endurance capacity, weightlifting performance, simple reaction time and memory. Factors, such as motivation, belief, and habitual CAF consumption habits influenced the response. In many instances, these effects were comparable to CAF pharmacology. Given these findings and the lack of consistency in the experimental design, future research acknowledging factors, such as habitual CAF consumption habits, habituated expectations, and the importance of subjective post-hoc analysis will help to advance knowledge within this area.

**Keywords:** Caffeine; placebo; sport; exercise; health; expectancy; cognitions

## 1. Introduction

Caffeine (CAF) is amongst the most frequently used psychoactive substances in the world [1–6]. Approximately 90% of adults consume CAF in their everyday eating/drinking patterns [7]. Furthermore, three out four British athletes consume CAF prior to competition [2]. CAF can be ingested from natural sources (e.g., coffee and chocolate beans, tea leaves, kola nuts, etc.) or can be artificially synthesized and included in food and drinks (e.g., energy drinks/gels) [2]. CAF may improve numerous cognitive and behavioural mechanisms that are associated with successful sport, exercise and cognitive performance, including: alertness, concentration, energy levels, and self-reported feelings of fatigue [8,9]. CAF has also been observed to improve sport, exercise and cognitive performance directly [2,3,7,10]. Typically, the ergogenic effects of CAF have been observed with doses ranging from 3–9 mg/kg/body mass (BM) [11]. However, some individuals may be liable to CAF's anxiogenic effects, whilst others are susceptible to its ability to induce sleep disturbances and insomnia [11–14], and these effects may have substantial ramifications on the quality of exercise recovery, training, and preparation for sports competitions or general training. CAF consumption has also been observed to increase blood pressure [15], heart rate [16], and the production of catecholamines, the latter of which have been reported to damage myocardial cells and increase the risk of myocardial infarctions, especially during exercise performance, whereby catecholamine total volume is already augmented [17].

It is likely that the effects of CAF are mediated by various interpersonal factors, such as age, the use of other drugs or medications (e.g., alosetron, adenosine, deferasirox etc.) that may interact with CAF's effects, circadian factors/time of ingestion, in some instances the development of CAF tolerances (whereby a greater dosage is required to elicit the same physiological effect, as previously consumed lower dosages), and genetic predispositions [18,19].

Genetic predispositions may influence the acute and chronic responses to CAF ingestion both directly and indirectly. For example, genetic transcription of the AA allele of the CYP1A2 gene and subsequent mobilisation of the enzyme p450 has been reported to increase CAF metabolism, whereas a single base change of A to C at position 734 within intron 1 may decrease enzyme inducibility [20–22]. As such, individuals with the AA allele are considered as fast metabolisers, whereas those with the AC and CC alleles are considered slow metabolisers [20,21]. Slower CAF metabolism may increase the plasma half-life of CAF, potentially augmenting the previously ascribed risks to exercise and health states [11–16,23,24]. In some populations, these genetic differences are significantly more prominent, for example, females exhibit reduced CYP1A2 activity versus men, and females who are taking the oral contraceptive pill may be at even greater risk due to the ability of both oestrogen and progesterone to inhibit CYP1A2 activity [25,26]. CAF half-life has also been observed to extend up to 16 h in pregnant females, which may pose a risk to foetus health and development [21]. Polymorphisms may often go unnoticed until the debilitative effects of slower CAF metabolism have already manifested, this is unless individuals are genetically screened or are made aware of such a condition [27,28].

The aforementioned health concerns are typically problematic following consumption of pharmacologically active caffeine. However, the psychological permutations (e.g., changes in motivation, determination, belief, mood states, etc.) that are associated with expectancy of oral caffeine consumption may influence sport, exercise and/or cognitive performance comparably versus caffeine pharmacology, but significantly reduce any risks to health [1,4,6]. Expectancy is closely associated, and in some instances assumed to have a direct relationship with, the placebo effect [29–31]. It is suggested by manipulating the degree of expectancy, subsequently placebo efficacy might increase [32,33]. According to expectancy theory, placebo effects are mediated by explicit (consciously accessible) expectations that are influenced by factors, such as verbal information and observational learning [31]. Positive and negative expectations may generally influence the effectiveness of an inert intervention by resulting in either a facilitative (placebo) or debilitative (nocebo) response [34,35], although some contradictory findings have been observed [4,36,37]. Expectations may also influence the magnitude of effect observed after administration of pharmacologically active agents. Indeed, previous research advocates when compared in isolation, the synergistic effect of the pharmacological and psychological influence of nutritional interventions lead to the greatest improvements in sport, exercise and cognitive performance [3,6,30]. Within the context of sport and exercise nutrition, expectancy has been implicated following deceptive administration of anabolic steroids [38,39], carbohydrates [40,41], amino acids [42], sodium bicarbonate [29,30], super oxygenated water [43], and creatine monohydrate [44].

At present, the psychological permutations that are associated with caffeine are largely unaddressed in most experimental designs but could be as important as caffeine pharmacology in understanding if/how CAF elicits an ergogenic response on sport, exercise, and/or cognitive performance. Furthermore, caffeine expectancies may represent an alternative to caffeine pharmacology, which could prove particularly useful to individuals predisposed to caffeine's debilitative health concerns. For individuals who are not predisposed to caffeine's debilitative health concerns, synergism of caffeine psychology, and pharmacology may present the greatest ergogenic benefit. However, in contrast to biological sensitivity that is associated with adenosine and/or ryanodine receptors, expectancies and beliefs may be trained and/or manipulated, which may further enhance any ergogenic benefit.

Therefore, the primary purpose and novelty of the current systematic review is to analyse and explore existing literature regarding the effects of CAF expectancies on sport, exercise, and cognitive tasks [45,46] (e.g., The Bakan vigilance task, congruent, incongruent stimulus tasks, card organisation tasks, rapid visual information processing tasks, etc.) that are considered to be important determinants of skills (including concentration levels, attentional focus, information recall, memory, simple motor speed performance, and many more [47,48]) associated with successful sport, exercise, and cognitive performance. These cognitions may also improve an individual's ability to learn psychological (imagery, self-talk, muscular relaxation methods etc.) and performance specific skills (passing, dribbling during soccer, etc.) [49–51].

The inclusion criteria for the current review entailed studies with a primary aim of exploring CAF expectancies across sport, exercise, and/or cognitive performance (i.e., participants are administered an experimental/inert intervention, whilst being informed correctly/incorrectly with respect of its purpose). Various databases were searched (i.e., Google Scholar, Sport Discus, Research Gate) with search criteria including terminology such as "caffeine expectancy", "caffeine placebos" and "caffeine deception". Where applicable secondary search criteria were included and consisted of terminology, such as "sport", "exercise", "cognitions", and "mental processing". If databases did not provide this option, then primary and secondary search terminology were amalgamated. Finally, the reference sections of select papers were also used to inform this process. In total, 17 studies fulfilled this criterion and were subsequently included. This review is therefore split into two sections; Section 1 explores CAF expectancies and sport and exercise performance (Table 1), whilst Section 2 explores CAF expectancies and cognitive performance (Table 2).

## 2. CAF Expectancies and Sport and Exercise Performance

Beedie et al. [45]

The improvements in cycling capacity following CAF expectancies in Beedie et al. [45] were comparable to the administration of CAF reported elsewhere. However, the study design that was employed did not entail CAF consumption therefore no direct comparisons were made. No significant differences were observed for any physiological variables which indicates the mechanisms underlying these results were not mediated by substantial changes in effort. To further explore the potential mechanisms, two semi-structured interviews nota bene (N.B.) before and after the experimental deception was revealed) were performed exploring participant expectancies, and they were subsequently analysed using inductive content analysis [52].

Four out of seven participants indicated that they believed CAF would positively influence their performance. Five participants reported changes in subjective perceptions associated with CAF, with dose-dependent increases in aggression, vigour, and energy following the consumption of CAF-LOW and CAF-HIGH, respectively. Some participants even misinterpreted better starts to exercise performance because of CAF ingestion, which augmented feelings of motivation and effort [6], with one participant suggesting 'oh great, well I'll press a little bit harder and I'll go a little bit faster' (page (p). 2161). Six participants provided perceived mechanisms that are associated with CAF. These included; reductions in pain perception, belief-behaviour relationships (enhanced expectations resulting in changes in behaviour), increased attentional and physiological arousal. Yet, no clear relationship between expectancies and performance effects emerged. This may be due to only 67% of participants believing that they had ingested CAF. Had a design been adopted that more effectively manipulated expectancies, then this figure would be closer to 100%. This may have been achieved through a double-dissociation design, which is considered to be the most suitable design when exploring CAF psychology and pharmacology [37,45]. The double dissociation design includes four groups representing a placebo (given placebo (PLA)/told PLA (GP/TP)) and the pharmacological (given CAF/told PLA (GC/TP)), psychological (given PLA/told CAF (GP/TC)) and synergistic effect(s) of CAF (given CAF/told CAF (GC/TC)) on the dependent variable(s) assessed. When compared to experimental designs non-inclusive of deceptive administration (e.g., traditional single-blind and double-blind protocols), participant beliefs are intentionally manipulated in accordance with the experimental purpose, which reduces the discrepancy of individuals guessing which supplement they have ingested. If uncontrolled, this might cause overlaps between pharmacology and expectancies, making it difficult to delineate the individual effects of these properties.

Foad et al. [36]

Foad et al. [36] suggest that the low magnitude of effect for GP/TC may be attributable to a lack of counterbalancing. Due to a clearly distinct taste in CAF containing saline solutions GC conditions always preceded GP. Therefore, the differences in taste and potential reductions in perceived side effects may have raised participant suspicions and lowered expectancies during GP. This issue may have been augmented as participants were considered moderate CAF consumers and may have consciously expected CAF associated symptoms [6]. Alternatively, the reduction in mean power output (MPO) following synergism of CAF belief and pharmacology could be attributed to reductions in conscious efforts that are associated with an overreliance on CAF's ergogenic effectiveness (this notion is later supported by Tallis et al. [37]). Unfortunately, post-hoc analysis was not performed therefore these explanations remain speculative. Implementation of post-hoc analysis is fundamental to gain a greater understanding of the mechanism(s) associated with expectancy. This can be achieved via the use of questionnaires [30], visual analogue scales [46], and verbal feedback mechanisms (e.g., interviews, private Dictaphone logs, etc.) [45]. Within the current review, only two studies [45,46] performed post-hoc analysis to subjectively explore these mechanisms.

Pollo et al. [53]

A greater placebo effect was observed following implementation of acute conditioning procedures, and this was likely mediated by greater reductions in perceptual fatigue. The authors suggest these results underline the role of learning during the placebo response, and the importance of habituated expectancies that may be influenced by previous CAF experiences. Unfortunately, only 4/17 studies explored habituated expectancies in the current review [37,54–56]. Alternatively, these results may have been influenced by methodological limitations that are associated with a between-subjects design. This design entails various inter-participant differences (e.g., genetics, age, gender, personality traits, etc.) that have been observed to influence CAF metabolism [25,26]. For example, while no significant differences were observed in anthropometric variables, weight lifted or 1 repetition max (RPM), personality differences were not accounted for and may have influenced placebo responsiveness [37]. Moreover, coffee contains over 1000 compounds, of which many have undergone negligible investigation regarding their influence on sport, exercise, and cognitive performance [57]. Therefore, there remains a potential for other ingredients to have impacted these results.

Duncan et al. [46]

In line with previous findings [6,36,45], ratings of perceived exertion (RPE) [58] was significantly greater during PLA versus CAF and control (CON) [46], which may indicate a nocebo effect. The nocebo effect has been observed to overestimate the placebo effect by causing greater disparity between expectancies and beliefs [59]. Future studies should aim to neutralise expectancies during PLA which may reduce the prevalence of nocebo responses and improve the reliability of comparisons. Moreover, the techniques used to manipulate expectancies are yet to be validated. Alternatively, these results may have been influenced by daily variation. A study by Smith et al. [60], devoid of any experimental manipulation observed similar deviations in repetitions performed (+4) during knee extension at an even greater exercise intensity (70% 1 RPM). Additional repetitions at higher exercise intensities may indicate greater daily variation at lower exercise intensities, due to enhanced fatigue resistance [37]. These results may have also been influenced by learning effects (as no familiarisation sessions were performed) and/or the provision of a minimum recovery period of 24 h. Bishop et al. [61] suggests resistance trained male individuals should be provided a minimum of 48 h recovery between sessions, with 72 and 96 h considered optimum. In contrast, participants in Duncan et al. [46] were provided between 24 and 72 h of recovery. However, an expectancy effect cannot be ruled out, as post-hoc analysis revealed 88% of participants expected CAF to have an ergogenic effect on exercise performance. Additionally, during CAF, all of the participants reported either CAF-related symptoms or performance effects (with some participants reporting both). This suggests, perceived CAF consumption resulted in relative psychosomatic symptoms, which could have augmented expectancies and subsequently improved exercise performance [6].

**Table 1.** Characteristics and findings of studies assessing caffeine expectancies on sport and exercise performance.

| Author(s) | Sample Characteristics | Experimental Design & Main Outcome Measure(s) | Intervention/Informed | Main Findings |
|---|---|---|---|---|
| Beedie et al. [45] | 7 well trained male cyclists (30 ± 11 years). Habitual caffeine consumption not reported | **Design** Deceptive administration, randomised, within-subjects and double-blind **Main outcome measure** 10 km cycle ergometer time trial | **Received** PLA during all trials. No treatment Control (CON) **Informed** Placebo (PLA) 4.5 mg/kg/BM caffeine (CAF-LOW) 9 mg/kg/BM (CAF-HIGH) CON **Expectancy manipulation** Literature detailing caffeine ergogenicity amongst elite cyclists. | Perceived placebo reduced mean power output by ~2.3 W vs. baseline. Perception of 4.5 mg/kg/BM and 9 mg/kg/BM caffeine increased mean power output by 4 and 9.3 W vs. baseline, respectively. |
| Foad et al. [26] | 14 male (43 ± 7 years), moderate caffeine consuming (310 ± 75 mg) recreational cyclists | **Design** Double-dissociation, within-subjects, non-randomised and single-blind **Main outcome measure** 40 km cycle ergometer time trial | **Received** Saline solutions (told for hydration purposes only) containing PLA or CAF (5 mg/kg/BM) **Informed** Given CAF told CAF (GC/TC) Given CAF/ told PLA (GC/TP) Given PLA/told CAF (GP/TC) Given PLA/told PLA (GP/TP) **Expectancy manipulation** Placebo capsule perceived to contain 5 mg/kg/BM CAF and a 90-min presentation displaying CAF benefits on cycling performance | Consumption (3.5 ± 2.0%) and belief of CAF (0.7–1.4%), respectively resulted in very likely and possibly beneficial increases in MPO. Following CAF consumption, individuals were 100%, 99% and 98% likely to display improvements in MPO equivalent to 0.5%, 1.0% and 1.5%, respectively. The chances of improved MPO following belief of CAF only, was 62%, 33% and 12%, respectively. Synergism of caffeine belief and pharmacology (2.6 ± 3.3%) indicated improvements following lower expectations. A possibly harmful placebo effect (−1.9% ± 2.2%) was observed for given PLA/told PLA. |
| Pollo et al. [53] | 44 male undergraduate students (22 ± 2 years). Habitual caffeine consumption N/A | **Design** Deceptive administration, between-subjects and single-blind **Main outcome measure** Knee extension exercise at 60% 1 repetition maximum (1 RPM) | **Received** PLA No treatment CON **Informed** 20 mL caffeinated coffee (CAF) CON **Expectancy manipulation** Literature displaying CAF benefits on resistance exercise. During study 2, two acute conditioning sessions were included, whereby exercise intensity was reduced to 45% 1 RPM but perceived as 60% 1 RPM | CAF increased PPO (11.8 ± 16.1%) and repetitions performed (2.53) versus baseline, however no effect was observed for a control. A greater placebo effect was observed during study 2 with more repetitions (4.82) performed and a greater improvement in PPO (22.1 ± 23.5%) for CAF versus baseline. CAF also reduced perceptual exertion (RPE) (~1) and this was for repetitions 3, 6, 9, 12 and 15 during study 2. |
| Duncan et al. [46] | 12 resistance trained male participants (23 ± 6 years). Habitual caffeine consumption not reported | **Design** Deceptive administration, within-subjects, randomised and double-blind **Main outcome measure** Single leg knee extension at 60% 1 RPM | **Received** 250 mL artificially sweetened water No treatment CON **Informed** CAF (3 mg/kg/BM) PLA CON **Expectancy manipulation** Literature displaying the benefits of CAF on resistance-based exercise performance | CAF increased the number of repetitions performed (20 ± 5) and weight lifted (weight x repetitions) (713 ± 121 kg) versus CON (16 ± 4,577 ± 101 kg) and PLA (18 ± 4, 656 ± 155 kg), respectively. RPE was ~1 unit lower for CAF versus PLA, but similar for CAF and CON. |

Table 1. *Cont.*

| Author(s) | Sample Characteristics | Experimental Design & Main Outcome Measure(s) | Intervention/Informed | Main Findings |
|---|---|---|---|---|
| Duncan et al. [62] | 12 male (24 ± 4 years) moderate caffeine consuming (250 mg per day) trained participants | **Design** Double-dissociation, randomised, within-subjects and single-blind. Main outcome measure 30 s Wingate test at a resistance equivalent to 7.5% BM | **Received** 250 mL artificially sweetened water combined with 5 mg/kg/BM or PLA. **Informed** GC/TC, GC/TP, GP/TC, GP/TP. **Expectancy manipulation** Literature reviewing the benefits of caffeine on high intensity exercise performance | GC/TC significantly increased PPO, MPO and lowered RPE, in comparison to all other conditions. No significant differences were observed for GP/TC GC/TP. However, both groups improved PPO (59.5 and 48.9 W) and RPE (−1 and −1), versus GP/TP, respectively. |
| Tallis et al. [37] | 14 male (21 ± 1 years) low caffeine consuming (92 ± 17 mg per day) participants | **Design** Double-dissociation, randomised, counterbalanced and single-blind. Main outcome measure Maximal voluntary concentric force and fatigue resistance of the knee flexors and extensors at velocities equivalent to 30° per second and 120° per second | **Received** Orange squash solutions (4 mL/kg/BM water and 1 mL/kg/BM sugar free orange squash) with or without 5 mg/kg/BM caffeine. **Informed** GC/TC, GC/TP, GP/TC, GP/TP. **Expectancy manipulation** Verbally informed TP orange squash solutions contained no caffeine. | Peak force produced for GC/TP and GC/TC was comparable, but significantly greater versus GP/TP at both 30° per second (12.8% and 15.8%) and 120° per second (6.8% and 11.2%, respectively). Only GC/TC produced significantly greater average force production versus GP/TP, at both 30° per second (18%) and 120° per second (14.4%), respectively. |
| Saunders et al. [6] | 42 male (37 ± years) moderate habitual caffeine consuming (195 ± 56 mg per day) trained cyclists | **Design** Randomised, counterbalanced, double-blind and within-subjects. Main outcome measures Cycle ergometer time trial at 85% peak power output. Questionnaire exploring which supplement participants believed they had ingested pre and post exercise | **Received** Capsules containing CAF (6 mg/kg/BM) or PLA. No treatment CON. **Informed** N/A CON. **Expectancy manipulation** N/A | Correct identification of CAF (n = 17) increased MPO by 4.5% (+10 W) versus CON. Three more participants correctly identified CAF post-exercise, this increased MPO by a further 1.3% (+3 W). MBI indicated 100% chance of beneficial effects after administration and correct identification of caffeine. Correct identification of PLA (n = 17) decreased MPO by −0.8% for PLA (−2 W) versus CON. One more participant identified PLA post-exercise, this decreased MPO by -a further 0.6% (−1 W) versus CON. The chance of harmful effects at pre-exercise and post-exercise was 31% and 47%, respectively. Expectation for CAF following PLA ingestion (n = 8) increased MPO by 2.5% (+5 W) versus CON. Three more participants incorrectly perceived PLA as CAF post-exercise, this increased MPO by a further 0.9% (+3 W) versus CON. The chance of beneficial effects at pre-exercise and post-exercise, was 66% and 87%, respectively. |

Duncan et al. [62]

Duncan et al. [62] explain that their results may be explained by reduced priori expectancies associated with GP/TP. However, only 3/12 participants correctly identified GP/TP, whereas seven correctly identified GC/TC. These differences are likely related to the perception of CAF symptoms. Saunders et al. [6] suggests habitual CAF users will likely display greater habituated expectancies versus CAF naive individuals, and the perception of side effects may catalyse beliefs to a greater extent in these individuals. This further supports a relationship between CAF pharmacology and psychology and explains why GC/TC conditions generally result in the greatest ergogenic benefit [3]. Alternatively, the aforementioned discrepancies also indicate an issue with the efficacy of expectancy manipulations, which are necessary to uphold the integrity of the double-dissociation design. Once more, this issue may be associated with a lack of validation for the techniques that are used to modulate expectancies. Moreover, these results may be due to learning effects associated with a lack of familiarisation sessions, or the use of a single blind study design, which has been observed to overestimate the placebo effect versus double blind administration due to experimenter bias [63,64].

Tallis et al. [37]

Using a 10-point Likert scale (−5 representing very negative and +5 very positive effect), all participants in Tallis et al. [37] expected CAF to improve performance at the beginning (mean +3.09 ± 0.44) and end of exercise (mean +3.18 ± 0.42). Interestingly, when participants perceived CAF to have a greater performance benefit, there was a negative association in peak force of the knee extensors at 120° per second for GP/TC versus GP/TP. These results suggest that a greater perceived benefit may deduce a smaller practical significance whereas lower perceived benefits may have greater practical significance. This theory is in contrast to Geers et al. [65], who concludes that perceived optimism or pessimism will facilitate a placebo or nocebo response, respectively. In contrast, Tallis et al. [37] suggest an inverse relationship between expectations and motivation with too positive an expectation resulting in over reliance of CAF ergogenicity and reductions in conscious effort. Therefore, for the greatest performance benefits expectations may need to be modulated to an optimum point (much like the inverted U-hypothesis proposed by Yerkes & Dodson [66]), and this point might differ individually (based on belief and concurrent level of motivation), temporally and experientially.

Saunders et al. [6]

In contrast to previous observations [36,45,62], the findings of Saunders et al. [6] suggest that the correct identification and subsequent expectation of a placebo does not influence exercise performance. The variances in these findings might be associated with differences in participant perceptions being associated with placebo efficacy. Like CAF expectancies, a relationship may be plausible between placebo expectancies and performance effects [67]. However, in the current review no studies explored placebo expectancies. Moreover, when assessing the influence of CAF psychology and pharmacology, post-exercise expectancies influenced by perceptions related to the experimental manipulation are often overlooked, but should be considered as significant as pre-exercise expectancies for subsequent bouts of exercise. This was evident through a relationship between CAF expectancies, perceived symptoms (e.g., tachycardia, alertness, trembling), and improvements in mood states during exercise, with participants feeling "better" and "less tired" (p.7). These perceptions may have been further influenced, as participants were considered aware of CAF's ergogenic impetus and may have anticipated CAF-related symptoms. Consequently, this may have enhanced expectancies and improved cycling performance. However, a relationship between habituated CAF consumption and expectancies should not be assumed and instead assessed independently as some contradictory findings have been observed [4,36,37].

## 3. CAF Expectancies and Cognitive Performance

Fillmore & Vogel-Sprott [56]

Four types of events are relative to the type of expectancy effects observed, these are; the stimuli that are associated with administration of the drug, the stimulus effect of the drug, the drugs effect on a symptom/sensation related to the activity, and the subsequent outcome [56,57,68]. Post-hoc analysis revealed that all participants in the current study believed they had received caffeinated coffee, and the expectation for a positive/negative performance effect generally correlated with the type of symptom/sensation experienced. For example, individuals with positive expectancies felt more alert, whereas individuals with negative expectancies felt less alert and more tense. Moreover, the differences in these perceptions were directly affiliated with successful/unsuccessful psychomotor performance [56]. These findings postulate that expectancies may mediate CAF-related symptoms/sensations, and these symptoms/sensations might be influenced by the direction of expectancy and the performance measure employed. The authors suggest that expectancy effects are more likely experienced by individuals who hold neutral habituated expectancies due to a greater responsiveness to expectancy manipulation techniques employed. More salient techniques may be required for individuals who hold greater habituated expectancies (e.g., false performance feedback, vicarious performance observations that are associated with CAF, etc.) [36,37].

Walach et al. [69]

The lack of expectancy effect observed by Walach et al. [69] might be explained by various methodological limitations. Firstly, the perception of a five-minute ingestion period may have been deemed insufficient by participants, especially as elevated CAF levels are detected in the blood stream between 20–120 min [70]. This issue may have been compounded as participants were considered regular CAF consumers and may have held habituated expectancies regarding CAF metabolism [6]. Post-hoc analysis revealed only 50% of participants believed the cover story used with 15% discovering the deception employed. Secondly, the consumption of exogenous CAF may have influenced these findings, especially as CAF half-life ranges from 1.5–9.5 h [71] and participants were asked to avoid CAF only 4 h prior to trials. This issue seems a reoccurring theme [55,72]. Thirdly, the concentration tasks that were deployed involved participation in video games on a desk computer. 1/6% of participants had no experience with video games and 28% did not work with a computer. Therefore, a lack of understanding for the tasks employed may have influenced these findings.

Walach et al. [54]

Subjective expectancies were considered to be neutral at baseline and they were not augmented by the experimental manipulation employed. The authors attribute this to the low suggested dose of CAF used (one cup of coffee). However, the low *a priori* expectation observed at baseline suggests that participants held neutral beliefs regarding CAF ergogenicity from the onset. In distinction to the postulate of Saunders et al. [6], these findings propose that habitual CAF consumption may not necessarily indicate habituated expectancies. Therefore, future research should explore habituated expectancies independently. Alternatively, these findings may have been influenced by the success of the expectancy manipulation employed with 16% of participants describing it as somewhat believable and 11% second guessing the true nature of the study. In contrast, Fillmore & Vogel-Sprott [56] observed performance effects across participants who displayed low *a priori* expectancies, however, a more successful expectancy manipulation procedure was confirmed. Finally, it is unclear whether the limitations that were described in Walach et al. [69] were addressed in this study.

Oei and Hartley [55]

It is unclear whether 'told CAF' refers to given CAF/placebo conditions. Likewise, it is difficult to interpret the information that was provided during 'given CAF' conditions. Yet, if told CAF conditions refer solely to expectancies, then the results of Oei and Hartley [55] suggest that positive habituated expectancies can improve sustained attention performance comparably versus CAF pharmacology. These findings are in contrast to Walach et al. [54] who observed no performance effect in individuals displaying low *a priori* expectancies. However, in the current study subjective expectancies were modulated through the use of verbal feedback and open preparation of solutions. The latter technique was also used by Fillmore & Vogel-Sprott [56] who also observed expectancy effects but in individuals displaying low *a priori* expectancies. This observation supports the notion that more salient manipulation techniques could exert greater expectancy effects. Habituated expectancies may significantly influence the ergogenicity of CAF expectancies, therefore further information regarding the origin of these beliefs is required, as it is likely personal and vicarious experiences associated with CAF, social factors (sports cultures etc.), and perceptions influenced by advertisement campaigns will likely prove influential here [6,73,74].

Schneider et al. [75]

The authors attributed the lack of expectancy effect that observed to the dose of CAF used, which may have been insufficient to stimulate central nervous activity or expectancies, especially if participants were accustomed to consuming greater quantities whereby a physiological tolerance may have been developed to lower dosages [76]. However, no information regarding habitual CAF consumption was provided, therefore this cannot be confirmed. This seems a reoccurring theme [45,46,53]. It is important for future research to explore participants' dietary habits and habituated expectancies to elucidate whether a relationship exists between these factors, and if so, why contradictory observations are prevalent [6,54,56]. This may be associated with the techniques used to manipulate expectancies. Similar to Walach et al. [54,69] who also observed no expectancy effect, the current study also used leaflets to describe CAF's ergogenic benefit. In contrast, when visual techniques (e.g., presentations, watching coffee brewed, etc.) were used, an expectancy effect was always observed [36,55–77] and successful expectancy manipulation was confirmed whenever this was explored.

Harrell and Juliano [4]

Harrell and Juliano [4] explored the effects of caffeine expectancies on reaction time, alertness and concentration which have been observed to enhance performance across a range of sports (e.g., soccer, rugby, boxing) [78–80]. The induction of side effects (e.g., episodes of headaches and negative somatic effects) and prevalence of CAF withdrawal symptoms were considered more substantive during "told impair" conditions. The authors suggest compensating for these debilitative perceptions and reverse any performance declines individuals may have increased conscious effort. Alternatively, participants in "told enhance" conditions may have become over confident resulting in reductions in effort [37]. In support of this notion, post-hoc analysis revealed that all participants believed the deception employed and general expectancies for improved cognitive performance were greater in "told enhance" versus "told impair" conditions. This observation is supported by Tallis et al. [37] and further contradicts the notion of a linear relationship between expectancies and performance [65].

Moreover, the benefit that is associated with CAF pharmacology may have been overestimated due to the potential reversal of withdrawal symptoms (N.B. participants were described as experiencing CAF withdrawal symptoms from the onset of this study) [81–83]. Interestingly, CAF only ameliorated these symptoms during "told enhance" conditions, with "given CAF/told impair", resulting in greater perceptual side effects and withdrawal symptoms versus all other conditions. It is unclear why similar effects were not observed for "given PLA/told impair". We speculate, during "told impair" conditions, CAF's stimulatory properties may have augmented the perception of side effects and withdrawal symptoms experienced and induced a reverse nocebo effect. This advocates an interesting relationship between beliefs and CAF side effects. However, further research is required.

Elliman et al. [3]

The findings of this study propose, when explored in isolation, neither CAF pharmacology nor psychology influenced reaction time. However, in combination performance improved which may further advocate a potential synergistic-relationship. For example, a possible lack of pharmacological stimulation associated with GP/TC may have induced suspicions and limited expectancies. Likewise, if the information that was relayed to participants during GC/TP was not kept neutral, any reduction in a *priori* expectancies may have reduced motivation and induced a nocebo response. Alternatively, it is possible that the performance benefits that are associated with GC/TC may also be related to the reversal of withdrawal effects, which are only applicable to habitual CAF consumers [3]. In line with Harrell and Juliano [4], this further supports the notion that CAF expectancies may influence the perception of symptoms/sensations associated with its use. However, this remains speculative, as subjective perceptions were not explored and no significant differences were observed across mood states.

Dawkins et al. [72]

The findings of Dawkins et al. [72] are in contrast to Elliman et al. [3], however, various methodological differences may account for these discrepancies. For example, participants in the present study were considered CAF abstinent only 2 h prior to trials which is considerably less than the 12 h in Elliman et al. [3]. Subsequently, expectancy effects would have been less likely masked by the reversal of CAF withdrawal. However, CAF abstinence 2 h prior to trials suggests exogenous CAF may have influenced these results, especially as consumption rates were not checked at any point. Moreover, participant body mass was undisclosed, but it is unlikely that the 75 mg dosage of CAF used fell within the previously defined ergogenic range (3–9 mg/kg/BM). Absolute doses of CAF also present difficulties in regulating subjective CAF intake, which may negate CAF pharmacology, especially if between-group anthropometry is not standardised. Furthermore, because this dosage represented habitual CAF consumption, the development of CAF tolerances cannot be ruled out [84]. Therefore, these results may indicate that CAF expectancies are not limited by the development of pharmacological tolerances and individuals may not need to increase habitual dosages. Moreover, the success of expectancy manipulations may partly depend on an individual's ability to perceive consumption of pharmacologically active CAF, which is less likely following lower dosages. This notion is supported in the current study as no participant guessed the true nature of the research. In contrast, the dose of CAF consumed was substantially greater during Elliman et al. [3], and the authors did not confirm successful expectancy manipulation. Finally, participants in TP conditions reported less vigour and greater depression from pre-drink to post-drink; therefore, a nocebo effect cannot be ruled out. The opposite was observed for TC conditions.

Denson et al. [77]

The strength model of self-regulation [84] explains that self-control and composure rely on executive control capacity, which during cognitively demanding tasks can be temporarily depleted. Once participants become depleted, they will be less able to control emotional impulses, which may inhibit mental function (e.g., decision making, awareness etc.) and subsequently impair sport, exercise, and cognitive performance [85].

Denson et al. [77] suggest caffeine expectancies provided participants a cognitive boost and increased motivation. However, it is unclear why similar results were not applicable to CAF. Alternatively, CAF may have increased physiological arousal through central nervous stimulation, which may have augmented feelings of aggression and subsequently reduced executive control capacity. This would support the findings of Harrell and Juliano [4] and it may represent a link between perceptions of side effects, the direction of expectancy, and the resulting benefit/lack of benefit on the outcome measure(s) assessed. To further assess the effect of CAF on executive control capacity, future studies should explore subjective perceptions and include a cognitively demanding outcome measure (e.g., Stroop task, Bakan vigilance task, BATAK, etc.). This would help to triangulate the link between expectancies, executive control capacity, and cognitive performance more effectively.

**Table 2.** Characteristics and findings of studies assessing caffeine expectancies on cognitive performance.

| Author(s) | Sample Characteristics | Experimental Design & Main Performance Measure(s) | Intervention/Informed | Main Findings |
|---|---|---|---|---|
| Fillmore & Vogel-Sprott [56] | 56 male (19–29 years) low caffeine consuming (2 ± 2 cups of coffee per day) undergraduate students | **Design** Deceptive administration, single-blind and between-subjects Main outcome measure Computerised pursuit rotor task adjudged by % time correctly following moving object | **Received** Decaffeinated coffee No treatment CON **Informed** Caffeinated coffee CON **Expectancy manipulation** 'Fairly strong dose of coffee' was prepared in front of participants. Groups were subsequently informed caffeine would positively (E+), negatively (E−) or not effect performance (E?) | Baseline psychomotor performance was similar between all groups. Additionally, all participants expected caffeine to have negligible influence. The expected effect of caffeine predicted the placebo response observed with E+ displaying the greatest performance benefits (67.5 ± 10.27%) vs. E− (49.17 ± 14.20%). E? (57.40 ± 11.78%) and CON (57.62 ± 9.98%). |
| Walach et al. [54] | 53 male and 104 female (28 ± 8 years) regular caffeine consuming (≥1 cup of coffee per day) undergraduate students | **Design** Deceptive administration, between-subjects and double-blind Main outcome measure Self-devised test (finding misprints in a text), and Wally the worm video game | **Received** Decaffeinated coffee No treatment CON **Informed** Caffeinated coffee Decaffeinated coffee Double-blind administration CON **Expectancy manipulation** Flyer describing caffeine's effects on concentration levels | No expectancy effect observed. |
| Walach et al. [69] | 44 male undergraduate students (22 ± 2 years). Habitual caffeine consumption not reported | **Design** Deceptive administration, between-subjects and double-blind Main outcome measure Self-devised test finding misprints in a text and clicking X on a computer when a previously denoted sequence of numbers appeared once more | **Received** Decaffeinated coffee No treatment CON **Informed** Caffeinated coffee Decaffeinated coffee. Double-blind administration CON **Expectancy manipulation** Flyer describing caffeine's effects on concentration levels | No expectancy effect observed. |
| Oei & Hartley [55] | 11 male and 21 female (25 ± 8 years) low caffeine consuming (≤120 mg per day or 2 cups of coffee per day) undergraduate students | **Design** Deceptive administration, mixed-factorial, between-subjects and single-blind Main outcome measure Sustained attention, memory, and delayed recall task | **Received** 250 mL caffeinated (~143 mg) or decaffeinated coffee **Informed** GC/TC GC/TP GP/TC GP/TP **Expectancy manipulation** Caffeinated coffee prepared in front of participants Participants were also allowed to inspect the jar that was perceived to contain caffeine | For sustained attention, more correct detections were observed for told caffeine (69.05 ± 0.97) and given caffeine (69.00 ± 1.23) versus placebo (66.48 ± 1.51 and 66.53 ± 1.21, respectively) for individuals displaying positive habituated expectancies only. Participants committed fewer false alarms for told caffeine (5.42 ± 0.78) and given caffeine (5.42 ± 0.68) versus placebo (7.11 ± 1.01 and 7.11 ± 1.08, respectively). |

**Table 2.** Cont.

| Author(s) | Sample Characteristics | Experimental Design & Main Performance Measure(s) | Intervention/Informed | Main Findings |
|---|---|---|---|---|
| Schneider et al. [76] | 20 males and 25 female German adults (27 ± 8 years) Habitual caffeine consumption not reported | Design<br>Deceptive administration, between-subjects and double-blind<br>Main outcome measure<br>The interactive test battery for attentional performance [75] | Received<br>250 mL caffeinated (2 mg/kg/BM) orange juice solution in all trials<br>Informed<br>Caffeinated orange juice solution<br>Non-caffeinated orange juice solution<br>Expectancy manipulation<br>Flyer describing caffeine's effects on the central nervous, cognitive and cardiovascular systems | No expectancy effect observed. |
| Harrell & Juliano [4] | 19 male and 41 female (23 years) regular caffeine consuming (463 ± 208 mg per day) adults | Design<br>Deceptive administration, between-subjects and single-blind<br>Main outcome measures<br>Rapid visual information processing (RVIP), and finger tapping tasks<br>Perceived motivation was explored prior to cognitive performance using a 4-point Likert scale (0—not at all, to 4—extremely) | Received<br>500 mL caffeinated (280 mg) coffee<br>500 mL decaffeinated (280 mg) coffee<br>Informed<br>Caffeinated coffee<br>Expectancy manipulation<br>Verbally informed caffeine would either enhance or impair performance | CAF consumption resulted in improvements across all performance measures versus PLA, however no significant differences were observed between told impair/enhance conditions.<br>Told enhance increased motivation for the RVIP (+0.58) and finger tapping task (+0.87) versus told impair.<br>However, given placebo/told impair resulted in greater improvements in reaction time (−10.08 ± 10.67 milliseconds (ms)) and RVIP hits (+2.67 ± 2.33) versus given placebo/told enhance. |
| Elliman et al. [3] | 6 male and 21 female (21 years) habitual caffeine consuming (>1 cup of coffee per day) undergraduate students | Design<br>Double-dissociation, within-subjects, counter-balanced and single blind<br>Main outcome measure<br>Bakan vigilance task | Received<br>200 mL caffeinated (200 mg) or decaffeinated coffee<br>Informed<br>GC/TC<br>GC/TP<br>GP/TC<br>GP/TP<br>Expectancy manipulation<br>Verbally informed decaffeinated coffee was administered in TP conditions | No effect was observed for mean correct and false hits for GC/TP (3.88 and 0.31 hits) versus GP/TC (3.72 and 0.32 hits), respectively. Neither group presented a meaningful improvement versus GP/TP.<br>Significant differences for correct hits were observed for GC/TC versus GC/TP (+0.24) and GP/TC (+0.40), respectively. |
| Dawkins et al. [72] | 44 male and 44 female habitual caffeine consuming 75 mg per day) undergraduate students | Design<br>Double-dissociation, between-subjects and single-blind<br>Main outcome measures<br>A card sorting task, 40 congruent (printed words and colours the same) and 40 incongruent stimulus tasks | Received<br>250 mL caffeinated (75 mg) or decaffeinated coffee<br>Informed<br>GC/TC<br>GC/TP<br>GP/TC<br>GP/TP<br>Expectancy manipulation<br>Verbally informed decaffeinated coffee was administered in TP conditions | GC/TC performed the best on all 3 performance measures, whilst GP/TP performed the worst.<br>GP/TC performed better on the congruent (39 versus 36 correct responses), incongruent (37 versus 35 correct responses) and card sorting task (10% faster) versus GC/TP. |

**Table 2.** *Cont.*

| Authors(s) | Sample Characteristics | Experimental Design & Main Performance Measure(s) | Intervention/Informed | Main Findings |
|---|---|---|---|---|
| Denson et al. [77] | 63 male and 61 female (27 ± 8 years) light caffeine consuming (≤1 cup of coffee per day) undergraduate students | Design<br>Deceptive administration, between-subjects and single-blind<br>Main outcome measures<br>The Taylor aggression paradigm following cognitive depletion (e.g., exhausting reading task and aggression provocation procedure) | Received<br>CAF tablets (200 mg)<br>PLA tablets<br>No tablet CON<br>Informed<br>CAF tablets<br>CON<br>Expectancy manipulation<br>Verbally informed CAF tablets were equivalent to 2 cups of coffee | Following cognitive depletion, PLA resulted in greater executive control capacity versus CON and CAF. No difference was observed for CAF vs CON. |
| Domotor et al. [56] | 42 male and 65 female (22 ± 4 years) habitual caffeine consuming (3 ± 1 cups of coffee per day) undergraduate students | Design<br>Deceptive administration, between-subjects and double-blind.<br>Main outcome measure<br>Simple reaction time using the PsychLabWin v.1.1 software (Informer technologies Inc., Washington, DC, USA). | Received<br>Caffeinated coffee (5 mg/kg/BM)<br>Decaffeinated coffee<br>No treatment CON<br>Informed<br>CON<br>Conditional placebo (Group 2)<br>Conditional caffeine (Group 3)<br>Deceived placebo (Group 4)<br>Caffeine (group 5)<br>Expectancy manipulation<br>Verbally informed CAF tablets were equivalent to 2 cups of coffee | No expectancy effect observed. |

Domotor et al. [86]

Knowledge of CAF consumption augmented general expectancies and reduced SBP (5 mmHg) and HR (3 bpm), versus uncertainty of CAF consumption. Reductions in physiological arousal have been observed to improve cognitive function and attention [87], however, it is unclear whether this was mediated by expectancies or another mechanism, as CAF is generally considered to be stimulatory in action. Alternatively, the concept of uncertainty in group 3 may have increased blood pressure, which could also help to explain this discrepancy [88]. Alternatively, these results may have been influenced by methodological limitations, including a between-subjects study design, a lack of counterbalancing, and familiarisation sessions.

## 4. Discussion

This review has addressed seven intervention studies relating to CAF expectancies within the sport and exercise literature, and a further 10 studies relating to measures of cognitive function that may be indirectly affiliated with sport and exercise performance. With respect to the 17 studies included, potential expectancy effects were implicated across 13 studies and these were for tasks including cycling [6,36,45,62], knee extension performance [37,46,53], attentional focus [55,72,77], simple reaction time [4,55], and cognitions [3,55,56,72,77]. This review advocates the importance for future studies to implement experimental designs that explore expectancies and the psychological permutations associated with CAF. This will provide further clarity regarding CAF's mechanism(s) of action. At present, these psychological permutations remain largely unaccounted for but may be as influential as CAF pharmacology [6,72].

Where applicable, we propose the use of a double dissociation design and a mixed methods approach for studies assessing caffeine expectancies and/or generic caffeine intervention studies. With respect of generic caffeine intervention studies, it is important to standardize expectancies to prevent overlaps between caffeine psychology and pharmacology. This will increase the reliability when attempting to denote the true magnitude of effect for caffeine pharmacology. A double dissociation design not only permits direct comparison of CAF pharmacology and psychology through the use of active placebos, but also the synergistic effect of both. Within the current review, during the adoption of a double-dissociation design, synergism of CAF pharmacology and psychology generally resulted in the greatest performance improvements. A relationship between these properties is plausible. However, at present, limited information is available here and further research is required. A mixed methods approach entails quantitative analysis of the performance parameters employed, but also qualitative exploration of the psychological permutations associated with CAF. As previously described this can be achieved via the use of questionnaires [30], visual analogue scales [46], and verbal feedback mechanisms [45].

Participant expectancies may be influenced by a host of experimental and non-experimental parameters and should therefore be considered dynamic in nature and explored across studies, as the experiences during one trial may affect subsequent trials. Additionally, perceptions have been observed to change from pre to post exercise [6]. Henceforth, the implementation of post-hoc analysis is important to understand the influence of expectancies across studies. Subjective post-hoc analysis could also provide further information regarding the influence that inter-personal differences may have on placebo responsiveness. To our knowledge, no studies have yet employed a double dissociation design in combination with subjective post-hoc analysis to explore expectancy mechanisms. We believe implementation of these methodological practices will help to elucidate further information regarding CAF expectancy.

Expectancy effects are likely mediated by a variety of factors. Within the current review examples included perceived side effects [3,6,56,72], habituated expectancies [37,45,46,54–56], confirmation of successful expectancy manipulation [4,36,37,45,46,56,72,75], pre-existing CAF consumption habits [37,55], and the mode of expectancy manipulation [36,62,72,75,86]. Visual stimuli were always correlated with an expectancy effect [36,55–77], irrespective of the performance measure assessed. In contrast, when literature describing CAF ergogenicity was employed, an expectancy effect was never observed during cognitive assessment [54,69,75], but always observed for sport and exercise performance [46,53,62]. Two studies exploring cognitions proposed issues with the success of expectancy manipulation [54,69], whilst the other did not explore this [75]. Verbal affirmations [3,4,37,45,72,86] resulted in an expectancy effect of 75% and 100% of the time, for cognitive and sport and exercise performance, respectively. Three studies [4,72,86] exploring cognitions confirmed successful expectancy manipulations following verbal affirmations; this is in contrast to the lack of success observed following the provision of literature. Only one study used multiple techniques to modulate expectancies, and an expectancy effect was observed alongside confirmation of successful expectancy manipulation here [36]. These findings suggest that, although expectancy effects were always modulated during sport and exercise performance, visual depiction of CAF ergogenicity might represent the greatest expectancy benefit during cognitive performance and this may be linked to greater saliency [36,37,56]. In contrast, the provision of reading material proved least influential. Future studies should confirm the success of expectancy manipulations and validate the efficacy of techniques used to modulate these expectancies. Moreover, a lack of validation and general consideration is also apparent when administering 'told placebo' conditions. Studies should aim to neutralise expectations here. If this issue is unaddressed a nocebo response may occur which may subsequently overestimate CAF expectancies [6,46]. Alternatively, inclusion of a 5th group (CON), which is not subjected to an experimental manipulation, might also assist with this issue.

Thirteen out of 17 studies used individuals who were considered habitual consumers and expectancy effects were apparent in 10/13. A trend was observed when habitual CAF consumption and positive habituated expectancies were correlated with 2/2 studies observing an expectancy effect [37,55]. However, when individuals displayed a low *priori* expectation (2/4) [54,56], expectancy effects were only observed following confirmation of successful expectancy manipulation. In comparison, four studies did not confirm participants' habitual CAF consumption habits [45,46,53,75], with three observing expectancy effects. Two of these studies did however confirm habituated expectancies for performance effects [45,46]. Future studies should acknowledge the potential relationship between habitual CAF consumption and habituated expectations. However, expectancy effects may also be observed in individuals with low *priori* expectations following successful expectancy manipulations. The relationship between habituated expectancies and consumption habits may also hold implications regarding health states. For example, in some populations, habitual CAF consumers are at an increased risk of the debilitative health concerns versus acute consumers. Yet, these individuals may also reap a greater expectancy benefit due to potentially greater habituated expectancies [6]. However, too great an expectation may prove debilitative to performance by potentially increasing motivation/confidence to a point of debilitation [4,36,37]. Practitioners may therefore wish to consider factors (e.g., personality characteristics, social factors, etc.) that might influence the placebo effect, and how these may be managed to optimise the effectiveness of interventions. The perception of side effects was correlated with an expectancy effect during four studies [3,4,56,72] with only one study observing no effect [74]. However, the direction of these effects seemed to depend on individual perceptions for a positive or negative performance benefit.

Within the current review 12 studies attempted to explain the mechanisms associated with expectancy effects. Some examples included: feelings of side effects and physiological arousal [4,45,46,86], changes in mood states [45,77], reductions in the perception of effort [45,46,53], changes in motivation [4,37,45,77], and the nature of habituated expectancies and beliefs [6,45,46,56,62]. However, only two studies [45,46] performed post-hoc analysis to subjectively explore these mechanisms further. These mechanisms may be multifactorial and depend on a range of subjective factors inclusive of advertisements, beliefs, living experiences, and social relationships [6,89]. However, it is likely that individuals who share similar personal and/or sport culture(s) may utilise comparable mechanisms due to aligned beliefs [6,73].

## 5. Conclusions

To conclude, 13 out of 17 studies in the current review indicated expectancy effects of varying magnitudes across a range of exercise tasks and cognitive skills. These results support the notion that the psychological permutations associated with oral caffeine consumption may significantly influence caffeine ergogenicity and it may be as significant as caffeine pharmacology. Given these findings, we encourage future studies exploring the influence of caffeine expectancies on sport, exercise, and/or cognitive performance, to utilize the double dissociation design that permits direct comparisons between caffeine pharmacology versus psychology and may inform caffeine's proposed mechanism(s) of action to a greater extent. This recommendation is also particularly relevant to generic caffeine intervention studies where at present caffeine's psychological permutations are largely overlooked, but it may significantly influence any ergogenic response. However, to effectively employ such comparisons, future studies should assess the success of expectancy manipulation, which is likely influenced by various inter-personal factors including habitual caffeine consumption, habituated expectancies, and the social profile of participants used. These factors may be explored through the use of questionnaires and/or interview procedures. Furthermore, the techniques used to modulate expectancies are also important to the success of expectancy manipulation, however, at present, these require validation. Finally, it is fundamental to employ qualitative analytical techniques, including the use of questionnaires and post-hoc analysis to gain a greater understanding how expectancies are modulated and more importantly how they may influence sport, exercise, and cognitive performance.

**Author Contributions:** Conceptualization, M.F.H., A.S. and A.H.; Writing—Original Draft Preparation, A.S.; Writing—Review & Editing, M.F.H., A.H. and J.T.

**Funding:** This research received no external funding.

**Conflicts of Interest:** The authors declare no conflict of interest.

## References

1. Beedie, C.J.; Foad, A.J. The placebo effect in sports performance: A brief review. *Sports Med.* **2009**, *39*, 313–329. [CrossRef] [PubMed]
2. Del Coso, J.; Munoz-Fernandez, V.E.; Munoz, G.; Fernandez-Elias, V.E.; Ortega, J.F.; Hamouti, N.; Barbero, C.J.; Munoz-Guerra, J. Effects of a caffeine-containing energy drink on simulated soccer performance. *PLoS ONE* **2012**, *7*, 1–8. [CrossRef] [PubMed]
3. Elliman, N.A.; Ash, J.; Green, M.W. Pre-existent expectancy effects in the relationship between caffeine and performance. *Appetite* **2010**, *55*, 355–358. [CrossRef] [PubMed]
4. Harrell, P.T.; Juliano, L.M. Caffeine expectancies influence the subjective and behavioral effects of caffeine. *Psychopharmacology* **2009**, *207*, 335–342. [CrossRef] [PubMed]
5. McDaniel, L.W.; McIntire, K.; Streitz, C.; Jackson, A.; Guadet, L. The Effects of caffeine On Athletic Performance. *Coll. Teach. Methods Styles J.* **2010**, *6*, 33–37. [CrossRef]
6. Saunders, B.; de Oliveira, L.F.; da Silva, R.P.; de Salles Painelli, V.; Goncalves, L.S.; Yamaguchi, G.; Mutti, T.; Maciel, E.; Roschel, H.; Artioli, G.G.; et al. Placebo in sports nutrition: A proof-of-principle study involving caffeine supplementation. *Scand. J. Med. Sci. Sports* **2017**, *27*, 1240–1247. [CrossRef] [PubMed]
7. Burke, L.M. Caffeine and sports performance. *Appl. Physiol. Nutr. Metab.* **2008**, *33*, 1319–1334. [CrossRef] [PubMed]
8. Huntley, J.; Juliano, L.M. Caffeine Expectancy Questionnaire (CaffEQ): Construction, psychometric properties, and associations with caffeine use, caffeine dependence, and other related variables. *Psychol. Assess.* **2012**, *24*, 592–607. [CrossRef] [PubMed]
9. Lieberman, H.R.; Tharion, W.J.; Shukitt-Hale, B.; Speckman, K.L.; Tulley, R. Effects of caffeine, sleep loss, and stress on cognitive performance and mood during U.S. Navy SEAL training. Sea-Air-Land. *Psychopharmacology* **2002**, *164*, 250–261. [CrossRef] [PubMed]
10. Foskett, A.; Ajmol, A.; Gant, N. Caffeine enhances cognitive function and skill performance during simulated soccer activity. *Int. J. Sport Nutr. Exerc. Metab.* **2009**, *19*, 410–423. [CrossRef] [PubMed]

11. Pickering, C.; Kiely, J. Are the Current Guidelines on caffeine Use in Sport Optimal for Everyone? Inter-individual Variation in caffeine Ergogenicity, and a Move Towards Personalised Sports Nutrition. *Sports Med.* **2018**, *48*, 7–16. [CrossRef] [PubMed]

12. Gilbert, R.M. Caffeine consumption. *Prog. Clin. Biol. Res.* **1984**, *158*, 185–213. [PubMed]

13. Bchir, F.; Dogui, M.; Ben Fradj, R.; Arnaud, M.J.; Saquern, S. Differences in pharmacokinetic and electroencephalographic responses to caffeine in sleep-sensitive and non-sensitive subjects. *C. R. Biol.* **2006**, *329*, 512–519. [CrossRef] [PubMed]

14. Silverman, K.; Evans, S.M.; Strain, E.C.; Griffiths, R.R. Withdrawal syndrome after the double-blind cessation of caffeine consumption. *N. Engl. J. Med.* **1992**, *327*, 1109–1114. [CrossRef] [PubMed]

15. Hartley, T.R.; Sung, B.H.; Pincomb, G.A.; Whitsett, T.L.; Wilson, M.F.; Lovallo, W.R. Hypertension risk status and effect of caffeine on blood pressure. *Hypertension* **2000**, *36*, 137–141. [CrossRef] [PubMed]

16. Green, P.J.; Kirby, R.; Suls, J. The effects of caffeine on blood pressure and heart rate: A review. *Ann. Behav. Med.* **1996**, *18*, 201–216. [CrossRef] [PubMed]

17. Abraham, J.; Mudd, J.O.; Kapur, N.K.; Klein, K.; Champion, H.C.; Wittstein, I.S. Stress cardiomyopathy after intravenous administration of catecholamines and beta-receptor agonists. *J. Am. Coll. Cardiol.* **2009**, *53*, 1320–1325. [CrossRef] [PubMed]

18. Chait, L.D. Factors influencing the subjective response to caffeine. *Behav. Pharmacol.* **1992**, *3*, 219–228. [CrossRef] [PubMed]

19. Kendler, K.S.; Schmitt, E.; Aggen, S.H.; Prescott, C.A. Genetic and environmental influences on alcohol, caffeine, cannabis, and nicotine use from early adolescence to middle adulthood. *Arch. Gen. Psychiatry* **2008**, *65*, 678–682. [CrossRef] [PubMed]

20. Sachse, C.; Bhambra, U.; Smith, G.; Lightfoot, T.J.; Barett, J.H.; Scollay, J.; Garner, R.C.; Boobis, A.R.; Wolf, C.R.; Gooderham, N.J.; Colorectal Cancer Study Group. Polymorphisms in the cytochrome P450 CYP1A2 gene (CYP1A2) in colorectal cancer patients and controls: Allele frequencies, linkage disequilibrium and influence on caffeine metabolism. *Br. J. Clin. Pharmacol.* **2003**, *55*, 68–76. [CrossRef] [PubMed]

21. Tracy, T.S.; Venkataramanan, R.; Glover, D.D.; Caritis, S.N.; National Institute for Child Health and Human Development Network of Maternal-Fetal-Medicine Units. Temporal changes in drug metabolism (CYP1A2, CYP2D6 and CYP3A Activity) during pregnancy. *Am. J. Obstet. Glynecol.* **2005**, *192*, 633–639. [CrossRef] [PubMed]

22. Djordjevic, N.; Ghotbi, R.; Jankovic, S.; Aklillu, E. Induction of CYP1A2 by heavy coffee consumption is associated with the CYP1A2 -163C > A polymorphism. *Eur. J. Clin. Pharmacol.* **2010**, *66*, 697–703. [CrossRef] [PubMed]

23. Cornelis, M.C.; El-Sohemy, A.; Kabagambe, E.K.; Campos, H. Coffee, CYP1A2 genotype, and risk of myocardial infarction. *JAMA* **2006**, *295*, 1135–1141. [CrossRef] [PubMed]

24. El-Sohemy, A.; Cornelis, M.C.; Kabagambe, E.K.; Campos, H. Coffe, CYP1A2 genotype and risk of myocardial infarction. *Genes Nutr.* **2007**, *2*, 155–156. [CrossRef] [PubMed]

25. Carrillo, J.A.; Benitez, J. CYP1A2 activity, gender and smoking, as variables influencing the toxicity of caffeine. *Br. J. Clin. Pharmacol.* **1996**, *41*, 605–608. [CrossRef] [PubMed]

26. Ou-Yang, D.-S.; Huang, S.-L.; Wang, W.; Xie, H.-G.; Xu, Z.-H.; Shu, Y.; Zhou, H.-H. Phenotypic polymorphism and gender-related differences of CYP1A2 activity in a Chinese population. *Br. J. Clin. Pharmacol.* **2000**, *49*, 145–151. [CrossRef] [PubMed]

27. Yang, A.; Palmer, A.A.; de Wit, H. Genetics of caffeine consumption and responses to caffeine. *Psychopharmacology* **2010**, *211*, 245–257. [CrossRef] [PubMed]

28. Soyama, A.; Saito, Y.; Hanioka, N.; Maekawa, K.; Komamura, K.; Kamakura, S.; Kitakaze, M.; Tomoike, H.; Ueno, K.; Goto, Y.; et al. Single nucleotide polymorphisms and haplotypes of CYP1A2 in a Japanese population. *Drug Metab. Pharmacokinet.* **2005**, *20*, 24–33. [CrossRef] [PubMed]

29. Higgins, M.F.; Shabir, A. Expectancy of ergogenicity from sodium bicarbonate ingestion increases high-intensity cycling capacity. *Appl. Physiol. Nutr. Metab.* **2016**, *41*, 405–410. [CrossRef] [PubMed]

30. McClung, M.; Collins, D. "Because I know it will!": Placebo effects of an ergogenic aid on athletic performance. *J. Sport Exerc. Physiol.* **2007**, *29*, 382–394. [CrossRef]

31. Kirsch, I. Response expectancy as a determinant of experience and behavior. *Am. Psychol.* **1985**, *40*, 1189–1202. [CrossRef]

32. Mark, A. Placebo: The Belief Effect. *J. R. Soc. Med.* **2003**, *96*, 199–200. [CrossRef]

33. Montgomery, G.H.; Kirsch, I. Classical conditioning and the placebo effect. *Pain* **1997**, *72*, 107–113. [CrossRef]

34. Benedetti, F.; Arduino, C.; Costa, S.; Vighetti, S.; Tarenzi, L.; Rainero, I.; Asteggiano, G. Loss of expectation-related mechanisms in Alzheimer's disease makes analgesic therapies less effective. *Pain* **2006**, *121*, 133–144. [CrossRef] [PubMed]

35. Ross, M.; Olson, J.M. An expectancy-attribution model of the effects of placebos. *Psychol. Rev.* **1981**, *88*, 408–437. [CrossRef] [PubMed]

36. Foad, A.J.; Beedie, C.J.; Coleman, D.A. Pharmacological and psychological effects of caffeine ingestion in 40-km cycling performance. *Med. Sci. Sports Exerc.* **2008**, *40*, 158–165. [CrossRef] [PubMed]

37. Tallis, J.; Muhammad, B.; Islam, M.; Duncan, M.J. Placebo effects of caffeine on maximal voluntary concentric force of the knee flexors and extensors. *Muscle Nerve* **2016**, *54*, 479–486. [CrossRef] [PubMed]

38. Ariel, G.; Saville, W. The effect of anabolic steroids on reflex components. *Med. Sci. Sports* **1972**, *4*, 124–126.

39. Maganaris, C.N.; Collins, D.; Sharp, M. Expectancy Effects and Strength Training: Do Steroids Make a Difference? *Sport Psychol.* **2000**, *14*, 272–278. [CrossRef]

40. Clark, V.R.; Hopkins, W.G.; Hawley, J.A.; Burke, L.M. Placebo effect of carbohydrate feedings during a 40-km cycling time trial. *Med. Sci. Sports Exerc.* **2000**, *32*, 1642–1647. [CrossRef] [PubMed]

41. Hulston, C.J.; Jeukendrup, A.E. No placebo effect from carbohydrate intake during prolonged exercise. *Int. J. Sport Nutr. Metab.* **2009**, *19*, 275–284. [CrossRef]

42. Kalasountas, V.; Reed, J.; Fitzpatrick, J. The Effect of Placebo-Induced Changes in Expectancies on Maximal Force Production in College Students. *J. Appl. Sport Psych.* **2004**, *19*, 116–124. [CrossRef]

43. Porcari, J.; Jennifer, O.; Heidi, F.; Richard, M.; Foster, C. The placebo effect on exercise performance. *J. Cardiopulm. Rehabil. Rev.* **2006**, *25*, 269. [CrossRef]

44. Wright, G.; Porcari, J.P.; Foster, C.; Felker, H.; Kosholek, J.; Otto, E.M.; Sorenson Udermann, B. Placebo effects on Exercise Performance. *Med. J.* **2009**, *6*, 3–7.

45. Beedie, C.J.; Stuart, E.M.; Coleman, D.A.; Foad, A.J. Placebo effects of caffeine on cycling performance. *Med. Sci. Sports Exerc.* **2006**, *38*, 2159–2164. [CrossRef] [PubMed]

46. Duncan, M.J.; Lyons, M.; Hankey, J. Placebo effects of caffeine on short-term resistance exercise to failure. *Int. J. Sports Physiol. Perform.* **2009**, *4*, 244–253. [CrossRef] [PubMed]

47. Scarpina, F.; Tangini, S. The Stroop Color and Word Test. *Front. Psychol.* **2017**, *8*, 557. [CrossRef] [PubMed]

48. Vakil, E.; Weisz, H.; Jedwab, L.; Groswasser, Z.; Aberbuch, S. Stroop color-word task as a measure of selective attention: Efficiency in closed-head-injured patients. *J. Clin. Exp. Neuropsychol.* **1995**, *17*, 335–342. [CrossRef] [PubMed]

49. Anuar, N.; Williams, S.E.; Cumming, J. Do the physical and environment PETTLEP elements predict sport imagery ability? *Eur. J. Sport Sci.* **2017**, *17*, 1319–1327. [CrossRef] [PubMed]

50. Verburgh, L.; Scherder, E.J.; Van Lange, P.A.; Oosterlaan, J. Do Elite and Amateur Soccer Players Outperform Non-Athletes on Neurocognitive Functioning? A Study Among 8-12 Year Old Children. *PLoS ONE* **2016**, *11*, 1–12. [CrossRef] [PubMed]

51. Wulf, G.; McConnel, N.; Gartner, M.; Schwarz, A. Enhancing the learning of sport skills through external-focus feedback. *J. Motor Behav.* **2002**, *34*, 171–182. [CrossRef] [PubMed]

52. Jackson, S.A. The growth of qualitative research in sport psychology. In *Sport Psychology: Theory, Applications and Issues*; Morris, T., Summers, J.J., Eds.; John Wiley: Milton, UK, 1995; pp. 575–591.

53. Pollo, A.; Carlino, E.; Benedetti, F. The top-down influence of ergogenic placebos on muscle work and fatigue. *Eur. J. Neurosci.* **2008**, *28*, 379–388. [CrossRef] [PubMed]

54. Walach, H.; Schmidt, S.; Dirhold, T.; Nosch, S. The effects of a caffeine placebo and suggestion on blood pressure, heart rate, well-being and cognitive performance. *Int. J. Psychophysiol.* **2002**, *43*, 247–260. [CrossRef]

55. Oei, A.; Hartley, L.R. The effects of caffeine and expectancy on attention and memory. *Hum. Psychopharmacol.* **2005**, *20*, 193–202. [CrossRef] [PubMed]

56. Fillmore, M.; Vogel-Sprott, M. Expected effect of caffeine on motor performance predicts the type of response to placebo. *Psychopharmacology* **1992**, *106*, 209–214. [CrossRef] [PubMed]

57. Tunnicliffe, J.M.; Erdman, K.A.; Reimer, R.A.; Lun, V.; Shearer, J. Consumption of dietary caffeine and coffee in physically active populations: Physiological interactions. *Appl. Physiol. Nutr. Metab.* **2008**, *33*, 1301–1310. [CrossRef] [PubMed]

58. Borg, G.A. Psychophysical bases of perceived exertion. *Med. Sci. Sports Exerc.* **1982**, *14*, 377–381. [CrossRef] [PubMed]

59. Chavarria, V.; Vian, J.; Pereira, C.; Data-Franco, J.; Fernandes, B.S.; Berk, M.; Dodd, S. The Placebo and Nocebo Phenomena: Their Clinical Management and Impact on Treatment Outcomes. *Clin. Ther.* **2017**, *39*, 477–486. [CrossRef] [PubMed]

60. Smith, C.M.; House, T.J.; Hill, E.C.; Schmidt, R.J.; Johnson, G.O. Time Course of Changes in Neuromuscular Responses at 30% versus 70% 1 Repetition Maximum during Dynamic Constant External Resistance Leg Extensions to Failure. *Int. J. Exerc. Sci.* **2017**, *10*, 365–375. [PubMed]

61. Bishop, P.A.; Jones, E.; Woods, A.K. Recovery from training: A brief review: Brief review. *J. Strength Cond. Res.* **2008**, *22*, 1015–1024. [CrossRef] [PubMed]

62. Duncan, M.J. Placebo effects of caffeine on anaerobic performance in moderately trained adults. *Serb. J. Sports Sci.* **2010**, *4*, 99–106.

63. Pollo, A.; Amanzio, M.; Arslanian, A.; Casadio, C.; Maggi, G.; Benedetti, F. Response expectancies in placebo analgesia and their clinical relevance. *Pain* **2001**, *93*, 77–84. [CrossRef]

64. Vase, L.; Riley, J.L., 3rd; Price, D.D. A comparison of placebo effects in clinical analgesic trials versus studies of placebo analgesia. *Pain* **2002**, *99*, 443–452. [CrossRef]

65. Geers, A.L.; Kosbab, K.; Helfer, S.G.; Weiland, P.E.; Wellman, J.A. Further evidence for individual differences in placebo responding: An interactionist perspective. *J. Psychosom. Res.* **2007**, *62*, 563–570. [CrossRef] [PubMed]

66. Yerkes, R.M.; Dodson, J.D. The relation of strength of stimulus to rapidity of habit-formation. *J. Comp. Neurol.* **1908**, *18*, 459–482. [CrossRef]

67. Beedie, C.J. Placebo Effects in Competitive Sport: Qualitative Data. *J. Sports Sci. Med.* **2007**, *6*, 21–28. [PubMed]

68. Vogel-Sprott, M.; Fillmore, M. The placebo response to alcohol: A three-expectancy problem? University of Waterloo: Waterloo, Canada, (found in [56]). 1990; unpublished work.

69. Walach, H.; Schmidt, S.; Bihr, Y.-M.; Wiesch, S. The effects of a caffeine placebo and experimenter expectation on blood pressure, heart rate, well-being and cognitive performance: a failure to reproduce. *J. Eur. Psychiatry* **2001**, *6*, 15–25. [CrossRef]

70. Sinclair, C.J.; Geiger, J.D. Caffeine use in sports. A pharmacological review. *J. Sports Med. Phys. Fitness* **2000**, *40*, 71–79. [PubMed]

71. Somani, S.M.; Gupta, P. Caffeine: A new look at an age-old drug. *Int. J. Clin. Pharmacol. Ther. Toxicol.* **1988**, *26*, 521–533. [PubMed]

72. Dawkins, L.; Shahzad, F.Z.; Ahmed, S.S.; Edmonds, C.J. Expectation of having consumed caffeine can improve performance and mood. *Appetite* **2011**, *57*, 597–600. [CrossRef] [PubMed]

73. Moerman, D.E.; Jonas, W.B. Deconstructing the placebo effect and finding the meaning response. *Ann. Intern. Med.* **2002**, *136*, 471–476. [CrossRef] [PubMed]

74. Zimmerman, B.J. Becoming a Self-Regulated Learner: An Overview. *Theory Pract.* **2002**, *41*, 67–70. [CrossRef]

75. Schneider, R.; Gruner, M.; Heiland, A.; Keller, M.; Kujanova, Z.; Peper, M.; Reigl, M.; Schmidt, S.; Volz, P.; Walach, H. Effects of expectation and caffeine on arousal, well-being, and reaction time. *Int. J. Behav. Med.* **2006**, *13*, 330–339. [CrossRef] [PubMed]

76. Evans, S.M.; Griffiths, R.R. Caffeine tolerance and choice in humans. *Psychopharmacology* **1992**, *108*, 51–59. [CrossRef] [PubMed]

77. Denson, T.F.; Jacobsen, M.; von Hippel, W.; Kemp, R.I.; Mak, T. CAF expectancies but not CAF reduce depletion-induced aggression. *Psychol. Addict. Behav.* **2012**, *26*, 140–144. [CrossRef] [PubMed]

78. Stuart, G.R.; Hopkins, W.G.; Cook, C.; Cairns, S.P. Multiple effects of caffeine on simulated high-intensity team-sport performance. *Med. Sci. Sports Exerc.* **2005**, *37*, 1998–2005. [CrossRef] [PubMed]

79. Golby, J.; Sheard, M. Mental toughness and hardiness at different levels of rugby league. *Pers. Indiv. Differ.* **2004**, *37*, 933–942. [CrossRef]

80. Guidetti, L.; Musulin, A.; Baldari, C. Physiological factors in middleweight boxing performance. *J. Sports Med. Phys. Fitness* **2002**, *42*, 309–314. [PubMed]

81. Rogers, J.P.; Dernoncourt, C. Regular caffeine consumption: A balance of adverse and beneficial effects for mood and psychomotor performance. *Pharmacol. Biochem. Behav.* **1998**, *59*, 1039–1045. [CrossRef]

82. Yeomans, M.R.; Ripley, T.; Davies, L.H.; Rusted, J.M.; Rogers, P.J. Effects of caffeine on performance and mood depend on the level of caffeine abstinence. *Psychopharmacology* **2002**, *164*, 241–249. [CrossRef] [PubMed]

83. James, J.E.; Rogers, P.J. Effects of caffeine on performance and mood: Withdrawal reversal is the most plausible explanation. *Psychopharmacology* **2005**, *182*, 1–8. [CrossRef] [PubMed]

84. Baumeister, R.F.; Vohs, K.D.; Tice, D.M. The Strength Model of Self-Control. *Curr. Dir. Psychol. Sci.* **2007**, *16*, 351–355. [CrossRef]

85. DeWall, C.N.; Baumeister, R.F.; Stillman, T.F.; Gailliot, M.T. Violence restrained: Effects of self-regulation and its depletion on aggression. *J. Exp. Soc. Psychol.* **2007**, *43*, 62–76. [CrossRef]

86. Domotor, Z.; Szemersky, R.; Koteles, F. Subjective and objective effects of coffee consumption—Caffeine or expectations? *Acta Physiol. Hung.* **2015**, *102*, 77–85. [CrossRef] [PubMed]

87. Cha, S.D.; Patel, H.P.; Hains, D.S.; Mahan, J.D. The Effects of Hypertension on Cognitive Function in Children and Adolescents. *Int. J. Pediatr.* **2012**, *2012*, 1–5. [CrossRef] [PubMed]

88. Peters, A.; McEwen, B.S.; Friston, K. Uncertainty and stress: Why it causes diseases and how it is mastered by the brain. *Prog. Neurobiol.* **2017**, *156*, 164–188. [CrossRef] [PubMed]

89. Bandura, A. Self-efficacy: Toward a unifying theory of behavioral change. *Psychol. Rev.* **1977**, *84*, 191–215. [CrossRef] [PubMed]

nutrients

MDPI

*Review*

# Impact of Genetic Variability on Physiological Responses to Caffeine in Humans: A Systematic Review

Jacob L. Fulton [1], Petros C. Dinas [2], Andres E. Carrillo [1,2], Jason R. Edsall [1], Emily J. Ryan [3] and Edward J. Ryan [1,*]

[1]  Department of Movement Science, Chatham University, Pittsburgh, PA 15232, USA;
    jacob.fulton@chatham.edu (J.L.F.); ACarrillo@chatham.edu (A.E.C.); J.Edsall@chatham.edu (J.R.E.)
[2]  FAME Laboratory, Department of Exercise Science, University of Thessaly, GR42100 Trikala, Greece;
    petros.cd@gmail.com
[3]  Department of Exercise Physiology, West Virginia University School of Medicine, West Virginia University,
    Morganton, WV 26506, USA; ejfickes@hsc.wvu.edu
*   Correspondence: ERyan@chatham.edu; Tel.: +1-412-365-1143

Received: 1 September 2018; Accepted: 23 September 2018; Published: 25 September 2018

**Abstract:** Emerging research has demonstrated that genetic variation may impact physiological responses to caffeine consumption. The purpose of the present review was to systematically recognize how select single nucleotide polymorphisms (SNPs) impact habitual use of caffeine as well as the ergogenic and anxiogenic consequences of caffeine. Two databases (PubMed and EBSCO) were independently searched using the same algorithm. Selected studies involved human participants and met at least one of the following inclusion criteria: (a) genetic analysis of individuals who habitually consume caffeine; (b) genetic analysis of individuals who underwent measurements of physical performance with the consumption of caffeine; (c) genetic analysis of individuals who underwent measurements of mood with the consumption of caffeine. We included 26 studies (10 randomized controlled trials, five controlled trials, seven cross-sectional studies, three single-group interventional studies and one case-control study). Single nucleotide polymorphisms in or near the cytochrome P450 (*CYP1A2*) and aryl hydrocarbon receptor (*AHR*) genes were consistently associated with caffeine consumption. Several studies demonstrated that the anxiogenic consequences of caffeine differed across adenosine 2a receptor (*ADORA2A*) genotypes, and the studies that investigated the effects of genetic variation on the ergogenic benefit of caffeine reported equivocal findings (*CYP1A2*) or warrant replication (*ADORA2A*).

**Keywords:** polymorphism; anxiety; ergogenic; adenosine receptor; cytochrome P450; caffeine; pharmacogenomics

## 1. Introduction

Caffeine (1,3,7-trimethylxanthine) is one of the most widely used drugs in the world and is available in many mediums for consumption. The pharmacokinetics and pharmacodynamics of caffeine have been well studied [1]. Caffeine metabolism occurs primarily in the liver via the cytochrome P450 system (CYP1A2) [2]. The CYP1A2 proteins are encoded by the *CYP1A2* gene, and CYP1A2 activity is induced when aromatic hydrocarbons bind the aryl hydrocarbon receptor [3]. Caffeine acts as an adenosine antagonist via competitive inhibition [4], and research in mice has demonstrated that blockade of adenosine 2a receptors (encoded via *ADORA2A* gene) may potentiate dopaminergic neurotransmission [5]. It is biologically plausible that variations in the *CYP1A2* and aryl hydrocarbon receptor (*AHR*) genes impact the metabolism of caffeine and thus subsequent physiological concentrations of caffeine achieved. Further, it can be hypothesized that variations in the *ADORA2A* gene may impact caffeine-adenosine 2a receptor binding characteristics and thus downstream dopaminergic neurotransmission. Recently, the effects of single nucleotide polymorphisms (SNPs) in the aforementioned genes on caffeine use and metabolism have been investigated [3,6,7].

With the widespread consumption of caffeine-containing beverages, the health consequences of these beverages are of particular interest to researchers. For example, the chronic consumption of coffee has been associated with cognitive performance and cardiovascular health [8,9]. The identification of predictors of habitual caffeine consumption may prove useful to epidemiologists and health professionals. To date, several SNPs, such as the *CYP1A2* (rs2472297) and *AHR* (rs4410790, rs6968554), have been implicated in habitual use [10]. Further, while caffeine is generally well tolerated, some individuals report feelings of anxiety following consumption [11]. Recent investigations have explored the effect of variations in the *ADORA2A* and *CYP1A2* genes as a potential explanation for caffeine's anxiogenic impact in some individuals [6,11,12].

Athletes have long utilized caffeine as an ergogenic aid [13]. Research has demonstrated that 3–6 mg kg$^{-1}$ of body mass mildly improves exercise/physical performance [14–16]. Nonetheless, investigators have reported equivocal findings, with some reporting interindividual variation in ergogenic responses to caffeine within their subject pools [17–19]. Earlier work has demonstrated that a Single Nucleotide Polymorphism (SNP) in the *CYP1A2* gene (rs762551) led to differing rates of caffeine metabolism across genotypes in smokers [3]. Recently, researchers have examined the influence of this specific SNP and select others on the ergogenic benefit of caffeine [20,21].

To our knowledge, investigators have not systematically recognized studies evaluating the effects of indexed and unknown SNPs in biologically plausible genes on physiological responses to caffeine across scholarly disciplines. Such a systematic review may provide a basis for further interdisciplinary approaches and future directions. Therefore, the purpose of the present review was to systematically investigate the impact of select SNPs on the ergogenic and anxiogenic consequences, and habitual use, of caffeine in humans.

## 2. Materials and Methods

### 2.1. Search Strategy

The Preferred Reporting Items for Systematic Review and Meta-Analyses (PRISMA) guidelines [22–24] were followed. Two databases [PubMed and Medline (EBSCO)] were independently searched by two investigators (J.L.F and P.C.D) up until 5 July 2018 using an appropriate algorithm (Figure S1). Any conflicts in the searching procedure were resolved through consensus, while the searching results were reviewed and sorted to identify relevant publications to the topic under review.

### 2.2. Selection Criteria

The studies included in this review involved human participants and met at least one of the following criteria: (a) genetic analysis of individuals who habitually consume caffeine; (b) genetic analysis of individuals who underwent measurements of physical performance with the consumption of caffeine; (c) genetic analysis of individuals who underwent measurements of mood with the consumption of caffeine. Included studies displayed outcomes regarding SNPs associated with habitual caffeine consumption, relationships between certain SNPs, and relationships between caffeine consumption and mood. We excluded animal studies, reviews, conference proceedings, and editorials; however, we screened the reference lists of such publications and of the retrieved articles for relevant papers. The list of the included studies (*n* = 26) is available in the data extraction table (Table 1), while the list of the excluded studies (*n* = 3512) is available in Figure S2.

Table 1. Characteristics of the studies included in the systematic review.

| First Author | Design | Participants | Main Outcome | Secondary Outcome |
|---|---|---|---|---|
| Algrain [25] | Controlled Trial | Male (M) = 13 Female (F) = 7 | Polymorphism in CYP1A2 gene (AA and C-allele carriers) did not impact ergogenic benefit of caffeine in recreational cyclists ($p > 0.05$) | |
| Alsene [11] | Randomized Controlled Trial | 94 healthy, infrequent caffeine users | 1976T/T and 2592Tins/Tins genotypes report greater increase in anxiety after caffeine administration ($p < 0.05$) | |
| Childs [6] | Randomized Controlled Trial | 102 healthy individuals (M = 51 and F = 51) who consumed less than 300 mg caffeine per week | ADORA2A TT genotype reported highest anxiety (VAS) ($4.6 \pm 1.9$) and ADORA2A CC ($-7.5 \pm 3.7$) reported the least anxiety after 150 mg of caffeine but was not significant when data for European-American participants were considered ($p = 0.1$); caffeine-induced anxiety was associated with dopamine receptor 2 gene (DRD2) polymorphism | |
| Cornelis [26] | Cross-sectional | $n = 2735$ | ADORA2A, but not CYP1A2, genotype was associated with different amounts of caffeine intake; compared to persons consuming <100 mg caffeine/day, odds ratios for having the ADORA2A TT genotype were 0.74, 0.63, and 0.57 for persons consuming 100–200, 200–400, and >400 mg caffeine/day, respectively | Association more pronounced among current smokers compared to nonsmokers |
| Cornelis [27] | Cross-sectional | 47,341 individuals of European descent | Two loci-7p21 ($p = 2.4 \times 10^{-19}$), near AHR, and 15q24 ($p = 5.2 \times 10^{-14}$), near CYP1A1 and CYP1A2; both candidates as CYP1A2 caffeine metabolizers | |
| Cornelis [10] | Cross-sectional | Coffee consumers of European ancestry $n = 91,462$ African American ancestry $n = 7964$ | Eight loci, six being novel, met genome-wide significance ($\log_{10}$Bayes factor >5.64); loci near genes potentially involved in pharmacokinetics (ABCG2, AHR, POR, and CYP1A2) and pharmacodynamics (BDNF and SLC6A4). Loci related to metabolic traits (GCKR and MLXIPL) | |
| Djordjevic [28] | Single-group interventional design | 126 Healthy Serbians, 64 nonsmoking (from their previous study) | Inducing effect of CYP1A2 activity with heavy coffee consumption among Serbian ($p = 0.022$) and Swedish ($p = 0.016$) participants carrying the CYP1A2-163 C > A polymorphism | |
| Domschke [29] | Controlled Trial | M = 56 and F = 54 healthy individuals | Startle magnitude highest for unpleasant pictures and lowest for pleasant pictures across ADORA2A genotypes ($p < 0.001$). TT (risk) genotype carriers had highest startle magnitude in the caffeine condition in response to unpleasant pictures, occurring mostly among females | Females of this group had higher startle magnitudes than males |
| Domschke [30] | Controlled Trial | 58 M and 66 F healthy proband | ADORA2A TT risk genotype carriers had significantly increased startle magnitude in response to neutral stimuli ($p = 0.02$) and a significant decrease in startle magnitude in response to unpleasant stimuli ($p = 0.02$) in caffeine compared to placebo condition; no change in AA/AT nonrisk genotype | |
| Gajewska [31] | Randomized Controlled Trial | 57 M and 57 F healthy individuals controlled for anxiety sensitivity | Prepulse inhibition was influenced by genetics (ADORA2A 1976C/T); impaired prepulse facilitation in anxiety sensitive ADORA2A TT group in response to caffeine compared to placebo ($t(56) = 2.16$, $p = 0.04$) | |
| Giersch [32] | Controlled Trial | 20 male subjects between age of 18–45 years | CYP1A2 C-allele carriers had higher serum caffeine one hour after caffeine ingestion (C-allele carriers = $14.2 \pm 1.8$ ppm, AA homozygotes = $11.7 \pm 1.7$ ppm, $p = 0.001$). No difference between genotypes in caffeine metabolites ($p > 0.05$). Main effect of caffeine on performance ($p = 0.03$); no caffeine by genotype interaction ($p > 0.05$) | |

**Table 1.** *Cont.*

| First Author | Design | Participants | Main Outcome | Secondary Outcome |
|---|---|---|---|---|
| Guest [33] | Randomized Controlled Trial | Competitive male athletes $n = 101$ | In *CYP1A2* AA genotype, cycling time decreased by 4.8% ($p = 0.0005$) and 6.8% ($p < 0.0001$) with 2 and 4 mg kg$^{-1}$ caffeine consumption, respectively; (2 and 4 mg kg$^{-1}$); in CC genotype, cycling time increased by 13.7% ($p = 0.04$) with caffeine consumption (4 mg kg$^{-1}$), no effects were observed among AC genotype | 4 mg kg$^{-1}$ caffeine decreased cycling time by 3% vs. placebo |
| Josse [34] | Cross-sectional | $n = 1639$ nonsmokers and $n = 884$ current smokers | Subjects who consumed >400 mg caffeine compared to who consumed <100 mg caffeine were more likely to be carriers of T, C, or T alleles for rs6968865, rs4410790, and rs2472297, respectively; corresponding Odds Ratios and 95% confidence intervals (CIs) were 1.41 (1.03, 1.93), 1.41 (1.04, 1.92), and 1.55 (1.01, 2.36) | |
| Loy [21] | Randomized Controlled Trial | Women with high self-reported caffeine sensitivity and low daily caffeine consumption, TT $n = 6$, CT/CC $n = 6$ | Caffeine proved to be ergogenic for *ADORA2A* TT allele homozygotes (6.85 ± 4.41 kJ) but not ergogenic for CT/CC alleles (−2.70 ± 5.64 kJ) (d = −1.89) | |
| Luciano [35] | Cross-sectional | 3808 Australian adult twin pairs ($n = 1799$ monozygous pairs and $n = 2009$ dizygous pairs) | Genes not typically associated with sleep disturbance were implicated in coffee-attributed insomnia | |
| McMahon [7] | Cross-sectional | 4460–7520 women | Caffeine consumption was associated with *CYP1A1* (Betas = 8.7 to 21.4, $p$-values = $1.59 \times 10^{-3}$ to $3.33 \times 10^{-10}$) and *AHR* (Betas = 4.0 to 14.6, $p$-values = $1.1510^{-1}$ to $3.34 \times 10^{-6}$) genotypes; association not strengthened with combined allelic score (1.28% of phenotypic variance) | |
| Pataky [36] | Randomized Controlled Trial | 25 M and 13 F recreational cyclists from James Madison University | *CYP1A2* AC heterozygotes experienced greater power output (6%) with caffeine ingestion. Caffeine ingestion favored AC heterozygotes compared to AA homozygotes when performance gains were compared to placebo (5.1 ± 6.1%, $p = 0.12$) | |
| Pirastu [37] | Cross-sectional | 370 individuals from Puglia, Italy and 843 individuals from Friuli Venezia Region, Italy | *PDSS2* gene shown in sample was linked to negative regulation of the expression of caffeine metabolism genes in several tissues (e.g., subcutaneous adipose tissue −0.27, skeletal muscle −0.52) | |
| Puente [38] | Randomized Controlled Trial | 10 men and 9 women elite basketball players | *CYP1A2* genotype (rs762551) AA improved Abalakov jump height ($p = 0.03$) with caffeine consumption, while C-allele carriers remained unchanged ($p = 0.33$); Sprint was not improved in either genotype with caffeine, while number of body impacts increased in both AA (4.1 ± 5.3%, $p = 0.02$) and C-allele carriers (3.3 ± 3.2%; $p = 0.01$) | |
| Rogers [12] | Randomized Controlled Trial | 162 non/low and 217 medium/high caffeine consumers | *ADORA2A* (rs5751876) TT genotype showed largest increase in anxiety after caffeine (mean ± standard error ) for caffeine = 1.65 ± 0.15 and for placebo = 0.95 ± 0.17, $p < 0.01$) | |

**Table 1.** *Cont.*

| First Author | Design | Participants | Main Outcome | Secondary Outcome |
|---|---|---|---|---|
| Sachse [3] | Single-group interventional design | 185 healthy Caucasian nonsmokers and 51 smokers | Among smokers ($n = 51$), subjects who possessed the *CYP1A2* (rs762551) *AA* genotype metabolized caffeine (100 mg) faster 1.37 (*AC* –0.88; and *CC* – 0.82) relative to C-allele carriers while utilizing a 5 h paraxanthine/caffeine ratio as the outcome measure ($p = 0.008$) | 31.3% of C-allele carriers reported increased nervousness after caffeine ingestion |
| Salinero [39] | Randomized Controlled Trial | 21 healthy active participants | Caffeine ingestion increased peak and mean power in both *AA* and C-allele carriers of the *CYP1A2* gene ($p > 0.05$); no difference in Wingate test performance between *AA* and C-allele carriers ($p > 0.05$) | |
| Soares [40] | Single-group interventional design | 37 individuals between ages of 19–50 | Systolic blood pressure (BP) increased with caffeine ingestion only among individuals with *CYP1A2 AA* genotype ($p < 0.05$); both *CYP1A2 AA* and *AC* had high diastolic BP after caffeine ingestion ($p < 0.05$); physical activity only modulated the BP responses to acute caffeine ingestion in *AC* individuals | |
| Thomas [41] | Controlled Trial | *CYP1A2 AA* (F = 4 and M = 7), C-allele carriers (F= 3 and M = 6) | No difference in heart rate variability between *CYP1A2*1F* polymorphisms (i.e., *AA* and C-allele carriers) measured at baseline and postexercise ($p > 0.05$) | |
| Urry [42] | Case–Control Study | 57 subjects with type 2 diabetes (T2D) and 146 non-T2D | *CYP1A2* enzyme activity was significantly higher in T2D compared to control group ($p = 0.004$) | |
| Womack [20] | Randomized Controlled Trial | Trained male cyclists $n = 35$ | Caffeine supplementation reduced 40 km time greater in *CYP1A2 AA* homozygotes (4.9%) than in C-allele carriers (1.8%) ($p < 0.05$) | |

*2.3. Risk of Bias Assessment*

Two reviewers (J.L.F. and P.C.D.) independently evaluated the risk of bias of the non- randomized controlled trials (RCT) via the 13-item tool developed by the Research Triangle Institute (RTI), Evidence-based Practice Center [24]. This tool has previously shown median interrater agreement of 75% [43] and 93.5% [44]. The risk of bias of the RCTs was assessed via the "Cochrane Collaboration's tool for assessing risk of bias" [45]. Conflicts in the risk of bias assessment were resolved by two independent referee investigators (E.J.R. and A.E.C.).

*2.4. Data Extraction and Analysis*

The results of the data extraction procedure are shown in Table 1. Data extraction was performed independently by two investigators (J.L.F. and E.J.R.), and conflicts were resolved by a referee investigator (A.E.C.). For all included studies, we extracted the first authors' name, year of publication, design of the studies, participants' characteristics (i.e., number, sex, age, health status, and intervention) and the main and secondary outcomes, including results from statistical analyses. A qualitative synthesis of the retrieved evidence was completed thereafter.

**3. Results**

The reporting of the available information in this systematic review is shown in a PRISMA checklist in Figure S3.

*3.1. Searching Procedure Results*

The entire search yielded 3532 records. Of these, 2115 were duplicates; therefore, 1417 were initially screened to exclude reviews, conferences and editorials ($n$ = 387). Consequently, 1033 records were assessed for eligibility, with 20 studies meeting the inclusion criteria. Finally, an additional six records were added manually. The searching outcome is presented in a PRISMA flow diagram (Figure S4).

*3.2. Characteristics of the Included Studies*

The characteristics and the results of the included studies are presented in Table 1. Of the included 26 studies, 10 were RCT (38%), five were controlled trials (CT) (19%), seven were cross-sectional studies (CSS) (27%), three were single-group interventional studies (13%), and one was a case–control study (CCS) (4%).

*3.3. Risk of Bias Assessment*

The risk of bias assessment results can be found in Tables 2 and 3, and a summary of the results are displayed in Figures 1 and 2. For the RTCs, eight showed an unclear risk of bias for random sequence generation [11,12,31,36,38,39] and two showed a low risk of bias [21,33]. For allocation concealment, six studies were classified as showing an unclear risk of bias [6,11,12,20,31,36,38,39] and two showed a low risk of bias [21,33]. Seven studies demonstrated an unclear risk of bias for the blinding of participants and researchers [6,11,12,20,31,33,36], although the studies stated that the participants were "blinded", and three displayed a low risk of bias [21,38,39]. Six studies were categorized as having an unclear risk of bias for the blinding of outcome assessment [6,12,21,31,38,39], and four were categorized as being a low risk of bias [11,20,33,36]. One out of the 10 RCTs [21] showed an unclear risk of bias for incomplete data, while the others displayed a low risk of bias. Finally, all 10 of the RCTs were categorized as having a low risk of bias for selective reporting and other bias.

*Nutrients* **2018**, *10*, 1373

**Table 2.** Risk of bias assessment using the Cochrane Collaboration's Tool.

| First Author | Random Sequence Generation | Allocation Concealment | Blinding of Participants and Researchers | Blinding of Outcome Assessment | Incomplete Outcome Data | Selective Reporting | Other Bias |
|---|---|---|---|---|---|---|---|
| | | | Randomized Controlled Trials (RCTs) | | | | |
| Gajewska [31] | ? | ? | ? | ? | + | + | + |
| Alsene [11] | ? | ? | ? | + | + | + | + |
| Pataky [36] | ? | ? | ? | + | + | + | + |
| Rogers [12] | ? | + | ? | ? | + | + | + |
| Puente [38] | + | + | + | ? | + | + | + |
| Guest [33] | + | + | ? | + | + | + | + |
| Salinero [39] | + | + | + | ? | ? | + | + |
| Loy [21] | + | ? | + | ? | + | + | + |
| Womack [20] | ? | + | ? | + | ? | + | + |
| Childs [6] | ? | ? | ? | ? | + | + | + |

Key: +: Low risk of bias (green); ?: Unclear risk of bias (yellow). RCTs: randomized controlled trials.

**Table 3.** Risk of bias assessment using the Research Triangle Institute (RTI) Item Bank.

| First Author | Selection Bias | Performance Bias | Detection Bias | Attrition Bias | Selective Outcome | Confounding |
|---|---|---|---|---|---|---|
| | | | Non-RCT | | | |
| Djordjevic [28] | - | + | ? | o | + | ? |
| McMahon [7] | o | o | ? | ? | + | + |
| Soares [40] | + | + | ? | o | + | + |
| Giersch [32] | + | + | + | o | + | + |
| Domschke [30] | + | + | ? | o | + | + |
| Sachse [5] | + | + | ? | o | + | + |
| Urry [42] | ? | + | ? | o | + | ? |
| Pirastu [37] | ? | + | + | o | + | ? |
| Josse [34] | + | + | ? | o | + | ? |
| Cornelis [27] | + | + | + | o | + | + |
| Domschke [29] | + | + | ? | o | + | ? |
| Cornelis [10] | + | + | + | o | + | + |
| Thomas [41] | + | + | + | o | + | + |
| Algrain [25] | + | + | + | o | + | + |
| Cornelis [26] | + | + | ? | o | + | ? |
| Luciano [35] | o | + | ? | o | + | ? |

Key: +: Low risk of bias (green); -: High risk of bias (red); ?: Unclear risk of bias (yellow); o: Non-applicable (blue).

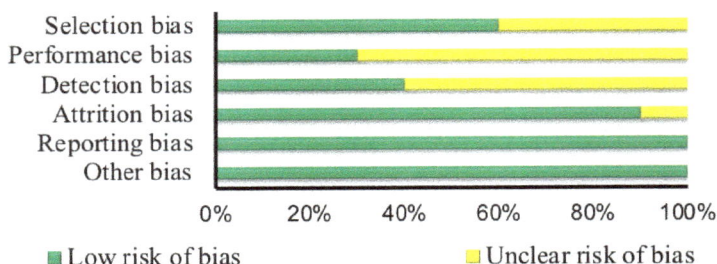

**Figure 1.** Summary of risk of bias assessment for randomized controlled trials ($n = 10$). Selection bias (random sequence generation, low risk ($n = 2$), unclear risk ($n = 8$) + allocation concealment, low risk ($n = 4$), unclear risk ($n = 6$)); Performance bias (blinding of participants and researchers, low risk ($n = 3$), unclear risk ($n = 7$)); Detection bias (blinding of outcome assessment, low risk ($n = 4$), unclear risk ($n = 6$)); Attrition bias (incomplete outcome data, low risk ($n = 9$), unclear risk ($n = 1$)); Reporting bias (selective reporting, low risk ($n = 10$)); Other bias, low risk ($n = 10$).

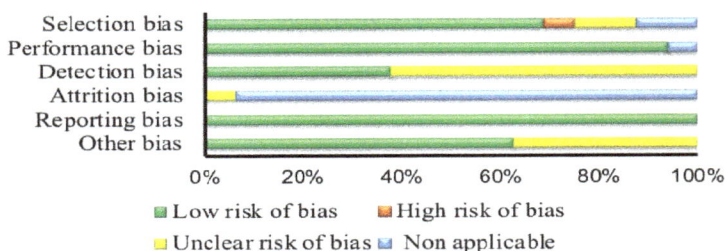

**Figure 2.** Summary of risk of bias assessment for non-randomized controlled trials. Selection bias, high risk ($n = 1$), low risk ($n = 11$), unclear risk ($n = 2$), non-applicable ($n = 2$); Performance bias, low risk ($n = 15$), non-applicable ($n = 1$); Detection bias, low risk ($n = 6$), unclear risk ($n = 10$); Attrition bias, unclear risk ($n = 1$), non-applicable ($n = 15$); Reporting bias (selective reporting, low risk ($n = 16$)); Other bias (confounding, low risk ($n = 10$), unclear risk ($n = 6$)).

Out of the 16 non-RCTs, one showed a high risk of bias [28], two were non-applicable to the category [7,35], two showed an unclear risk of bias [37,42], and the other 11 showed a low risk of bias [3,10,25–27,29,30,32,34,40,41] for selection bias. Of these studies, only one was non-applicable to the category [7], while the other 15 showed a low risk of bias for performance bias. For detection bias, six studies showed a low risk of bias [25,29,30,32,37,41], and the other 10 studies showed an unclear risk of bias. One of the 16 studies showed an unclear risk of bias for attrition bias [7], and the other 15 studies were non-applicable to the category. All 16 of the studies displayed a low risk of bias for selective outcome. For the confounding category, six of the 16 studies displayed an unclear risk of bias [10,27,28,34,35,37], and the remaining 10 displayed a low risk of bias.

*3.4. Reporting of the Outcomes*

3.4.1. Habitual Use

Six of the included studies reported genetic variation associated with habitual use [7,10,26,27, 34,37]. In a CSS, Cornelis et al. [26] examined how polymorphisms in the *CYP1A2* (rs762551) and *ADORA2A* (rs5751876) genes were associated with caffeine intake as measured via a validated food frequency questionnaire. These data demonstrated that the *CYP1A2* genotype was not associated with caffeine intake, but the *ADORA2A TT* genotype was associated with lower caffeine intake in smokers ($p = 0.008$) and nonsmokers ($p = 0.011$). Further work conducted by Cornelis et al. [27] demonstrated associations between caffeine consumption and genetic loci near the *AHR* (rs4410790, $p = 2.4 \times 10^{-19}$) and *CYP1A2* (rs2472304, $p = 2.5 \times 10^{-7}$) genes in 47,341 subjects of European descent. A CSS confirmed these associations in a distinct Costa Rican population (rs4410790, Odds Ratio = 1.41 high versus low consumers; rs2472304, Odd Ratio = 1.55 high versus low consumers) [34], and an additional CSS reported similar associations in the *AHR* gene (rs6968865, *p* range = $1.15 \times 10^{-1}$ to $3.34 \times 10^{-6}$ [7]. More recently, Cornelis et al. [10] demonstrated associations between caffeine consumption and several indexed SNPs (rs1260326, $Log_{10}$ Bayes-Factor (BF) = 6.48; rs1481012, $Log_{10}BF$ = 6.08; rs7800944, $Log_{10}BF$ = 8.83; rs17685, $Log_{10}BF$ = 15.12; rs6265, $Log_{10}BF$ = 5.76; rs9902453, $Log_{10}BF$ = 6.29) in individuals of European and African-American ancestry. Further, Pirastu et al. [37] implicated a novel gene (*PDSS2*) that encodes for coenzyme Q10 in caffeine consumption.

### 3.4.2. Anxiogenic Consequences

Eight of the included studies reported data on genetic variation and anxiety/side effects of caffeine [6,11,12,29–31,38,39]. Two of the included studies investigated the effects of an SNP in the *CYP1A2* (rs762551) gene on self-reported side effects of caffeine following consumption in basketball players [38,39]. The results of these studies demonstrated that self-reported feelings of anxiety were not different across genotypes. Alsene et al. [11] investigated the impact of genetic variation in the *ADORA2A* gene on anxiety following caffeine consumption in caffeine-naive subjects via a double-blind RCT. The data demonstrated that two polymorphisms (rs5751876 and rs35060421) were associated with self-reported anxiety, with the *TT* and *2592Tins/Tins* genotypes reporting higher anxiety, respectively. In an additional double blind RCT, Childs et al. [6] demonstrated that genetic variation in the *ADORA2A* (rs5751876, rs2298383, rs4822492) and dopamine receptor *DRD2* (rs1110976) genes were associated with anxiety in 102 non-to-moderate caffeine users. Further supporting the findings that individuals with the rs5751876 *TT* genotype may be prone to anxiety with caffeine, Gajewska et al. [31] and Domschke et al. [29] reported that subjects with the rs5751876 *TT* genotype exhibited impaired prepulse inhibition (female subjects) and an increased startle reflex (particularly female subgroup) with caffeine, respectively. Nonetheless, one study [12] demonstrated that the anxiogenic effect of caffeine was only apparent in subjects with the rs5751876 *TT* genotype that were caffeine naive. The authors concluded that tolerance to the anxiogenic impact of caffeine is observed when individuals habitually consume moderate to large doses [12]. Additionally, one study reported that variation in the Neuropeptide S receptor gene (rs324981) may (in conjunction with *ADORA2A* (rs5751876)) further impact the anxiogenic effects of caffeine [30].

### 3.4.3. Ergogenic Consequences

Eight of the included studies investigated the effects of genetic variability on the ergogenic consequences of caffeine [20,21,25,32,33,36,39,46]. Five of these studies [20,25,32,33,36] assessed the impact of the *CYP1A2* (rs762551) SNP on the ergogenic consequences of caffeine using cycling time trials as a performance measure with disparate findings. Womack et al. [20] reported that male cyclists (*n* = 35) with the *CYP1A2* (rs762551) *AA* genotype demonstrated greater improvements in cycling performance (40 km time trial) versus C-allele carriers following caffeine consumption (6 mg kg$^{-1}$ anhydrous caffeine). Similarly, Guest et al. [33] reported that male cyclists (*n* = 101) with the *CYP1A2* (rs762551) *AA* genotype demonstrated greater improvements in cycling performance (10 km time trial) relative to those with the *CYP1A2* (rs762551) *CC* genotype following caffeine treatment (2 and 4 mg kg$^{-1}$). Equivocally, Pataky et al. [36] demonstrated that recreational cyclists (*n* = 38) with the *CYP1A2* (rs762551) *AC* genotype derived a more robust ergogenic benefit (3 km time trial) following caffeine treatment (6 mg kg$^{-1}$ and 6 mg kg$^{-1}$ plus caffeinated mouth rinse) relative to *CYP1A2* (rs762551) *AA* homozygotes. Algrain et al. [25] demonstrated that subject groups (*AA* vs. C-allele carriers) responded similarly to caffeine treatment (300 mg in gum vs. placebo gum) with performance measured via a 15-min performance ride. Giersch et al. [32] found that the ergogenic consequences (3 km time trial) of caffeine (6 mg kg$^{-1}$ anhydrous caffeine) did not differ across *CYP1A2* (rs762551) genotype groups (*AA* vs. C-allele carriers). Algrain et al. [25], Pataky et al. [36], and Giersch et al. [32] all cited methodological differences as a potential explanation for the disparate findings in the literature. Studies investigating the impact of the *CYP1A2* (rs762551) SNP on the erogenicity of caffeine utilizing sports-related outcome measures (Wingate test, reaction time, basketball-specific skills) have reported that genotype groups responded equally to caffeine [38,39]. One of the included studies [21] examined the impact of the adenosine receptor *ADORA2A* (rs5751876) SNP on ergogenic responses to caffeine in females (*n* = 12) with performance assessed via a 10-min cycling time trial. These data demonstrated that subjects with the *ADORA2A* (rs5751876) *TT* genotype derived a larger ergogenic benefit from caffeine relative to heterozygotes or *CC* homozygotes [21].

*3.5. Other Outcomes*

Three of the included studies investigated the effect of the *CYP1A2* (rs762551) genotype on caffeine metabolism [3,25,32]. Sachse et al. [3] reported that, among smokers (*n* = 51), subjects who possessed the *CYP1A2* (rs762551) *AA* genotype metabolized caffeine (100 mg) faster relative to C-allele carriers while utilizing a 5 h paraxanthine/caffeine ratio as the outcome measure. The authors concluded that the *CYP1A2* (rs762551) *AA* genotype may confer high inducibility and CYP1A2 activity. More recently, confirming the findings above in 20 nonsmokers, Giersch et al. [32] reported that, 1 h following the administration of 6 mg kg$^{-1}$ anhydrous caffeine, circulating caffeine concentrations were lower in subjects possessing the *CYP1A2* (rs762551) *AA* genotype relative to C-allele carriers. Equivocally, Algrain et al. [25] reported comparable circulating caffeine concentrations over time in nonsmokers across *CYP1A2* (rs762551) genotypes. Two of the included studies investigated the effects of the *CYP1A2* (rs762551) SNP on the cardiovascular consequences of caffeine [40,41]. Thomas et al. [41] reported that changes in postexercise heart rate variability with caffeine treatment (300 mg) were similar between *CYP1A2* genotype groups (*AA* vs. C-allele carriers). Soares et al. [40] reported that *CYP1A2* heterozygotes demonstrated increases in systolic blood pressure, while subjects with the *CYP1A2 AA* genotype did not following acute caffeine ingestion (6 mg kg$^{-1}$).

Three additional studies were included in the present review [28,35,42]. Djordjevic [28] explored the association of multiple *CYP1A2* polymorphisms with the induction of CYP1A2 enzyme activity resultant from heavy caffeine consumption. These data demonstrated that high *CYP1A2* enzyme activity was associated with heavy coffee consumption only in subjects possessing the *CYP1A2* (rs762551) *AA* genotype. In a classical twin design study, Luciano et al. [35] demonstrated that genes not typically associated with sleep disturbance were implicated in coffee-attributed insomnia. In a CCS, Urry et al. [42] demonstrated that subjects with type 2 diabetes mellitus exhibited higher estimated *CYP1A2* enzyme activity relative to control subjects.

**4. Discussion**

The purpose of the present review was to systematically recognize how select SNPs impact habitual use of caffeine as well as the ergogenic and anxiogenic consequences of caffeine. The primary findings of our work will be discussed in the subsections below.

*4.1. Habitual Use*

Genome-wide association scans have implicated several indexed SNPs in caffeine consumption. The aforementioned SNPs occur in genes known to be involved in the pharmacokinetics and pharmacodynamics of caffeine. The results of the present review indicate that SNPs in the *CYP1A2* gene and near the *AHR* gene have been consistently associated with caffeine consumption [7,10,27,34]. Further, less conclusive evidence suggests that SNPs in the *ADORA2A* gene are associated with caffeine consumption [10,26,27]. Recently, novel genes have been implicated in caffeine consumption, with authors calling for replication of the findings and postulating biological plausibility [10,37].

*4.2. Anxiogenic Consequences*

Our search provided strong evidence that SNPs in the *ADORA2A* gene (primarily rs5751876) are associated with the anxiogenic impact of caffeine [6,11,29,31]. Particularly, caffeine-naïve females possessing the *ADORA2A* (rs5751876) *TT* genotype may be especially prone to experiencing anxiety following caffeine consumption [29,31]. Interestingly, one study demonstrated that self-reported anxiety with caffeine was not apparent in subjects possessing *ADORA2A* (rs5751876) *TT* who habitually consume large to moderate doses. Thus, the available evidence suggests that habitual use may lead to tolerance to the anxiogenic consequences of caffeine regardless of select genetic variations.

*4.3. Ergogenic Consequences*

The studies included in this section of the review focused primarily on the *CYP1A2* (rs762551) SNP and reported equivocal findings. Two studies reported that the *CYP1A2* (rs762551) *AA* genotype resulted in a more robust ergogenic benefit of caffeine [20,33] relative to C-allele carriers, while one study reported opposing findings [36], and two studies reported comparable ergogenic responses across genotype groups [25,32]. Two studies reported that the *CYP1A2* (rs762551) SNP did not influence sport-specific outcomes with and without caffeine [38,39]. Our search included one pilot study that investigated the effects of the *ADORA2A* (rs5751876) SNP on the erogenicity of caffeine; Loy et al. [21] reported that females with the *ADORA2A* (rs5751876) *TT* genotype derived a larger ergogenic benefit with caffeine relative to C-allele carriers. In general, the studies included in this section cited some methodological constraints, such as low sample size and capricious outcome measures. We recommend that future studies increase sample size, utilize more standardized outcome measures, and examine a multitude of biologically plausible SNPs to further elucidate the impact genetic variation has on the ergogenic consequences of caffeine.

*4.4. Quality of Evidence, Limitations, and Potential Biases in the Review Process*

Based on the studies selected for the aim of the current systematic review, we may form adequate conclusions regarding the impact of select SNPs on the ergogenic and anxiogenic consequences and habitual use of caffeine in humans. This is because we identified enough available evidence in the area. The included RCTs displayed unclear and low risk of bias in the selection, performance, and detection biases, while they mostly displayed low risk of bias in the attrition, reporting, and other biases. Similarly, the non-RCTs displayed mostly low risk of bias in the selection, performance, reporting, and other biases, while in the attrition bias, most studies were non-applicable for the category. This indicates that both the RCTs and the non-RCTs may provide fairly good quality evidence.

Our systematic review has a number of strengths. We used the PRISMA guidelines [22–24] and appropriate databases and algorithms with standardized indexing terms for our searching procedure. We also used well-established tools to evaluate the included [24,43–45] studies. Furthermore, to minimize bias in our systematic review process, two investigators worked independently on the searching and screening procedure, data extraction, and risk of bias assessment. Finally, we have not excluded studies based on language.

A possible limitation of the current systematic review is that we avoided the use of non-peer-review data (grey literature) and conference papers to test our research question. However, the inclusion of non-peer-review data may have itself introduced bias, given that there was available peer-reviewed evidence [47]. Another possible limitation is the small number of the included studies, especially the RCTs, which indicates the need for additional research of this topic in the future.

## 5. Conclusions

In conclusion, our search has provided evidence that the *CYP1A2*, *AHR*, and *ADORA2A* genes are associated with habitual consumption, and further exploration is warranted to clarify how these genes directly or indirectly impact physiological and/or psychological mechanisms responsible for the

*Nutrients* **2018**, *10*, 1373

variability in consumption of caffeine across individuals. The literature also demonstrates that gender, habitual caffeine consumption, and variability in the *ADORA2A* gene collectively influence individual susceptibility to the anxiogenic consequences of caffeine and that variability in the *CYP1A2* gene, in conjunction with environmental factors (heavy coffee drinking, smoking), impact the metabolism of caffeine. Future work is warranted to elucidate the effects of variability in the *CYP1A2* and *ADORA2A* genes on the ergogenic impact of caffeine. We recommend all future studies in this area utilize an interdisciplinary approach as the physiological consequences of caffeine in humans are likely dependent on a complex interaction of genetic, physiological, and behavioral factors.

**Supplementary Materials:** The following are available online at http://www.mdpi.com/2072-6643/10/10/1373/s1, Figure S1: Algorithm, Figure S2: Excluded Studies; Figure S3: PRISMA Checklist; Figure S4: PRISMA Flowchart.

**Author Contributions:** The following authors contributed to conceptualization: Conceptualization J.L.F., E.J.R. (Edward J. Ryan), P.C.D., A.E.C., J.R.E., E.J.R. (Emily J. Ryan) The methodology was formed by J.L.F., E.J.R. (Edward J. Ryan), P.C.D., A.E.C., J.R.E., E.J.R. (Emily J. Ryan), and the software included contributions from J.L.F., E.J.R. (Edward J. Ryan), P.C.D., A.E.C. Validations were done by J.L.F., E.J.R. (Edward J. Ryan), P.C.D., A.E.C., J.R.E., E.J.R. (Emily J. Ryan), and formal analysis were done by J.L.F., E.J.R. (Edward J. Ryan), P.C.D., A.E.C. The investigation was done by J.L.F., E.J.R. (Edward J. Ryan), P.C.D., A.E.C., J.R.E., E.J.R. (Emily J. Ryan), and the resources were done by J.L.F., E.J.R. (Edward J. Ryan), P.C.D., A.E.C., J.R.E., E.J.R. (Emily J. Ryan). Contributions to the data curation were from J.L.F., E.J.R. (Edward J. Ryan), P.C.D., A.E.C. For writing—original draft preparation, the contributions were from J.L.F., E.J.R. (Edward J. Ryan), P.C.D., A.E.C. Writing—Review and editing was done by J.L.F., E.J.R. (Edward J. Ryan), P.C.D., A.E.C., J.R.E., E.J.R (Emily J. Ryan), and visualization was done by J.L.F, E.J.R. (Edward J. Ryan), P.C.D., A.E.C., J.R.E., E.J.R (Emily J. Ryan). Supervision was done by P.C.D., E.J.R. (Edward J. Ryan), A.E.C., and project administration was done by J.L.F., E.J.R. (Emily J. Ryan), P.C.D., A.E.C.

**Funding:** This research received no external funding.

**Conflicts of Interest:** The authors declare no conflict of interest.

## References

1. Magkos, F.; Kavouras, S.A. Caffeine use in sports, pharmacokinetics in man, and cellular mechanisms of action. *Crit. Rev. Food Sci. Nutr.* **2005**, *45*, 535–562. [CrossRef] [PubMed]
2. Lelo, A.; Birkett, D.J.; Robson, R.A.; Miners, J.O. Comparative pharmacokinetics of caffeine and its primary demethylated metabolites paraxanthine, theobromine and theophylline in man. *Br. J. Clin. Pharmacol.* **1986**, *22*, 177–182. [CrossRef] [PubMed]
3. Sachse, C.; Brockmoller, J.; Bauer, S.; Roots, I. Functional significance of a C→A polymorphism in intron 1 of the cytochrome p450 cyp1a2 gene tested with caffeine. *Br. J. Clin. Pharmacol.* **1999**, *47*, 445–449. [CrossRef] [PubMed]
4. Fredholm, B.B. Astra award lecture. Adenosine, adenosine receptors and the actions of caffeine. *Pharmacol. Toxicol.* **1995**, *76*, 93–101. [CrossRef] [PubMed]
5. Salmi, P.; Chergui, K.; Fredholm, B.B. Adenosine-dopamine interactions revealed in knockout mice. *J. Mol. Neurosci.* **2005**, *26*, 239–244. [CrossRef]
6. Childs, E.; Hohoff, C.; Deckert, J.; Xu, K.; Badner, J.; de Wit, H. Association between adora2a and drd2 polymorphisms and caffeine-induced anxiety. *Neuropsychopharmacology* **2008**, *33*, 2791–2800. [CrossRef] [PubMed]
7. McMahon, G.; Taylor, A.E.; Davey Smith, G.; Munafò, M.R. Phenotype refinement strengthens the association of ahr and cyp1a1 genotype with caffeine consumption. *PLoS ONE* **2014**, *9*, e103448. [CrossRef] [PubMed]
8. Eskelinen, M.H.; Kivipelto, M. Caffeine as a protective factor in dementia and alzheimer's disease. *J. Alzheimers Dis.* **2010**, *20*, S167–S174. [CrossRef] [PubMed]
9. Bohn, S.K.; Ward, N.C.; Hodgson, J.M.; Croft, K.D. Effects of tea and coffee on cardiovascular disease risk. *Food Funct.* **2012**, *3*, 575–591. [CrossRef] [PubMed]
10. Cornelis, M.C.; Byrne, E.M.; Esko, T.; Nalls, M.A.; Ganna, A.; Paynter, N.; Monda, K.L.; Amin, N.; Fischer, K.; Renstrom, F.; et al. Genome-wide meta-analysis identifies six novel loci associated with habitual coffee consumption. *Mol. Psychiatry* **2015**, *20*, 647–656. [CrossRef] [PubMed]
11. Alsene, K.; Deckert, J.; Sand, P.; de Wit, H. Association between a2a receptor gene polymorphisms and caffeine-induced anxiety. *Neuropsychopharmacology* **2003**, *28*, 1694–1702. [CrossRef] [PubMed]

12. Rogers, P.J.; Hohoff, C.; Heatherley, S.V.; Mullings, E.L.; Maxfield, P.J.; Evershed, R.P.; Deckert, J.; Nutt, D.J. Association of the anxiogenic and alerting effects of caffeine with adora2a and adora1 polymorphisms and habitual level of caffeine consumption. *Neuropsychopharmacology* **2010**, *35*, 1973–1983. [CrossRef] [PubMed]

13. Graham, T.E. Caffeine and exercise: Metabolism, endurance and performance. *Sports Med.* **2001**, *31*, 785–807. [CrossRef] [PubMed]

14. Ganio, M.S.; Klau, J.F.; Casa, D.J.; Armstrong, L.E.; Maresh, C.M. Effect of caffeine on sport-specific endurance performance: A systematic review. *J. Strength Cond. Res.* **2009**, *23*, 315–324. [CrossRef]

15. Doherty, M.; Smith, P.M. Effects of caffeine ingestion on exercise testing: A meta-analysis. *Int. J. Sport Nutr. Exerc. Metab.* **2004**, *14*, 626–646. [CrossRef]

16. Goldstein, E.R.; Ziegenfuss, T.; Kalman, D.; Kreider, R.; Campbell, B.; Wilborn, C.; Taylor, L.; Willoughby, D.; Stout, J.; Graves, B.S.; et al. International society of sports nutrition position stand: Caffeine and performance. *J. Int. Soc. Sports Nutr.* **2010**, *7*, 5. [CrossRef]

17. Jenkins, N.T.; Trilk, J.L.; Singhal, A.; O'Connor, P.J.; Cureton, K.J. Ergogenic effects of low doses of caffeine on cycling performance. *Int. J. Sport Nutr. Exerc. Metab.* **2008**, *18*, 328–342. [CrossRef]

18. Meyers, B.M.; Cafarelli, E. Caffeine increases time to fatigue by maintaining force and not by altering firing rates during submaximal isometric contractions. *J. Appl. Physiol.* **2005**, *99*, 1056–1063. [CrossRef]

19. Doherty, M.; Smith, P.M.; Davison, R.C.; Hughes, M.G. Caffeine is ergogenic after supplementation of oral creatine monohydrate. *Med. Sci. Sports Exerc.* **2002**, *34*, 1785–1792. [CrossRef]

20. Womack, C.J.; Saunders, M.J.; Bechtel, M.K.; Bolton, D.J.; Martin, M.; Luden, N.D.; Dunham, W.; Hancock, M. The influence of a cyp1a2 polymorphism on the ergogenic effects of caffeine. *J. Int. Soc. Sports Nutr.* **2012**, *9*, 7. [CrossRef]

21. Loy, B.D.; O'Connor, P.J.; Lindheimer, J.B.; Covert, S.F. Caffeine is ergogenic for adenosine a2a receptor gene (adora2a) t allele homozygotes: A pilot study. *J. Caffeine Res.* **2015**, *5*, 2. [CrossRef]

22. Harris, J.D.; Quatman, C.E.; Manring, M.M.; Siston, R.A.; Flanigan, D.C. How to write a systematic review. *Am. J. Sports Med.* **2014**, *42*, 2761–2768. [CrossRef]

23. Davey, J.; Turner, R.M.; Clarke, M.J.; Higgins, J.P. Characteristics of meta-analyses and their component studies in the cochrane database of systematic reviews: A cross-sectional, descriptive analysis. *BMC Med. Res. Methodol.* **2011**, *11*, 160. [CrossRef] [PubMed]

24. Viswanathan, M.; Berkman, N.D.; Dryden, D.M.; Hartling, L. Assessing Risk of Bias and Confounding in Observational Studies of Interventions or Exposure: Further Development of the RTI Item Bank. Available online: http://www.effectivehealthcare.ahrq.gov/reports/final.cfm (accessed on 22 August 2018).

25. Algrain, H.A.; Thomas, R.M.; Carrillo, A.E.; Ryan, E.J.; Kim, C.-H.; Lettanll, R.B.; Ryan, E.J. The effects of a polymorphism in the cytochrome p450 cyp1a2 gene on performance enhancement with caffeine in recreational cyclists. *J. Caffeine Res.* **2015**, *6*, 1. [CrossRef]

26. Cornelis, M.C.; El-Sohemy, A.; Campos, H. Genetic polymorphism of the adenosine a2a receptor is associated with habitual caffeine consumption. *Am. J. Clin. Nutr.* **2007**, *86*, 240–244. [CrossRef] [PubMed]

27. Cornelis, M.C.; Monda, K.L.; Yu, K.; Paynter, N.; Azzato, E.M.; Bennett, S.N.; Berndt, S.I.; Boerwinkle, E.; Chanock, S.; Chatterjee, N.; et al. Genome-wide meta-analysis identifies regions on 7p21 (ahr) and 15q24 (cyp1a2) as determinants of habitual caffeine consumption. *PLoS Genet.* **2011**, *7*, e1002033. [CrossRef]

28. Djordjevic, N.; Ghotbi, R.; Bertilsson, L.; Jankovic, S.; Aklillu, E. Induction of cyp1a2 by heavy coffee consumption in serbs and swedes. *Eur. J. Clin. Pharmacol.* **2008**, *64*, 381–385. [CrossRef]

29. Domschke, K.; Gajewska, A.; Winter, B.; Herrmann, M.J.; Warrings, B.; Muhlberger, A.; Wosnitza, K.; Glotzbach, E.; Conzelmann, A.; Dlugos, A.; et al. Adora2a gene variation, caffeine, and emotional processing: A multi-level interaction on startle reflex. *Neuropsychopharmacology* **2012**, *37*, 759–769. [CrossRef] [PubMed]

30. Domschke, K.; Klauke, B.; Winter, B.; Gajewska, A.; Herrmann, M.J.; Warrings, B.; Muhlberger, A.; Wosnitza, K.; Dlugos, A.; Naunin, S.; et al. Modification of caffeine effects on the affect-modulated startle by neuropeptide s receptor gene variation. *Psychopharmacology* **2012**, *222*, 533–541. [CrossRef] [PubMed]

31. Gajewska, A.; Blumenthal, T.D.; Winter, B.; Herrmann, M.J.; Conzelmann, A.; Muhlberger, A.; Warrings, B.; Jacob, C.; Arolt, V.; Reif, A.; et al. Effects of adora2a gene variation and caffeine on prepulse inhibition: A multi-level risk model of anxiety. *Prog. Neuropsychopharmacol. Biol. Psychiatry* **2013**, *40*, 115–121. [CrossRef] [PubMed]

32. Giersch, G.E.; Boyett, J.C.; Hargens, T.A.; Luden, N.D.; Saunders, L.J.; Daley, H.M.; Hughey, C.A.; El-Sohemy, A.; Womack, C.J. The effect of the cyp1a2 163 C>A polymorphism on caffeine metabolism and subsequent cycling performance. *J. Caffeine Adenosine Res.* **2018**, *8*, 2. [CrossRef]

33. Guest, N.; Corey, P.; Vescovi, J.; El-Sohemy, A. Caffeine, cyp1a2 genotype, and endurance performance in athletes. *Med. Sci. Sports Exerc.* **2018**, *50*, 1570–1578. [CrossRef] [PubMed]

34. Josse, A.R.; Da Costa, L.A.; Campos, H.; El-Sohemy, A. Associations between polymorphisms in the ahr and cyp1a1-cyp1a2 gene regions and habitual caffeine consumption. *Am. J. Clin. Nutr.* **2012**, *96*, 665–671. [CrossRef] [PubMed]

35. Luciano, M.; Zhu, G.; Kirk, K.M.; Gordon, S.D.; Heath, A.C.; Montgomery, G.W.; Martin, N.G. "No thanks, it keeps me awake": The genetics of coffee-attributed sleep disturbance. *Sleep* **2007**, *30*, 1378–1386. [CrossRef]

36. Pataky, M.W.; Womack, C.J.; Saunders, M.J.; Goffe, J.L.; D'Lugos, A.C.; El-Sohemy, A.; Luden, N.D. Caffeine and 3-km cycling performance: Effects of mouth rinsing, genotype, and time of day. *Scand. J. Med. Sci. Sports* **2016**, *26*, 613–619. [CrossRef]

37. Pirastu, N.; Kooyman, M.; Robino, A.; van der Spek, A.; Navarini, L.; Amin, N.; Karssen, L.C.; Van Duijn, C.M.; Gasparini, P. Non-additive genome-wide association scan reveals a new gene associated with habitual coffee consumption. *Sci. Rep.* **2016**, *6*, 31590. [CrossRef]

38. Puente, C.; Abián-Vicén, J.; Del Coso, J.; Lara, B.; Salinero, J.J. The cyp1a2 -163C>A polymorphism does not alter the effects of caffeine on basketball performance. *PLoS ONE* **2018**, *13*, e0195943. [CrossRef] [PubMed]

39. Salinero, J.J.; Lara, B.; Ruiz-Vicente, D.; Areces, F.; Puente-Torres, C.; Gallo-Salazar, C.; Pascual, T.; Del Coso, J. Cyp1a2 genotype variations do not modify the benefits and drawbacks of caffeine during exercise: A pilot study. *Nutrients* **2017**, *9*, 269. [CrossRef]

40. Soares, R.N.; Schneider, A.; Valle, S.C.; Schenkel, P.C. The influence of cyp1a2 genotype in the blood pressure response to caffeine ingestion is affected by physical activity status and caffeine consumption level. *Vascul. Pharmacol.* **2018**, *106*, 67–73. [CrossRef]

41. Thomas, R.M.; Algrain, H.A.; Ryan, E.J.; Popojas, A.; Carrigan, P.; Abdulrahman, A.; Carrillo, A.E. Influence of a cyp1a2 polymorphism on post-exercise heart rate variability in response to caffeine intake: A double-blind, placebo-controlled trial. *Ir. J. Med. Sci.* **2017**, *186*, 285–291. [CrossRef]

42. Urry, E.; Jetter, A.; Landolt, H.P. Assessment of cyp1a2 enzyme activity in relation to type-2 diabetes and habitual caffeine intake. *Nutr. Metab.* **2016**, *13*, 66. [CrossRef] [PubMed]

43. Margulis, A.V.; Pladevall, M.; Riera-Guardia, N.; Varas-Lorenzo, C.; Hazell, L.; Berkman, N.D.; Viswanathan, M.; Perez-Gutthann, S. Quality assessment of observational studies in a drug-safety systematic review, comparison of two tools: The Newcastle-Ottawa scale and the RTI item bank. *Clin. Epidemiol.* **2014**, *6*, 359–368. [CrossRef] [PubMed]

44. Al-Saleh, M.A.; Armijo-Olivo, S.; Thie, N.; Seikaly, H.; Boulanger, P.; Wolfaardt, J.; Major, P. Morphologic and functional changes in the temporomandibular joint and stomatognathic system after transmandibular surgery in oral and oropharyngeal cancers: Systematic review. *J. Otolaryngol. Head Neck Surg.* **2012**, *41*, 345–360. [PubMed]

45. Higgins, J.P.; Altman, D.G.; Gotzsche, P.C.; Juni, P.; Moher, D.; Oxman, A.D.; Savovic, J.; Schulz, K.F.; Weeks, L.; Sterne, J.A.; et al. The cochrane collaboration's tool for assessing risk of bias in randomised trials. *BMJ* **2011**, *343*, d5928. [CrossRef] [PubMed]

46. Puente, C.; Abian-Vicen, J.; Salinero, J.J.; Lara, B.; Areces, F.; Del Coso, J. Caffeine improves basketball performance in experienced basketball players. *Nutrients* **2017**, *9*, 1033. [CrossRef] [PubMed]

47. Higgins, J.P.; Green, S. *Cochrane Handbook for Systematic Reviews of Interventions*; Version 5.1.0; John Wiley & Sons, Inc.: Hoboken, NJ, USA, 2011.

*nutrients*

MDPI

*Review*

# The Role of Genetics in Moderating the Inter-Individual Differences in the Ergogenicity of Caffeine

Kyle Southward [1], Kay Rutherfurd-Markwick [2,3], Claire Badenhorst [1,3] and Ajmol Ali [1,3,*]

[1] School of Sport, Exercise and Nutrition, Massey University, North Shore Mail Centre, Private Bag 102 904, Auckland 0745, New Zealand; K.A.Southward@massey.ac.nz (K.S.); C.Badenhorst@massey.ac.nz (C.B.)
[2] School of Health Sciences, Massey University, Auckland 0745, New Zealand; K.J.Rutherfurd@massey.ac.nz
[3] Centre for Metabolic Health Research, Massey University, Auckland 0745, New Zealand
* Correspondence: a.ali@massey.ac.nz; Tel.: +64-(0)9-213-6414

Received: 17 August 2018; Accepted: 17 September 2018; Published: 21 September 2018

**Abstract:** Caffeine use is widespread among athletes following its removal from the World Anti-Doping Agency banned list, with approximately 75% of competitive athletes using caffeine. While literature supports that caffeine has a small positive ergogenic effect for most forms of sports and exercise, there exists a significant amount of inter-individual difference in the response to caffeine ingestion and the subsequent effect on exercise performance. In this narrative review, we discuss some of the potential mechanisms and focus on the role that genetics has in these differences. CYP1A2 and ADORA2A are two of the genes which are thought to have the largest impact on the ergogenicity of caffeine. CYP1A2 is responsible for the majority of the metabolism of caffeine, and ADORA2A has been linked to caffeine-induced anxiety. The effects of CYP1A2 and ADORA2A genes on responses to caffeine will be discussed in detail and an overview of the current literature will be presented. The role of these two genes may explain a large portion of the inter-individual variance reported by studies following caffeine ingestion. Elucidating the extent to which these genes moderate responses to caffeine during exercise will ensure caffeine supplementation programs can be tailored to individual athletes in order to maximize the potential ergogenic effect.

**Keywords:** CYP1A2; ADORA2A; time trial performance; caffeine metabolism; pharmacological ergogenic aid

## 1. Introduction

Caffeine was placed on the World Anti-Doping Agency's (WADA) banned list in 1984 and remained there until 2004 when it was removed from this list and placed on the monitoring program after it was determined that it no longer satisfied two of the three criteria needed to be on the banned list. Since the removal of caffeine from the banned list, the use of caffeine as an ergogenic aid amongst athletes has become widespread around the world with one study reporting approximately 73.8% of athletes will have consumed caffeine shortly before or during an event, with a higher prevalence in endurance athletes [1].

The first study to show the ergogenic potential of caffeine was published in 1907 [2], there was no further research in this area until the work of Ivy and Costill in the late 1970s [2–4]. Ivy and Costill [3,4] suggested that the ergogenic effect of caffeine was due to increased lipid mobilisation and utilisation through both direct action on fat stores as well as through stimulating the release of cortisol and norepinephrine, thus increasing lipolysis [5–8]. An increase in lipid mobilisation was hypothesised to increase glycogen sparing and thus, delayed fatigue during endurance exercise [3,4]. However, since then, other studies have shown that caffeine does not significantly increase lipid metabolism during exercise [9–13]. Evidence now indicates the ergogenicity of caffeine is most likely due to its effect as a potent adenosine receptor antagonist, whereby it blocks the actions of adenosine primarily in the brain [9,14,15].

The effects of adenosine on the central nervous system (CNS) have been well documented [16,17]. Adenosine has been shown to decrease feelings of arousal, alertness and vigilance, which increases central fatigue and negatively impacts on exercise performance [16,17]. Normally, adenosine slowly accumulates throughout the day, as well as during exercise when there is insufficient oxygen to regenerate adenosine triphosphate (ATP). Adenosine concentrations decrease during rest or sleep when ATP stores are regenerated [18,19]. Adenosine has been shown to down regulate various neurotransmitters such as dopamine, serotonin, glutamate, acetylcholine and norepinephrine [16,17]. Dopamine is a key neurotransmitter in parts of the brain which regulates behavioral activation and effort-based behavioral processes; thus, decreases in dopamine concentrations can lead to a reduced effort during exercise and a reduction in overall exercise performance [20]. This is more evident in endurance exercise where central fatigue plays an important role in moderating exercise performance, compared to strength and speed-based exercise such as sprinting, where peripheral fatigue may have a larger impact on overall performance.

The effects of adenosine are inhibited through competition by caffeine at the adenosine receptor sites. Caffeine, as well as theophylline and paraxanthine, caffeine metabolites, have a similar structure to adenosine, which allows them to bind to adenosine receptors throughout the body. It is the antagonism of adenosine by caffeine and theophylline molecules which likely has the biggest impact on the ergogenicity of caffeine, particularly during endurance exercise by reducing the effects of adenosine and ultimately decreasing feelings of tiredness and improving vigilance, arousal and a willingness to exert effort during exercise [15,16,21].

Despite well-documented overall improvements in exercise performance following caffeine ingestion [9,15,22,23], there exists a significant variation, both between individuals and between studies, in the responses to caffeine ingestion [24]. Studies which have reported individual data have shown that a number of individuals either do not respond to caffeine supplementation, such that their performance is unchanged between placebo and caffeine trials, or performance is decreased following caffeine supplementation (Table 1), as opposed to the majority of individuals who show improved exercise performance following caffeine ingestion. These studies show that approximately 33% of individuals in these investigations did not improve their endurance time-trial performance following caffeine ingestion. While this shows evidence of the variance between individuals and studies, the cause of the variance is not yet fully understood. A recent publication by Pickering and Kiely [25], provided some discussion around the inter-individual variation in the ergogenicity of caffeine, however, the present paper looks to expand upon this further with specific reference to endurance sports as well as taking a more in-depth look at the genes associated with caffeine ergogenicity.

**Table 1.** Time-trial studies investigating ergogenicity of caffeine which reported individual data.

| Study | Caffeine Dose | Number of Individuals Who Performed Worse in Caffeine Trials Compared to Placebo |
|---|---|---|
| Acker-Hewitt et al. [26] | $6 \text{ mg·kg}^{-1}$ | 2/10 |
| Astorino et al. [27] | $5 \text{ mg·kg}^{-1}$ | 3/16 |
| Astorino et al. [28] | $5 \text{ mg·kg}^{-1}$ | 1/9 |
| Beaumont & James [29] | $6 \text{ mg·kg}^{-1}$ | 1/8 |
| Christensen et al. [30] | $3 \text{ mg·kg}^{-1}$ | 4/12 |
| Church et al. [31] | $3 \text{ mg·kg}^{-1}$ | 8/20 |
| Desbrow et al. [32] | $3 \text{ mg·kg}^{-1}$ | 3/9 |
| Desbrow et al. [33] | $6 \text{ mg·kg}^{-1}$ | 4/16 |
| De Souza Goncalves et al. [34] | $6 \text{ mg·kg}^{-1}$ | 6/40 |
| Graham-Paulson et al. [35] | $4 \text{ mg·kg}^{-1}$ | 1/11 |
| Guest et al. [36] | $2 \text{ mg·kg}^{-1}$ | 38/101 [1] |
| Guest et al. [36] | $4 \text{ mg·kg}^{-1}$ | 32/101 [1] |

Table 1. Cont.

| Study | Caffeine Dose | Number of Individuals Who Performed Worse in Caffeine Trials Compared to Placebo |
|---|---|---|
| O'Rourke et al. [37] | 5 mg·kg$^{-1}$ | 3/30 |
| Pitchford et al. [38] | 3 mg·kg$^{-1}$ | 2/9 |
| Roelands et al. [39] | 6 mg·kg$^{-1}$ | 4/8 |
| Santos et al. [40] | 5 mg·kg$^{-1}$ | 2/8 |
| Skinner et al. [41] | 6 mg·kg$^{-1}$ | 1/14 |
| Stadheim et al. [42] | 6 mg·kg$^{-1}$ | 2/10 |
| Stadheim et al. [43] | 4.5 mg·kg$^{-1}$ | 4/13 |
| Womack et al. [44] | 6 mg·kg$^{-1}$ | 3/35 |
| Total | | 124/379 (33%) |

<sup>1</sup> Same group of participants.

## 2. The Effect of CYP1A2 on Inter-Individual Differences in Ergogenicity of Caffeine

The response to caffeine and the potency of the effects of caffeine seem to be multifaceted. Caffeine habituation [7,45], metabolism of caffeine [9,46], method of caffeine ingestion [47], caffeine dosage [9,48], training status [49,50], and timing of caffeine ingestion [41] have all been identified as having an effect on the ergogenicity of caffeine. Similarly, oral contraceptives, pregnancy, ethnicity, age, and smoking have all been suggested to affect the metabolism of caffeine [51]. Genetic variation in specific genes, namely CYP1A2 (rs762551) and ADORA2A (rs5751876), have also been suggested to have a significant effect on the responses to caffeine ingestion and the metabolism of caffeine [51].

The cytochrome P450s (CYP) are a family of haemoprotein enzymes which are responsible for approximately 75% of all drug metabolism [52]. The cytochrome P450 family 1 subfamily A member 2 (CYP1A2) is predominantly found in the liver and metabolises many clinical drugs and endogenous compounds [53]. The CYP1A2 enzyme is responsible for >90% of caffeine metabolism, breaking it down into the three metabolites: paraxanthine (81.5%), theobromine (10%) and theophylline (5.4%), while CYP2E1 is responsible for the majority of the transformation between caffeine to theophylline and theobromine [54,55].

As the majority of caffeine metabolism is determined by CYP1A2 enzyme activity, variations to the gene encoding for the CYP1A2 enzyme will alter its inducibility, thus significantly impacting the metabolism of caffeine. A single nucleotide polymorphism ($-163$ C > A) at position 734 within intron 1 (rs762551) has been identified as the major source of inducibility of CYP1A2 and thus caffeine metabolism [46,51,53]. Individuals with the homozygous A/A allele show enhanced caffeine metabolism and have been classified as "fast metabolisers", whereas C allele carriers (A/C and C/C) have a reduced caffeine metabolism and are known as "slow metabolisers" [51,53]. Therefore, slow metabolisers are likely to have a prolonged caffeine half-life compared to fast metabolisers. Half-lives of paraxanthine, theobromine and theophylline are ~3 h, ~6 h, and ~7 h respectively [56]. On average, 40% of the general population carry the A/A genotype while 50% and 10% carry the A/C and C/C genotypes, respectively [36,46]. A slower caffeine metabolism would seem to be beneficial for endurance performance as the effects of caffeine would be longer lasting as well as potentially more pronounced, however, some studies have reported the opposite effect [36,44].

Relative to the amount of research on caffeine and performance *per se*, there is limited information on the effects of CYP1A2 genotypes on the ergogenicity of caffeine (Table 2). Two studies found greater performance improvements in individuals with the A/A genotypes compared to C allele carriers [36,44]; two studies found no differences between genotypes in time-trial performance [57,58]; and one study found individuals with the A/C genotype performed better in a 3 km time-trial compared to individuals with the A/A genotype [59]. Other studies [60–62] found no effect of CYP1A2 genotype on the ergogenicity of caffeine using their respective protocols.

It is evident that the results of investigations into the effects of CYP1A2 genotypes on the ergogenicity of caffeine remain equivocal. This may be largely due to the protocols used by the majority of these studies and in some instances the protocol may be counter-productive to the aim of the study. Caffeine as an ergogenic aid has been shown to be most effective and reliable when used for endurance exercise [63], therefore, using a short duration protocol such as the 30 s Wingate test or sport-specific skill tests will inherently lead to less reliable results. Using a long duration endurance protocol to investigate the effects of CYP1A2 on the ergogenicity of caffeine would be beneficial as it would allow for larger changes to be seen in caffeine metabolism between fast and slow metabolizers as caffeine would have had more time to breakdown within the body. It would therefore be more specific and applicable to the endurance sports it is most commonly consumed in, thus having a greater impact. Caffeine is most commonly consumed for endurance sports such as triathlons and marathons [1] which typically last between 1–3 h for half-marathons, marathons, sprint and standard distance triathlons, and between 4–15 h for half and full iron-man events. Therefore, while exercise protocols which last longer than 5 min could technically be considered endurance exercise, endurance events usually last significantly longer than many of the protocols used in the current literature. The protocols used to examine the ergogenicity of caffeine on endurance exercise in future studies should attempt to use exercise protocols lasting longer than 1 h.

**Table 2.** Studies investigating the effects of CYP1A2 genotype on time-trial performance following caffeine ingestion.

| Study | Sample | Caffeine Dose and Timing Prior to Exercise | Protocol | Results |
|---|---|---|---|---|
| Algrain et al. [57] | 13 male and 7 female recreational cyclists | 300 mg caffeinated chewing gum; 10 min | 15 min@70% $VO_{2max}$ followed by 10 min rest and 15 min performance cycle ride | No effect of genotype on performance ride performance |
| Giersch et al. [58] | 20 male cyclists | 6 mg·kg$^{-1}$; 60 min | 3 km TT cycle | No effect of genotype on 3 km TT performance |
| Guest et al. [36] | 101 male competitive cyclists | 2 mg·kg$^{-1}$; 75 min | 10 km TT cycle | Improved A/A genotype performance by 4.8%; No significant difference in A/C and C/C genotypes |
| Guest et al. [36] | 101 male competitive cyclists | 4 mg·kg$^{-1}$; 75 min | 10 km TT cycle | Improved A/A genotype 10 km TT performance 6.8%; Decreased C/C genotype 10 km TT performance by 13.7% |
| Klein et al. [60] | 8 male and 8 female tennis players | 6 mg·kg$^{-1}$; 60 min | 30 min intermittent treadmill running followed by tennis skill test | No effect of genotypes on tennis skill test |
| Pataky et al. [59] | 25 male and 13 females | 6 mg·kg$^{-1}$; 60 min; 25 mL 1.14% caffeinated mouth rinse | 3 km TT cycle | Greater improvements in 3 km TT in A/C genotypes compared to A/A genotypes |
| Puente et al. [62] | 10 males and 9 female elite basketball players | 3 mg·kg$^{-1}$; 60 min | 10 repetitions of: Abalakov jump test and change of direction and acceleration test; 20 min simulated basketball game | No effect of genotype on tests performance |
| Salinero et al. [61] | 14 male and 7 females recreationally active | 3 mg·kg$^{-1}$; 60 min | 30 s Wingate test | No effect of genotypes on Wingate performance |
| Womack et al. [44] | 35 recreationally competitive male cyclists | 6 mg·kg$^{-1}$; 60 min | 40 km TT cycle | Improved cycling TT performance to a greater degree in A/A genotypes compared to C allele carriers |

TT: Time-trial.

Furthermore, protocols used for a novel area of research, such as the effects of genetics on the ergogenicity of caffeine, should be controlled to provide the best chance of obtaining a favorable outcome as proof of concept. Using a short duration test (strength or power tests), or sport-specific skill tests, which have been shown to produce less reliable improvements following caffeine ingestion [63], are unlikely to provide conclusive results when examining the effect of CYP1A2 genotypes on the ergogenicity of caffeine using these protocols. Conversely, endurance exercise protocols have shown much more reliable results compared to short duration exercises and would aid in isolating the effect of CYP1A2, as well as other genes, on the ergogenicity of caffeine. However, well-designed and well-controlled studies using shorter duration protocols should still be conducted as there are many sports, which do not last longer than 1 h, where caffeine may still have an ergogenic effect.

Of the studies included in Table 2, only those by Algrain [57] and Giersch [58] measured caffeine pharmacokinetics and in both cases found no difference in caffeine metabolism between individuals with the A/A allele and C allele carriers. However, Giersch et al. [58] measured caffeine metabolites for 1 h post caffeine ingestion (6 mg·kg$^{-1}$ anhydrous caffeine pill), and Algrain et al. [57] recorded caffeine metabolites for 65 min post ingestion (300 mg caffeinated gum). When using anhydrous caffeine in pill form, peak caffeine concentration usually occurs around 1 h post ingestion [9] and around 30 min after using caffeinated gum [64]. Therefore, measuring caffeine metabolites for 1 h post-caffeine ingestion (via a capsule) would mostly measure caffeine absorption, which is not determined by CYP1A2, rather than metabolism. Thus, studies investigating the effects of CYP1A2 on the ergogenicity of caffeine should focus on using protocols which last >1 h, which allows caffeine metabolism to be measured over a 2-h period should caffeine be administered 1 h prior to exercise.

While it is expected that the CYP1A2 genotype affects caffeine metabolism, some studies have reported no difference in the rate of caffeine metabolism between fast and slow metabolizers in healthy adults [46,57,58]. CYP1A2 genotype affects the inducibility of the CYP1A2 enzyme, which results in changes to the metabolism of caffeine. Individuals with the A/A allele have a higher CYP1A2 inducibility compared to the A/C and C/C alleles [46,51]. The results of these studies may be partly explained by environmental factors which have also been shown to affect the inducibility of CYP1A2, including consumption of cruciferous foods (broccoli and Brussels sprouts), heavy exercise, tobacco smoke, oral contraceptives and various medicines (fluvoxamine, omeprazole) [65]. Combining multiple factors could lead to a greatly increased or reduced activity of CYP1A2, potentially up to 60-fold [65]. Therefore, individuals with the A/C or C/C CYP1A2 allele may have a similar caffeine metabolism to an individual with the A/A allele due to increased inducibility from environmental factors. These factors should be kept in mind when examining studies which have reported effects of CYP1A2 genotype on caffeine ergogenicity and metabolism, and future studies should control or record the use of potential potent inducers and inhibitors of CYP1A2 during caffeine trials.

While the current literature is equivocal as to whether CYP1A2 has an effect on the ergogenicity of caffeine, it is likely that it still has a role in mediating the effects of caffeine on exercise performance. The effects of CYP1A2 on the ergogenicity of caffeine may be more evident in endurance events lasting longer than 1 h, where the metabolism of caffeine may have a more pronounced effect, as those who metabolize caffeine faster would not maintain high circulating levels of caffeine throughout the event compared to those with a slower metabolism of caffeine, unless further ingestion occurred. It should be noted that the half-life of caffeine is 2–12 h and on average 4–6 h [66] in most adults; and it is not yet known to what degree caffeine metabolism is altered between fast and slow metabolizers. Therefore, it is unknown at what time point there would be a large enough difference in the circulating levels of caffeine between fast and slow metabolizers to have a significant impact on the ergogenicity of caffeine.

It has been suggested that fast metabolizers may receive a greater ergogenic effect from caffeine due to the faster metabolism of caffeine into the metabolites, theophylline and paraxanthine [37]. While it has been shown that the concentrations of the metabolites are likely too small to have an ergogenic effect [67], it should be noted that paraxanthine and theophylline are both adenosine antagonists as well [9]. Although the concentrations of paraxanthine and theophylline may be too small in isolation, together with caffeine, they may still provide a noticeable ergogenic effect. The effects of the CYP1A2 genotype on caffeine ergogenicity may be dependent on the duration of exercise, where longer duration exercise (>1 h) may be more suited to slow metabolizers and short-term high intensity exercise may be more suited to faster metabolizers of caffeine. However, this may be highly dependent on the difference between metabolism rates which currently have not been thoroughly explored and reported. In a letter to the editor, Pickering [68] echoed similar sentiments and stated that while several recent studies have found C allele carriers were not reported to have an ergogenic effect from caffeine, it may be due to the time at which caffeine was ingested, advocating for an earlier time of caffeine ingestion to test this hypothesis. Furthermore, studies should record and report caffeine pharmacokinetics before, during and after exercise when investigating CYP1A2 genotypes on caffeine ergogenicity in order to determine the magnitude of the effect CYP1A2 has on the metabolism of caffeine.

## 3. The Effect of ADORA2A on Inter-Individual Differences in Ergogenicity of Caffeine

The ADORA2A (C→T) gene encodes for the adenosine receptor A2A found predominantly in the brain and has a role in the down-regulation of dopamine and glutamate release [16,17]. Genotype distributions are varied between studies, however, approximately 45% of people carry the C/T allele while the T/T and C/C carriers range between 20–30% [69,70].

A number of studies have investigated the effects of ADORA2A gene variation on responses to caffeine ingestion [69,71–75]. The studies by Alsene et al. [75] and Childs et al. [74] both reported greater increases in self-reported caffeine-induced anxiety in individuals with the T/T allele compared to those with the C alleles. However, Retey et al., [71] reported a greater proportion of C/C genotype in individuals who rated themselves as caffeine "sensitive", and a greater prevalence of T/T genotype in individuals who rated themselves as caffeine "insensitive". The ADORA2A genotype may also affect habitual caffeine intake, as a study reported ADORA2A knockout mice self-administered less caffeinated solution compared to wild-type mice [76]. This suggests that ADORA2A may have a regulating role in the appetitive properties of caffeine, which is likely to influence habitual caffeine intake. A later study in humans [69] reported that CYP1A2 genotype had no effect on caffeine intake, however, individuals with the ADORA2A T/T allele had lower habitual caffeine consumption compared with the C allele carriers. This would suggest that individuals with the T/T genotype may have reduced habitual self-administered caffeine due to negative feedback. This supports the work of others [74,75] who reported greater caffeine-induced anxiety in T/T genotypes compared to C/C genotypes. A recent study [73] reported an increase in caffeine-induced anxiety in individuals with the T/T genotypes but not the C/T and C/C genotypes [73]. However, these results were mediated by habitual caffeine consumption, as caffeine-induced anxiety was greater in low and non-users of caffeine compared to medium and high users of caffeine. Caffeine-induced anxiety and habitual caffeine consumption are important factors which play a role in the ergogenic effects of caffeine, both of which have been associated with ADORA2A polymorphisms. Individuals with high or very low caffeine-induced anxiety, due to ADORA2A genotype may experience ergolytic or no ergogenic effects from caffeine consumption, respectively. Similarly, ADORA2A may influence habitual caffeine consumption which may attenuate the ergogenic effect of caffeine which is important for caffeine supplementation in days prior to competitions.

Homodimerisation of ADORA1A and ADORA2A has also been suggested to affect caffeine concentration as the binding of caffeine to the ADORA1A and ADORA2A receptors increases the affinity for a second caffeine molecule to bind to the ADORA1A dimer [77]. The increased uptake in caffeine by the adenosine receptors potentially leads to a decrease in local caffeine concentration and may partially explain the biphasic effects of caffeine on locomotors activation. After caffeine has binded to the adenosine receptor dimer it increases the affinity of adenosine and caffeine molecules to bind to the remaining receptor. Therefore, at low doses of caffeine ($\sim$<2 mg·kg$^{-1}$) there is not sufficient caffeine to bind to both adenosine receptors of the dimer, thus the effects of caffeine are attenuated due to adenosine binding to the remaining receptor site. However, at high doses of caffeine, there is sufficient caffeine that both adenosine receptor sites can be occupied by caffeine rather than a caffeine and adenosine molecule.

To the authors' knowledge, only one study [78] has investigated the effects of ADORA2A genotype on exercise performance following caffeine ingestion. Loy et al. [78] examined the effects of ADORA2A genotype on caffeine ergogenicity during a 30 min cycle in a randomized cross-over design. Twelve females performed 20 min of moderate-intensity cycling followed by a maximal 10 min cycle time-trial. Results of the 10 min time-trial revealed only one individual with the C/T and C/C group ($n$ = 6) improved time-trial cycle performance following caffeine ingestion, whereas all participants in the T/T group (*n* = 6) showed improvements following caffeine ingestion. This is interesting as the T/T genotype has previously been associated with increased anxiety following caffeine ingestion [73–75]. While anxiety is generally thought to be ergolytic, it has not been shown to have a strong relationship with exercise performance [79]. Therefore, individuals with the ADORA2A T/T genotype may be more sensitive to the effects of caffeine as caffeine-induced anxiety may also be perceived as increasing arousal leading to potential ergogenic effects. Further research should be conducted using similar protocols to verify the results found by Loy et al. [78]. Moreover, different exercise modalities, as well as the use of perceptual and mood measures, should be utilised to determine the effects of ADORA2A genotype on the ergogenicity of caffeine during various forms of sports and exercise and any potential contributing mechanisms.

While CYP1A2 and ADORA2A remain the most researched genes with regards to the ergogenicity of caffeine, AHR (aryl hydrocarbon receptor) has been shown to affect caffeine metabolism through detecting polycyclic hydrocarbons such as those found in roasted coffee and induces the transcription of CYP1A2 in the liver [69,72,80,81]. While no studies have investigated the effects of AHR on the ergogenicity of caffeine, studies investigating the effects of CYP1A2 on the ergogenicity of caffeine should include AHR genotyping to elucidate the role it may have in caffeine metabolism. Together AHR, CYP1A2 and ADORA2A have all been associated with caffeine consumption. The variability of CYP1A2 and AHR has been associated with a 42% increase in coffee intake and a cooperative action between these two genes may exist for moderating caffeine consumption [82]. This adds support to the hypothesis that "individuals adjust their dietary caffeine consumption to maintain biological exposure levels of caffeine that elicit optimal stimulant effects" [83]. ADORA1A has also been reported to be largely responsible for the anxiogenic effects of caffeine as well as influencing the disruptions to sleep following caffeine ingestion [73]. Thus, ADORA1A could be another potential gene to investigate with caffeine supplementation, particularly for its disruptions to sleep and increased anxiety prior to and during competitions [84].

ADORA2A and CYP1A2 may also lead to unique interaction effects in response to caffeine ingestion. For example, an individual with the ADORA2A T/T and CYP1A2 C/C genotypes might have a greater ergogenic effect from caffeine when competing in a longer duration event (>1 h) as the slow metabolism may be beneficial in maintaining biologically active levels of caffeine in the body. Furthermore, an individual with ADORA2A T/T and CYP1A2 A/A may have a greater ergogenic benefit during a high-intensity, short-duration activity (such as a sprinting) through increased arousal and not requiring high levels of circulating caffeine for long periods of time. Therefore, individuals may require different levels of circulating caffeine to receive an equitable effect based on their genetics. Similarly, individuals will require more or less frequent consumption of caffeine to maintain their optimal circulating level of caffeine and respective ergogenic effect based on their genetics. However, factors such as habitual caffeine consumption, additional caffeine intake during the event, and individual reactions to caffeine may yet further influence the overall ergogenic effect. Further research is needed to examine and quantify the effect genetics may have on the ergogenicity of caffeine, and potentially identify any other genes associated with responses to caffeine ingestion.

## 4. Limitations and Future Considerations

The key limitation to this area of research is the relatively few number of studies investigating the effects of CYP1A2 and ADORA2A genotype on the ergogenicity of caffeine. There is a greater need to verify results of published studies rather than adding novel elements to the very small pool of studies currently available. This will be more useful to the end users such as coaches and athletes who need to know what the ergogenic effects of caffeine are likely to be, based on their particular combination of CYP1A2, AHR, ADORA2A, and ADORA1A genotypes. Therefore, studies should test participants for ADORA2A, ADORA1A, AHR and CYP1A2 genotypes as the combination of these genes may partly explain the equivocal results when investigating the ergogenicity of caffeine. Additionally, future studies should record caffeine and caffeine metabolite pharmacokinetics when administering caffeine and testing for CYP1A2 genotypes to gain a greater understanding of the differences between fast and slow metabolizers. This would also provide evidence for differences in caffeine metabolism for CYP1A2 genotypes which is the main premise for studies investigating the effects of CYP1A2 on the ergogenicity of caffeine. Studies investigating the effects of CYP1A2 on the ergogenicity of caffeine and caffeine metabolism should use longer duration protocols which last more than 1 h to ensure that differences between fast and slow metabolizers can more easily be determined. Currently, it cannot be stated conclusively whether fast or slow metabolizers perform better following caffeine ingestion because their different metabolism rates have not been measured and may in fact be similar between these individuals.

## 5. Conclusions

It is clear that various factors affect the ergogenic effects of caffeine and contribute to inter-individual variability in response to caffeine ingestion. Even with limited research, it appears that genetics plays a key role. Future research should further investigate which genes may affect the ergogenicity of caffeine as well as the mechanisms by which it is achieved. This would enable practitioners and coaches to tailor individualised caffeine supplementation regimes for athletes to achieve the maximum possible ergogenic effect in their specific sport.

**Author Contributions:** Conceptualization, K.S., A.A., K.R.-M. and C.B.; Writing, K.S.; Writing-review and proof-reading, A.A., K.R.-M. and C.B.

**Conflicts of Interest:** The authors declare no conflict of interest.

## References

1. Del Coso, J.; Muñoz, G.; Muñoz-Guerra, J. Prevalence of caffeine use in elite athletes following its removal from the world anti-doping agency list of banned substances. *Appl. Physiol. Nutr. Metab.* **2011**, *36*, 555–561. [CrossRef] [PubMed]
2. Rivers, W.H.R.; Webber, H.N. The action of caffeine on the capacity for muscular work. *J. Physiol.* **1907**, *36*, 33–47. [CrossRef] [PubMed]
3. Ivy, J.L.; Costill, D.L.; Fink, W.J.; Lower, R.W. Influence of caffeine and carbohydrate feedings on endurance performance. *Med. Sci. Sports* **1979**, *11*, 6–11. [CrossRef] [PubMed]
4. Costill, D.L.; Dalsky, G.P.; Fink, W.J. Effects of caffeine ingestion on metabolism and exercise performance. *Med. Sci. Sports* **1978**, *10*, 155–158. [PubMed]
5. Van Soeren, M.H.; Sathasivam, P.; Spriet, L.L.; Graham, T.E. Caffeine metabolism and epinephrine responses during exercise in users and nonusers. *J. Appl. Physiol.* **1993**, *75*, 805–812. [CrossRef] [PubMed]
6. Jackman, M.; Wendling, P.; Friars, D.; Graham, T.E. Metabolic, catecholamine, and endurance responses to caffeine during intense exercise. *J. Appl. Physiol.* **1996**, *81*, 1658–1663. [CrossRef] [PubMed]
7. Van Soeren, M.H.; Graham, T.E. Effect of caffeine on metabolism, exercise endurance, and catecholamine responses after withdrawal. *J. Appl. Physiol.* **1998**, *85*, 1493–1501. [CrossRef] [PubMed]
8. Lovallo, W.R.; Farag, N.H.; Vincent, A.S.; Thomas, T.L.; Wilson, M.F. Cortisol responses to mental stress, exercise, and meals following caffeine intake in men and women. *Pharmacol. Biochem. Behav.* **2006**, *83*, 441–447. [CrossRef] [PubMed]
9. Graham, T.E. Caffeine and exercise-Metabolism, endurance and performance. *Sport. Med.* **2001**, *31*, 785–807. [CrossRef]
10. Graham, T.E.; Battram, D.S.; Dela, F.; El-Sohemy, A.; Thong, F.S.L. Does caffeine alter muscle carbohydrate and fat metabolism during exercise? *Appl. Physiol. Nutr. Metab.* **2008**, *33*, 1311–1318. [CrossRef] [PubMed]
11. Graham, T.E.; Helge, J.W.; MacLean, D.A.; Kiens, B.; Richter, E.A. Caffeine ingestion does not alter carbohydrate or fat metabolism in human skeletal muscle during exercise. *J. Physiol.* **2000**, *529*, 837–847. [CrossRef] [PubMed]
12. Spriet, L.L. Caffeine and Performance. *Int. J. Sport Nutr.* **1995**, *5*, S84–S99. [CrossRef] [PubMed]

13. Southward, K.; Rutherfurd-Markwick, K.J.; Ali, A. Correction to: The effect of acute caffeine ingestion on endurance performance: A systematic review and meta-analysis. *Sport. Med.* **2018**, 1–17. [CrossRef] [PubMed]

14. Nehlig, A.; Daval, J.-L.; Debry, G. Caffeine and the central nervous system: Mechanisms of action, biochemical, metabolic and psychostimulant effects. *Brain Res. Rev.* **1992**, *17*, 139–170. [CrossRef]

15. Spriet, L.L. Caffeine and exercise performance. In *Sports Nutrition*; Maughan, R.J., Ed.; Comite International Olympique (International Olympic Committee): Lausanne, Switzerland, 2014.

16. Meeusen, R.; Roelands, B.; Spriet, L.L. Caffeine, exercise and the brain. *Nestle Nutr. Inst. Workshop Ser.* **2013**, *76*, 1–12. [PubMed]

17. Fredholm, B.B. Adenosine, adenosine receptors and the actions of caffeine. *Pharmacol. Toxicol.* **1995**, *76*, 93–101. [CrossRef] [PubMed]

18. Porkka-Heiskanen, T.; Strecker, R.E.; McCarley, R.W. Brain site-specificity of extracellular adenosine concentration changes during sleep deprivation and spontaneous sleep: An in vivo microdialysis study. *Neuroscience* **2000**, *99*, 507–517. [CrossRef]

19. Marshall, J.M. The roles of adenosine and related substances in exercise hyperaemia. *J. Physiol.* **2007**, *583*, 835–845. [CrossRef] [PubMed]

20. Salamone, J.D.; Farrar, A.M.; Font, L.; Patel, V.; Schlar, D.E.; Nunes, E.J.; Collins, L.E.; Sager, T.N. Differential actions of adenosine A1 and A2A antagonists on the effort-related effects of dopamine D2 antagonism. *Behav. Brain Res.* **2009**, *201*, 216–222. [CrossRef] [PubMed]

21. Shearer, J.; Graham, T.E. Performance effects and metabolic consequences of caffeine and caffeinated energy drink consumption on glucose disposal. *Nutr. Rev.* **2014**, *72*, 121–136. [CrossRef] [PubMed]

22. Ganio, M.S.; Klau, J.F.; Casa, D.J.; Armstrong, L.E.; Maresh, C.M. Effect of caffeine on sport-specific endurance performance: A systematic review. *J. Strength Cond. Res.* **2009**, *23*, 315–324. [CrossRef] [PubMed]

23. Souza, D.B.; Del Coso, J.; Casonatto, J.; Polito, M.D. Acute effects of caffeine-containing energy drinks on physical performance: A systematic review and meta-analysis. *Eur. J. Nutr.* **2017**, *56*, 13–27. [CrossRef] [PubMed]

24. Doherty, M.; Smith, P.M.; Davison, R.C.R.; Hughes, M.G. Caffeine is ergogenic after supplementation of oral creatine monohydrate. *Med. Sci. Sports Exerc.* **2002**, *34*, 1785–1792. [CrossRef] [PubMed]

25. Pickering, C.; Kiely, J. Are the current guidelines on caffeine use in sport optimal for everyone? Inter-individual variation in caffeine ergogenicity, and a move towards personalised sports nutrition. *Sport. Med.* **2018**, *48*, 7–16. [CrossRef] [PubMed]

26. Acker-Hewitt, T.L.; Shafer, B.M.; Saunders, M.J.; Goh, Q.; Luden, N.D. Independent and combined effects of carbohydrate and caffeine ingestion on aerobic cycling performance in the fed state. *Appl. Physiol. Nutr. Metab.* **2012**, *37*, 276–283. [CrossRef] [PubMed]

27. Astorino, T.A.; Cottrell, T.; Lozano, A.T.; Aburto-Pratt, K.; Duhon, J. Ergogenic effects of caffeine on simulated time-trial performance are independent of fitness level. *J. Caffeine Res.* **2011**, *1*, 179–185. [CrossRef]

28. Astorino, T.A.; Roupoli, L.R.; Valdivieso, B.R. Caffeine does not alter RPE or pain perception during intense exercise in active women. *Appetite* **2012**, *59*, 585–590. [CrossRef] [PubMed]

29. Beaumont, R.E.; James, L.J. Effect of a moderate caffeine dose on endurance cycle performance and thermoregulation during prolonged exercise in the heat. *J. Sci. Med. Sport* **2016**, *20*, 1024–1028. [CrossRef] [PubMed]

30. Christensen, P.M.; Petersen, M.H.; Friis, S.N.; Bangsbo, J. Caffeine, but not bicarbonate, improves 6 min maximal performance in elite rowers. *Appl. Physiol. Nutr. Metab.* **2014**, *39*, 1058–1063. [CrossRef] [PubMed]

31. Church, D.D.; Hoffman, J.R.; LaMonica, M.B.; Riffe, J.J.; Hoffman, M.W.; Baker, K.M.; Varanoske, A.N.; Wells, A.J.; Fukuda, D.H.; Stout, J.R. The effect of an acute ingestion of Turkish coffee on reaction time and time trial performance. *J. Int. Soc. Sports Nutr.* **2015**, *12*, 37. [CrossRef] [PubMed]

32. Desbrow, B.; Barrett, C.M.; Minahan, C.L.; Grant, G.D.; Leveritt, M.D.; Coker, R.H. Caffeine, cycling performance and exogenous, CHO oxidation: A dose-response study-Comment. *Med. Sci. Sports Exerc.* **2009**, *41*, 1744–1751. [CrossRef] [PubMed]

33. Desbrow, B.; Biddulph, C.; Devlin, B.; Grant, G.D.; Anoopkumar-Dukie, S.; Leveritt, M.D. The effects of different doses of caffeine on endurance cycling time trial performance. *J. Sports Sci.* **2012**, *30*, 115–120. [CrossRef] [PubMed]

34. Gonçalves, L.S.; Painelli, V.S.; Yamaguchi, G.; de Oliveira, L.F.; Saunders, B.; da Silva, R.P.; Maciel, E.; Artioli, G.G.; Roschel, H.; Gualano, B.; et al. Dispelling the myth that habitual caffeine consumption influences the performance response to acute caffeine supplementation. *J. Appl. Physiol.* **2017**. [CrossRef] [PubMed]

35. Graham-Paulson, T.S.; Perret, C.; Watson, P.; Goosey-Tolfrey, V.L. Improvement of sprint performance in wheelchair sportsmen with caffeine supplementation. *Int. J. Sports Physiol. Perform.* **2016**, *11*, 214–220. [CrossRef] [PubMed]

36. Guest, N.; Corey, P.; Vescovi, J.; El-Sohemy, A. Caffeine, *CYP1A2* genotype, and endurance performance in athletes. *Med. Sci. Sports Exerc.* **2018**. [CrossRef] [PubMed]

37. O'Rourke, M.P.; O'Brien, B.J.; Knez, W.; Paton, C.D. Caffeine has a small effect on 5-km running performance of well-trained and recreational runners. *J. Sci. Med. Sport* **2008**, *11*, 231–233. [CrossRef] [PubMed]

38. Pitchford, N.W.; Fell, J.W.; Leveritt, M.D.; Desbrow, B.; Shing, C.M. Effect of caffeine on cycling time-trial performance in the heat. *J. Sci. Med. Sport* **2014**, *17*, 445–449. [CrossRef] [PubMed]

39. Roelands, B.; Buyse, L.; Pauwels, F.; Delbeke, F.; Deventer, K.; Meeusen, R. No effect of caffeine on exercise performance in high ambient temperature. *Eur. J. Appl. Physiol.* **2011**, *111*, 3089–3095. [CrossRef] [PubMed]

40. De Alcantara Santos, R.; Peduti Dal Molin Kiss, M.A.; Silva-Cavalcante, M.D.; Correia-Oliveira, C.R.; Bertuzzi, R.; Bishop, D.J.; Lima-Silva, A.E.; Kiss, M.A.P.D.M.; Silva-Cavalcante, M.D.; Correia-Oliveira, C.R.; et al. Caffeine alters anaerobic distribution and pacing during a 4000-m cycling time trial. *PLoS ONE* **2013**, *8*, e75399.

41. Skinner, T.L.; Jenkins, D.G.; Taaffe, D.R.; Leveritt, M.D.; Coombes, J.S. Coinciding exercise with peak serum caffeine does not improve cycling performance. *J. Sci. Med. Sport* **2013**, *16*, 54–59. [CrossRef] [PubMed]

42. Stadheim, H.K.; Kvamme, B.; Olsen, R.; Drevon, C.A.; Ivy, J.L.; Jensen, J. Caffeine increases performance in cross-country double-poling time trial exercise. *Med. Sci. Sport. Exerc.* **2013**, *45*, 2175–2183. [CrossRef] [PubMed]

43. Stadheim, H.K.; Nossum, E.M.; Olsen, R.; Spencer, M.; Jensen, J. Caffeine improves performance in double poling during acute exposure to 2000-m altitude. *J. Appl. Physiol.* **2015**, *119*, 1501–1509. [CrossRef] [PubMed]

44. Womack, C.J.; Saunders, M.J.; Bechtel, M.K.; Bolton, D.J.; Martin, M.; Luden, N.D.; Dunham, W.; Hancock, M. The influence of a CYP1A2 polymorphism on the ergogenic effects of caffeine. *J. Int. Soc. Sports Nutr.* **2012**, *9*, 7. [CrossRef] [PubMed]

45. Bell, D.G.; McLellan, T.M. Exercise endurance 1, 3, and 6 h after caffeine ingestion in caffeine users and nonusers. *J. Appl. Physiol.* **2002**, *93*, 1227–1234. [CrossRef] [PubMed]

46. Sachse, C.; Brockmöller, J.; Bauer, S.; Roots, I. Functional significance of a C→A polymorphism in intron 1 of the cytochrome P450 *CYP1A2* gene tested with caffeine. *Br. J. Clin. Pharmacol.* **1999**, *47*, 445–449. [CrossRef] [PubMed]

47. Astorino, T.A.; Roberson, D.W. Efficacy of acute caffeine ingestion for short-term high-intensity exercise performance: A systematic review. *J. Strength Cond. Res.* **2010**, *24*, 257–265. [CrossRef] [PubMed]

48. Talanian, J.L.; Spriet, L.L. Low and moderate doses of caffeine late in exercise improve performance in trained cyclists. *Appl. Physiol. Nutr. Metab.* **2016**, *41*, 850–855. [CrossRef] [PubMed]

49. Collomp, K.; Ahmaidi, S.; Chatard, J.C.; Audran, M.; Préfaut, C. Benefits of caffeine ingestion on sprint performance in trained and untrained swimmers. *Eur. J. Appl. Physiol. Occup. Physiol.* **1992**, *64*, 377–380. [CrossRef] [PubMed]

50. Boyett, J.C.; Giersch, G.E.W.; Womack, C.J.; Saunders, M.J.; Hughey, C.A.; Daley, H.M.; Luden, N.D. Time of day and training status both impact the efficacy of caffeine for short duration cycling performance. *Nutrients* **2016**, *8*, 639. [CrossRef] [PubMed]

51. Yang, A.; Palmer, A.A.; de Wit, H. Genetics of caffeine consumption and responses to caffeine. *Psychopharmacology* **2010**, *211*, 245–257. [CrossRef] [PubMed]

52. Guengerich, F.P. Cytochrome P450 and chemical toxicology. *Chem. Res. Toxicol.* **2008**, *21*, 70–83. [CrossRef] [PubMed]

53. Koonrungsesomboon, N.; Khatsri, R.; Wongchompoo, P.; Teekachunhatean, S. The impact of genetic polymorphisms on *CYP1A2* activity in humans: A systematic review and meta-analysis. *Pharmacogenomics J.* **2017**. [CrossRef] [PubMed]

54. Gu, L.; Gonzalez, F.J.; Kalow, W.; Tang, B.K. Biotransformation of caffeine, paraxanthine, theobromine and theophylline by cDNA-expressed human *CYP1A2* and *CYP2E1*. *Pharmacogenetics* **1992**, *2*, 73–77. [CrossRef] [PubMed]

55.  Perera, V.; S. Gross, A.; J. McLachlan, A. Current drug metabolism. In *Measurement of CYP1A2 Activity: A Focus on Caffeine as a Probe*; Bentham Science Publishers: Sharjah, UAE, 2012; Volume 13, pp. 667–678.

56.  Lelo, A.; Birkett, D.; Robson, R.; Miners, J. Comparative pharmacokinetics of caffeine and its primary demethylated metabolites paraxanthine, theobromine and theophylline in man. *Br. J. Clin. Pharmacol.* **1986**, *22*, 177–182. [CrossRef] [PubMed]

57.  Algrain, H.A.; Thomas, R.M.; Carrillo, A.E.; Ryan, E.J.; Kim, C.-H.; Lettan, R.B.; Ryan, E.J. The effects of a polymorphism in the cytochrome P450 *CYP1A2* gene on performance enhancement with caffeine in recreational cyclists. *J. Caffeine Res.* **2016**, *6*, 34–39. [CrossRef]

58.  Giersch, G.E.W.; Boyett, J.C.; Hargens, T.A.; Luden, N.D.; Saunders, M.J.; Daley, H.; Hughey, C.A.; El-Sohemy, A.; Womack, C.J. The effect of the *CYP1A2*-163 C > A polymorphism on caffeine metabolism and subsequent cycling performance. *J. Caffeine Adenosine Res.* **2018**, *8*, 65–70. [CrossRef]

59.  Pataky, M.W.; Womack, C.J.; Saunders, M.J.; Goffe, J.L.; D'Lugos, A.C.; El-Sohemy, A.; Luden, N.D. Caffeine and 3-km cycling performance: Effects of mouth rinsing, genotype, and time of day. *Scand. J. Med. Sci. Sport.* **2016**, *26*, 613–619. [CrossRef] [PubMed]

60.  Klein, C.S.; Clawson, A.; Martin, M.; Saunders, M.J.; Flohr, J.A.; Bechtel, M.K.; Dunham, W.; Hancock, M.; Womack, C.J. The effect of caffeine on performance in collegiate tennis players. *J. Caffeine Res.* **2012**, *2*, 111–116. [CrossRef]

61.  Salinero, J.; Lara, B.; Ruiz-Vicente, D.; Areces, F.; Puente-Torres, C.; Gallo-Salazar, C.; Pascual, T.; Del Coso, J. *CYP1A2* genotype variations do not modify the benefits and drawbacks of caffeine during exercise: A pilot study. *Nutrients* **2017**, *9*, 269. [CrossRef] [PubMed]

62.  Puente, C.; Abián-Vicén, J.; Del Coso, J.; Lara, B.; Salinero, J.J. The *CYP1A2*-163 C > A polymorphism does not alter the effects of caffeine on basketball performance. *PLoS ONE* **2018**, *13*, e0195943. [CrossRef] [PubMed]

63.  Doherty, M.; Smith, P.M. Effects of caffeine ingestion on exercise testing: A meta-analysis. *Int. J. Sport Nutr. Exerc. Metab.* **2004**, *14*, 626–646. [CrossRef] [PubMed]

64.  Kamimori, G.H.; Karyekar, C.S.; Otterstetter, R.; Cox, D.S.; Balkin, T.J.; Belenky, G.L.; Eddington, N.D. The rate of absorption and relative bioavailability of caffeine administered in chewing gum versus capsules to normal healthy volunteers. *Int. J. Pharm.* **2002**, *234*, 159–167. [CrossRef]

65.  Gunes, A.; Dahl, M.-L. Variation in *CYP1A2* activity and its clinical implications: Influence of environmental factors and genetic polymorphisms. *Pharmacogenomics* **2008**, *9*, 625–637. [CrossRef] [PubMed]

66.  Benowitz, N.L. Clinical Pharmacology of Caffeine. *Annu. Rev. Med.* **1990**, *41*, 277–288. [CrossRef] [PubMed]

67.  Magkos, F.; Kavouras, S.A. Caffeine use in sports, pharmacokinetics in man, and cellular mechanisms of action. *Crit. Rev. Food Sci. Nutr.* **2005**, *45*, 535–562. [CrossRef] [PubMed]

68.  Pickering, C. Caffeine, *CYP1A2* genotype, and sports performance: Is timing important? *Ir. J. Med. Sci.* **2018**. [CrossRef] [PubMed]

69.  Cornelis, M.C.; El-Sohemy, A.; Campos, H. Genetic polymorphism of the adenosine A2A receptor is associated with habitual caffeine consumption. *Am. J. Clin. Nutr.* **2007**, *86*, 240–244. [CrossRef] [PubMed]

70.  Dhaenens, C.-M.; Burnouf, S.; Simonin, C.; Van Brussel, E.; Duhamel, A.; Defebvre, L.; Duru, C.; Vuillaume, I.; Cazeneuve, C.; Charles, P.; et al. A genetic variation in the *ADORA2A* gene modifies age at onset in Huntington's disease. *Neurobiol. Dis.* **2009**, *35*, 474–476. [CrossRef] [PubMed]

71.  Rétey, J.V.; Adam, M.; Khatami, R.; Luhmann, U.F.O.; Jung, H.H.; Berger, W.; Landolt, H.-P. A genetic variation in the adenosine $A_{2A}$ receptor gene (*ADORA2A*) contributes to individual sensitivity to caffeine effects on sleep. *Clin. Pharmacol. Ther.* **2007**, *81*, 692–698. [CrossRef] [PubMed]

72.  Cornelis, M.C.; Monda, K.L.; Yu, K.; Paynter, N.; Azzato, E.M.; Bennett, S.N.; Berndt, S.I.; Boerwinkle, E.; Chanock, S.; Chatterjee, N.; et al. Genome-wide meta-analysis identifies regions on 7p21 (*AHR*) and 15q24 (*CYP1A2*) as determinants of habitual caffeine consumption. *PLoS Genet.* **2011**, *7*, e1002033. [CrossRef] [PubMed]

73.  Rogers, P.J.; Hohoff, C.; Heatherley, S.V.; Mullings, E.L.; Maxfield, P.J.; Evershed, R.P.; Deckert, J.; Nutt, D.J. Association of the anxiogenic and alerting effects of caffeine with *ADORA2A* and *ADORA1* polymorphisms and habitual level of caffeine consumption. *Neuropsychopharmacology* **2010**, *35*, 1973–1983. [CrossRef] [PubMed]

74.  Childs, E.; Hohoff, C.; Deckert, J.; Xu, K.; Badner, J.; de Wit, H. Association between *ADORA2A* and *DRD2* polymorphisms and caffeine-induced anxiety. *Neuropsychopharmacology* **2008**, *33*, 2791–2800. [CrossRef] [PubMed]

75. Alsene, K.; Deckert, J.; Sand, P.; de Wit, H. Association between A$_{2a}$ receptor gene polymorphisms and caffeine-induced anxiety. *Neuropsychopharmacology* **2003**, *28*, 1694–1702. [CrossRef] [PubMed]

76. El Yacoubi, M.; Ledent, C.; Parmentier, M.; Costentin, J.; Vaugeois, J.-M. Reduced appetite for caffeine in adenosine A$_{2A}$ receptor knockout mice. *Eur. J. Pharmacol.* **2005**, *519*, 290–291. [CrossRef] [PubMed]

77. Gracia, E.; Moreno, E.; Cortés, A.; Lluís, C.; Mallol, J.; McCormick, P.J.; Canela, E.I.; Casadó, V. Homodimerization of adenosine A1 receptors in brain cortex explains the biphasic effects of caffeine. *Neuropharmacology* **2013**, *71*, 56–69. [CrossRef] [PubMed]

78. Loy, B.D.; O'Connor, P.J.; Lindheimer, J.B.; Covert, S.F. Caffeine is ergogenic for adenosine A$_{2A}$ receptor gene (*ADORA2A*) T allele homozygotes: A pilot study. *J. Caffeine Res.* **2015**, *5*, 73–81. [CrossRef]

79. Craft, L.L.; Magyar, T.M.; Becker, B.J.; Feltz, D.L. The relationship between the competitive state anxiety inventory-2 and sport performance: A meta-analysis. *J. Sport Exerc. Psychol.* **2003**, *25*, 44–65. [CrossRef]

80. Sulem, P.; Gudbjartsson, D.F.; Geller, F.; Prokopenko, I.; Feenstra, B.; Aben, K.K.H.; Franke, B.; den Heijer, M.; Kovacs, P.; Stumvoll, M.; et al. Sequence variants at *CYP1A1–CYP1A2* and *AHR* associate with coffee consumption. *Hum. Mol. Genet.* **2011**, *20*, 2071–2077. [CrossRef] [PubMed]

81. Nehlig, A. Interindividual differences in caffeine metabolism and factors driving caffeine consumption. *Pharmacol. Rev.* **2018**, *70*, 384–411. [CrossRef] [PubMed]

82. Nordestgaard, A.T.; Nordestgaard, B.G. Coffee intake, cardiovascular disease and all-cause mortality: Observational and Mendelian randomization analyses in 95,000–223,000 individuals. *Int. J. Epidemiol.* **2016**, *45*, 1938–1952. [CrossRef] [PubMed]

83. Cornelis, M.C.; Kacprowski, T.; Menni, C.; Gustafsson, S.; Pivin, E.; Adamski, J.; Artati, A.; Eap, C.B.; Ehret, G.; Friedrich, N.; et al. Genome-wide association study of caffeine metabolites provides new insights to caffeine metabolism and dietary caffeine-consumption behavior. *Hum. Mol. Genet.* **2016**, *25*, 5472–5482. [CrossRef] [PubMed]

84. Wei, C.J.; Li, W.; Chen, J.-F. Normal and abnormal functions of adenosine receptors in the central nervous system revealed by genetic knockout studies. *Biochim. Biophys. Acta Biomembr.* **2011**, *1808*, 1358–1379. [CrossRef] [PubMed]

nutrients

MDPI

*Review*

# Mendelian Randomization Studies of Coffee and Caffeine Consumption

**Marilyn C. Cornelis [1],\* and Marcus R. Munafo [2]**

[1] Department of Preventive Medicine, Northwestern University Feinberg School of Medicine, Chicago, IL 60611, USA

[2] MRC Integrative Epidemiology Unit (IEU) at the University of Bristol, UK Centre for Tobacco and Alcohol Studies, School of Psychological Science, University of Bristol, Bristol BS8 1TU, UK; marcus.munafo@bristol.ac.uk

\* Correspondence: marilyn.cornelis@northwestern.edu; Tel.: +1-312-503-4548

Received: 31 August 2018; Accepted: 17 September 2018; Published: 20 September 2018

**Abstract:** Habitual coffee and caffeine consumption has been reported to be associated with numerous health outcomes. This perspective focuses on Mendelian Randomization (MR) approaches for determining whether such associations are causal. Genetic instruments for coffee and caffeine consumption are described, along with key concepts of MR and particular challenges when applying this approach to studies of coffee and caffeine. To date, at least fifteen MR studies have investigated the causal role of coffee or caffeine use on risk of type 2 diabetes, cardiovascular disease, Alzheimer's disease, Parkinson's disease, gout, osteoarthritis, cancers, sleep disturbances and other substance use. Most studies provide no consistent support for a causal role of coffee or caffeine on these health outcomes. Common study limitations include low statistical power, potential pleiotropy, and risk of collider bias. As a result, in many cases a causal role cannot confidently be ruled out. Conceptual challenges also arise from the different aspects of coffee and caffeine use captured by current genetic instruments. Nevertheless, with continued genome-wide searches for coffee and caffeine related loci along with advanced statistical methods and MR designs, MR promises to be a valuable approach to understanding the causal impact that coffee and caffeine have in human health.

**Keywords:** Mendelian Randomization; coffee; caffeine; behavior; causality; genetic epidemiology; epidemiological methods

## 1. Introduction

Coffee is one of the most widely consumed beverages in the world. Consumption patterns vary by country with larger per capita consumptions reported for Nordic countries, such as Finland (12.2 kg), Sweden (10.1 kg) and Norway (8.7 kg) compared to other countries such Brazil (5.9 kg), Netherlands (5.3 kg), USA (4.5 kg), Australia (4.0 kg), Russia (1.7 kg), China (0.8 kg) and Turkey (0.7 kg) [1]. For most populations, regular coffee is the primary dietary source of caffeine; a psychostimulant also present in tea, cola, and cocoa products. Absorption and exposure to caffeine from these different sources is similar although a slight delay in absorption has been reported for cola and chocolate [2–4]. Roasted coffee also contains unique polyphenols (i.e., chlorogenic acid) and melanoidins that are major contributors to antioxidants in diet [5,6]. Boiled or unfiltered coffee contains diterpenoids, including cafestol and kahweol [7]. Trigonelline, magnesium, potassium, niacin, lignans, as well as heterocyclic amines and acrylamide have also been detected in the beverage [8–12]. With widespread popularity and availability of coffee, there is increasing public and scientific interest in the potential health consequences of its regular consumption. Traditional epidemiology has been fundamental to our increased knowledge on habitual coffee intake and health; but while a highly efficient and relevant approach, it has several limitations that warrant consideration when interpreting the results [13].

Among these is establishing causal associations. The current perspective focuses on Mendelian Randomization (MR) approaches for determining a causal role of habitual coffee and caffeine intake on health. Because coffee and dietary caffeine intake are highly correlated we focus on both exposures. We first provide a brief review of coffee, caffeine and health. We follow with key concepts of the MR approach and particular challenges when applying it to studies of coffee and caffeine. Recent MR studies of coffee, caffeine and health are discussed, and we conclude with future directions for the field.

## 2. Coffee, Dietary Caffeine and Health

A recent umbrella review considered data from 201 meta-analysis of epidemiological studies of 67 unique health outcomes, and concluded that coffee likely has a beneficial role in reducing risk of type 2 diabetes (T2D), cardiovascular diseases (CVD), several cancers and Parkinson's disease (PD), but that high caffeine intake is likely harmful on pregnancy outcomes, such as low birth weight and pregnancy loss [14]. Overall, coffee consumption seems generally safe within usual levels of intake (i.e., at 3 to 4 cups a day) and more likely to benefit health than harm [14]. Rigorous reviews of caffeine toxicity conclude that consumption of up to 400 mg caffeine/day (equivalent to ~4 cups of coffee) in healthy adults, or 2.5 mg/kg/day for children and adolescents is not associated with overt adverse effects [15] and thus generally support the overall findings on habitual coffee intake and health [14]. Meanwhile, the Diagnostic and Statistical Manual of Mental Disorders (DSM-V) lists caffeine intoxication and withdrawal as disorders, and have added 'caffeine use disorder' to 'Conditions for Further Study' [16]. Much of our knowledge pertaining to habitual coffee and caffeine intake on risk of chronic disease has been limited to observational research [14,15]. Inferring causality from observational data is difficult, due to potential residual confounding and reverse causality [17]. For example, in some populations coffee consumption is highly correlated with disease risk factors, such as smoking. Participants might acknowledge their true coffee behavior, but underreport their smoking behavior. As a consequence the coffee intake variable will continue to convey information about smoking even after adjustment for measures of smoking [18]. Coffee drinkers may also have reduced their coffee intake in light of disease symptoms or diagnosis, which might result in an apparent, but non-causal protective association between coffee and the disease [19]. Observational studies also provide no insight to mechanisms linking coffee to health. Coffee contains caffeine, but also hundreds of other chemicals that might benefit or harm health via different biological pathways [9]. Randomized trials of coffee consumption and disease outcomes would require long-term adherence to high or no coffee consumption, which is challenging given strong coffee consumption habits [20].

## 3. Mendelian Randomization (MR)

MR is a method of using the association of variation in genes with biomarkers or modifiable exposures to examine the causal effect of these biomarkers and exposures on disease outcomes in observational studies. The underlying principle of MR is that if a genetic variant alters the level of an exposure of interest, then this genetic variant should also be associated with disease risk and to the extent predicted by the effect of the genetic variant on the exposure [21,22]. According to Mendel's Law of Inheritance, alleles segregate randomly from parents to offspring. Thus, offspring genotypes are unlikely to be associated with confounders in the population. Moreover, germ-line genotypes are fixed at conception and so precede the observed variables, avoiding issues of reverse causation [23]. MR studies are often described as natural RCTs, but there are important differences [24]. For example, RCTs are usually of short duration while an individual's genetics generally reflect life-long exposures [21,24,25].

MR relies on a number of assumptions, in particular that the genetic variants(s): (1) Is associated with the modifiable exposure of interest, (2) is not associated with confounders of the exposure to outcome association and (3) only influences the outcome through the exposure of interest [17]. The first assumption is the only one that can be formally tested, but MR methods and study designs have advanced much over the last few years and now include methods that are robust to potential violations of assumptions (2) and (3). It is increasingly widely used as a causal inference method in epidemiology. One-sample (genetics, exposure and outcome measured in the same sample) and two-sample (exposure and outcome measured in different samples) are the most common MR study designs. The latter is advantageous in situations where it is difficult to measure exposure and outcome in the same sample and can also be performed on publicly available genome-wide association study (GWAS) data (summary-level data). When possible, an instrument (genetic marker of exposure) that combines the effects of many SNPs is used to boost power while also addressing MR assumption violations (see below). The basic method for summary-level data, inverse-variance weighted (IVW), uses a fixed effects meta-analysis approach to combine the Wald ratio estimates of the causal effect (SNP-outcome effect divided by the SNP-exposure effect [26]) obtained from different SNPs, but assumes all SNPs are valid instruments or are invalid in such a way that the overall bias is zero [27,28]. The IVW is generally equivalent to the two-stage least squares estimate commonly used with individual level data.

## 4. Genetic Determinants of Coffee and Caffeine Consumption

Opportunities for MR studies of coffee and health have been made possible by the success of GWAS, which have identified multiple genetic variants associated with self-reported habitual coffee and caffeine consumption (Table 1) [29–33]. Loci near *ADORA2A*, *BDNF* and *SLC6A4* likely act directly on coffee drinking behavior by modulating the acute psychostimulant and rewarding properties of caffeine; driving factors for coffee drinking and caffeine use [34]. However, loci near *AHR*, *CYP1A2*, *POR*, and *ABCG2* generally present with the largest effect sizes and likely impact drinking behavior indirectly by altering the metabolism of caffeine and thus the physiological levels of this compound available for its psychostimulant effects. Only one locus is implicated in the sensory properties of coffee (*OR8U8*). Others have no obvious role in coffee or caffeine consumption, but have previously been associated with other traits in GWAS notably obesity, glucose and lipids [35–38]. GWAS and smaller follow-up studies have linked these loci to consumption of regular coffee, decaffeinated coffee, tea, total caffeine and water [31,39,40]. A subsequent GWAS of circulating caffeine metabolite levels further informed the roles of these loci in coffee and caffeine consumption behavior, but also identified variants near *CYP2A6* associated with paraxanthine-to-caffeine ratio (index for caffeine metabolism), that were nominally associated with drinking behavior [41]. Importantly, genetic variants leading to increased coffee/caffeine consumption associate with lower circulating caffeine levels and higher paraxanthine-to-caffeine ratio suggesting a fast caffeine metabolism phenotype. Thus, many of the loci affecting coffee and caffeine drinking behavior do so by modulating the physiological levels of caffeine.

## 5. Key Challenges to MR Studies of Coffee and Caffeine

Despite progress in the identification of robust genetic variants for coffee and caffeine consumption, efforts to apply these variants to MR studies of coffee and caffeine have been met with challenges, such as trait heterogeneity, pleiotropy and collider bias as discussed below. Limitations in the conduct and interpretation of MR studies more generally, along with potential solutions, have been reviewed in detail elsewhere [23,25,42], and include weak instrument bias, lack of reliable genetic instruments, population stratification, low statistical power (and therefore wide confidence intervals around causal estimates), linkage disequilibrium (LD) and the Winner's Curse phenomenon (i.e., the tendency for effect sizes in initial studies to be inflated).

Table 1. Genetic determinants of coffee and caffeine consumption [29–33].

| Locus (Index SNP, Coffee/Caffeine Increasing Allele) | Closest Gene(s) | Encoded Protein(s): Function [UniProtKb] | Assoc. with Caffeine Metabolites * | Assoc. with Other Traits † | Hypothesized Link to Caffeine or Coffee Consumption |
|---|---|---|---|---|---|
| 1q25.2 (rs574367, T) | SEC16B | SEC16 Homolog B, Endoplasmic Reticulum Export Factor: Required for secretory cargo traffic from the endoplasmic reticulum to the Golgi apparatus and for normal transitional endoplasmic reticulum organization. | $p > 0.05$ | Y | None |
| 2p25.3 (rs10865548, G) | TMEM18 | Transmembrane Protein 18: Transcription repressor. Sequence-specific ssDNA and dsDNA binding protein, with preference for GCT end CTG repeats. Cell migration modulator, which enhances the glioma-specific migration ability of neural stem cells and neural precursor cells. | $p > 0.05$ | Y | None |
| 2p23.3 (rs1260326,C) | GCKR | Glucokinase regulatory protein (GKRP): Inhibits glucokinase by forming an inactive complex with this enzyme. | $\downarrow\downarrow$ $p < 1 \times 10^{-5}$ | Y | Response to caffeine/coffee: May function in the glucose-sensing process of the brain that may influence central pathways responding to caffeine/coffee. Metabolism of caffeine: Inferred by association with caffeine metabolites |
| 4q22 (rs1481012, A) | ABCG2 | ATP-binding cassette sub-family G member 2: High-capacity urate exporter. Plays a role in porphyrin homeostasis and cellular export of hemin and heme. May play an important role in the exclusion of xenobiotics from the brain. Implicated in the efflux of numerous drugs and xenobiotics. | $\uparrow$ $p < 0.05$ | Y | Metabolism of caffeine: Caffeine/metabolite efflux transporter. |
| 7p21 (rs4410790 C, rs6968554, G) | AHR | Aryl hydrocarbon receptor: Ligand-activated transcriptional activator. Activates the expression of multiple phase I and II xenobiotic metabolizing enzymes. Involved in cell-cycle regulation and likely plays a role in the development/maturation of many tissues. | $\downarrow\downarrow$ $p < 5 \times 10^{-8}$ | N | Metabolism of caffeine: Regulates CYP1A2 expression. |
| 7q11.23 (rs7800944, C) | MLXIPL | Carbohydrate-responsive element-binding protein: Transcriptional repressor. | | Y | Response to caffeine/coffee: May regulate transcription of genes (e.g., GCKR) implicated in the response to caffeine. |

**Table 1.** *Cont.*

| Locus (Index SNP, Coffee/Caffeine Increasing Allele) | Closest Gene(s) | Encoded Protein(s): Function [UniProtKb] | Assoc. with Caffeine Metabolites * | Assoc. with Other Traits † | Hypothesized Link to Caffeine or Coffee Consumption |
|---|---|---|---|---|---|
| 7q11.23 (rs17685, A) | POR | NADPH-cytochrome P450 reductase: Required for electron transfer from NADP to cytochrome P450 in microsomes and can also facilitate electron transfer to heme oxygenase and cytochrome B5. | $\downarrow$ $p < 0.05$ | N | Metabolism of caffeine: Required for CYP1A2 catalytic activity. |
| 11p13 (rs6265, C) | BDNF | Brain-derived neurotrophin factor: During development, promotes survival and differentiation of selected neuronal populations of the PNS and CNS. Major regulator of synaptic transmission and plasticity at adult synapses in many regions of the CNS. | $p > 0.05$ | Y | Response to caffeine: Modulates neurotransmitters potentially mediating the rewarding response to caffeine. |
| 11q12.1 (rs597045, A) | OR8U8 | Olfactory Receptor Family 8 Subfamily U Member 8: Odorant receptor | $p > 0.05$ | N | Smell/taste perception of coffee |
| 14q12 (rs1956218, G) | AKAP6 | A-Kinase Anchoring Protein 6: Binds to type II regulatory subunits of protein kinase A and anchors/targets them to the nuclear membrane or sarcoplasmic reticulum. May act as an adapter for assembling multiprotein complexes. | $p > 0.05$ | N | None |
| 15q24 (rs2470893 T, rs2472297, T) | CYP1A1, CYP1A2 | Cytochrome P450 1A1/2: Cytochromes P450 are a group of enzymes involved in NADPH-dependent electron transport pathways. They oxidize a variety of compounds, including steroids, fatty acids, and xenobiotics. | $\downarrow\downarrow$ $p < 5 \times 10^{-8}$ | N | Metabolism of caffeine: CYP1A2 metabolizes >95% of caffeine. |
| 17q11.2 (rs9902453, G) | EFCAB5, SLC6A4 | EF-hand calcium-binding domain-containing protein 5: Unknown Sodium-dependent serotonin transporter: In CNS, regulates serotonergic signaling via transport of serotonin molecules from the synaptic cleft back into the presynaptic terminal for reuse. | $p > 0.05$ | N | Response to caffeine/coffee: Serotonin may mediate the rewarding response to caffeine. |
| 18q21.32 (rs66723169, A) | MC4R | Melanocortin 4 Receptor: Receptor specific to the heptapeptide core common to adrenocorticotropic hormone and alpha-, beta-, and gamma-MSH. Plays a central role in energy homeostasis and somatic growth. | $p > 0.05$ | Y | None |
| 22q11.23 (rs2330783, G) | SPECC1L-ADORA2A | Adenosine A2a Receptor: Receptor for adenosine. The activity of this receptor is mediated by G proteins, which activate adenylyl cyclase. | $\uparrow$ $p < 0.05$ | N | Response to caffeine/coffee: Caffeine blocks this receptor, which mediates some of the psychostimulant effects of caffeine. |

* SNP is associated with (i) higher blood levels of caffeine ($\uparrow$); (ii) lower blood levels of caffeine ($\downarrow$); or (iii) lower blood levels of caffeine and higher paraxanthine-to-caffeine ratio ($\downarrow\downarrow$).
† GWAS (genome-wide association study) catalogue traits unrelated to caffeine or coffee. Y, Yes; N, No.

*5.1. Trait Heterogeneity*

The most comprehensive (and therefore powerful) genetic instrument employed in an MR study of coffee will reflect multiple aspects of coffee drinking behavior (Table 2), such as caffeine metabolism, reward-response and potentially taste. Such heterogeneity does not preclude causal inference, but it does limit the ability to infer causality for particular dimensions of coffee (e.g., caffeine vs non-caffeine) and makes interpretation of MR analyses more difficult [23,25]. An instrumental variable (IV) that narrows in on a particular aspect of coffee drinking might also face issues of interpretation. For example, genetically-inferred 'fast' and 'slow' caffeine metabolizers may consume different amounts of the same type of coffee, but their circulating caffeine levels may not be different. However, circulating levels of non-caffeine constituents of coffee *will* differ. Alternatively, given the same amount and type of coffee consumed, slow caffeine metabolizers will, on average, have higher circulating caffeine levels than fast caffeine metabolizers. Circulating levels of non-caffeine constituents will generally be the same. Because most of the SNPs associate with caffeine intake, and not exclusively coffee intake, the genetic instrument for coffee might also reflect exposure to other dietary sources of caffeine, which might confound or mask any causal relationship between coffee and outcome [43]. Although MR studies are thought to be relatively protected against exposure measurement error, this is less likely to be the case for an MR study of coffee or caffeine [20]. For example, the genetic predisposition to drink coffee, due to an increased caffeine metabolism might also impact preference for regular strong coffee over other coffee types. Taken together, it is important to specify the hypothesis being tested a priori, select the optimal IV and sample for analysis, and consider alternate explanations for positive or null results.

Table 2. Mendelian Randomization (MR) studies of coffee and caffeine consumption.

| Study | Outcome | Instrumental Variable (IV) | Design & Approach | Results | Interpretation | Limitations Reported |
|---|---|---|---|---|---|---|
| Nordestgaard et al. 2015 [44] | Obesity, metabolic syndrome, T2D and related measures (BMI, WC, height, weight, SBP, DBP, TGs, TC, HDL, glucose) | 5-SNPs AHR, CYP1A2 Score and single SNPs | One-sample Individual-level data 2SLS $n \leq 93,179$ Copenhagen General Population Study (CGPS) and the Copenhagen City Heart Study (CCHS). Summary-level data Wald ratio, IVW T2D only DIAGRAM ($n \leq 78,021$) | Observational: Coffee significantly reduced risk of obesity, metabolic syndrome and T2D Coffee significantly increased BMI, WC, weight, height, SBP, DBP, TGs, and TC and decreased HDL SNP-outcome: NS Similar results when individuals were stratified into coffee drinkers and coffee abstainers however, among those without coffee intake, blood pressure was lower with higher coffee-intake allele score | No evidence supporting a causal relationship between coffee and outcomes | Underpowered IV Pleiotropy Collider Bias |
| Nordestgaard & Nordestgaard, 2016 [43] | CVD (IHD, IS, IVD) All-cause and CVD mortality | 2-SNPs AHR, CYP1A2 Score and single SNPs | One-sample Individual-level data 2SLS $n \leq 112,509$ CGPS, CCHS and Copenhagen Ischaemic Heart Disease Study (CIHDS) 3822 IHD cases 4971 IS cases 1708 IVD cases 971 CVD deaths 5422 total deaths Summary-level data Wald ratio, IVW IHD only Cardiogram ($n = 80,517$) and C4D ($n = 30,433$) | Observational: U-shaped association between coffee intake and IHD. IS, IVD and all-cause mortality. Lowest risk with medium coffee intake compared with no coffee intake. SNP-outcome: NS Similar results when individuals were stratified into coffee abstainers, coffee drinkers, coffee drinkers excluding tea and cola drinkers. | No evidence supporting a causal relationship between coffee and outcomes | Underpowered IV Pleiotropy Collider Bias (stratified analysis) Confounding by other caffeine containing-beverages Cannot rule out non-linear effects of coffee on outcomes |
| Kwok et al., 2016 [45] | T2D, IHD, depression, Alzheimer's disease, lipids, glycemic traits, adiposity or adiponectin | 9-SNPs AHR, CYP1A2(2), GCKR, MLXIPL, POR, EFCAB5, BDNF, ABCG2 5 SNPs AHR, CYP1A2(2), POR, EFCAB5 3 SNPs AHR, CYP1A2(2) | Two-sample Summary-level data Multiple published GWAS WME | 9 SNPs: ↑T2D, ↓TGs, ↑BMI, ↑WHR, ↑IR 5 SNPs: NS 3 SNPs: NS | No evidence supporting a causal relationship between coffee and outcomes | Confounding (Population stratification) Pleiotropy Cannot rule out non-linear effects of coffee on outcomes |

**Table 2.** *Cont.*

| Study | Outcome | Instrumental Variable (IV) | Design & Approach | Results | Interpretation | Limitations Reported |
|---|---|---|---|---|---|---|
| Treur et al., 2016 [46] | Smoking behavior Coffee intake Caffeine use | 1-SNP for smoking heaviness (CHRNA3) 8-SNP score for coffee intake AHR, CYP1A2, GCKR, MLXIPL, POR, EFCAB5, BDNF, ABCG2 | Individual-level data Bivariate genetic modelling (SEM) n = 10,368 current smoking (y/n) caffeine use (high/low) coffee use (high/low) Bidirectional MR Regression analyses n = 12,319 Self-reported caffeine use (mg/day), coffee use (cups/day), cigs/day, smoking initiation and persistence Summary-level data LD score regression CCGC Tobacco, Alcohol and Genetics Consortium (TAG): cigs/day, smoking initiation and persistence n ≤ 38,181 | Bivariate genetic modelling Current smoking-coffee intake: G r = 0.47, E r = 0.30 Current smoking-caffeine use: G r = 0.44, E r = 0.00 MR: NS LD score regression Smoking heaviness- coffee intake: r = 0.44 Smoking initiation-coffee intake: r = 0.28 Smoking persistence-coffee intake: r = 0.25 | Genetic factors explain most of the association between smoking and caffeine consumption. Quitting smoking may be more difficult for heavy caffeine consumers, given their genetic susceptibility. | Underpowered Pleiotropy |
| Taylor et al., 2017 [47] | Prostate cancer (PC) risk and progression | 2-SNPs AHR, CYP1A2 | Individual-level data Two-sample MR Regression analyses + meta-analysis Practical consortium (n = 46,687) 4 studies GS-coffee GS-tea GS-(tea + coffee) 23 studies GS-PC GS-PC stage GS-PC grade GS-mortality | Significant GS-coffee, GS-tea and GS-(tea + coffee) GS-PC grade (p = 0.02) | No clear evidence supporting a causal relationship between coffee and outcomes | Between-study heterogeneity in case definition Imprecise IV Pleiotropy Underpowered |

**Table 2.** *Cont.*

| Study | Outcome | Instrumental Variable (IV) | Design & Approach | Results | Interpretation | Limitations Reported |
|---|---|---|---|---|---|---|
| Ware et al., 2017 [48] | Smoking heaviness, cigs/day | 8-SNP GS AHR, CYP1A2, GCKR, MLXIPL, POR, EFCAB5, BDNF, ABCG2; 6-SNP GS AHR, CYP1A2, GCKR, MLXIPL, POR, EFCAB5; 2-SNP GS AHR, CYP1A2 | 2-sample MR Summary-level data IVW, WME CCGC TAG GWAS Cotinine levels (n = 4548) [in vitro experiments] Individual-level data (replication, n = 8072 smokers who drink coffee) IVW, WME | Each cup of coffee/day lead to a decrease in 1.5 (8 SNPs), 1.7 (6 SNPs) or 2.0 (2 SNPs) cigs/day. Coffee did not influence cotinine levels. Coffee did not influence cigs/day in replication sample. | Coffee intake is unlikely to have a major causal impact on cigarette smoking | Pleiotropy Underpowered replication Underpowered IV |
| Bjorngaard et al., 2017 [49] | Coffee intake (cups/day, sensitivity analysis: Any vs. none) Tea intake (cups/day, sensitivity analysis: Any vs. none) Smoking status (never, former, current) Smoking heaviness (cigs/day) | 1-SNP (CHRNA3) for smoking heaviness 2-SNPs (AHR, CYP1A2) for coffee intake GS | Individual-level data Bidirectional MR Regression analyses + meta-analysis UK biobank (n ≤ 114,029) HUNT (n ≤ 56,664) CGPS (n ≤ 78,650) coffee or tea drinkers only | Observational Former & current smoking associated with higher coffee consumption (not tea) vs. never smokers. Among smokers: Each cig/day increased coffee and tea intake; stronger for coffee MR SMK-SNP associated with coffee intake in current or ever smokers only Coffee-SNP not associated with smoking behavior | Higher cigarette consumption causally increases coffee intake. | Underpowered to rule out causal coffee → smoking association. UK Biobank non-representative sample Collider bias: (i) if selection into the sample is related to both coffee and smoking (ii) via smoking stratification Phenotype measurement error |
| Larsson et al., 2017 [50] | Alzheimer's Disease (AD) | 5-SNP GS AHR, CYP1A2, MLXIPL, POR, EFCAB5 (coffee and 23 other exposures tested) | Summary-level data 2-sample MR IVW, WME, MR Egger CCGC International Genomics of Alzheimer's Project (n = 17,009 cases, 37,154 controls) | Suggestive association between coffee GS and increased risk of AD ($p = 0.01$) | Suggestive causal relationship between coffee and AD risk, but in opposite direction to that expected based on observational studies. | None. |
| Verweij et al., 2018 [51] | Causal associations between nicotine, alcohol, caffeine, and cannabis use | Polygenic scores ($p < 5 \times 10^{-8}$ or $p < 1 \times 10^{-5}$) for each exposure | Summary-level data two-sample bidirectional MR IVW, Wald ratio Multiple published GWAS | Smoking cigs/day—caffeine use ($p = 0.01$) Alcohol use: Smoking initiation ($p = 0.03$) | Little evidence for causal relationships between nicotine, alcohol, caffeine, and cannabis use, but may suggest a common liability model (shared genetics) | Imprecise IV GWAS sample overlap (bias to null) |

**Table 2.** *Cont.*

| Study | Outcome | Instrumental Variable (IV) | Design & Approach | Results | Interpretation | Limitations Reported |
|---|---|---|---|---|---|---|
| Ong et al., 2017 [52] | Epithelial ovarian cancer | 4-SNP GS (coffee IV) ABCG2, AHR, CYP1A2, POR 2-SNP GS (caffeine IV) AHR, CYP1A2 | Summary-level data Two-sample MR Wald-type ratio estimator CCGC Ovarian Cancer Association Consortium ($n$ = 44,062, 20,683 cases) | NS | No evidence supporting a causal relationship between coffee/caffeine and outcome | MR Assumption 3 not confirmed Not generalizable to non-European populations. Underpowered or imprecise IV Cannot rule out non-linear effects of coffee/caffeine on cancer |
| Larsson et al., 2018 [53] | Gout | 5-SNPs AHR, CYP1A2, MLXIPL, POR, EFCAB5 | Summary-level data 2-sample MR IVW, WME, MR Egger CCGS Serum Uric acid GWAS ($n$ = 110,347) Gout GWAS (2115 cases and 67,259 controls). | CYP1A2 and MLXIPL SNPs inversely associated with uric acid Combined MR: significant inverse relationship ($p = 7.9 \times 10^{-6}$) All but AHR SNP associated with lower gout risk. Combined MR: significant inverse relationship ($p = 0.005$) | Supports causal inverse association between coffee intake and risk of gout. | None |
| Treur et al., 2018 [54] | Sleep behaviors (sleep duration, chronotype and insomnia complaints) | IV threshold $p < 5 \times 10^{-8}$ 4 SNPs (POR, AHR, CYP1A2, MXLIPL) $p < 5 \times 10^{-5}$ 4-SNPs plus 23 SNPs | Summary-level data Two-sample bidirectional MR IVW, LD score regression CCGC Caffeine metabolite GWAS Sleep GWAS | MR: NS LD score regression: NS | No evidence for causal relationship between habitual coffee intake and sleep behaviors. | Underpowered LD score regression using caffeine metabolite GWAS Phenotype measurement error |
| Noyce et al., 2018 [55] | Parkinson's Disease (PD) | Morning person primary exposure (15 SNPs) coffee secondary exposure (4-SNPs, AHR, BDNF, POR, CYP1A2) | Summary-level data Two-sample MR IVW CCGC Morning person GWAS ($n$ = 89,283) PD GWAS (13,708 cases, 95,282 controls) | Morning person MR: $p = 0.01$ Coffee MR: NS | Along with published RCT results, findings suggest that caffeine may neither prevent PD occurring nor be of benefit in those with the condition. | Use of summary-level data does not allow adjustment for potential confounding factors. |

**Table 2.** *Cont.*

| Study | Outcome | Instrumental Variable (IV) | Design & Approach | Results | Interpretation | Limitations Reported |
|---|---|---|---|---|---|---|
| Zhou et al. 2018 [56] | Cognitive function composite global cognition and memory scores | 2-SNPs *AHR, CYP1A2* Other SNPs (secondary analysis) | Individual-level data $n = 415,530$ (300,760 coffee drinkers) from 10 meta-analyzed European ancestry cohorts. Genetic analysis performed under different levels of habitual coffee intake (1–4 and $\geq$4 cups/day. Negative control: Non-coffee drinkers. | Observational: No overall association between coffee intake and global cognition and memory. SNP-outcome: NS | Study provides no evidence to support beneficial or adverse long-term effects of coffee intake on global cognition or memory. | Pleiotropy. Caution when interpreting coffee IV |
| Lee, 2018 [57] | Osteoarthritis | 4 SNPs, *POR, CYP1A2, NRCAM, NCALD* | Summary-level data Two-sample MR IVW, WME, MR-Egger regression CCGC + Amin et al. 2012 ($n = 18,176$) Osteoarthritis GWAS (7410 cases, 11,009 controls) | IVW: $p = 0.03$ WME: $p = 0.05$ MR Egger: NS (however, no pleiotropy was evident) | Results suggest that coffee consumption is causally associated with an increased risk of osteoarthritis. | Underpowered or imprecise IV Results limited to populations of European ancestry and limited to osteoarthritis in the knee and hip |

AD—Alzheimer's disease; BMI—body mass index; CCGC—Coffee and Caffeine Genetics Consortium; DBP—diastolic blood pressure; DIAGRAM—Diabetes Genetics Replication and Meta-analysis; GS—genetic (SNP) score; HDL—high-density lipoprotein; IHD—ischaemic heard disease; IS—ischaemic stroke, IVD—ischaemic vascular disease, IVW—inverse-variance weighted meta-analysis, NS—non-significant; PC—prostate cancer; PD—Parkinson's Disease; SBP—systolic blood pressure; T2D—type 2 diabetes; TC—total cholesterol); TGs—triglycerides; WC—waist circumference; WME—weighted median estimate.

## 5.2. Pleiotropy

Pleiotropy can violate MR assumption 3, which requires that the genetic variant only influences the outcome through the exposure of interest. Vertical pleiotropy does not violate MR assumption 3 and occurs when the genetic variant is associated with a factor on the pathway between the exposure and outcome, but only because of its effect on the exposure [58]. Horizontal (or biological) pleiotropy occurs when a genetic variant is associated with multiple exposures or traits and is therefore a violation of MR assumption 3 [17,58]. Seven of the fourteen loci associated with coffee or caffeine consumption are also associated with other traits based on GWAS [35] (Table 1). Whether this results from horizontal pleiotropy or a true causal relationship between coffee and these other traits is unclear. Nevertheless, since it is not possible to prove assumption 3 holds for all SNPs in an MR study its becoming common practice to implement extensions of the basic MR methodology that detect the presence of pleiotropy and account for it in causal estimates of the exposure [59]. Random effects IVW or weighted generalized linear regressions are simple options [22,60,61], but common methods that explicitly account for pleiotropy include MR-Egger regression [62], and the weighted-median estimate [63]. Newer methods include MR-PRESSO [64] and generalized summary MR (GSMR) [65]. Each approach relies on different (and largely uncorrelated) assumptions, and therefore the use of multiple approaches allows triangulation; if all provide consistent causal estimates we can be more confident that a true causal effect exists.

## 5.3. Collider Bias

When individual-level data are available, a common strategy is to restrict SNP-outcome analysis to coffee drinkers arguing that the SNPs are associated with coffee drinking (heaviness) and thus causal relationships should only be observed among coffee drinkers (a form of gene-environment interaction) [43,44,48,49,56,59,66]. SNP-outcome associations among non-drinkers ('negative control sample') would suggest a violation in at least one of the assumptions [59,66]. However, this strategy introduces potential for collider bias given that several loci associated with coffee intake also distinguish between non-drinker and heavy coffee drinkers [31]. Collider bias occurs when the exposure and outcome of interest independently influence a third risk factor, and this third risk factor is conditioned upon, either through statistical adjustment or stratification [67–69]. This bias will also apply to the genetic correlates of the exposure and outcome. Indeed, MR studies of coffee intake among the Copenhagen population provided evidence for collider bias [43,44]. For example, among coffee-abstainers, the genetic IV for coffee intake was inversely associated with age. Since age was a risk factor for the outcome and was strongly associated with coffee intake, but among coffee consumers only, the IV-age association in the 'negative control sample' likely arises from collider bias [43].

## 6. MR Studies of Coffee, Caffeine and Health

Table 2 summarizes all MR studies of coffee or caffeine and health outcomes published to-date. Studies are in descending order by date of publication (column 1). For each study we extracted the outcome of interest (column 2), the genetic variants used as the IV (column 3), the basic design and approach (column 4), main results (column 5), interpretation or overarching conclusion of the study (column 6) and limitations as acknowledged by study authors (column 7). With one exception [57], all study IVs included at least SNPs near *CYP1A2* and *AHR*—the strongest and most robust variants linked to coffee drinking behavior and caffeine metabolite levels (Table 1). Primary analysis was conducted using predominately regression analyses or IVW meta-analysis for multi-SNP analysis. These were generally followed by weighted median estimates and MR-Egger regressions to address potential assumption violations. In most studies, the exposure of interest was simply defined as coffee consumption or caffeine use. Data from the GWAS of coffee consumption among 91,462 coffee drinkers in the Coffee and Caffeine Genetics Consortium (CCGC) [31] were used in all summary-level data analysis.

Epidemiological studies report a consistent inverse linear association between coffee consumption and T2D [14], which extends to decaffeinated coffee. This is typically interpreted as evidence for non-caffeine constituents of coffee underlying the coffee-T2D relationship [14]. Two studies, using individual-level and summary-level data for up to ~170,000 participants (26,000 T2D cases) provided no evidence in support of a causal association between coffee intake and T2D risk [44,45], which also extended to measures of adiposity, blood pressure, lipid and glucose metabolism [44,45]. Nordestgaard and colleagues [44] additionally examined a BMI IV (SNPs in/near *FTO*, *MC4R* and *TMEM18*) to examine potential reverse causation from BMI to coffee intake, and as a positive control for risk of T2D. The coffee-intake IV was not linked to BMI, but the BMI-IV was positively associated with coffee intake. Interestingly, SNPs included in the BMI-IV were recently shown to associate with coffee consumption in GWAS (Table 1) [33] and so possibly relate to reward mechanisms (the causal pathway) relevant to coffee drinking behavior and obesity and not adiposity per se [33].

Epidemiological studies also suggest coffee intake may reduce risk of CVD, CVD-mortality and all-cause mortality, but with greatest risk reduction with 3 to 5 cups/day (i.e., a non-linear association) [14]. Nordestgaard and Nordestgaard [43] examined all three of these outcomes in 112,509 Danes and observed a similar pattern of benefits associated with coffee consumption over a 6 year follow-up, but no evidence for causality. In the subgroup of coffee drinkers they noted strong positive and plausible LDL-SNP and HRT-SNP associations, but could not rule-out that such associations could have resulted from collider bias [43].

Caffeine, nicotine, alcohol, and cannabis use are highly correlated behaviors [70]. Potential mechanisms include shared genetic and/or shared environmental factors (i.e., common liability) or a causal influence of one on the other [71]. The co-occurrence of coffee/caffeine use with other substance use behaviors has been investigated in four MR studies [46,48,49,51]. Three of these studies employed bidirectional MR [46,49,51], in which IVs for each substance use were used to evaluate causal effects and their direction [23,72]. The first study focused on the association between smoking and caffeine using three approaches: Bivariate genetic modelling in a twin sample, LD score regression with summary level-data and bidirectional MR analysis using individual-levels data [46]. The results suggested shared genetic factors for caffeine/coffee intake and smoking behavior, rather than a causal influence of one behavior on the other. Ware and colleagues [48] specifically focused on the causal role of coffee consumption on smoking heaviness. Two-sample MR analyses indicated that heavier coffee consumption might lead to *reduced* heaviness of smoking. However, their in vitro experiments, and attempt to replicate in the UK Biobank sample of smokers who drank coffee, did not support these initial causal findings, and overall were not consistent with the direction of association reported in observational analysis. Bjorngaard and colleagues [49] also examined coffee and tea drinkers from three population studies using bidirectional MR and provided evidence for a causal relationship of smoking heaviness on coffee and tea intake, but not vice versa. Finally, Verweij and colleagues [51] examined causal relationships among caffeine, smoking, as well as alcohol, and cannabis use with a variation of bidirectional MR that used 'polygenic scores'. The latter relaxes the significance threshold for GWAS to produce a stronger instrument, but also runs the risk of vertical pleiotropy [59]. Their findings did not support the hypothesis that causal relationships explain the co-occurrence of use of different substances, but are consistent with a common liability model [51].

Alzheimer's Disease (AD) was investigated by Kwok and colleagues [45], and Larsson and colleagues [50], using the same summary-level data, but employed different multi-SNP IVs. Larsson and colleagues [50] used an IV with SNPs for *AHR*, *CYP1A2*, *MLXIPL*, *POR* and *EFCAB* and reported a suggestive causal relationship between coffee and AD risk, but in the opposite direction to that expected based on observational data. Kwok and colleagues [45], whom did not include the *MLXIPL* SNP in their IV, reported no evidence for a causal relationship. A causal relationship between coffee and cognitive function was also not supported by a separate MR [56]. The latter accounted for the potential non-linear association between coffee and cognitive function by conducting analysis by different levels of coffee intake. An association among non-coffee consumers served as a negative control sample. While collider bias was not acknowledged as a limitation, they noted caution when interpreting their results as the instruments indexing greater caffeine consumption may reflect a faster rate of caffeine clearance, and hence a lower (rather than higher) circulating level of bioactive caffeine [56].

Although data are limited, coffee intake has been linked to lower risk of gout [14]. Larsson and colleagues [53] examined the causal association between coffee and gout, as well as uric acid, a related biomarker. The five SNP-IV (excluding the *ABCG2* SNP, which associates with uric acid) was inversely related to both gout risk and uric acid levels, supporting a causal relationship between coffee drinking and gout.

MR studies have failed to support a causal association between coffee/caffeine intake and epithelial ovarian cancer [52], prostate cancer [47], sleep behaviors [54] and Parkinson's disease (PD) [55]. The latter finding is in marked contrast to consistent observational and animal experimental data suggesting coffee and caffeine are protective for PD, but rather align with RCTs and suggest "caffeine may neither prevent PD occurring nor be of benefit in those with the condition" [55]. The authors nevertheless noted that potentially causal effects of coffee may not occur exclusively through caffeine [55], suggesting their IV aimed to capture caffeine exposure rather than coffee drinking per se. The most recent coffee MR was applied to osteoarthritis [57] and supported a causal positive relationship between coffee and this outcome. However, the selection of SNPs for the study was unclear and no human observational study has examined coffee and osteoarthritis, so that the findings are largely hypothesis-generating.

Taken together, at least fifteen studies to date have investigated the causal role of coffee or caffeine use in T2D, CVD, AD and cognition, PD, gout, osteoarthritis, cancers, sleep and other substance use behaviors. Single studies investigated and provided support for a causal role of coffee in reducing risk of gout [14] and increasing risk of osteoarthritis [57]. Four studies examined the co-occurrence of caffeine use and other substances with conflicting results [46,48,49,51]. For the remaining outcomes, studies did not provide clear support for a causal role of coffee or caffeine, but often acknowledged limitations (such as low statistical power, pleiotropy and collider bias), such that a causal role cannot yet be ruled out.

## 7. Future Directions

There is continued enthusiasm for understanding the causal role of coffee and caffeine in health. Thus far, most outcomes of interest have been investigated by single studies and thus the significant and null findings warrant confirmation in independent studies. Many outcomes, for which coffee and caffeine have been implicated, have yet to be investigated [14]. Methodological challenges, such as insufficient power, pleiotropy and collider bias are commonly acknowledged. However, conceptual challenges arising from the different aspects of coffee/caffeine use captured by genetic instruments warrant careful consideration going forward. With continued investment in GWAS it may be possible to parse variants related to non-caffeine aspects of coffee from those related to caffeine providing opportunities to identify the causal elements of coffee per se, rather than coffee drinking behavior. The increasing availability of large individual-level data sets and advanced statistical methods means that more sophisticated MR designs might also be considered. For example, the use of polygenic scores might be optimized using the MR robust adjusted profile score (MR-RAPS) method, which weights each variant differently based on effect size and precision of the SNP-exposure association [62]. Given the co-occurrence of coffee drinking and smoking, a factorial MR may be an attractive approach to study the combined causal effects (i.e., interaction) of these behaviors on disease [22]. Individuals can be allocated into either a high or low-SNP score for coffee and then each group further allocated into either a high or low-SNP score for smoking. The causal estimates for each of the resulting four groups on disease could then be determined. A two-step MR may also be used to assess whether an intermediate trait, say a biomarker or metabolite, acts as a causal mediator between coffee drinking and an outcome [73,74]. An IV for coffee drinking is first used to estimate the causal effect of coffee drinking on the potential mediator (step 1). IVs for the potential mediator are then used to assess the causal effect of the mediator on the outcome (step 2). Evidence of association in both steps implies some degree of mediation of the association between coffee drinking and the outcome by the intermediate variable. Finally, multivariable MRs allow multiple exposures to be examined simultaneously, and provide an effect estimate of one conditional on the other (e.g., effects of coffee consumption conditional on circulating caffeine levels) [75]. These alternate MR designs will still require careful attention to challenges and limitations discussed above.

Multiple statistical methods to accommodate different MR violations combined with replication studies and other mechanistic studies will be necessary to support stronger causal relationship between coffee or caffeine intake and health [59]. GWAS of more refined coffee drinking behaviors, and circulating metabolite markers of coffee intake will also be important, but the collection of such data on a large scale will be needed first. Nevertheless, in light of the rapid pace, in which advancements are being made in these areas, MR promises to be an increasingly valuable approach to understanding the causal impact that coffee and caffeine have in human health.

**Author Contributions:** M.C.C. and M.R.M. conceptualized the paper. M.C.C. wrote the first draft of the paper. All authors revised and approved the final the manuscript.

**Acknowledgments:** This work was funded by the National Institute on Aging (K01AG053477 to M.C.C.) and a Benjamin Meaker Visiting Professorship (to M.C.C.).

**Conflicts of Interest:** The authors declare no conflict of interest.

## References

1. International Coffee Organization. Trade Statistics. Available online: http://www.ico.org/profiles_e.asp (accessed on 1 August 2018).
2. Marks, V.; Kelly, J. Absorption of caffeine from tea, coffee, and coca cola. *Lancet* **1973**, *301*, 827. [CrossRef]
3. Nehlig, A. Interindividual differences in caffeine metabolism and factors driving caffeine consumption. *Pharmacol. Rev.* **2018**, *70*, 384–411. [CrossRef] [PubMed]
4. White, J.R., Jr.; Padowski, J.M.; Zhong, Y.; Chen, G.; Luo, S.; Lazarus, P.; Layton, M.E.; McPherson, S. Pharmacokinetic analysis and comparison of caffeine administered rapidly or slowly in coffee chilled or hot versus chilled energy drink in healthy young adults. *Clin. Toxicol.* **2016**, *54*, 308–312. [CrossRef] [PubMed]
5. Scalbert, A.; Williamson, G. Dietary intake and bioavailability of polyphenols. *J. Nutr.* **2000**, *130*, 2073S–2085S. [CrossRef] [PubMed]
6. Yanagimoto, K.; Ochi, H.; Lee, K.G.; Shibamoto, T. Antioxidative activities of fractions obtained from brewed coffee. *J. Agric. Food Chem.* **2004**, *52*, 592–596. [CrossRef] [PubMed]
7. Urgert, R. Levels of the cholesterol-elevating diterpenes cafestol and kahweol in various coffee brews. *J. Agric. Food Chem.* **1995**, *43*, 2167–2172. [CrossRef]
8. Milder, I.E.; Arts, I.C.; van de Putte, B.; Venema, D.P.; Hollman, P.C. Lignan contents of dutch plant foods: A database including lariciresinol, pinoresinol, secoisolariciresinol and matairesinol. *Br. J. Nutr.* **2005**, *93*, 393–402. [CrossRef] [PubMed]
9. Spiller, M.A. The chemical components of coffee. In *Caffeine*; Spiller, G.A., Ed.; CRC: Boca Raton, FL, USA, 1998; pp. 97–161.
10. Andrzejewski, D.; Roach, J.A.; Gay, M.L.; Musser, S.M. Analysis of coffee for the presence of acrylamide by lc-ms/ms. *J. Agric. Food Chem.* **2004**, *52*, 1996–2002. [CrossRef] [PubMed]
11. Minamisawa, M.; Yoshida, S.; Takai, N. Determination of biologically active substances in roasted coffees using a diode-array hplc system. *Anal. Sci.* **2004**, *20*, 325–328. [CrossRef] [PubMed]
12. Ludwig, I.A.; Mena, P.; Calani, L.; Cid, C.; Del Rio, D.; Lean, M.E.; Crozier, A. Variations in caffeine and chlorogenic acid contents of coffees: What are we drinking? *Food Funct.* **2014**, *5*, 1718–1726. [CrossRef] [PubMed]
13. Cornelis, M.C. Toward systems epidemiology of coffee and health. *Curr. Opin. Lipidol.* **2015**, *26*, 20–29. [CrossRef] [PubMed]
14. Poole, R.; Kennedy, O.J.; Roderick, P.; Fallowfield, J.A.; Hayes, P.C.; Parkes, J. Coffee consumption and health: Umbrella review of meta-analyses of multiple health outcomes. *BMJ* **2017**, *359*, j5024. [CrossRef] [PubMed]

15. Wikoff, D.; Welsh, B.T.; Henderson, R.; Brorby, G.P.; Britt, J.; Myers, E.; Goldberger, J.; Lieberman, H.R.; O'Brien, C.; Peck, J. Systematic review of the potential adverse effects of caffeine consumption in healthy adults, pregnant women, adolescents, and children. *Food Chem. Toxicol.* **2017**, *109*, 585–648. [CrossRef] [PubMed]

16. American Psychiatric Association. *Diagnostic and Statistical Manual of Mental Disorders*, 5th ed.; American Psychiatric Publishing: Arlington, VA, USA, 2013.

17. Davey Smith, G.; Hemani, G. Mendelian randomization: Genetic anchors for causal inference in epidemiological studies. *Hum. Mol. Genet.* **2014**, *23*, 89–98. [CrossRef] [PubMed]

18. Leviton, A. Coffee consumption and residual confounding. *Epidemiology* **1996**, *7*, 110. [CrossRef] [PubMed]

19. Soroko, S.; Chang, J.; Barrett-Connor, E. Reasons for changing caffeinated coffee consumption: The rancho bernardo study. *J. Am. Coll. Nutr.* **1996**, *15*, 97–101. [CrossRef] [PubMed]

20. van Dam, R. Can 'omics' studies provide evidence for causal effects of coffee consumption on risk of type 2 diabetes? *J. Int. Med.* **2018**, *283*, 588–590. [CrossRef] [PubMed]

21. Katan, M.B. Apolipoprotein e isoforms, serum cholesterol, and cancer. *Lancet* **1986**, *1*, 507–508. [CrossRef]

22. Davey Smith, G.; Ebrahim, S. "Mendelian randomization": Can genetic epidemiology contribute to understanding environmental determinants of disease? *Int. J. Epidemiol.* **2003**, *32*, 1–22. [CrossRef]

23. Zheng, J.; Baird, D.; Borges, M.-C.; Bowden, J.; Hemani, G.; Haycock, P.; Evans, D.M.; Smith, G.D. Recent developments in Mendelian randomization studies. *Curr. Epidemiol. Rep.* **2017**, *4*, 330–345. [CrossRef] [PubMed]

24. Swanson, S.A.; Tiemeier, H.; Ikram, M.A.; Hernan, M.A. Nature as a trialist?: Deconstructing the analogy between Mendelian randomization and randomized trials. *Epidemiology* **2017**, *28*, 653–659. [CrossRef] [PubMed]

25. Holmes, M.V.; Ala-Korpela, M.; Smith, G.D. Mendelian randomization in cardiometabolic disease: Challenges in evaluating causality. *Nat. Rev. Cardiol.* **2017**, *14*, 577. [CrossRef] [PubMed]

26. Wald, A. The fitting of straight lines if both variables are subject to error. *Ann. Math. Stat.* **1940**, *11*, 284–300. [CrossRef]

27. Lawlor, D.A.; Harbord, R.M.; Sterne, J.A.; Timpson, N.; Davey Smith, G. Mendelian randomization: Using genes as instruments for making causal inferences in epidemiology. *Stat. Med.* **2008**, *27*, 1133–1163. [CrossRef] [PubMed]

28. Burgess, S.; Butterworth, A.; Thompson, S.G. Mendelian randomization analysis with multiple genetic variants using summarized data. *Genet. Epidemiol.* **2013**, *37*, 658–665. [CrossRef] [PubMed]

29. Amin, N.; Byrne, E.; Johnson, J.; Chenevix-Trench, G.; Walter, S.; Nolte, I.M.; kConFab, I.; Vink, J.M.; Rawal, R.; Mangino, M. Genome-wide association analysis of coffee drinking suggests association with cyp1a1/cyp1a2 and nrcam. *Mol. Psychiatry* **2012**, *17*, 1116–1129. [CrossRef] [PubMed]

30. Cornelis, M.C.; Monda, K.L.; Yu, K.; Paynter, N.; Azzato, E.M.; Bennett, S.N.; Berndt, S.I.; Boerwinkle, E.; Chanock, S.; Chatterjee, N. Genome-wide meta-analysis identifies regions on 7p21 (ahr) and 15q24 (cyp1a2) as determinants of habitual caffeine consumption. *PLoS Genet.* **2011**, *7*, e1002033. [CrossRef] [PubMed]

31. Coffee and Caffeine Genetics Consortium; Cornelis, M.C.; Byrne, E.M.; Esko, T.; Nalls, M.A.; Ganna, A.; Paynter, N.; Monda, K.L.; Amin, N.; Fischer, K. Genome-wide meta-analysis identifies six novel loci associated with habitual coffee consumption. *Mol. Psychiatry* **2015**, *20*, 647–656. [CrossRef] [PubMed]

32. Sulem, P.; Gudbjartsson, D.F.; Geller, F.; Prokopenko, I.; Feenstra, B.; Aben, K.K.; Franke, B.; den Heijer, M.; Kovacs, P.; Stumvoll, M. Sequence variants at cyp1a1-cyp1a2 and ahr associate with coffee consumption. *Hum. Mol. Genet.* **2011**, *20*, 2071–2077. [CrossRef] [PubMed]

33. Zhong, V.; Kuang, A.; Danning, R.; Kraft, P.; van Dam, R.; Chasman, D.; Cornelis, M.C. A Genome-Wide Association Study of Habitual Bitter and Sweet Beverage Consumption. 2018, submitted for publication.

34. Fredholm, B.B.; Battig, K.; Holmen, J.; Nehlig, A.; Zvartau, E.E. Actions of caffeine in the brain with special reference to factors that contribute to its widespread use. *Pharmacol. Rev.* **1999**, *51*, 83–133. [PubMed]

35. Hindorf, L.; MacArthur, J.; Morales, J.; Junkins, H.; Hall, P.; Klemm, A.; Manolio, T. Catalogue of Published Genome-Wide Association Studies. Available online: https://www.ebi.ac.uk/gwas/ (accessed on 1 August 2018).

36. Locke, A.E.; Kahali, B.; Berndt, S.I.; Justice, A.E.; Pers, T.H.; Day, F.R.; Powell, C.; Vedantam, S.; Buchkovich, M.L.; Yang, J. Genetic studies of body mass index yield new insights for obesity biology. *Nature* **2015**, *518*, 197–206. [CrossRef] [PubMed]

37. Teslovich, T.M.; Musunuru, K.; Smith, A.V.; Edmondson, A.C.; Stylianou, I.M.; Koseki, M.; Pirruccello, J.P.; Ripatti, S.; Chasman, D.I.; Willer, C.J. Biological, clinical and population relevance of 95 loci for blood lipids. *Nature* **2010**, *466*, 707–713. [CrossRef] [PubMed]

38. Manning, A.K.; Hivert, M.F.; Scott, R.A.; Grimsby, J.L.; Bouatia-Naji, N.; Chen, H.; Rybin, D.; Liu, C.T.; Bielak, L.F.; Prokopenko, I. A genome-wide approach accounting for body mass index identifies genetic variants influencing fasting glycemic traits and insulin resistance. *Nat. Genet.* **2012**, *44*, 659–669. [CrossRef] [PubMed]

39. Taylor, A.E.; Davey Smith, G.; Munafò, M.R. Associations of coffee genetic risk scores with consumption of coffee, tea and other beverages in the uk biobank. *Addiction* **2018**, *113*, 148–157. [CrossRef] [PubMed]

40. McMahon, G.; Taylor, A.E.; Smith, G.D.; Munafo, M.R. Phenotype refinement strengthens the association of ahr and cyp1a1 genotype with caffeine consumption. *PLoS ONE* **2014**, *9*, e103448. [CrossRef] [PubMed]

41. Cornelis, M.C.; Kacprowski, T.; Menni, C.; Gustafsson, S.; Pivin, E.; Adamski, J.; Artati, A.; Eap, C.B.; Ehret, G.; Friedrich, N. Genome-wide association study of caffeine metabolites provides new insights to caffeine metabolism and dietary caffeine-consumption behavior. *Hum. Mol. Genet.* **2016**, *25*, 5472–5482. [PubMed]

42. Davies, N.M.; Holmes, M.V.; Smith, G.D. Reading Mendelian randomisation studies: A guide, glossary, and checklist for clinicians. *BMJ* **2018**, *362*, k601. [CrossRef] [PubMed]

43. Nordestgaard, A.T.; Nordestgaard, B.G. Coffee intake, cardiovascular disease and all-cause mortality: Observational and Mendelian randomization analyses in 95,000–223,000 individuals. *Int. J. Epidemiol.* **2016**, *45*, 1938–1952. [CrossRef] [PubMed]

44. Nordestgaard, A.T.; Thomsen, M.; Nordestgaard, B.G. Coffee intake and risk of obesity, metabolic syndrome and type 2 diabetes: A Mendelian randomization study. *Int. J. Epidemiol.* **2015**, *44*, 551–565. [CrossRef] [PubMed]

45. Kwok, M.K.; Leung, G.M.; Schooling, C.M. Habitual coffee consumption and risk of type 2 diabetes, ischemic heart disease, depression and Alzheimer's disease: A Mendelian randomization study. *Sci. Rep.* **2016**, *6*, 36500. [CrossRef] [PubMed]

46. Treur, J.L.; Taylor, A.E.; Ware, J.J.; Nivard, M.G.; Neale, M.C.; McMahon, G.; Hottenga, J.J.; Baselmans, B.M.; Boomsma, D.I.; Munafò, M.R. Smoking and caffeine consumption: A genetic analysis of their association. *Addict. Biol.* **2017**, *22*, 1090–1102. [CrossRef] [PubMed]

47. Taylor, A.E.; Martin, R.M.; Geybels, M.S.; Stanford, J.L.; Shui, I.; Eeles, R.; Easton, D.; Kote-Jarai, Z.; Amin Al Olama, A.; Benlloch, S. Investigating the possible causal role of coffee consumption with prostate cancer risk and progression using mendelian randomization analysis. *Int. J. Cancer* **2017**, *140*, 322–328. [CrossRef] [PubMed]

48. Ware, J.J.; Tanner, J.A.; Taylor, A.E.; Bin, Z.; Haycock, P.; Bowden, J.; Rogers, P.J.; Davey Smith, G.; Tyndale, R.F.; Munafò, M.R. Does coffee consumption impact on heaviness of smoking? *Addiction* **2017**, *112*, 1842–1853. [CrossRef] [PubMed]

49. Bjørngaard, J.H.; Nordestgaard, A.T.; Taylor, A.E.; Treur, J.L.; Gabrielsen, M.E.; Munafò, M.R.; Nordestgaard, B.G.; Åsvold, B.O.; Romundstad, P.; Davey Smith, G. Heavier smoking increases coffee consumption: Findings from a Mendelian randomization analysis. *Int. J. Epidemiol.* **2017**, *46*, 1958–1967. [CrossRef] [PubMed]

50. Larsson, S.C.; Traylor, M.; Malik, R.; Dichgans, M.; Burgess, S.; Markus, H.S. Modifiable pathways in alzheimer's disease: Mendelian randomisation analysis. *BMJ* **2017**, *359*, j5375. [CrossRef] [PubMed]

51. Verweij, K.J.; Vinkhuyzen, A.A.; Benyamin, B.; Lynskey, M.T.; Quaye, L.; Agrawal, A.; Gordon, S.D.; Montgomery, G.W.; Madden, P.A.; Heath, A.C. The genetic aetiology of cannabis use initiation: A meta-analysis of genome-wide association studies and a snp-based heritability estimation. *Addict. Biol.* **2013**, *18*, 846–850. [CrossRef] [PubMed]

52. Ong, J.-S.; Hwang, L.-D.; Cuellar-Partida, G.; Martin, N.G.; Chenevix-Trench, G.; Quinn, M.C.; Cornelis, M.C.; Gharahkhani, P.; Webb, P.M.; MacGregor, S. Assessment of moderate coffee consumption and risk of epithelial ovarian cancer: A Mendelian randomization study. *Int. J. Epidemiol.* **2017**, *47*, 450–459. [CrossRef] [PubMed]

53. Larsson, S.C.; Carlström, M. Coffee consumption and gout: A Mendelian randomisation study. *Ann. Rheum. Dis.* **2018**, *77*, 1544–1546. [CrossRef] [PubMed]

54. Treur, J.L.; Gibson, M.; Taylor, A.E.; Rogers, P.J.; Munafo, M.R. Investigating genetic correlations and causal effects between caffeine consumption and sleep behaviours. *J. Sleep Res.* **2018**, *3*, e12695. [CrossRef] [PubMed]

55. Noyce, A.J.; Kia, D.; Heilbron, K.; Jepson, J.; Hemani, G.; Hinds, D.; Lawlor, D.A.; Smith, G.D.; Hardy, J.; Singleton, A. Tendency towards being a "morning person" increases risk of Parkinson's disease: Evidence from mendelian randomisation. *bioRxiv* **2018**. [CrossRef]

56. Zhou, A.; Taylor, A.E.; Karhunen, V.; Zhan, Y.; Rovio, S.P.; Lahti, J.; Sjögren, P.; Byberg, L.; Lyall, D.M.; Auvinen, J. Habitual coffee consumption and cognitive function: A Mendelian randomization meta-analysis in up to 415,530 participants. *Sci. Rep.* **2018**, *8*, 7526. [CrossRef] [PubMed]

57. Lee, Y.H. Investigating the possible causal association of coffee consumption with osteoarthritis risk using a Mendelian randomization analysis. *Clin. Rheumatol.* **2018**, in press. [CrossRef] [PubMed]

58. Burgess, S.; Thompson, S.G. Multivariable Mendelian randomization: The use of pleiotropic genetic variants to estimate causal effects. *Am. J. Epidemiol.* **2015**, *181*, 251–260. [CrossRef] [PubMed]

59. Hemani, G.; Bowden, J.; Davey Smith, G. Evaluating the potential role of pleiotropy in Mendelian randomization studies. *Hum. Mol. Genet.* **2018**, *27*, R195–R208. [CrossRef] [PubMed]

60. Burgess, S.; Bowden, J.; Fall, T.; Ingelsson, E.; Thompson, S.G. Sensitivity analyses for robust causal inference from Mendelian randomization analyses with multiple genetic variants. *Epidemiology* **2017**, *28*, 30. [CrossRef] [PubMed]

61. Burgess, S.; Dudbridge, F.; Thompson, S.G. Combining information on multiple instrumental variables in Mendelian randomization: Comparison of allele score and summarized data methods. *Stat. Med.* **2016**, *35*, 1880–1906. [CrossRef] [PubMed]

62. Bowden, J.; Davey Smith, G.; Burgess, S. Mendelian randomization with invalid instruments: Effect estimation and bias detection through egger regression. *Int. J. Epidemiol.* **2015**, *44*, 512–525. [CrossRef] [PubMed]

63. Bowden, J.; Davey Smith, G.; Haycock, P.C.; Burgess, S. Consistent estimation in Mendelian randomization with some invalid instruments using a weighted median estimator. *Genet. Epidemiol.* **2016**, *40*, 304–314. [CrossRef] [PubMed]

64. Verbanck, M.; Chen, C.-Y.; Neale, B.; Do, R. Detection of widespread horizontal pleiotropy in causal relationships inferred from Mendelian randomization between complex traits and diseases. *Nat. Genet.* **2018**, *50*, 693–698. [CrossRef] [PubMed]

65. Zhu, Z.; Zheng, Z.; Zhang, F.; Wu, Y.; Trzaskowski, M.; Maier, R.; Robinson, M.R.; McGrath, J.J.; Visscher, P.M.; Wray, N.R. Causal associations between risk factors and common diseases inferred from gwas summary data. *Nat. Commun.* **2018**, *9*, 224. [CrossRef] [PubMed]

66. Cho, Y.; Shin, S.-Y.; Won, S.; Relton, C.L.; Smith, G.D.; Shin, M.-J. Alcohol intake and cardiovascular risk factors: A Mendelian randomisation study. *Sci. Rep.* **2015**, *5*, 18422. [CrossRef] [PubMed]

67. Munafò, M.R.; Tilling, K.; Taylor, A.E.; Evans, D.M.; Davey Smith, G. Collider scope: When selection bias can substantially influence observed associations. *Int. J. Epidemiol.* **2017**, *47*, 226–235. [CrossRef] [PubMed]

68. Paternoster, L.; Tilling, K.; Smith, G.D. Genetic epidemiology and Mendelian randomization for informing disease therapeutics: Conceptual and methodological challenges. *PLoS Genet.* **2017**, *13*, e1006944. [CrossRef] [PubMed]

69. Glymour, M.M.; Tchetgen, E.J.; Robins, J.M. Credible Mendelian randomization studies: Approaches for evaluating the instrumental variable assumptions. *Am. J. Epidemiol.* **2012**, *175*, 332–339. [CrossRef] [PubMed]

70. Kendler, K.S.; Schmitt, E.; Aggen, S.H.; Prescott, C.A. Genetic and environmental influences on alcohol, caffeine, cannabis, and nicotine use from early adolescence to middle adulthood. *Arch. Gen. Psychiatry* **2008**, *65*, 674–682. [CrossRef] [PubMed]

71. Vanyukov, M.M.; Tarter, R.E.; Kirillova, G.P.; Kirisci, L.; Reynolds, M.D.; Kreek, M.J.; Conway, K.P.; Maher, B.S.; Iacono, W.G.; Bierut, L. Common liability to addiction and "gateway hypothesis": Theoretical, empirical and evolutionary perspective. *Drug Alcohol Depend.* **2012**, *123* (Suppl. 1), S3–S17. [CrossRef]

72. Haycock, P.C.; Burgess, S.; Wade, K.H.; Bowden, J.; Relton, C.; Davey Smith, G. Best (but oft-forgotten) practices: The design, analysis, and interpretation of Mendelian randomization studies. *Am. J. Clin. Nutr.* **2016**, *103*, 965–978. [CrossRef] [PubMed]

73. Relton, C.L.; Davey Smith, G. Two-step epigenetic Mendelian randomization: A strategy for establishing the causal role of epigenetic processes in pathways to disease. *Int. J. Epidemiol.* **2012**, *41*, 161–176. [CrossRef] [PubMed]

74. Burgess, S.; Daniel, R.M.; Butterworth, A.S.; Thompson, S.G.; Consortium, E.-I. Network Mendelian randomization: Using genetic variants as instrumental variables to investigate mediation in causal pathways. *Int. J. Epidemiol.* **2014**, *44*, 484–495. [CrossRef] [PubMed]

75. Sanderson, E.; Smith, G.D.; Windmeijer, F.; Bowden, J. An examination of multivariable Mendelian randomization in the single sample and two-sample summary data settings. *bioRxiv* **2018**. [CrossRef]

nutrients

MDPI

*Review*

# Biases Inherent in Studies of Coffee Consumption in Early Pregnancy and the Risks of Subsequent Events

Alan Leviton

Boston Children's Hospital and Harvard Medical School, 1731 Beacon Street, Brookline, MA 02445, USA; alan.leviton@childrens.harvard.edu; Tel.: +1-617-485-7187

Received: 24 July 2018; Accepted: 21 August 2018; Published: 23 August 2018

**Abstract:** Consumption of coffee by women early in their pregnancy has been viewed as potentially increasing the risk of miscarriage, low birth weight, and childhood leukemias. Many of these reports of epidemiologic studies have not acknowledged the potential biases inherent in studying the relationship between early-pregnancy-coffee consumption and subsequent events. I discuss five of these biases, recall bias, misclassification, residual confounding, reverse causation, and publication bias. Each might account for claims that attribute adversities to early-pregnancy-coffee consumption. To what extent these biases can be avoided remains to be determined. As a minimum, these biases need to be acknowledged wherever they might account for what is reported.

**Keywords:** epidemiology; bias; causation; coffee; pregnancy

## 1. Introduction

Maternal consumption of coffee during early pregnancy has been viewed as increasing the risk of miscarriage [1–4], fetal growth restriction [2,5–11], and childhood leukemias [12–20]. Unfortunately, many of the epidemiologic studies have not acknowledged the potential biases that appear to have influenced these perceptions of risk. The list of potential biases is long [21].

In this essay, I review five of these biases, namely recall bias, misclassification, residual confounding, reverse causation, and publication bias. Each of these biases might account for some of what has been reported. Unfortunately, eliminating these biases can sometimes be extraordinarily difficult, if not impossible. Indeed, a Cochrane Review concluded, "There is insufficient evidence to confirm or refute the effectiveness of caffeine avoidance on birthweight or other pregnancy outcomes. There is a need to conduct high-quality, double-blinded random clinical trials (RCTs) to determine whether caffeine has any effect on pregnancy outcome." [22]. In essence, observational studies are probably not able to overcome some of the biases. I know of only two clinical trials and they have shown no adverse effect of caffeine consumption on the risk of low birth weight [23], or miscarriage [24].

## 2. Bias 1: Recall/Respondent Bias

Recall or respondent bias occurs when the person interviewed does not fully report what is asked or tends to remember the past differently than others. Perhaps the most common form of recall bias occurs when respondents want to present themselves in an idealized light. For example, based on a review of 67 studies that examined the relationship between self-reported smoking and smoking confirmed by cotinine (a metabolite of nicotine) measurement in saliva or urine, the authors concluded, "Overall, the data show trends of underestimation when smoking prevalence is based on self-report." [25]. Indeed, approximately 20% of pregnant women who report that they are not smokers have smoker-level cotinine concentrations in blood or saliva [26,27]. A review of 34 papers concluded that obese adults tend to significantly under-report their food intake [28]. These reports document that people do not always report the truth.

One of the explanations offered for much recall/respondent bias is social desirability [29]. As applied to answering questionnaires, social desirability is seen as having two components [30]. One, identified as 'impression management,' is the conscious tendency to deceive others, while the other, labeled, 'self-deception,' is the unconscious tendency to believe one's own positive self-reports. Either way, those who try to get accurate information are thwarted by social desirability [31]. whether they want to study hand washing [32], or tobacco consumption [25,33].

Another form of recall bias occurs when some respondents try harder than others to remember the past. For example, when asked to remember exposures during early pregnancy, the mothers of children who developed leukemia are more likely to report higher coffee consumption than the mothers of children selected from the same community, or the mothers of children hospitalized with acute orthopedic trauma [12–20,34–36]. How well do people remember what they drank years before? The time between the consumption and the query is not the only influence on the accuracy of the information provided.

Compared to the mothers of healthy newborns, mothers of children with a major congenital malformation diagnosed soon after birth tend to recall more exposures or characteristics during the index pregnancy [37]. This led to the inference that mothers of malformed babies are more likely to try hard to account for what happened than mothers of children who do not have obvious malformations. Preferential recall was also raised by the authors of one study when fathers of children who had leukemia reported levels of cigarette smoking similar to those reported by fathers of ontrols, but mothers of children with leukemia cases reported higher exposure levels to passive smoking than did the mothers of controls [38].

The authors of a meta-analysis of studies that evaluated the relationship between maternal coffee consumption and the risk of childhood leukemia acknowledged the possibility that mothers of children who had leukemia might recall exposures during the index pregnancy differently than community controls ("the possibility of a recall bias could not be precluded") [39]. On the other hand, another meta-analysis "noted the positive association between coffee consumption and childhood ALL and childhood AML among studies using interviewing techniques, but not among studies using self-administered questionnaire" [40]. The differential recall implies bias somewhere along the information-gathering process.

In light of these phenomena, strategies to minimize recall bias take several forms. "Cohort studies are generally regarded as providing stronger evidence than case-control studies for causality because they satisfy the temporality criterion that the measurement of exposure precede the ascertainment of the outcome." [41]. Not surprisingly then, that some tobacco-related exposures (including coffee consumption) are not associated with tobacco-related malignancies in cohort studies (dependent on exposure data collected before recognition of the disorder) [42–44], but are reported as associated in case-control studies (dependent on exposure data collected after recognition of the disorder) [45,46].

Because of the potential recall bias even when the exposure was recent, some studies of the relationship between caffeine consumption and miscarriage assessed consumption prior to pregnancy [47–50]. "Overall, while most of these studies were small, the majority showed that pre-pregnancy consumption of caffeine was not associated with increased risk of spontaneous abortion." [51].

One way to minimize recall bias that might have contributed to the association between maternal gestational coffee consumption and childhood leukemias would be to choose controls who also have a potentially fatal illness that might have antenatal origins. This strategy of selecting controls who have another disorder that prompts the mother to search her memory especially thoroughly [52], has yet to be applied to the study of childhood leukemia. It would be reasonable to do so if the malignancies of controls each had a relatively unique risk profile.

## 3. Bias 2: Misclassification

The most obvious misclassification that has the potential to distort our perception of truth about relationships between coffee drinking and any disorder is inappropriately quantifying exposure [53].

What is a cup of coffee? 5 ounces (150 mL)? 8 ounces (240 mL)? Is a mug 8 ounces (240 mL)? 10 ounces (300 mL)? 12 ounces (360 mL)? Similar concerns apply to the 'strength of the brew', as well as to additives (e.g., sugar, non-nutritive sweeteners, milk, cream).

Misclassification bias is potentially high in studies that assess the effects of caffeine as the exposure of interest. Almost invariably, authors make assumptions based on reports of caffeine content of coffee, tea, other beverages and foods [54–56], and about attributing to a population, the caffeine content as estimated by self-report [57,58].

## 4. Bias 3: Residual Confounding

Confounding defines the distortion of our perception of the relationship between an exposure (coffee consumption) and a disorder (e.g., childhood leukemia, miscarriage).

This distortion occurs when a variable that is a potential confounder is not considered in the analysis. A potential confounder has to be associated with the disorder and the exposure, but must not be on the causal pathway between the exposure and the disorder [59].

Tobacco smoke induces cytochrome P450 1A2 (CYP1A2), the main enzyme involved in caffeine metabolism, thereby increasing the rate of caffeine metabolism, and shortening the half-life of caffeine [60–62]. One consequence is that the duration of desired behavioral effects of caffeine is shortened, prompting smokers to consume more coffee than non-smokers [63]. Among Norwegian pregnant women, the average daily caffeine consumption varied with smoking. For example, never-smokers consumed 54 mg of caffeine daily, while occasional smokers consumed 109 mg daily, and daily smokers consumed on average 143 mg each day [8]. Therefore, tobacco is a potential confounder of the relationship between a mother's coffee consumption and her child's risk of childhood leukemia. This can be minimized to some extent by "adjusting" for tobacco exposure.

Residual confounding occurs when efforts to minimize confounding are not adequate.

In the most extreme examples, investigators classify as "smokers" all women who smoked during pregnancy, even though these women varied considerably in their level of tobacco consumption, and classify all others as "non-smokers".

Tobacco smoke exposure is a known carcinogen [64]. Some studies have reported that maternal tobacco exposure during pregnancy is associated with increased risk of the offspring developing childhood leukemia [13,19,65–67], whereas others report that paternal tobacco exposure during pregnancy (a source of second-hand smoke for the mother) is associated with the child's heightened risk of childhood leukemia [34,38,68].

Successful adjustment in multivariable models of the risk of a disorder depends on high-quality exposure data. All the adjusting in the world cannot eliminate distortions due to "social desirability responding," such as that which occurs when respondents are truthful about their coffee consumption, but not about their tobacco exposure.

The statement, "Only at the very highest level of pre-pregnancy intake (e.g., >900 mg/day, a consumption level rarely seen in people who do not smoke) was caffeine consumption associated with increased risk of miscarriage" [49]. This raises the possibility that such high consumptions reflect the influence of tobacco exposure, which leads to the inference that the increased risk of miscarriage might also reflect residual confounding of tobacco [69]. Cigarette smoking is also a risk factor for low birth weight [70] and placenta dysfunction [71]. These associations again raise the possibility of residual confounding in studies of coffee consumption during early pregnancy and low birth weight [8,72,73], fetal growth restriction [74], and perhaps even epigenetic effects, such as childhood overweight [75].

Another challenge to eliminating confounding is posed by polymorphisms of multiple genes that influence caffeine and/or coffee consumption [76–84]. Some of these polymorphisms also influence the risk of diseases associated with caffeine and/or coffee consumption [85–90].

A common strategy to disentangle the contribution of genetic propensity to consume coffee/caffeine is to stratify the sample by possession of each gene variant. In essence, this amounts to exploring the caffeine/coffee association in those with and without a specific variant. However, this can be considerably more complex and pose analysis challenges. For example, alleles near genes associated with high coffee consumption are associated with adiposity, cigarette smoking, high levels of fasting insulin and glucose, low risk of hypertension, as well as favorable lipid, inflammatory, and liver enzyme profiles [82].

## 5. Bias 4: Reverse Causation

### 5.1. Coffee Consumption Changes during Early Pregnancy

Even before some women realize they are pregnant, they decrease their coffee consumption. Coffee consumption by women tends to decline as early as the 4th and 5th weeks of normal pregnancy [91,92] (see Figure 1 of [91]; and Figures 1 and 2 of [92]).

Perhaps the first signal of a viable pregnancy is the sensitivity to odors, which can be accompanied by a diminished desire for coffee and the aromas associated with it [93]. As the pregnancy signal intensifies, nausea and overt aversion to odors become increasingly evident [94].

Because women who have early nausea are at a lower risk of early fetal loss (miscarriage) than women who do not experience nausea [95–97], a strong pregnancy signal is seen as an indicator of a viable pregnancy, and the absence of a pregnancy signal is seen as an indicator that the situation might be suboptimal.

The decline in coffee consumption early in pregnancy among women who apparently did not intend to reduce their coffee consumption has been attributed to epiphenomena, including "aversion to tastes and smells ordinarily well tolerated." [98]. Subsequently, the term "pregnancy signal" was used to describe some of the earliest physiologic changes associated with pregnancy, including food aversions, and (hyper)sensitivities to aromas, including those of brewed coffee and perfume [99–102]. Some now use the term, 'pregnancy awareness' [103].

More than half a century ago, the pregnancy signal was attributed to the high-estrogen-content of the first commercially-available oral contraceptives [104,105]. Two decades later, the pregnancy signal was linked to elevated (early morning) urine concentrations of estrone-3-glucuronide and human chorionic gonadotropin [106]. "The number of potential contributors to maternal recognition of pregnancy continues to grow and this highlights our limited appreciation of the complexity of the key molecules and signal transduction pathways that intersect during these key developmental processes." [107]. And indeed, the number of potential contributors does continue to grow [108].

The hormonal characteristics linked to coffee consumption during pregnancy to some extent also appear to apply to consumption when women are not pregnant. For example, the lower the peak estradiol level among women prior to in vitro fertilization, the higher their caffeine consumption [109]. A similar phenomenon occurs in premenopausal women [110].

### 5.2. Inferences That Follow from a Weak Pregnancy Signal

If a weak pregnancy signal is an indicator of a placenta not able to produce the high concentrations of hormones and growth factors needed for fetal wellbeing and optimal growth, the fetus is at increased risk of death and limited growth. If a weak pregnancy signal also allows the gravida to continue her normal coffee consumption, then coffee will be blamed (inappropriately) for increasing the risk of miscarriage and lower birth weight. The blame is inappropriate because the level of coffee consumption is influenced by the very process that will result in potentially dire consequences. In essence, the same placental deficiencies that contribute to the adversities also fail to reduce coffee consumption. Continued pre-pregnancy level of coffee consumption is a consequence, and not a cause, of the placental deficiencies.

This is an example of "reverse causation," which refers to situations where an antecedent is a consequence rather than a cause of illness [50,111–128]. Another example has occurred in some studies that have found that people whose weight (or body mass index) is low were at heightened risk of death [129]. Low weight can be a consequence of disease that results in a loss of appetite [130]. In such situations the processes that lead to death also lead to weight loss, rather than low weight contributing to mortality risk [131].

"Reverse causation" also applies to the situation where a limited-function placenta is more likely to allow a woman to continue her usual levels of coffee consumption throughout pregnancy than is the healthy placenta that prompts a woman to reduce her coffee consumption. As a result, coffee consumption is associated with the consequences of a limited-function placenta precisely because a limited-function placenta allows higher coffee consumption than does a full-function placenta.

The limited-function placenta is associated with fetal growth restriction [132–134]. So is coffee/caffeine consumption [9], even if only by reverse causation.

Among the risk factors for childhood leukemias are two pregnancy phenomena, prior pregnancy loss ("fetal wastage") [135–138] and low birth weight [139–143]. Both of these have been associated with continued normal (pre-pregnancy) level of coffee consumption [6,8,11,91,144–147]. To some extent, each of these (i.e., fetal wastage, low birth weight, and continued coffee consumption during pregnancy at pre-pregnancy levels) is a correlate of impaired implantation of the placenta, and a weaker pregnancy signal than occurs following a healthy implantation. Might the association between gestational coffee consumption and childhood leukemia reflect "reverse causation"?

## 6. Bias 5: Publication Bias

Publication (or dissemination) bias has been defined as the selective publication of studies [148,149]. This appears to happen most commonly when reviewers and editors view "positive" findings as more attractive for publication than "negative" (or non-significant) findings [150]. Publication bias can also reflect self-censorship by authors who are reluctant to continue to battle editors about the need to publish reasonably-powered "negative" studies [151]. So many other persistent influences contribute to publication bias [152–154] that some do not consider elimination of this bias to be feasible [155,156]. Consequently, a negative finding, such as no relationship between early pregnancy coffee consumption and risk of miscarriage, is unlikely to be attractive to editors in light of the plethora of studies reporting a positive relationship. The result is publication bias [157,158], which is especially distorting in meta-analyses [159]. These include distortion of the standardized mean difference plotted against the standard error, which can be severe when the primary studies are small [160]. In addition, asymmetry of the funnel plot might not accurately indicate publication bias [161].

## 7. Conclusions

I have provided comments about biases that might account for associations between maternal coffee consumption early in pregnancy and subsequent events. All of the reports of detrimental effects of coffee consumption during early pregnancy can be explained by one or more of the biases mentioned above. To what extent these biases explain away the associations between maternal early-pregnancy coffee consumption and subsequent events remains to be determined. Obviously, the more these biases can be avoided, the closer we will come to the truth. A laudable goal, but one that is difficult to achieve.

**Funding:** This research received no external funding.

**Acknowledgments:** Preparation of this report was supported by the National Coffee Association of U.S.A.

**Conflicts of Interest:** The authors declare no conflict of interest.

# References

1. Li, J.; Zhao, H.; Song, J.M.; Zhang, J.; Tang, Y.L.; Xin, C.M. A meta-analysis of risk of pregnancy loss and caffeine and coffee consumption during pregnancy. *Int. J. Gynaecol. Obstet.* **2015**, *130*, 116–122. [CrossRef] [PubMed]

2. Chen, L.W.; Wu, Y.; Neelakantan, N.; Chong, M.F.; Pan, A.; van Dam, R.M. Maternal caffeine intake during pregnancy and risk of pregnancy loss: A categorical and dose-response meta-analysis of prospective studies. *Public Health Nutr.* **2016**, *19*, 1233–1244. [CrossRef] [PubMed]

3. Lyngso, J.; RamLau-Hansen, C.H.; Bay, B.; Ingerslev, H.J.; Hulman, A.; Kesmodel, U.S. Association between coffee or caffeine consumption and fecundity and fertility: A systematic review and dose-response meta-analysis. *Clin. Epidemiol.* **2017**, *9*, 699–719. [CrossRef] [PubMed]

4. Gaskins, A.J.; Rich-Edwards, J.W.; Williams, P.L.; Toth, T.L.; Missmer, S.A.; Chavarro, J.E. Pre-pregnancy caffeine and caffeinated beverage intake and risk of spontaneous abortion. *Eur. J. Nutr.* **2018**, *57*, 107–117. [CrossRef] [PubMed]

5. Hoyt, A.T.; Browne, M.; Richardson, S.; Romitti, P.; Druschel, C. Maternal caffeine consumption and small for gestational age births: Results from a population-based case-control study. *Matern. Child. Health J.* **2014**, *18*, 1540–1551. [CrossRef] [PubMed]

6. Voerman, E.; Jaddoe, V.W.; Gishti, O.; Hofman, A.; Franco, O.H.; Gaillard, R. Maternal caffeine intake during pregnancy, early growth, and body fat distribution at school age. *Obesity* **2016**, *24*, 1170–1177. [CrossRef] [PubMed]

7. Xue, F.; Willett, W.C.; Rosner, B.A.; Forman, M.R.; Michels, K.B. Parental characteristics as predictors of birthweight. *Hum. Reprod.* **2008**, *23*, 168–177. [CrossRef] [PubMed]

8. Sengpiel, V.; Elind, E.; Bacelis, J.; Nilsson, S.; Grove, J.; Myhre, R.; Haugen, M.; Meltzer, H.M.; Alexander, J.; Jacobsson, B.; et al. Maternal caffeine intake during pregnancy is associated with birth weight but not with gestational length: Results from a large prospective observational cohort study. *BMC Med.* **2013**, *11*, 42. [CrossRef] [PubMed]

9. Rhee, J.; Kim, R.; Kim, Y.; Tam, M.; Lai, Y.; Keum, N.; Oldenburg, C.E. Maternal caffeine consumption during pregnancy and risk of low birth weight: A dose-response meta-analysis of observational studies. *PLoS ONE* **2015**, *10*, e0132334. [CrossRef] [PubMed]

10. Greenwood, D.C.; Thatcher, N.J.; Ye, J.; Garrard, L.; Keogh, G.; King, L.G.; Cade, J.E. Caffeine intake during pregnancy and adverse birth outcomes: A systematic review and dose-response meta-analysis. *Eur. J. Epidemiol.* **2014**, *29*, 725–734. [CrossRef] [PubMed]

11. CARE Study Group. Maternal caffeine intake during pregnancy and risk of fetal growth restriction: A large prospective observational study. *BMJ* **2008**, *337*, a2332. [CrossRef] [PubMed]

12. Petridou, E.; Trichopoulos, D.; Kalapothaki, V.; Pourtsidis, A.; Kogevinas, M.; Kalmanti, M.; Koliouskas, D.; Kosmidis, H.; Panagiotou, J.P.; Piperopoulou, F.; et al. The risk profile of childhood leukaemia in Greece: A nationwide case-control study. *Br. J. Cancer* **1997**, *76*, 1241–1247. [CrossRef] [PubMed]

13. Clavel, J.; Bellec, S.; Rebouissou, S.; Menegaux, F.; Feunteun, J.; Bonaiti-Pellie, C.; Baruchel, A.; Kebaili, K.; Lambilliotte, A.; Leverger, G.; et al. Childhood leukaemia, polymorphisms of metabolism enzyme genes, and interactions with maternal tobacco, coffee and alcohol consumption during pregnancy. *Eur. J. Cancer Prev.* **2005**, *14*, 531–540. [CrossRef] [PubMed]

14. Menegaux, F.; Steffen, C.; Bellec, S.; Baruchel, A.; Lescoeur, B.; Leverger, G.; Nelken, B.; Philippe, N.; Sommelet, D.; Hemon, D.; et al. Maternal coffee and alcohol consumption during pregnancy, parental smoking and risk of childhood acute leukaemia. *Cancer Detect. Prev.* **2005**, *29*, 487–493. [CrossRef] [PubMed]

15. Petridou, E.; Ntouvelis, E.; Dessypris, N.; Terzidis, A.; Trichopoulos, D. Maternal diet and acute lymphoblastic leukemia in young children. *Cancer Epidemiol. Biomark. Prev.* **2005**, *14*, 1935–1939. [CrossRef] [PubMed]

16. Menegaux, F.; Ripert, M.; Hemon, D.; Clavel, J. Maternal alcohol and coffee drinking, parental smoking and childhood leukaemia: A French population-based case-control study. *Paediatr. Perinat. Epidemiol.* **2007**, *21*, 293–299. [CrossRef] [PubMed]

17. Milne, E.; Royle, J.A.; Bennett, L.C.; de Klerk, N.H.; Bailey, H.D.; Bower, C.; Miller, M.; Attia, J.; Scott, R.J.; Kirby, M.; et al. Maternal consumption of coffee and tea during pregnancy and risk of childhood ALL: Results from an Australian case-control study. *Cancer Causes Control* **2011**, *22*, 207–218. [CrossRef] [PubMed]

18. Bonaventure, A.; Rudant, J.; Goujon-Bellec, S.; Orsi, L.; Leverger, G.; Baruchel, A.; Bertrand, Y.; Nelken, B.; Pasquet, M.; Michel, G.; et al. Childhood acute leukemia, maternal beverage intake during pregnancy, and metabolic polymorphisms. *Cancer Causes Control* **2013**, *24*, 783–793. [CrossRef] [PubMed]

19. Orsi, L.; Rudant, J.; Ajrouche, R.; Leverger, G.; Baruchel, A.; Nelken, B.; Pasquet, M.; Michel, G.; Bertrand, Y.; Ducassou, S.; et al. Parental smoking, maternal alcohol, coffee and tea consumption during pregnancy, and childhood acute leukemia: The ESTELLE study. *Cancer Causes Control* **2015**, *26*, 1003–1017. [CrossRef] [PubMed]

20. Milne, E.; Greenop, K.R.; Petridou, E.; Bailey, H.D.; Orsi, L.; Kang, A.Y.; Baka, M.; Bonaventure, A.; Kourti, M.; Metayer, C.; et al. Maternal consumption of coffee and tea during pregnancy and risk of childhood ALL: A pooled analysis from the childhood leukemia international consortium. *Cancer Causes Control* **2018**, *29*, 539–550. [CrossRef] [PubMed]

21. Sackett, D.L. Bias in analytic research. *J. Chronic Dis.* **1979**, *32*, 51–63. [CrossRef]

22. Jahanfar, S.; Jaafar, S.H. Effects of Restricted Caffeine Intake by Mother on Fetal, Neonatal And Pregnancy Outcome. *Cochrane Database Syst. Rev.* **2013**, *2*, CD006965.

23. Bech, B.H.; Obel, C.; Henriksen, T.B.; Olsen, J. Effect of reducing caffeine intake on birth weight and length of gestation: Randomised controlled trial. *BMJ* **2007**, *334*, 409. [CrossRef] [PubMed]

24. Howards, P.P.; Hertz-Picciotto, I.; Bech, B.H.; Nohr, E.A.; Andersen, A.M.; Poole, C.; Olsen, J. Spontaneous abortion and a diet drug containing caffeine and ephedrine: A study within the Danish national birth cohort. *PLoS ONE* **2012**, *7*, e50372. [CrossRef] [PubMed]

25. Connor Gorber, S.; Schofield-Hurwitz, S.; Hardt, J.; Levasseur, G.; Tremblay, M. The accuracy of self-reported smoking: A systematic review of the relationship between self-reported and cotinine-assessed smoking status. *Nicotine Tob. Res.* **2009**, *11*, 12–24. [CrossRef] [PubMed]

26. Spencer, K.; Cowans, N.J. Accuracy of self-reported smoking status in first trimester aneuploidy screening. *Prenat. Diagn.* **2013**, *33*, 245–250. [CrossRef] [PubMed]

27. Dietz, P.M.; Homa, D.; England, L.J.; Burley, K.; Tong, V.T.; Dube, S.R.; Bernert, J.T. Estimates of nondisclosure of cigarette smoking among pregnant and nonpregnant women of reproductive age in the United States. *Am. J. Epidemiol.* **2011**, *173*, 355–359. [CrossRef] [PubMed]

28. Wehling, H.; Lusher, J. People With a Body Mass Index 30 Under-Report Their Dietary Intake: A Systematic Review. 2017. Available online: http://journals.sagepub.com/doi/abs/10.1177/1359105317714318 (accessed on 22 August 2018).

29. Crowne, D.P.; Marlowe, D. A new scale of social desirability independent of psychopathology. *J. Consult. Psychol.* **1960**, *24*, 349–354. [CrossRef] [PubMed]

30. Paulus, D.L. Two-component models of socially desirable responding. *J. Personal. Soc. Psychol.* **1984**, *46*, 598–609. [CrossRef]

31. Tracey, T.J. A note on socially desirable responding. *J. Counsel. Psychol.* **2016**, *63*, 224–232. [CrossRef] [PubMed]

32. Contzen, N.; De Pasquale, S.; Mosler, H.J. Over-Reporting in handwashing self-reports: Potential explanatory factors and alternative measurements. *PLoS ONE* **2015**, *10*, e0136445. [CrossRef] [PubMed]

33. Biglan, M.; Gilpin, E.A.; Rohrbach, L.A.; Pierce, J.P. Is there a simple correction factor for comparing adolescent tobacco-use estimates from school- and home-based surveys? *Nicotine Tob. Res.* **2004**, *6*, 427–437. [CrossRef] [PubMed]

34. Milne, E.; Greenop, K.R.; Scott, R.J.; Bailey, H.D.; Attia, J.; Dalla-Pozza, L.; de Klerk, N.H.; Armstrong, B.K. Parental prenatal smoking and risk of childhood acute lymphoblastic leukemia. *Am. J. Epidemiol.* **2012**, *175*, 43–53. [CrossRef] [PubMed]

35. Greenop, K.R.; Miller, M.; Attia, J.; Ashton, L.J.; Cohn, R.; Armstrong, B.K.; Milne, E. Maternal consumption of coffee and tea during pregnancy and risk of childhood brain tumors: Results from an Australian case-control study. *Cancer Causes Control.* **2014**, *25*, 1321–1327. [CrossRef] [PubMed]

36. Melchior, M.; Moffitt, T.E.; Milne, B.J.; Poulton, R.; Caspi, A. Why do children from socioeconomically disadvantaged families suffer from poor health when they reach adulthood? A life-course study. *Am. J. Epidemiol.* **2007**, *166*, 966–974. [CrossRef] [PubMed]

37. Werler, M.M.; Pober, B.R.; Nelson, K.; Holmes, L.B. Reporting accuracy among mothers of malformed and nonmalformed infants. *Am. J. Epidemiol.* **1989**, *129*, 415–421. [CrossRef] [PubMed]

38. Farioli, A.; Legittimo, P.; Mattioli, S.; Miligi, L.; Benvenuti, A.; Ranucci, A.; Salvan, A.; Rondelli, R.; Conter, V.; Magnani, C. Tobacco smoke and risk of childhood acute lymphoblastic leukemia: Findings from the SETIL case-control study. *Cancer Causes Control.* **2014**, *25*, 683–692. [CrossRef] [PubMed]

39. Thomopoulos, T.P.; Ntouvelis, E.; Diamantaras, A.A.; Tzanoudaki, M.; Baka, M.; Hatzipantelis, E.; Kourti, M.; Polychronopoulou, S.; Sidi, V.; Stiakaki, E.; et al. Maternal and childhood consumption of coffee, tea and cola beverages in association with childhood leukemia: A meta-analysis. *Cancer Epidemiol.* **2015**, *39*, 1047–1059. [CrossRef] [PubMed]

40. Cheng, J.; Su, H.; Zhu, R.; Wang, X.; Peng, M.; Song, J.; Fan, D. Maternal coffee consumption during pregnancy and risk of childhood acute leukemia: A metaanalysis. *Am. J. Obstet. Gynecol.* **2014**, *210*, 151 e1–151 e10. [CrossRef] [PubMed]

41. The Health Consequences of Smoking—50 Years of Progress: 2014. Available online: https://www.surgeongeneral.gov/library/reports/50-years-of-progress/full-report.pdf (accessed on 29 April 2018).

42. Yu, X.; Bao, Z.; Zou, J.; Dong, J. Coffee consumption and risk of cancers: A meta-analysis of cohort studies. *BMC Cancer* **2011**, *11*, 96. [CrossRef] [PubMed]

43. Huang, T.B.; Guo, Z.F.; Zhang, X.L.; Zhang, X.P.; Liu, H.; Geng, J.; Yao, X.D.; Zheng, J.H. Coffee consumption and urologic cancer risk: A meta-analysis of cohort studies. *Int. Urol. Nephrol.* **2014**, *46*, 1481–1493. [CrossRef] [PubMed]

44. Sugiyama, K.; Sugawara, Y.; Tomata, Y.; Nishino, Y.; Fukao, A.; Tsuji, I. The association between coffee consumption and bladder cancer incidence in a pooled analysis of the miyagi cohort study and ohsaki cohort study. *Eur. J. Cancer Prev.* **2017**, *26*, 125–130. [CrossRef] [PubMed]

45. Bae, J.M.; Kim, E.H. Hormonal replacement therapy and the risk of lung cancer in women: An adaptive meta-analysis of cohort studies. *J. Prev. Med. Public health* **2015**, *48*, 280–286. [CrossRef] [PubMed]

46. Lee, P.N.; HamLing, J.S. Environmental tobacco smoke exposure and risk of breast cancer in nonsmoking women. An updated review and meta-analysis. *Inhal. Toxicol.* **2016**, *28*, 431–454. [CrossRef] [PubMed]

47. Pollack, A.Z.; Buck Louis, G.M.; Sundaram, R.; Lum, K.J. Caffeine consumption and miscarriage: A prospective cohort study. *Fertil. Steril.* **2010**, *93*, 304–306. [CrossRef] [PubMed]

48. Wen, W.; Shu, X.O.; Jacobs, D.R., Jr.; Brown, J.E. The associations of maternal caffeine consumption and nausea with spontaneous abortion. *Epidemiology* **2001**, *12*, 38–42. [CrossRef] [PubMed]

49. Tolstrup, J.S.; Kjaer, S.K.; Munk, C.; Madsen, L.B.; Ottesen, B.; Bergholt, T.; Gronbaek, M. Does caffeine and alcohol intake before pregnancy predict the occurrence of spontaneous abortion? *Hum. Reprod.* **2003**, *18*, 2704–2710. [CrossRef] [PubMed]

50. Hahn, K.A.; Wise, L.A.; Rothman, K.J.; Mikkelsen, E.M.; Brogly, S.B.; Sorensen, H.T.; Riis, A.H.; Hatch, E.E. Caffeine and caffeinated beverage consumption and risk of spontaneous abortion. *Hum. Reprod.* **2015**, *30*, 1246–1255. [CrossRef] [PubMed]

51. Gaskins, A.J.; Toth, T.L.; Chavarro, J.E. Prepregnancy nutrition and early pregnancy outcomes. *Curr. Nutr. Rep.* **2015**, *4*, 265–272. [CrossRef] [PubMed]

52. Werler, M.M.; Louik, C.; Mitchell, A.A. Case-control studies for identifying novel teratogens. *Am. J. Med. Genet. C Semin. Med. Genet.* **2011**, *157C*, 201–208. [CrossRef] [PubMed]

53. Barone, J.J.; Roberts, H.R. Caffeine consumption. *Food chem. toxicol.* **1996**, *34*, 119–129. [CrossRef]

54. Rudolph, E.; Farbinger, A.; Konig, J. Determination of the caffeine contents of various food items within the Austrian market and validation of a caffeine assessment tool (CAT). *Food Addit. Contam.* **2012**, *29*, 1849–1860. [CrossRef] [PubMed]

55. Lisko, J.G.; Lee, G.E.; Kimbrell, J.B.; Rybak, M.E.; Valentin-Blasini, L.; Watson, C.H. Caffeine concentrations in coffee, tea, chocolate, and energy drink flavored e-liquids. *Nicotine Tob. Res.* **2017**, *19*, 484–492. [CrossRef] [PubMed]

56. Sanchez, J.M. Methylxanthine content in commonly consumed foods in Spain and determination of its intake during consumption. *Foods* **2017**, *6*, 109. [CrossRef] [PubMed]

57. Bracken, M.B.; Triche, E.; Grosso, L.; Hellenbrand, K.; Belanger, K.; Leaderer, B.P. Heterogeneity in assessing self-reports of caffeine exposure: Implications for studies of health effects. *Epidemiology* **2002**, *13*, 165–171. [CrossRef] [PubMed]

58. Ludwig, I.A.; Mena, P.; Calani, L.; Cid, C.; Del Rio, D.; Lean, M.E.; Crozier, A. Variations in caffeine and chlorogenic acid contents of coffees: What are we drinking? *Food Funct.* **2014**, *5*, 1718–1726. [CrossRef] [PubMed]

59. Jager, K.J.; Zoccali, C.; Macleod, A.; Dekker, F.W. Confounding: What it is and how to deal with it. *Kidney Int.* **2008**, *73*, 256–260. [CrossRef] [PubMed]

60. Plowchalk, D.R.; Rowland Yeo, K. Prediction of drug clearance in a smoking population: Modeling the impact of variable cigarette consumption on the induction of CYP1A2. *Eur. J. Clin. Pharmacol.* **2012**, *68*, 951–960. [CrossRef] [PubMed]

61. Hukkanen, J.; Jacob, P., 3rd; Peng, M.; Dempsey, D.; Benowitz, N.L. Effect of nicotine on cytochrome P450 1A2 activity. *Br. J. Clin. Pharmacol.* **2011**, *72*, 836–838. [CrossRef] [PubMed]

62. Bjorngaard, J.H.; Nordestgaard, A.T.; Taylor, A.E.; Treur, J.L.; Gabrielsen, M.E.; Munafo, M.R.; Nordestgaard, B.G.; Asvold, B.O.; Romundstad, P.; Davey Smith, G. Heavier smoking increases coffee consumption: Findings from a Mendelian randomization analysis. *Int. J. Epidemiol.* **2017**, *46*, 1958–1967. [CrossRef] [PubMed]

63. De Leon, J.; Diaz, F.J.; Rogers, T.; Browne, D.; Dinsmore, L.; Ghosheh, O.H.; Dwoskin, L.P.; Crooks, P.A. A pilot study of plasma caffeine concentrations in a US sample of smoker and nonsmoker volunteers. *Prog. Neuro-Psychopharmacol. Biol. Psych.* **2003**, *27*, 165–171. [CrossRef]

64. Hecht, S.S.; Szabo, E. Fifty years of tobacco carcinogenesis research: From mechanisms to early detection and prevention of lung cancer. *Cancer Prev. Res.* **2014**, *7*, 1–8. [CrossRef] [PubMed]

65. Metayer, C.; Zhang, L.; Wiemels, J.L.; Bartley, K.; Schiffman, J.; Ma, X.; Aldrich, M.C.; Chang, J.S.; Selvin, S.; Fu, C.H.; et al. Tobacco smoke exposure and the risk of childhood acute lymphoblastic and myeloid leukemias by cytogenetic subtype. *Cancer Epidemiol. Biomark. Prev.* **2013**, *22*, 1600–1611. [CrossRef] [PubMed]

66. Ferreira, J.D.; Couto, A.C.; Pombo-de-Oliveira, M.S.; Koifman, S. Pregnancy, maternal tobacco smoking, and early age leukemia in Brazil. *Front. Oncol.* **2012**, *2*, 151. [CrossRef] [PubMed]

67. Infante-Rivard, C.; Krajinovic, M.; Labuda, D.; Sinnett, D. Parental smoking, CYP1A1 genetic polymorphisms and childhood leukemia (Quebec, Canada). *Cancer Causes Control* **2000**, *11*, 547–553. [CrossRef] [PubMed]

68. Liu, R.; Zhang, L.; McHale, C.M.; Hammond, S.K. Paternal smoking and risk of childhood acute lymphoblastic leukemia: Systematic review and meta-analysis. *J. Oncol.* **2011**, *2011*, 16. [CrossRef] [PubMed]

69. Pineles, B.L.; Park, E.; Samet, J.M. Systematic review and meta-analysis of miscarriage and maternal exposure to tobacco smoke during pregnancy. *Am. J. Epidemiol.* **2014**, *179*, 807–823. [CrossRef] [PubMed]

70. Pereira, P.P.; Da Mata, F.A.; Figueiredo, A.C.; de Andrade, K.R.; Pereira, M.G. Maternal active smoking during pregnancy and low birth weight in the americas: A systematic review and meta-analysis. *Nicotine Tob. Res.* **2017**, *19*, 497–505. [CrossRef] [PubMed]

71. Huuskonen, P.; Amezaga, M.R.; Bellingham, M.; Jones, L.H.; Storvik, M.; Hakkinen, M.; Keski-Nisula, L.; Heinonen, S.; O'Shaughnessy, P.J.; Fowler, P.A.; et al. The human placental proteome is affected by maternal smoking. *Reprod. Toxicol.* **2016**, *63*, 22–31. [CrossRef] [PubMed]

72. Eskenazi, B.; Stapleton, A.L.; Kharrazi, M.; Chee, W.Y. Associations between maternal decaffeinated and caffeinated coffee consumption and fetal growth and gestational duration. *Epidemiology* **1999**, *10*, 242–249. [CrossRef] [PubMed]

73. Rondo, P.H.; Rodrigues, L.C.; Tomkins, A.M. Coffee consumption and intrauterine growth retardation in Brazil. *Eur. J. Clin. Nutr.* **1996**, *50*, 705–709. [PubMed]

74. Fortier, I.; Marcoux, S.; Beaulac-Baillargeon, L. Relation of caffeine intake during pregnancy to intrauterine growth retardation and preterm birth. *Am. J. Epidemiol.* **1993**, *137*, 931–940. [CrossRef] [PubMed]

75. Papadopoulou, E.; Botton, J.; Brantsaeter, A.L.; Haugen, M.; Alexander, J.; Meltzer, H.M.; Bacelis, J.; Elfvin, A.; Jacobsson, B.; Sengpiel, V. Maternal caffeine intake during pregnancy and childhood growth and overweight: Results from a large Norwegian prospective observational cohort study. *BMJ Open* **2018**, *8*, e018895. [CrossRef] [PubMed]

76. Nehlig, A. Interindividual differences in caffeine metabolism and factors driving caffeine consumption. *Pharmacol. Rev.* **2018**, *70*, 384–411. [CrossRef] [PubMed]

77. Taylor, A.E.; Davey Smith, G.; Munafo, M.R. Associations of coffee genetic risk scores with consumption of coffee, tea and other beverages in the UK Biobank. *Addiction* **2018**, *113*, 148–157. [CrossRef] [PubMed]

78. Denden, S.; Bouden, B.; Haj Khelil, A.; Ben Chibani, J.; Hamdaoui, M.H. Gender and ethnicity modify the association between the CYP1A2 rs762551 polymorphism and habitual coffee intake: Evidence from a meta-analysis. *Genet. Mol. Res.* **2016**, *15*. [CrossRef] [PubMed]

79. McMahon, G.; Taylor, A.E.; Davey Smith, G.; Munafo, M.R. Phenotype refinement strengthens the association of AHR and CYP1A1 genotype with caffeine consumption. *PLoS ONE* **2014**, *9*, e103448. [CrossRef] [PubMed]

80. Pirastu, N.; Kooyman, M.; Robino, A.; van der Spek, A.; Navarini, L.; Amin, N.; Karssen, L.C.; Van Duijn, C.M.; Gasparini, P. Non-additive genome-wide association scan reveals a new gene associated with habitual coffee consumption. *Sci. Rep.* **2016**, *6*, 31590. [CrossRef] [PubMed]

81. Pirastu, N.; Kooyman, M.; Traglia, M.; Robino, A.; Willems, S.M.; Pistis, G.; d'Adamo, P.; Amin, N.; d'Eustacchio, A.; Navarini, L.; et al. Association analysis of bitter receptor genes in five isolated populations identifies a significant correlation between TAS2R43 variants and coffee liking. *PLoS ONE* **2014**, *9*, e92065. [CrossRef] [PubMed]

82. Cornelis, M.C.; Byrne, E.M.; Esko, T.; Nalls, M.A.; Ganna, A.; Paynter, N.; Monda, K.L.; Amin, N.; Fischer, K.; Renstrom, F.; et al. Genome-wide meta-analysis identifies six novel loci associated with habitual coffee consumption. *Mol. Psych.* **2015**, *20*, 647–656. [CrossRef] [PubMed]

83. Cornelis, M.C.; Kacprowski, T.; Menni, C.; Gustafsson, S.; Pivin, E.; Adamski, J.; Artati, A.; Eap, C.B.; Ehret, G.; Friedrich, N.; et al. Genome-wide association study of caffeine metabolites provides new insights to caffeine metabolism and dietary caffeine-consumption behavior. *Hum. Mol. Genet.* **2016**, *25*, 5472–5482. [CrossRef] [PubMed]

84. Cornelis, M.C.; Monda, K.L.; Yu, K.; Paynter, N.; Azzato, E.M.; Bennett, S.N.; Berndt, S.I.; Boerwinkle, E.; Chanock, S.; Chatterjee, N.; et al. Genome-wide meta-analysis identifies regions on 7p21 (AHR) and 15q24 (CYP1A2) as determinants of habitual caffeine consumption. *PLoS Genet.* **2011**, *7*, e1002033. [CrossRef] [PubMed]

85. Kokaze, A.; Yoshida, M.; Ishikawa, M.; Matsunaga, N.; Karita, K.; Ochiai, H.; Shirasawa, T.; Nanri, H.; Mitsui, K.; Hoshimo, H.; et al. Mitochondrial DNA 5178 C/A polymorphism modulates the effects of coffee consumption on elevated levels of serum liver enzymes in male Japanese health check-up examinees: An exploratory cross-sectional study. *J. Physiol. Anthropol.* **2016**, *35*, 15. [CrossRef] [PubMed]

86. Chuang, Y.H.; Lill, C.M.; Lee, P.C.; Hansen, J.; Lassen, C.F.; Bertram, L.; Greene, N.; Sinsheimer, J.S.; Ritz, B. Gene-environment interaction in parkinson's disease: Coffee, ADORA2A, and CYP1A2. *Neuroepidemiology* **2016**, *47*, 192–200. [CrossRef] [PubMed]

87. Casiglia, E.; Tikhonoff, V.; Albertini, F.; Favaro, J.; Montagnana, M.; Danese, E.; Finatti, F.; Benati, M.; Mazza, A.; Dal Maso, L.; et al. Caffeine intake and abstract reasoning among 1374 unselected men and women from general population. Role of the −163C>A polymorphism of CYP1A2 gene. *Clin. Nutr. ESPEN* **2017**, *20*, 52–59. [CrossRef] [PubMed]

88. Wang, T.; Huang, T.; Kang, J.H.; Zheng, Y.; Jensen, M.K.; Wiggs, J.L.; Pasquale, L.R.; Fuchs, C.S.; Campos, H.; Rimm, E.B.; et al. Habitual coffee consumption and genetic predisposition to obesity: Gene-diet interaction analyses in three US prospective studies. *BMC Med.* **2017**, *15*, 97. [CrossRef] [PubMed]

89. Platt, D.E.; Ghassibe-Sabbagh, M.; Salameh, P.; Salloum, A.K.; Haber, M.; Mouzaya, F.; Gauguier, D.; Al-Sarraj, Y.; El-Shanti, H.; Zalloua, P.A.; et al. Caffeine impact on metabolic syndrome components is modulated by a CYP1A2 variant. *Ann. Nutr. Metabol.* **2016**, *68*, 1–11. [CrossRef] [PubMed]

90. Palatini, P.; Benetti, E.; Mos, L.; Garavelli, G.; Mazzer, A.; Cozzio, S.; Fania, C.; Casiglia, E. Association of coffee consumption and CYP1A2 polymorphism with risk of impaired fasting glucose in hypertensive patients. *Eur. J. Epidemiol.* **2015**, *30*, 209–217. [CrossRef] [PubMed]

91. Cnattingius, S.; Signorello, L.B.; Anneren, G.; Clausson, B.; Ekbom, A.; Ljunger, E.; Blot, W.J.; McLaughlin, J.K.; Petersson, G.; Rane, A.; et al. Caffeine intake and the risk of first-trimester spontaneous abortion. *N. Engl. J. Med.* **2000**, *343*, 1839–1845. [CrossRef] [PubMed]

92. Lawson, C.C.; LeMasters, G.K.; Wilson, K.A. Changes in caffeine consumption as a signal of pregnancy. *Reprod. Toxicol.* **2004**, *18*, 625–633. [CrossRef] [PubMed]

93. Hook, E.B. Dietary cravings and aversions during pregnancy. *Am. J. Clin. Nutr.* **1978**, *31*, 1355–1362. [CrossRef] [PubMed]

94. Weigel, M.M.; Coe, K.; Castro, N.P.; Caiza, M.E.; Tello, N.; Reyes, M. Food aversions and cravings during early pregnancy: Association with nausea and vomiting. *Ecol. Food Nutr.* **2011**, *50*, 197–214. [CrossRef] [PubMed]

95. Weigel, R.M.; Weigel, M.M. Nausea and vomiting of early pregnancy and pregnancy outcome. A meta-analytical review. *Br. J. Obstet. Gynaecol.* **1989**, *96*, 1312–1318. [CrossRef] [PubMed]

96. Sayle, A.E.; Wilcox, A.J.; Weinberg, C.R.; Baird, D.D. A prospective study of the onset of symptoms of pregnancy. *J. Clin. Epidemiol.* **2002**, *55*, 676–680. [CrossRef]

97. Chan, R.L.; Olshan, A.F.; Savitz, D.A.; Herring, A.H.; Daniels, J.L.; Peterson, H.B.; Martin, S.L. Severity and duration of nausea and vomiting symptoms in pregnancy and spontaneous abortion. *Hum. Reprod.* **2010**, *25*, 2907–2912. [CrossRef] [PubMed]

98. Stein, Z.; Susser, M. Miscarriage, caffeine, and the epiphenomena of pregnancy: The causal model. *Epidemiology* **1991**, *2*, 163–167. [PubMed]

99. Geisert, R.D.; Ross, J.W.; Ashworth, M.D.; White, F.J.; Johnson, G.A.; DeSilva, U. Maternal recognition of pregnancy signal or endocrine disruptor: The two faces of oestrogen during establishment of pregnancy in the pig. *Soc. Reprod. Fertil. Suppl.* **2006**, *62*, 131–145. [PubMed]

100. Wollenhaupt, K.; Brussow, K.P.; Tiemann, U.; Tomek, W. The embryonic pregnancy signal oestradiol influences gene expression at the level of translational initiation in porcine endometrial cells. *Reprod. Domest. Anim.* **2007**, *42*, 167–175. [CrossRef] [PubMed]

101. Brent, R.L.; Christian, M.S.; Diener, R.M. Evaluation of the reproductive and developmental risks of caffeine. *Birth Defects Res.* **2011**, *92*, 152–187. [CrossRef] [PubMed]

102. Porciuncula, L.O.; Sallaberry, C.; Mioranzza, S.; Botton, P.H.; Rosemberg, D.B. The Janus face of caffeine. *Neurochem. Int.* **2013**, *63*, 594–609. [CrossRef] [PubMed]

103. Peacock, A.; Hutchinson, D.; Wilson, J.; McCormack, C.; Bruno, R.; Olsson, C.A.; Allsop, S.; Elliott, E.; Burns, L.; Mattick, R.P. Adherence to the caffeine intake guideline during pregnancy and birth outcomes: A prospective cohort study. *Nutrients* **2018**, *10*, 319. [CrossRef] [PubMed]

104. Russell, M.; Ramcharan, S. Oral contraceptive estrogen content and adverse effects. *Can. Fam. Phys. Med.* **1987**, *33*, 445–460.

105. Speroff, L. The formulation of oral contraceptives: Does the amount of estrogen make any clinical difference? *Johns. Hopkins med. J.* **1982**, *150*, 170–176. [PubMed]

106. Lawson, C.C.; LeMasters, G.K.; Levin, L.S.; Liu, J.H. Pregnancy hormone metabolite patterns, pregnancy symptoms, and coffee consumption. *Am. J. Epidemiol.* **2002**, *156*, 428–437. [CrossRef] [PubMed]

107. Spencer, T.E.; Johnson, G.A.; Bazer, F.W.; Burghardt, R.C.; Palmarini, M. Pregnancy recognition and conceptus implantation in domestic ruminants: Roles of progesterone, interferons and endogenous retroviruses. *Reprod. Fertil. Dev.* **2007**, *19*, 65–78. [CrossRef] [PubMed]

108. Brooks, K.; Burns, G.; Spencer, T.E. Conceptus elongation in ruminants: Roles of progesterone, prostaglandin, interferon tau and cortisol. *J. Anim. Sci. Biotechnol.* **2014**, *5*, 53. [CrossRef] [PubMed]

109. Choi, J.H.; Ryan, L.M.; Cramer, D.W.; Hornstein, M.D.; Missmer, S.A. Effects of caffeine consumption by women and men on the outcome of in vitro fertilization. *J. Caff. Res.* **2011**, *1*, 29–34. [CrossRef] [PubMed]

110. Sisti, J.S.; Hankinson, S.E.; Caporaso, N.E.; Gu, F.; Tamimi, R.M.; Rosner, B.; Xu, X.; Ziegler, R.; Eliassen, A.H. Caffeine, coffee, and tea intake and urinary estrogens and estrogen metabolites in premenopausal women. *Cancer Epidemiol. Biomark. Prev.* **2015**, *24*, 1174–1183. [CrossRef] [PubMed]

111. Flegal, K.M.; Graubard, B.I.; Williamson, D.F.; Cooper, R.S. Reverse causation and illness-related weight loss in observational studies of body weight and mortality. *Am. J. Epidemiol* **2011**, *173*, 1–9. [CrossRef] [PubMed]

112. Kim, T.J.; von dem Knesebeck, O. Income and obesity: What is the direction of the relationship? A systematic review and meta-analysis. *BMJ open* **2018**, *8*, e019862. [PubMed]

113. Maselko, J.; Hayward, R.D.; Hanlon, A.; Buka, S.; Meador, K. Religious service attendance and major depression: A case of reverse causality? *Am. J. Epidemiol.* **2012**, *175*, 576–583. [CrossRef] [PubMed]

114. Kalantar-Zadeh, K.; Block, G.; Horwich, T.; Fonarow, G.C. Reverse epidemiology of conventional cardiovascular risk factors in patients with chronic heart failure. *J. Am. Coll. Cardiol.* **2004**, *43*, 1439–1444. [CrossRef] [PubMed]

115. Stokes, A.; Preston, S.H. Smoking and reverse causation create an obesity paradox in cardiovascular disease. *Obesity* **2015**, *23*, 2485–2490. [CrossRef] [PubMed]

116. Kerger, B.D.; Scott, P.K.; Pavuk, M.; Gough, M.; Paustenbach, D.J. Re-analysis of Ranch Hand study supports reverse causation hypothesis between dioxin and diabetes. *Crit. Rev. toxicol.* **2012**, *42*, 669–687. [CrossRef] [PubMed]

117. Januar, V.; Desoye, G.; Novakovic, B.; Cvitic, S.; Saffery, R. Epigenetic regulation of human placental function and pregnancy outcome: Considerations for causal inference. *Am. J. Obstet. Gynecol.* **2015**, *213*, S182–S196. [CrossRef] [PubMed]

118. Ahiadeke, C.; Gurak, D.T.; Schwager, S.J. Breastfeeding behavior and infant survival with emphasis on reverse causation bias: Some evidence from Nigeria. *Soc. Biol.* **2000**, *47*, 94–113. [CrossRef] [PubMed]

119. Fussman, C.; Todem, D.; Forster, J.; Arshad, H.; Urbanek, R.; Karmaus, W. Cow's milk exposure and asthma in a newborn cohort: Repeated ascertainment indicates reverse causation. *J. Asthma* **2007**, *44*, 99–105. [CrossRef] [PubMed]

120. Grassi, M.; Assanelli, D.; Pezzini, A. Direct, reverse or reciprocal causation in the relation between homocysteine and ischemic heart disease. *Thromb. Res.* **2007**, *120*, 61–69. [CrossRef] [PubMed]

121. Kummeling, I.; Thijs, C. Reverse causation and confounding-by-indication: Do they or do they not explain the association between childhood antibiotic treatment and subsequent development of respiratory illness? *Clin. Exp. Allergy* **2008**, *38*, 1249–1251. [CrossRef] [PubMed]

122. Lodge, C.J.; Lowe, A.J.; Dharmage, S.C. Is reverse causation responsible for the link between duration of breastfeeding and childhood asthma? *Am. J. Respir. Crit. Care Med.* **2008**, *178*, 994. [CrossRef] [PubMed]

123. Luciano, M.; Marioni, R.E.; Gow, A.J.; Starr, J.M.; Deary, I.J. Reverse causation in the association between C-reactive protein and fibrinogen levels and cognitive abilities in an aging sample. *Psychosom. Med.* **2009**, *71*, 404–409. [CrossRef] [PubMed]

124. Kusunoki, T.; Morimoto, T.; Nishikomori, R.; Yasumi, T.; Heike, T.; Mukaida, K.; Fujii, T.; Nakahata, T. Breastfeeding and the prevalence of allergic diseases in schoolchildren: Does reverse causation matter? *Pediatr. Allergy Immunol* **2010**, *21*, 60–66. [CrossRef] [PubMed]

125. Barbui, C.; Gastaldon, C.; Cipriani, A. Benzodiazepines and risk of dementia: True association or reverse causation? *Epidemiol. Psychiatr. Sci.* **2013**, *22*, 307–308. [CrossRef] [PubMed]

126. Brunner, E.J.; Shipley, M.J.; Britton, A.R.; Stansfeld, S.A.; Heuschmann, P.U.; Rudd, A.G.; Wolfe, C.D.; Singh-Manoux, A.; Kivimaki, M. Depressive disorder, coronary heart disease, and stroke: Dose-response and reverse causation effects in the Whitehall II cohort study. *Eur. J. Prev. Cardiol.* **2014**, *21*, 340–346. [CrossRef] [PubMed]

127. Duncan, G.E.; Mills, B.; Strachan, E.; Hurvitz, P.; Huang, R.; Moudon, A.V.; Turkheimer, E. Stepping towards causation in studies of neighborhood and environmental effects: How twin research can overcome problems of selection and reverse causation. *Health Place* **2014**, *27*, 106–111. [CrossRef] [PubMed]

128. Belfort, M.B.; Kuban, K.C.; O'Shea, T.M.; Allred, E.N.; Ehrenkranz, R.A.; Engelke, S.C.; Leviton, A. Weight status in the first 2 years of life and neurodevelopmental impairment in extremely low gestational age newborns. *J. Pediatr.* **2016**, *168*, 30–35. [CrossRef] [PubMed]

129. Niedziela, J.; Hudzik, B.; Niedziela, N.; Gasior, M.; Gierlotka, M.; Wasilewski, J.; Myrda, K.; Lekston, A.; Polonski, L.; Rozentryt, P. The obesity paradox in acute coronary syndrome: A meta-analysis. *Eur. J. Epidemiol.* **2014**, *29*, 801–812. [CrossRef] [PubMed]

130. Kalantar-Zadeh, K.; Horwich, T.B.; Oreopoulos, A.; Kovesdy, C.P.; Younessi, H.; Anker, S.D.; Morley, J.E. Risk factor paradox in wasting diseases. *Curr. Opin. Clin. Nutr. Metabol. Care* **2007**, *10*, 433–442. [CrossRef] [PubMed]

131. Habbu, A.; Lakkis, N.M.; Dokainish, H. The obesity paradox: Fact or fiction? *Am. J. Cardiol.* **2006**, *98*, 944–948. [CrossRef] [PubMed]

132. Krishna, U.; Bhalerao, S. Placental insufficiency and fetal growth restriction. *J. Obstet. Gynaecol. India* **2011**, *61*, 505–511. [CrossRef] [PubMed]

133. Parra-Saavedra, M.; Simeone, S.; Triunfo, S.; Crovetto, F.; Botet, F.; Nadal, A.; Gratacos, E.; Figueras, F. Correlation between histological signs of placental underperfusion and perinatal morbidity in late-onset small-for-gestational-age fetuses. *Ultrasound Obstet. Gynecol.* **2015**, *45*, 149–155. [CrossRef] [PubMed]

134. Kingdom, J.C.; Audette, M.C.; Hobson, S.R.; Windrim, R.C.; Morgen, E. A placenta clinic approach to the diagnosis and management of fetal growth restriction. *Am. J. Obstet. Gynecol.* **2018**, *218*, S803–S817. [CrossRef] [PubMed]

135. Karalexi, M.A.; Skalkidou, A.; Thomopoulos, T.P.; Belechri, M.; Biniaris-Georgallis, S.I.; Bouka, E.; Baka, M.; Hatzipantelis, E.; Kourti, M.; Polychronopoulou, S.; et al. History of maternal fetal loss and childhood leukaemia risk in subsequent offspring: Differentials by miscarriage or stillbirth history and disease subtype. *Paediatr. Perinat. Epidemiol.* **2015**, *29*, 453–461. [CrossRef] [PubMed]

136. Rudant, J.; Amigou, A.; Orsi, L.; Althaus, T.; Leverger, G.; Baruchel, A.; Bertrand, Y.; Nelken, B.; Plat, G.; Michel, G.; et al. Fertility treatments, congenital malformations, fetal loss, and childhood acute leukemia: The ESCALE study (SFCE). *Pediatr. Blood Cancer* **2013**, *60*, 301–308. [CrossRef] [PubMed]

137. Ajrouche, R.; Rudant, J.; Orsi, L.; Petit, A.; Baruchel, A.; Nelken, B.; Pasquet, M.; Michel, G.; Bergeron, C.; Ducassou, S.; et al. Maternal reproductive history, fertility treatments and folic acid supplementation in the risk of childhood acute leukemia: The ESTELLE study. *Cancer Causes Control* **2014**, *25*, 1283–1293. [CrossRef] [PubMed]

138. Yeazel, M.W.; Buckley, J.D.; Woods, W.G.; Ruccione, K.; Robison, L.L. History of maternal fetal loss and increased risk of childhood acute leukemia at an early age. A report from the Childrens Cancer Group. *Cancer* **1995**, *75*, 1718–1727. [CrossRef]

139. O'Neill, K.A.; Murphy, M.F.; Bunch, K.J.; Puumala, S.E.; Carozza, S.E.; Chow, E.J.; Mueller, B.A.; McLaughlin, C.C.; Reynolds, P.; Vincent, T.J.; et al. Infant birthweight and risk of childhood cancer: International population-based case control studies of 40,000 cases. *Int. J. Epidemiol.* **2015**, *44*, 153–168. [CrossRef] [PubMed]

140. Gruhn, B.; Taub, J.W.; Ge, Y.; Beck, J.F.; Zell, R.; Hafer, R.; Hermann, F.H.; Debatin, K.M.; Steinbach, D. Prenatal origin of childhood acute lymphoblastic leukemia, association with birth weight and hyperdiploidy. *Leuk* **2008**, *22*, 1692–1697. [CrossRef] [PubMed]

141. O'Neill, K.A.; Bunch, K.J.; Vincent, T.J.; Spector, L.G.; Moorman, A.V.; Murphy, M.F. Immunophenotype and cytogenetic characteristics in the relationship between birth weight and childhood leukemia. *Pediatr. Blood Cancer* **2012**, *58*, 7–11. [CrossRef] [PubMed]

142. Caughey, R.W.; Michels, K.B. Birth weight and childhood leukemia: A meta-analysis and review of the current evidence. *Int. J. Cancer. J. Int. Du cancer* **2009**, *124*, 2658–2670. [CrossRef] [PubMed]

143. Hjalgrim, L.L.; Rostgaard, K.; Hjalgrim, H.; Westergaard, T.; Thomassen, H.; Forestier, E.; Gustafsson, G.; Kristinsson, J.; Melbye, M.; Schmiegelow, K. Birth weight and risk for childhood leukemia in Denmark, Sweden, Norway, and Iceland. *J. Nati. Cancer Inst.* **2004**, *96*, 1549–1556. [CrossRef] [PubMed]

144. Kuzniewicz, M.W.; Wi, S.; Qian, Y.; Walsh, E.M.; Armstrong, M.A.; Croen, L.A. Prevalence and neonatal factors associated with autism spectrum disorders in preterm infants. *J. Pediatr.* **2014**, *164*, 20–25. [CrossRef] [PubMed]

145. Chen, L.W.; Wu, Y.; Neelakantan, N.; Chong, M.F.; Pan, A.; van Dam, R.M. Maternal caffeine intake during pregnancy is associated with risk of low birth weight: A systematic review and dose-response meta-analysis. *BMC Med.* **2014**, *12*, 174. [CrossRef] [PubMed]

146. Bakker, R.; Steegers, E.A.; Obradov, A.; Raat, H.; Hofman, A.; Jaddoe, V.W. Maternal caffeine intake from coffee and tea, fetal growth, and the risks of adverse birth outcomes: The Generation R Study. *Am. J. Clin. Nutr.* **2010**, *91*, 1691–1698. [CrossRef] [PubMed]

147. Barry, D. Differential recall bias and spurious associations in case/control studies. *Stat. Med.* **1996**, *15*, 2603–2616. [CrossRef]

148. Muller, K.F.; Briel, M.; D'Amario, A.; Kleijnen, J.; Marusic, A.; Wager, E.; Antes, G.; von Elm, E.; Lang, B.; Motschall, E.; et al. Defining publication bias: Protocol for a systematic review of highly cited articles and proposal for a new framework. *Syst. Rev.* **2013**, *2*, 34. [CrossRef] [PubMed]

149. Ioannidis, J.P.; Munafo, M.R.; Fusar-Poli, P.; Nosek, B.A.; David, S.P. Publication and other reporting biases in cognitive sciences: Detection, prevalence, and prevention. *Trends Cogn. Sci.* **2014**, *18*, 235–241. [CrossRef] [PubMed]

150. Dwan, K.; Gamble, C.; Williamson, P.R.; Kirkham, J.J. Systematic review of the empirical evidence of study publication bias and outcome reporting bias—An updated review. *PLoS ONE* **2013**, *8*, e66844. [CrossRef] [PubMed]

151. Connor, J.T. Positive reasons for publishing negative findings. *Am. J. Gastroenterol.* **2008**, *103*, 2181–2183. [CrossRef] [PubMed]

152. Post, R.M. Biased public health perspective on depression treatment: Media bias on publication bias. *Am. J. Psych.* **2009**, *166*, 934–935. [CrossRef] [PubMed]

153. Mathew, S.J.; Charney, D.S. Publication bias and the efficacy of antidepressants. *Am. J. Psych.* **2009**, *166*, 140–145. [CrossRef] [PubMed]

154. Bowden, J.; Jackson, D.; Thompson, S.G. Modelling multiple sources of dissemination bias in meta-analysis. *Stat. Med.* **2010**, *29*, 945–955. [CrossRef] [PubMed]

155. Meerpohl, J.J.; Schell, L.K.; Bassler, D.; Gallus, S.; Kleijnen, J.; Kulig, M.; La Vecchia, C.; Marusic, A.; Ravaud, P.; Reis, A.; et al. Evidence-informed recommendations to reduce dissemination bias in clinical research: Conclusions from the OPEN (Overcome failure to Publish nEgative fiNdings) project based on an international consensus meeting. *BMJ Open* **2015**, *5*, e006666. [CrossRef] [PubMed]

156. Carroll, H.A.; Toumpakari, Z.; Johnson, L.; Betts, J.A. The perceived feasibility of methods to reduce publication bias. *PLoS ONE* **2017**, *12*, e0186472. [CrossRef] [PubMed]

157. Nissen, S.B.; Magidson, T.; Gross, K.; Bergstrom, C.T. Publication bias and the canonization of false facts. *eLife* **2016**, *5*, e21451. [CrossRef] [PubMed]

158. Young, N.S.; Ioannidis, J.P.; Al-Ubaydli, O. Why current publication practices may distort science. *PLoS Med.* **2008**, *5*, e201. [CrossRef] [PubMed]

159. Kicinski, M.; Springate, D.A.; Kontopantelis, E. Publication bias in meta-analyses from the Cochrane database of systematic reviews. *Stat. Med.* **2015**, *34*, 2781–2793. [CrossRef] [PubMed]

160. Zwetsloot, P.P.; Van Der Naald, M.; Sena, E.S.; Howells, D.W.; IntHout, J.; De Groot, J.A.; Chamuleau, S.A.; MacLeod, M.R.; Wever, K.E. Standardized mean differences cause funnel plot distortion in publication bias assessments. *eLife* **2017**, *6*, e24260. [CrossRef] [PubMed]

161. Lau, J.; Ioannidis, J.P.; Terrin, N.; Schmid, C.H.; Olkin, I. The case of the misleading funnel plot. *BMJ* **2006**, *333*, 597–600. [CrossRef] [PubMed]

![nutrients logo] *nutrients*

MDPI

*Review*

# Effects of Caffeine on Myocardial Blood Flow: A Systematic Review

**Randy van Dijk** [1,2], **Daan Ties** [1,2], **Dirkjan Kuijpers** [1,3], **Pim van der Harst** [1,2] and **Matthijs Oudkerk** [1,*]

1    Center for Medical Imaging, University Medical Center Groningen, University of Groningen, 9713 GZ Groningen, The Netherlands; r.van.dijk02@umcg.nl (R.v.D.); d.ties@umcg.nl (D.T.); t.kuijpers@haaglandenmc.nl (D.K.); p.van.der.harst@umcg.nl (P.v.d.H.)
2    Department of Cardiology, University Medical Center Groningen, University of Groningen, 9713 GZ Groningen, The Netherlands
3    HMC-Bronovo, Haaglanden Medisch Centrum, Department of Radiology, Haaglanden Medisch Centrum-Bronovo, 2597 AX The Hague, The Netherlands
*    Correspondence: m.oudkerk@umcg.nl

Received: 6 July 2018; Accepted: 10 August 2018; Published: 13 August 2018

**Abstract:** Background. Caffeine is one of the most widely consumed stimulants worldwide. It is a well-recognized antagonist of adenosine and a potential cause of false-negative functional measurements during vasodilator myocardial perfusion. The aim of this systematic review is to summarize the evidence regarding the effects of caffeine intake on functional measurements of myocardial perfusion in patients with suspected coronary artery disease. Pubmed, Web of Science, and Embase were searched using a predefined electronic search strategy. Participants—healthy subjects or patients with known or suspected CAD. Comparisons—recent caffeine intake versus no caffeine intake. Outcomes—measurements of functional myocardial perfusion. Study design—observational. Fourteen studies were deemed eligible for this systematic review. There was a wide range of variability in study design with varying imaging modalities, vasodilator agents, serum concentrations of caffeine, and primary outcome measurements. The available data indicate a significant influence of recent caffeine intake on cardiac perfusion measurements during adenosine and dipyridamole induced hyperemia. These effects have the potential to affect the clinical decision making by re-classification to different risk-categories.

**Keywords:** caffeine; myocardial perfusion; coronary artery disease; adenosine; regadenoson; dipyridamole

## 1. Introduction

Noninvasive and invasive functional measurements are increasingly used to assess myocardial perfusion in both research and the clinical setting. To unmask relevant myocardial perfusion defects, it is essential to achieve maximal hyperemia during these measurements. The most widely used vasodilator agents used to achieve this hyperemic effect are adenosine, regadenoson, and dipyridamole. The hyperemic effect is primarily caused by binding to the adenosine $A_{2A}$-receptor on arteriolar vascular smooth muscle cells [1].

### 1.1. Vasodilator Agent Mechanisms of Action

Both adenosine and regadenoson act by directly binding to adenosine receptors. Adenosine is a nonselective adenosine receptor agonist and binds to all the different adenosine subtypes, including the adenosine $A_{2B}$-receptor subtype [2]. Binding to this receptor causes bronchospasm in patients with hypersensitive airways. Therefore, adenosine cannot be used in patients with either asthma or chronic obstructive pulmonary disease (COPD) [2–4]. Regadenoson can safely be used in patients with hypersensitive airways, due to its selective binding to the $A_{2A}$-receptor [5–8]. Dipyridamole acts as an adenosine re-uptake inhibitor. The inhibited uptake of adenosine by cells increases the extra-cellular adenosine concentration, increasing the amount of adenosine that is available for binding to adenosine receptors.

## 1.2. Caffeine Antagonism

When adenosine binds to the G-protein coupled $A_{2a}$-receptor, located on cardiac vascular smooth muscle cells, intra-cellular production of cAMP and activation of protein kinase increase, resulting in hyperpolarisation and consequently relaxation of vascular smooth muscle cells. Caffeine is a well-recognized antagonist of adenosine [9]. The competitive antagonistic nature of caffeine for the $A_{2A}$-receptor is a potential cause of achieving insufficient hyperemia, resulting in false-negative functional perfusion measurements [10]. Caffeine limits the binding of adenosine to the receptor and consequently possibly limits cardiac vasodilation and stress adequacy. Regadenoson is a potent selective $A_{2A}$-receptor agonist that is possibly less influenced by caffeine due to the stronger affinity for the receptor. Figure 1 is a simplified graphical illustration of the effect of the vasodilator agents and caffeine on the $A_{2a}$-receptor.

The effects of caffeine on different vasodilator myocardial perfusion measurements remains unclear, and conflicting results have been published. By conducting a review of current literature, two recent debate articles have attempted to shed light on the possible influence of caffeine on myocardial perfusion imaging and its clinical impact [11,12]. However, these papers both fail to provide a complete, unbiased systematic overview of current available evidence on the effects of caffeine on vasodilator myocardial perfusion measurements.

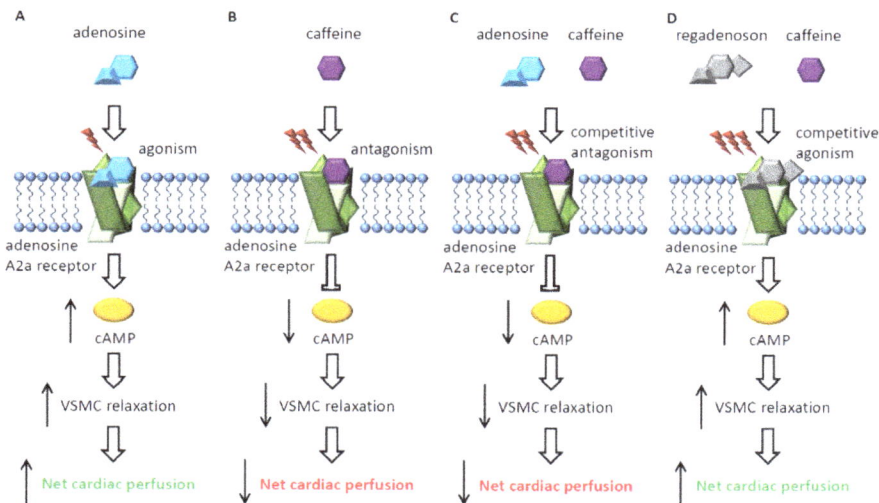

**Figure 1.** Suggested molecular $A_{2a}$-receptor effects showing adenosine agonism (**A**), caffeine antagonism (**B**), competitive antagonism of caffeine on adenosine (**C**), and competitive agonism of regadenoson on caffeine (**D**).

## 1.3. Caffeine Consumption

Caffeine is one of the most widely consumed stimulants worldwide and is present in a wide range of substances such as coffee, soft drinks, energy drinks, tea, and chocolate [13]. The European Food Safety Authority (EFSA) recently published a scientific opinion paper regarding the safety of caffeine [14]. The papers conducted extensive surveys in 22 European countries. They report a wide variability of mean daily caffeine intake per country. The daily intake ranged from 21.8–416.8 mg per day in individuals ≥18 years old, with coffee being the predominant caffeine containing beverage consumed [14]. An average cup of coffee contains approximately 85 mg of caffeine [14]. When taking the average amount of caffeine per cup, the reported coffee intake in the ESFA database roughly translates to a mean coffee intake of 0.25–5 cups. However, it should be recognized that the caffeine dose varies extensively depending on several factors, for example the type of coffee bean and brewing method [15].

### 1.4. Clinical Practice

In clinical practice, patients are generally instructed to refrain from consumption of caffeine containing substances for a period ranging from 12–24 h prior to myocardial perfusion testing. However, the adherence rate of patients to this advice is unclear and serum concentrations of caffeine are not routinely measured in the period preceding the functional measurement. In a study by Banko et al., 36/190 (19%) of patients who screened negative for recent caffeine ingestion by interview still had detectable serum caffeine levels prior to the examination [16]. It is also debatable whether 12 or 24 h caffeine abstinence prior to MPI should be recommended. Carlsson et al., compared coronary flow reserve (CFR) on MRI measured 12 and 24 h after study-induced caffeine ingestion, and showed a significantly lower coronary flow reserve after 12 h caffeine abstinence compared to 24 h caffeine abstinence [17].

### 1.5. Aim of the Study

This systematic review will summarize the evidence regarding the effects of caffeine intake on functional measurements of myocardial perfusion in patients with suspected coronary artery disease (CAD).

## 2. Methods and Results

### 2.1. Protocol and Registration

This systematic review was performed in concordance with the Preferred Reporting Items for Systematic Reviews and Meta-analyses (PRISMA) statement and was registered at PROSPERO under registration number CRD42018092187.

### 2.2. Eligibility Criteria

Participants—healthy subjects or patients with known or suspected CAD. Comparisons—recent caffeine intake versus no caffeine intake. Outcomes—measurements of functional myocardial perfusion. Study design—observational.

### 2.3. Search Strategy

Pubmed, Web of Science, and Embase were searched using a specific electronic search strategy. The following search strategy was used in Pubmed: ("Coffee"[Mesh] OR "Caffeine"[Mesh] OR caffeine [tiab] OR coffee [tiab] OR coffea [tiab]) AND ("Heart" [Mesh] OR "Myocardial Ischemia" [Mesh] OR Myocardi* [tiab] OR Cardiac [tiab] OR cardiovas* [tiab] OR heart [tiab] OR coronar* [tiab]) AND (perfusion* [tiab] OR "Perfusion Imaging" [Mesh] OR "Magnetic Resonance Imaging" [Mesh] OR Magnetic Resonance [tiab] OR Magnetic-resonance [tiab] OR MR [tiab] OR CMR [tiab] OR MRI [tiab] OR "Tomography, X-ray Computed" [Mesh] OR Computed tomograph* [tiab] OR CT [tiab] OR "Positron-Emission Tomography" [Mesh] OR Positron Emission Tomograp* [tiab] OR PET [tiab] OR "Single Photon Emission Computed Tomography Computed Tomography" [Mesh] OR photon Emission Computed Tomograph* [tiab] OR SPECT [tiab] OR "Fractional Flow Reserve, Myocardial" [Mesh] OR "Coronary Angiography" [Mesh] OR Fractional Flow Reserve [tiab] OR FFR [tiab] OR coronary angiograph* [tiab] OR coronary-angiograph* [tiab]). The search strategy for Web of Science and Embase was adjusted according to requirements and preferences of the different databases.

*2.4. Study Selection and Data Collection*

The results from the systematic search of the different databases were collected in Mendeley. Duplicates were removed by using the automatic "check for duplicates" function within Mendeley and an additional manual check for duplicates. Two reviewers independently screened the articles for eligibility using the title and abstract. After title and abstract screening, the results from the reviewers were compared and consensus was achieved in case of discrepancies. The remaining articles were read in full text independently by both reviewers and screened for inclusion. Results of full text screening were compared and discussed afterwards. The data extraction of eligible articles was performed with the use of a predefined template.

*2.5. Search Results*

In total, 702 articles were identified. After duplicate removal, 512 articles remained. After title–abstract screening, 38 articles were identified for eligibility for full text screening. Final number of studies included in the systematic review after full text screening $n = 14$.

*2.6. Study Characteristics*

An overview of the patient characteristics and study design details is shown in Tables 1 and 2 respectively. The imaging modality of choice for myocardial perfusion assessment was SPECT ($n = 5$), PET ($n = 2$), MRI ($n = 3$), or ICA ($n = 4$). The vasodilator agent used was dipyridamole ($n = 3$), adenosine triphosphate (ATP) ($n = 2$), adenosine ($n = 9$), or regadenoson ($n = 3$).

*2.7. Study Quality*

Study quality was assessed with a method based on the Quality Assessment of Diagnostic Accuracy Studies (QUADAS) forms. For the purpose of this systematic review on the effects of caffeine on myocardial perfusion measurements, the following study design components were assessed and graded as either low, high, or unclear risk of bias or applicability concern: 1. Patient selection (low: Prospective patients without inappropriate exclusion, high: (Pre-)selection based on imaging results or measurements, unclear: Not specified); 2. Intervention (low: Serum caffeine level was >4 mg/L, high: Serum caffeine levels <4 mg/L, unclear: Not specified); 3. Analysis (low: Analysis was interpreted without knowledge of the intervention, high: Analysis was interpreted without adequate blinding of the intervention status, unclear: Not specified); 4. Time interval between caffeine intervention and analysis (low: >30 min between caffeine intervention and analysis, high: <30 min between caffeine intervention and analysis, unclear: Not specified). The results of the study quality assessment are shown in Table 3. All studies included in this systematic review are at high risk of selection-bias either due to pre-selection of the study population based on imaging results (presence/absence of ischemia, presence of significant stenosis) or due to inclusion of only healthy volunteers.

**Table 1.** Patient characteristics of the studies included in the systematic review. Variables either presented as n, mean ± SD or n (%). CAD: coronary artery disease; N: number of patients; BMI: Body mass index; MPI: Myocardial perfusion imaging. * Intervention/control: Number of patients with caffeine intervention/number of patients without caffeine intervention.

| | N | Intervention/Controls * | Study Population | Age | Male | BMI |
|---|---|---|---|---|---|---|
| **SPECT** | | | | | | |
| Smits 1991 | 8 | 8/0 | Ischemia on baseline MPI | 60 ± 7 | 3(38) | 28 ± 4 |
| Zoghbi 2006 | 30 | 30/0 | Ischemia on baseline MPI | 64 ± 9 | 22(73) | NS |
| Reyes 2008 | 30 | 12/0 | Ischemia on baseline MPI | 66 ± 6 | NS | 29 ± 4 |
| | | 18/0 | Ischemia on baseline MPI | 64 ± 7 | NS | 27 ± 3 |
| Lee 2012 | 30 | 30/0 | Ischemia on baseline MPI | 70 ± 8 | 21(70) | NS |
| Tejani 2014 | 207 | 0/66 70/0 71/0 | Ischemia on baseline MPI | 68 ± 10.0/65.7 ± 11/69.4 ± 8.2 | 55(83.3)/58(82.9/51(71.8) | NS |
| **PET** | | | | | | |
| Böttcher 1995 | 12 | 12/0 | healthy volunteers | 27 ± 6 | 7(58) | NS |
| Kubo 2004 | 10 | 10/0 | healthy volunteers | 31 ± 6 | 10(100) | NS |
| | 10 | 10/0 | healthy volunteers | 31 ± 6 | 10(100) | NS |
| **MRI** | | | | | | |
| Carlsson 2015 | 16 | 16/0 | healthy volunteers | 41 ± 3 | 8(50) | NS |
| Greulich 2017 | 30 | 30/0 | Ischemia on baseline MPI | 68 ± 8 | 25(83) | NS |
| van Dijk 2017 | 98 | 15/50 | suspected of CAD | 65 ± 11 | 46(49) | NS |
| | | 9/24 | suspected of CAD | 65 ± 11 | 46(49) | NS |
| **ICA** | | | | | | |
| Matsumoto 2014 | 42 | 28/14 | Intermediate stenosis | 70 ± 8/69 ± 10 | 21(75)/11(79) | 24 ± 3/23 ± 4 |
| | 42 | 28/14 | Intermediate stenosis | 70 ± 8/69 ± 10 | 21(75)/11(79) | 24 ± 3/23 ± 4 |
| | 42 | 28/14 | Intermediate stenosis | 70 ± 8/69 ± 10 | 21(75)/11(79) | 24 ± 3/23 ± 4 |
| Mutha 2014 | 10 | 10/0 | Intermediate stenosis | 60 ± 9 | 8(80) | NS |
| Aqel 2004 | 10 | 10/0 | patients with CAD | 53 ± 8 | 10(100) | NS |
| Nakayama 2018 | 30 | 15/15 | patients with significant CAD | 69 ± 10 | 25(83) | 24 ± 3 |
| | 30 | 15/15 | patients with significant CAD | 69 ± 10 | 25(83) | 24 ± 3 |

**Table 2.** Study design details concerning vasodilator agent, caffeine, and main findings. * Timing of caffeine intervention prior to the examination. Continuous variables either reported as mean ± standard deviation or median (interquartile range).

| | Vasodilator | Dosage | Caffeine Dosage | Serum Concentration | Timing * | Main Finding | p-Value |
|---|---|---|---|---|---|---|---|
| **SPECT** | | | | | | | |
| Smits 1991 | Dipyridamole | 0.56 mg/kg | 4 mg/kg i.v. | 9.7 ± 1.3 mg/L | 30 min | Redistribution score caffeine 2.0 ± 1.1 vs. 9.0 ± 0.9 baseline | <0.05 |
| Zoghbi 2006 | Adenosine | 140 μg/kg/min | 8 oz cup of coffee | 3.1 ± 1.6 mg/L | 1 h | SDS caffeine 3.9 ± 2.3 vs. 3.8 ± 1.9 without caffeine | 0.8 |
| Reyes 2008 | Adenosine | 140 μg/kg/min | 2 shots espresso | 6.2 ± 2.6 | 1 h | SDS caffeine 4.1 ± 2.1 vs. baseline 12.0 ± 4.4 | <0.001 |
| | Adenosine | 210 μg/kg/min | 2 shots espresso | 5.7 ± 2.0 | 1 h | SDS caffeine 7.8 ± 4.2 vs. baseline 7.7 ± 4.0 | 0.7 |
| Lee 2012 | Adenosine | 140 μg/kg/min | one cup of coffee | 3.4 mg/L range 0.7–10.4 | 1 h | mean difference stress percent defect −1.6 | 0.3 |
| Tejani 2014 | Regadenoson | 400 μg | placebo, 200 mg or 400 mg caffeine orally | NS | 1.5 h | mean difference number of ischemic segments after 200 mg −0.61 ± 1.097, 400 mg −0.62 ± 1.367, placebo −0.12 ± 0.981 | <0.001 |
| **PET** | | | | | | | |
| Böttcher 1995 | Dipyridamole | 560 μg/kg | 1–2 cups of coffee | range 0–8 mg/L | 1–4 h | Flow reserve caffeine 2.3 ± 0.7 vs. 3.4 ± 0.8 | <0.001 |
| Kubo 2004 | Dipyridamole | 560 μg/kg | 2–3 cups of coffee | 3.3 ± 1.3 mg/L | 1.5 h | MFR caffeine 2.25 ± 0.94 vs. baseline 4.11 ± 1.44 | <0.005 |
| | ATP | 160 μg/kg/min | 2–3 cups of coffee | 3.1 ± 1.6 mg/L | 1.5 h | MFR caffeine 2.44 ± 0.88 vs. baseline 5.15 ± 1.64 | <0.005 |
| **MRI** | | | | | | | |
| Carlsson 2015 | Adenosine | 140 μg/kg/min | minimal 6 g instant coffee | NS | 12 vs. 24 h | CsFR 12 h 4.31 ± 0.57 vs. 24 h 5.32 ± 0.76 | 0.03 |
| Greulich 2017 | Adenosine | 140 μg/kg/min | 200 mg orally | 4.6 ± 2.2 mg/L | 1 h | Ischemic burden 6.9 ± 3.5 caffeine vs. 7.9 ± 3.5 baseline | <.001 |
| van Dijk 2017 | Adenosine | 140 μg/kg/min | 1–2 cups of coffee | NS | <4 h | T1 reactivity caffeine −7.8 ± 5.0 vs. control 4.3 ± 2.8 | <0.001 |
| | Regadenoson | 400 μg | 1–2 cups of coffee | NS | <4 h | T1 reactivity caffeine 4.4 ± 3.2 vs. control 5.4 ± 2.4 | 0.4 |

**Table 2.** *Cont.*

| | Vasodilator | Dosage | Caffeine Dosage | Serum Concentration | Timing * | Main Finding | *p*-Value |
|---|---|---|---|---|---|---|---|
| ICA | | | | | | | |
| Matsumoto 2014 | Adenosine | 140 µg/kg/min | 20 patients 100 or 200 mg orally | 2.9[1.8–4.6] mg/L | NS | FFR caffeine 0.81 ± 0.09 vs. 0.78 ± 0.09 papaverine | <0.001 |
| | Adenosine | 170 µg/kg/min | 20 patients 100 or 200 mg orally | 2.9[1.8–4.6] mg/L | NS | FFR caffeine 0.81 ± 0.09 vs. 0.78 ± 0.09 papaverine | <0.01 |
| | Adenosine | 210 µg/kg/min | 20 patients 100 or 200 mg orally | 2.9[1.8–4.6] mg/L | NS | FFR caffeine 0.79 ± 0.09 vs. 0.78 ± 0.09 papaverine | 0.01 |
| Mutha 2014 | Adenosine | 140 µg/kg/min | 4 mg/kg i.v. | 16.4 ± 5.5 mg/L | 7 min | FFR caffeine 0.82 ± 0.11 vs. 0.79 ± 0.07 baseline | 0.15 |
| Aqel 2004 | Adenosine | 30–50 µg bolus i.c. | 4 mg/kg i.v. | 3.8 ± 1.3 mg/L | 5 min | FFR caffeine 0.75 ± 0.14 vs. 0.76 ± 0.13 | 0.7 |
| Nakayama 2018 | ATP | 140 µg/kg/min | 222 mg orally | 7.3 ± 2.0 mg/L | 2 min | FFR caffeine 0.78 ± 0.12 vs. FFR papaverine 0.75 ± 0.14 | 0.002 |
| | ATP | 170 µg/kg/min | 222 mg orally | 7.3 ± 2.0 mg/L | 2 h | FFR caffeine 0.77 ± 0.12 vs. FFR papaverine 0.75 ± 0.14 | 0.007 |

**Table 3.** Study quality assessment. Red: High, Low: Green, Orange: Unclear risk of either bias (patient selection bias, analysis bias) or applicability concerns (intervention and timing interval possibly not reflection of clinical practice).

| | Patient Selection | Intervention | Analysis | Timing Interval |
|---|---|---|---|---|
| **SPECT** | | | | |
| Smits 1991 | | | | |
| Zoghbi 2006 | | | | |
| Reyes 2008 | | | | |
| Lee 2012 | | | | |
| Tejani 2014 | | | | |
| **PET** | | | | |
| Böttcher 1995 | | | | |
| Kubo 2004 | | | | |
| **CMR** | | | | |
| Carlsson 2015 | | | | |
| Greulich 2017 | | | | |
| van Dijk 2017 | | | | |
| **ICA** | | | | |
| Matsumoto 2014 | | | | |
| Mutha 2004 | | | | |
| Aqel 2004 | | | | |
| Nakayama 2018 | | | | |

## 3. Discussion

The competitive nature of caffeine for the adenosine receptor poses a threat to the validity of all myocardial perfusion modalities, irrespective of the vasodilator being used. In the past three decades, several publications have attempted to assess the impact of (recent) caffeine ingestion on the perfusion examinations. This systematic review aims to provide an overview of the available data and discusses the impact of caffeine ingestion on the different perfusion modalities and vasodilator agents.

Currently, the fractional flow reserve (FFR), as measured during invasive coronary angiography (ICA), is regarded as the reference standard for the functional assessment of myocardial perfusion. The guidelines state that FFR measurements should be performed in case of uncertainty regarding the significance of a coronary stenosis and that FFR measurements should be performed in case of intermediate stenosis (40–70%) [18]. Other imaging modalities that have the potential to provide functional information on myocardial pefusion are SPECT, PET, CT, and MRI. A recent meta-analysis focusing on the diagnostic accuracy of these cardiac perfusion imaging modalities showed a superior diagnostic accuracy of PET, MRI, and CT as compared to SPECT perfusion imaging [19]. SPECT imaging suffers from a limited spatial resolution and as a result, subtle differences in myocardial perfusion are more likely to be missed.

### 3.1. SPECT

Three out of five SPECT studies included in this systematic review reported a non-significant effect of recent caffeine ingestion on the functional perfusion measurement [20–22]. The studies by Lee et al. and Zoghbi et al. selected patients with ischemia on baseline SPECT and performed a second SPECT after caffeine ingestion [20,21]. They reported no significant effect of caffeine ingestion on MPI. However, both report a relatively low serum concentration of caffeine prior to performing the second MPI, possibly underestimating the effect of caffeine. The study by Reyes et al. selected patients with ischemia on baseline SPECT and performed a second SPECT after caffeine intervention (200 mg orally) with either the standard dosage of 140 µg/kg/min ($n$ = 12) or increased dosage of 210 µg/kg/min ($n$ = 18) [22]. The reported serum concentration of caffeine in this study was higher in both groups when compared to Lee et al. and Zoghbi et al. A significant effect of caffeine on the functional perfusion measurement in the group with standard adenosine dosage, but no significant effect in the group with the increased adenosine dosage, was detected, suggesting that the effect of caffeine can be overcome by an increased dosage of the vasodilator agent. Smits et al., report a significantly lower redistribution score as measured on dipyridamole-SPECT after intravenous injection of caffeine compared to baseline SPECT [23]. The serum caffeine concentration in this study was relatively high, potentially securing a maximal effect of the caffeine intervention. The placebo-controlled study by Tejani et al. with large sample size showed a significant decrease in the number of ischemic segments by caffeine measured during regadenoson-SPECT as compared to placebo [24]. When considering the study quality assessment, the two studies reporting no significant effects of the caffeine intervention on the perfusion measurement score worse as compared to the studies reporting significant effects, primarily driven by a lower serum concentration of caffeine.

### 3.2. PET

The two PET studies that report on the effects of caffeine on the functional perfusion measurements show a significant reduction in the myocardial flow reserve and myocardial blood flow, all at relatively low serum concentrations of caffeine during either dipyridamole or ATP induced hyperemia [25,26]. However, it should be noted that both studies included healthy individuals, making translation to clinical practice difficult.

## 3.3. MRI

All three studies that report on the effects of caffeine on adenosine MRI indicate a significant effect on the perfusion measurements [17,27,28]. Greulich et al., showed that caffeine one hour before the perfusion measurement at a serum level of 4.6 ± 2.2 mg/L caused a significant decrease in the ischemic burden [27]. In the other two studies, serum caffeine concentration is not reported. However, both studies report a significant effect on the Coronary Sinus Flow Reserve (CsFR) and T1-reactivity, respectively [17,28]. In the study by our research group, the T1-reactivity appeared unaffected by recent caffeine intake in patients that underwent regadenoson perfusion MRI [28].

## 3.4. ICA

The four studies assessing the effects of recent caffeine intake on the FFR used either ATP or adenosine as the vasodilator agent. The study by Nakayama et al., showed a significantly higher mean FFR value after caffeine ingestion at a "low" (140 µg/kg/min) and "high" (170 µg/kg/min) dose of ATP [29]. Matsumoto et al., also indicate a significant effect of recent caffeine ingestion on the FFR measurement at adenosine dosages of 140 µg/kg/min, 170 µg/kg/min, and 210 µg/kg/min [30]. Mutha et al. and Aqel et al. both report a non-significant effect of intravenous administration of caffeine 5–10 min before the FFR measurement [31,32]. Interestingly, the mean FFR values in the study by Mutha et al. do suggest a significant effect [31]. The lack of significance in this study might be due to the small study population, as they only included ten patients. The study does report that in 2 out of the 10 patients, the FFR value changed from significant (≤0.8) to non-significant (>0.8) after caffeine administration [31]. The study by Aqel et al., shows no significant effect of intravenous caffeine administration at a low serum concentration of caffeine and also in a small study population of only ten patients [32]. Additionally, the time interval between the coffee intervention and the perfusion measurement in the studies by Mutha et al. and Aqel et al. was only several minutes, which is possibly insufficient time for the caffeine to cause a maximal effect. The short time interval between caffeine intervention and the perfusion measurement is also not a good representation of clinical practice.

## 3.5. Contributing Factors

When summarizing the presented data on the potential effects of recent caffeine ingestion on functional perfusion measurements, several study design details appear to have an effect on the outcome. First of all, the different vasodilator agents appear to have a different sensitivity for recent caffeine ingestion, which also seems to be dose dependent. Almost all of the PET, MRI, and ICA studies reporting on the effects of caffeine on adenosine perfusion imaging at the standard dosage of 140 µg/kg/min show a significant effect on the perfusion parameter [17,25–28,30,31]. Only two ICA studies report non-significant effects [29,30]. As discussed in the previous section, possible explanations for the lack of a significant effect in these studies are the small study population, the timing of caffeine intervention, and the low serum concentrations of caffeine, which does not reflect the clinical setting. Additionally, the SPECT studies with a reasonable time interval between caffeine intervention and the perfusion measurement also indicate a significant effect on the perfusion measurement [23,24]. The SPECT and PET studies reporting on dipyridamole perfusion imaging show a significant effect of recent caffeine ingestion [23,25,26]. The effects of recent caffeine ingestion on regadenoson perfusion imaging remain unclear. Only two papers included in this systematic review report on the possible effects of caffeine on regadenoson, and these papers show contradictory results without a clear indication for the difference [24,28]. Regadenoson is increasingly used as the vasodilator agent of choice for perfusion measurement, and further research should be conducted to better understand the influence of caffeine intake on regadenoson perfusion.

*3.6. Clinical Relevance*

For translation to clinical practice, it is essential to investigate if the effects of caffeine on the perfusion measurements change clinical decision making. Current guidelines of the European Society of Cardiology (ESC) indicate an area of ischemia $\geq 10\%$ as high risk and a class IB indication for revascularization for both improvement of prognosis and persisting symptoms under Optical Medical Therapy (OMT) [33]. For MRI, this roughly translates to $\geq 2$ segments with new perfusion defects. The SDS score is used in SPECT analysis, with a score of >8 indicating "severe ischemia" [34]. During ICA, a cut-off value of $\leq 0.80$ is used to indicate stenosis with guideline based indication of revascularization [33].

Only a few articles included in this review provide information that can be used to make a statement on the possible clinical relevance. The MRI study by Greulich et al. reports that no conversion of a positive to a negative stress study occurred on a per patient basis, although the mean ischemic burden was significantly reduced by one segment after caffeine administration [27]. However, it must be noted that the study population consisted of a relatively diseased population with a high baseline mean number of ischemic segments (7.9 ± 3.5), meaning that in this specific population, re-classification as a result of caffeine ingestion would only occur if the detected ischemic burden would be reduced with $\geq 6$ segments by caffeine. Especially at the lower ranges of ischemic burden, the reduction of a small amount of segments by caffeine ingestion might result in re-classification. The SPECT study by Reyes et al. shows a re-classification of the SDS of patients from severe to mild-moderate at the standard adenosine dosage of 140 µg/kg/min [22]. With strict adherence to the guidelines, this would mean that these patients would not be referred for further treatment based on their ischemia burden. Both Zoghbi et al. and Lee et al. report the presence of non-significant ischemia at baseline without a change of classification after caffeine administration [20,21]. These results are to be expected, as the general hypothesis is that caffeine administration might potentially lower the amount of detected ischemia and not increase it, making "re-classification" in these studies impossible. The ICA study by Matsumoto et al. indicates a possible clinical relevant effect of caffeine on FFR measurements [30]. As stated previously, the current cut-off value of FFR for indicating relevant myocardial ischemia is $\leq 0.80$. The mean FFR value at baseline in their study population with adenosine dosage of either 140 µg/kg/min, 170 µg/kg/min, or 210 µg/kg/min indicates significant disease with an FFR $\leq 0.8$ (FFR 0.78 ± 0.09 papaverine). After caffeine ingestion, the mean FFR values in the 140 µg/kg/min and 170 µg/kg/min groups change from significant to non-significant >0.8 (FFR after caffeine administration 0.81 ± 0.09), clearly indicating the potential of recent caffeine ingestion to cause re-classification during ICA.

*3.7. Stress Adequacy*

The T1-reactivity can be used as an imaging biomarker for the assessment of stress adequacy during vasodilator perfusion MRI. It is useful to measure stress adequacy either before the perfusion acquisition or retrospectively during image post-processing and evaluation [35]. We believe that reporting the T1-reactivity will aid in the proper interpretation of MRI perfusion images and that the imaging biomarker should be used as a quality check for stress adequacy.

## 4. Conclusions

When considering the studies with high study quality, the available data indicate a significant influence of recent caffeine intake on cardiac perfusion measurements during adenosine and dipyridamole induced hyperemia in SPECT, PET, MRI, and ICA. Recent caffeine ingestion prior to functional perfusion measurements has the potential to affect clinical decision making by re-classification to different risk-categories.

## Implications of Key Findings

Caffeine intake prior to perfusion measurements should be discouraged and in case of recent caffeine intake, rescheduling of the procedure or switching to regadenoson as the vasodilator agent of choice should be considered. During vasodilator perfusion MRI, the T1-reactivity can be used as a biomarker to assess stress adequacy and to indicate patients at risk of false-negative perfusion results.

**Author Contributions:** Conceptualization, R.v.D., D.T., D.K. Formal Analysis, R.v.D., D.T.; Investigation, R.v.D., D.T.; Writing–Original Draft Preparation, R.v.D. Writing–Review & Editing, R.v.D., D.T., M.O., D.K., P.v.d.H. Supervision, D.K., M.O., P.v.d.H.

**Conflicts of Interest:** The authors declare no conflict of interest.

## References

1. Buhr, C.; Gössl, M.; Erbel, R.; Eggebrecht, H. Regadenoson in the detection of coronary artery disease. *Vasc. Health Risk Manag.* **2008**, *4*, 337–340. [PubMed]
2. Spicuzza, L.; Di Maria, G.; Polosa, R. Adenosine in the airways: Implications and applications. *Eur. J. Pharmacol.* **2005**, *533*, 77–88. [CrossRef] [PubMed]
3. Spicuzza, L.; Bonfiglio, C.; Polosa, R. Research applications and implications of adenosine in diseased airways. *Trends Pharmacol. Sci.* **2003**, *24*, 409–413. [CrossRef]
4. Cerqueira, M.D. The future of pharmacologic stress: Selective $A_{2A}$ adenosine receptor agonists. *Am. J. Cardiol.* **2004**, *94*, 33D–40D. [CrossRef] [PubMed]
5. Shryock, J.C.; Snowdy, S.; Baraldi, P.G.; Cacciari, B.; Spalluto, G.; Monopoli, A.; Ongini, E.; Baker, S.P.; Belardinelli, L. $A_{2A}$-Adenosine receptor reserve for coronary vasodilation. *Circulation* **1998**, *98*, 711–718. [CrossRef] [PubMed]
6. Jaroudi, W.A.; Iskandrian, A.E. Regadenoson: A new myocardial stress agent. *J. Am. Coll. Cardiol.* **2009**, *54*, 1123–1130. [CrossRef] [PubMed]
7. Hendel, R.C.; Bateman, T.M.; Cerqueira, M.D.; Iskandrian, A.E.; Leppe, J.A.; Blackburn, B.; Mahmarian, J.J. Initial clinical experience with regadenoson, a novel selective $A_{2A}$ Agonist for pharmacologic stress single-photon emission computed tomography myocardial perfusion imaging. *J. Am. Coll. Cardiol.* **2005**, *46*, 2069–2075. [CrossRef]
8. Thomas, G.S.; Tammelin, B.R.; Schiffman, G.L.; Marquez, R.; Rice, D.L.; Milikien, D.; Mathur, V. Safety of regadenoson, a selective adenosine $A_{2A}$ agonist, in patients with chronic obstructive pulmonary disease: A randomized, double-blind, placebo-controlled trial (RegCOPD trial). *J. Nucl. Cardiol.* **2008**, *15*, 319–328. [CrossRef] [PubMed]
9. Fredholm, B.B. Adenosine, adenosine receptors and the actions of caffeine. *Pharmacol. Toxicol.* **1995**, *76*, 93–101. [CrossRef] [PubMed]
10. Kidambi, A.; Sourbron, S.; Maredia, N.; Motwani, M.; Brown, J.M.; Nixon, J.; Everett, C.C.; Plein, S.; Greenwood, J.P. Factors associated with false-negative cardiovascular magnetic resonance perfusion studies: A clinical evaluation of magnetic resonance imaging in coronary artery disease (CE-MARC) substudy. *J. Magn. Reson. Imaging* **2016**, *43*, 566–573. [CrossRef] [PubMed]
11. Reyes, E. Caffeine reduces the sensitivity of vasodilator MPI for the detection of myoardial ischaemia: Pro. *J. Nucl. Cardiol.* **2016**, *23*, 447–453. [CrossRef]
12. Saab, R.; Bajaj, N.S.; Hage, F.G. Caffeine does not significantly reduce the sensitivity of vasodilator stress myocardial perfusion imaging. *J. Nucl. Cardiol.* **2015**, *23*, 442–446. [CrossRef] [PubMed]
13. Lapeyre, A.C.; Goraya, T.Y.; Johnston, D.L.; Gibbons, R.J. The impact of caffeine on vasodilator stress perfusion studies. *J. Nucl. Cardiol.* **2004**, *11*, 506–511. [CrossRef] [PubMed]
14. European Food Safety Authority (EFSA). Scientific opinion on the safety of caffeine1 EFSA panel on dietetic products, nutrition and allergies (NDA). *ESFA J.* **2015**, *13*, 4102.
15. McCusker, R.R.; Goldberger, B.A.; Cone, E.J. Caffeine content of specialty coffees. *J. Anal. Toxicol.* **2003**, *27*, 520–522. [CrossRef] [PubMed]
16. Banko, L.T.; Haq, S.A.; Rainaldi, D.A.; Klem, I.; Siegler, J.; Fogel, J.; Sacchi, T.J.; Heitner, J.F. Incidence of caffeine in serum of patients undergoing dipyridamole myocardial perfusion stress test by an intensive versus routine caffeine history screening. *Am. J. Cardiol.* **2009**, *105*, 1474–1479. [CrossRef] [PubMed]

17. Carlsson, M.; Jögi, J.; Bloch, K.M.; Hedén, B.; Ekelund, U.; Ståhlberg, F.; Arheden, H. Submaximal adenosine-induced coronary hyperaemia with 12 h caffeine abstinence: Implications for clinical adenosine perfusion imaging tests. *Clin. Physiol. Funct. Imaging* **2015**, *35*, 49–56. [CrossRef] [PubMed]

18. Windecker, S.; Kolh, P.; Alfonso, F.; Collet, J.P.; Cremer, J.; Falk, V.; Filippatos, G.; Hamm, C.; Head, S.J.; Jüni, P.; et al. 2014 ESC/EACTS Guidelines onmyocardial revascularization. The Task Force on Myocardial Revascularization of the European Society of Cardiology (ESC) and the European Association for Cardio-Thoracic Surgery (EACTS). *Eur. Heart J.* **2014**, *35*, 2541–2619. [PubMed]

19. Takx, R.A.; Blomberg, B.A.; El Aidi, H.; Habets, J.; de Jong, P.A.; Nagel, E.; Hoffmann, U.; Leiner, T. Diagnostic accuracy of stress myocardial perfusion imaging compared to invasive coronary angiography with fractional flow reserve meta-analysis. *Circ. Cardiovasc. Imaging* **2015**, *8*, e002666. [CrossRef] [PubMed]

20. Lee, J.C.; Fraser, J.F.; Barnett, A.G.; Johnson, L.P.; Wilson, M.G.; McHenry, C.M.; Walters, D.L.; Warnholtz, C.R.; Khafagi, A. Effect of caffeine on adenosine-induced reversible perfusion defects assessed by automated analysis. *J. Nucl. Cardiol.* **2012**, *19*, 474–481. [CrossRef] [PubMed]

21. Zoghbi, G.J.; Htay, T.; Aqel, R.; Blackmon, L.; Heo, J.; Iskandrian, A.E. Effect of caffeine on ischemia detection by adenosine single-photon emission computed tomography perfusion imaging. *J. Am. Coll. Cardiol.* **2006**, *47*, 2296–2302. [CrossRef] [PubMed]

22. Reyes, E.; Loong, C.Y.; Harbinson, M.; Donovan, J.; Anagnostopoulos, C.; Underwood, S.R. High-dose adenosine overcomes the attenuation of myocardial perfusion reserve caused by caffeine. *J. Am. Coll. Cardiol.* **2008**, *52*, 2008–2016. [CrossRef]

23. Smits, P.; Corstens, F.H.; Aengevaeren, W.R.; Wackers, F.J.; Thien, T. False-negative dipyridamole-thallium-201 myocardial imaging after caffeine infusion. *J. Nucl. Med.* **1991**, *32*, 1538–1541.

24. Tejani, F.H.; Thompson, R.C.; Kristy, R.; Bukofzer, S. Effect of caffeine on SPECT myocardial perfusion imaging during regadenoson pharmacologic stress: A prospective, randomized, multicenter study. *Int. J. Cardiovasc. Imaging* **2014**, *30*, 979–989. [CrossRef] [PubMed]

25. Böttcher, M.; Czernin, J.; Sun, K.T.; Phelps, M.E.; Schelbert, H.R. Effect of caffeine on myocardial blood flow at rest and during pharmacological vasodilation. *J. Nucl. Med.* **1995**, *36*, 2016–2021.

26. Kubo, S.; Tadamura, E.; Toyoda, H.; Mamede, M.; Yamamuro, M.; Magata, Y.; Mukai, T.; Kitano, H.; Tamaki, N.; Konishi, J. Effect of caffeine intake on myocardial hyperemic flow induced by adenosine triphosphate and dipyridamole. *J. Nucl. Med.* **2004**, *45*, 730–738. [PubMed]

27. Greulich, S.; Kaesemann, P.; Seitz, A.; Birkmeier, S.; Abu-Zaid, E.; Vecchio, F.; Sechtem, U.; Mahrholdt, H. Effects of caffeine on the detection of ischemia in patients undergoing adenosine stress cardiovascular magnetic resonance imaging. *J. Cardiovasc. Magn. Reson.* **2017**, *19*, 103. [CrossRef]

28. Van Dijk, R.; Kuijpers, D.; Kaandorp, T.A.M.; van Dijkman, P.R.M.; Vliegenthart, R.; van der Harst, P.; Oudkerk, M. Effects of caffeine intake prior to stress cardiac magnetic resonance perfusion imaging on regadenoson-versus adenosine-induced hyperemia as measured by T1 mapping. *Int. J. Cardiovasc. Imaging* **2017**, *33*, 1753–1759. [CrossRef] [PubMed]

29. Nakayama, M.; Chikamori, T.; Uchiyama, T.; Kimura, Y.; Hijikata, N.; Ito, R.; Yuhara, M.; Sato, H.; Kobori, Y.; Yamashina, A. Effects of caffeine on fractional flow reserve values measured using intravenous adenosine triphosphate. *Cardiovasc. Interv. Ther.* **2018**, *33*, 116–124. [CrossRef]

30. Matsumoto, H.; Nakatsuma, K.; Shimada, T.; Ushimaru, S.; Mikuri, M.; Yamazaki, T.; Matsuda, T. Effect of caffeine on intravenous adenosine-induced hyperemia in fractional flow reserve measurement. *J. Invasive Cardiol.* **2014**, *26*, 580–585. [PubMed]

31. Mutha, V.; Ul Haq, M.A.; Van Gaal, W.J. Effects of intravenous caffeine on fractional flow reserve measurements in coronary artery disease. *Open Heart* **2014**, *1*, e000060. [CrossRef] [PubMed]

32. Aqel, R.A.; Zoghbi, G.J.; Trimm, J.R.; Baldwin, S.A.; Iskandrian, A.E. Effect of caffeine administered intravenously on intracoronary-administered adenosine-induced coronary hemodynamics in patients with coronary artery disease. *Am. J. Cardiol.* **2004**, *93*, 343–346. [CrossRef]

33. Montalescot, G.; Sechtem, U.; Achenbach, S.; Andreotti, F.; Arden, C.; Montalescot, G.; Arden, C.; Budaj, A.; Bugiardini, R.; Crea, F.; et al. 2013 ESC guidelines on the management of stable coronary artery disease: The Task Force on the management of stable coronary artery disease of the European Society of Cardiology. *Eur. Heart J.* **2013**, *34*, 2949–3003. [CrossRef] [PubMed]

34. Hachamovitch, R.; Berman, D.S.; Shaw, L.J.; Kiat, H.; Cohen, I.; Cabico, J.A.; Friedman, J.; Diamond, G.A. Incremental prognostic value of myocardial perfusion single photon emission computed tomography for the prediction of cardiac death: Differential stratification for risk of cardiac death and myocardial infarction. *Circulation* **1998**, *97*, 535–543. [CrossRef]

35. Leiner, T. Hold off on that shot of Java: More evidence that caffeine intake leads to false negative adenosine stress myocardial perfusion. *Int. J. Cardiovasc. Imaging* **2017**, *33*, 97–99. [CrossRef] [PubMed]

![nutrients logo] *nutrients*

MDPI

*Review*

# Caffeine-Related Deaths: Manner of Deaths and Categories at Risk

Simone Cappelletti [1], Daria Piacentino [2], Vittorio Fineschi [1,*], Paola Frati [1], Luigi Cipolloni [1] and Mariarosaria Aromatario [1]

1   Department of Anatomical, Histological, Forensic Medicine and Orthopedic Sciences,
    Sapienza University of Rome, 00161 Rome, Italy; simone.cappelletti@uniroma1.it (S.C.);
    paola.frati@uniroma1.it (P.F.); luigi.cipolloni@uniroma1.it (L.C.);
    mariarosaria.aromatario@uniroma1.it (M.A.)
2   NESMOS (Neuroscience, Mental Health, and Sensory Organs) Department,
    Sapienza University of Rome, 00161 Rome, Italy; daria.piacentino@uniroma1.it
*   Correspondence: vfinesc@tin.it; Tel.: +39-0649912722

Received: 14 April 2018; Accepted: 9 May 2018; Published: 14 May 2018

**Abstract:** Caffeine is the most widely consumed psychoactive compound worldwide. It is mostly found in coffee, tea, energizing drinks and in some drugs. However, it has become really easy to obtain pure caffeine (powder or tablets) on the Internet markets. Mechanisms of action are dose-dependent. Serious toxicities such as seizure and cardiac arrhythmias, seen with caffeine plasma concentrations of 15 mg/L or higher, have caused poisoning or, rarely, death; otherwise concentrations of 3–6 mg/kg are considered safe. Caffeine concentrations of 80–100 mg/L are considered lethal. The aim of this systematic review, performed following the Preferred Reporting Items for Systematic Review and Meta-Analyses (PRISMA) statement for the identification and selection of studies, is to review fatal cases in which caffeine has been recognized as the only cause of death in order to identify potential categories at risk. A total of 92 cases have been identified. These events happened more frequently in infants, psychiatric patients, and athletes. Although caffeine intoxication is relatively uncommon, raising awareness about its lethal consequences could be useful for both clinicians and pathologists to identify possible unrecognized cases and prevent related severe health conditions and deaths.

**Keywords:** accidental death; caffeine; caffeine intoxication; intoxication; Suicide

---

## 1. Introduction

In recent years, the risk of caffeine intoxication has increased due to the more widespread availability of analgesics, CNS stimulant medicine and dietary supplements at shops, health stores and e-markets. Nonetheless, lethal cases from caffeine intoxication are quite uncommon. The first paper about lethal caffeine intoxication was published by Jokela et al. in 1959 [1], and it described the accidental death of a young woman following intravenous administration of caffeine.

The pharmacological effects of caffeine include central nervous system and cardiac stimulation and usually occur at plasma concentrations of 15 mg/L or higher. Common features of caffeine intoxication, also known as "caffeinism" (i.e., a state of chronic toxicity from excessive caffeine consumption), include anxiety, agitation, restlessness, insomnia, gastrointestinal disturbances, tremors, psychomotor agitation, and, in some cases, death. Symptoms of caffeine intoxication can mimic those of anxiety and other affective disorders [2]. The cardiovascular effects include supraventricular and ventricular tachyarrhythmias. The direct cause of death is often described as ventricular fibrillation.

Generally, life-threatening caffeine overdoses entail the ingestion of caffeine-containing medications, rather than caffeinated foods or beverages [3], and have been associated with blood concentrations in excess of 80 mg/L [4].

Up to now, there has been limited detailed research regarding caffeine fatalities and there have been sporadic reports about it, although complete reviews have been published on the topic of caffeine [5–7]. The aim of this systematic review is to summarize data regarding caffeine lethal intoxications and try to identify possible categories at risk for it; data obtained from our study could support both clinicians and pathologists in identifying possible unrecognized cases and render possible a better and further comprehension of an ever-growing phenomenon.

## 2. Methods

### 2.1. Eligibility Criteria

The present systematic review was carried out according to the Preferred Reporting Items for Systematic Review and Meta-Analyses (PRISMA) standards [8]. Studies examining caffeine-related deaths, paying particular attention to victims of pure caffeine intoxications, were included. Study designs comprised case reports, case series, retrospective and prospective studies, letters to the editors, and reviews. The latter were downloaded to search their reference lists similarly to other papers, but yielded no other potentially eligible paper. The search was limited to human studies.

### 2.2. Search Criteria and Critical Appraisal

A systematic literature search and a critical appraisal of the collected studies were conducted. An electronic search of PubMed, Science Direct Scopus, and Excerpta Medica Database (EMBASE) from the inception of these databases to the 22th of March 2018 was performed.

Search terms were ("caffeine" OR "coffee") AND ("toxicology" OR "death" OR "decease" OR "fatal intoxication" OR "fatality") in title, abstract, and keywords. Cases in which death has been related to the consumption of energy drinks or caffeinated drinks were excluded because they do not represent "pure" caffeine-related deaths as they are the results of a combination of more substances such as caffeine and alcohol, or caffeine and other caffeine-like substances that may have additional mechanisms of action on cardiovascular and neurological system.

The bibliographies of all located papers were examined and cross-referenced for further relevant literature.

Methodological appraisal of each study was conducted according to the PRISMA standards, including evaluation of bias. Data collection entailed study selection and data extraction. Two researchers (D.P., S.C.) independently examined those papers whose title or abstract appeared to be relevant and selected the ones that analyzed deaths due to caffeine intoxication. Disagreements concerning eligibility between the three researchers were resolved by consensus process. No unpublished or grey literature was searched. Data extraction was performed by one investigator (M.A.) and verified by another investigator (V.F.). This study was exempt from institutional review board approval as it did not involve human subjects.

## 3. Results

### 3.1. Search Results and Included Studies

An appraisal based on titles and abstracts as well as a hand search of reference lists was carried out. The reference lists of all located articles were reviewed to detect still unidentified literature. Figure 1 illustrate our search strategy.

A total of 36 studies fulfilled the inclusion criteria, producing a pooled dataset of 92 individuals. The reviewed studies involved a sample size ranging from 1 (i.e., case reports) to 22 individuals (i.e., a retrospective study), with a mean of 2.6 and a median of 1, indicating skewness towards smaller samples.

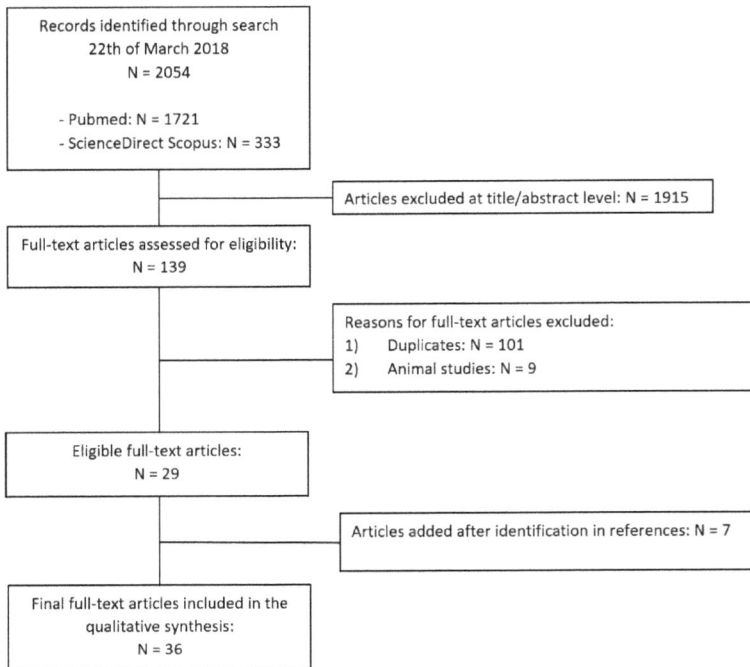

**Figure 1.** Search strategy

### 3.2. Study Characteristics

The following data were extracted from the included studies: study source; age and sex of participants in the study; toxicological data (if reported); way of administration. An exhaustive summary of the literature, including extracted data, is shown in Table 1.

**Table 1.** Caffeine-related fatalities.

| Author (Year) | Caffeine Blood Level (mg/L) | Age | Gender | Manner of Death | Route of Administration (Source) |
|---|---|---|---|---|---|
| Jokela et al. (1959) [1] | - | 35 | F | Accidental | Intravenous |
| Farago et al. (1968) [9] | 1040 mg/L | 15 months | - | Child abuse | Intravenous |
| Alstott et al. (1973) [10] | - | 27 | M | Suicide | Oral (pills) |
| Grusz-Hardy (1973) [11] | 79 mg/L | 45 | F | Accidental | Oral (pills) |
| Dimaio et al. (1974) [12] | 158.5 mg/L | 5 | F | Accidental | Oral (pills) |
| Turner et al. (1977) [13] | 106 mg/L | 34 | F | Uncertain | Oral (pills) |
| McGee (1980) [14] | 181 mg/L | 19 | F | Accidental | Oral (pills) |
| Bryant (1981) [15] | 113.5 mg/L | 42 | F | Suicide | Oral (pills) |
| Chaturvedi et al. (1983) [16] | 62 mg/L | 21 | M | Suicide | Oral (pills) |
| Garriott et al. (1985) [17] | 129.9 mg/L | 19 | F | Suicide | Oral (pills) |
|  | 147 mg/L | 21 | M | Suicide | Oral (pills) |
|  | 343.9 mg/L | 21 | M | Suicide | Oral (pills) |
|  | 184.1 mg/L | 23 | M | Accidental | Oral (pills) |
|  | 251 mg/L | 21 | F | Suicide | Oral (pills) |

**Table 1.** *Cont.*

| Author (Year) | Caffeine Blood Level (mg/L) | Age | Gender | Manner of Death | Route of Administration (Source) |
|---|---|---|---|---|---|
| Winek et al. (1985) [18] | 240 mg/L | 21 | F | Suicide | Oral (pills) |
| Hanzlick et al. (1986) [19] | 264 mg/L | 44 | F | suicide | Oral (pills) |
| | 182 mg/L | 20 | F | accidental | Oral (pills) |
| Morrow (1987) [20] | 117.3 mg/L | 14 months | - | Child abuse | Oral (pills) |
| Mrvos et al. (1989) [21] | 1560 mg/L | 22 | F | Accidental | Oral (pills) |
| Takayasu et al. (1993) [22] | 177.0 μg/g | 20 | F | Suicide | Oral (pills) |
| Rivenes et al. (1997) [23] | 117 mg/L | 5 weeks | M | Child abuse | Oral (pills) |
| Shum et al. (1997) [24] | 108 mcg/dL | 15 | F | Accidental | Oral (pills) |
| | 30 mcg/dL | 32 | M | Accidental | Oral (pills) |
| Riesselmann et al. (1999) [25] | 220 mg/L | 19 | F | Accidental | Oral (pills) |
| | 190 mg/L | 81 | F | Suicide | Not reported |
| Watson et al. (2004) [26] | - | 17 | - | Suicide | Oral (pills) |
| Holmgren et al. (2004) [27] | 173 mg/L | 54 | M | Uncertain | Oral (pills) |
| | 210 mg/L | 21 | M | Suicide | Oral (pills) |
| | 153 mg/L | 31 | M | Suicide | Oral (pills) |
| | 200 mg/L | 47 | F | Uncertain | Oral (pills) |
| Watson et al. (2005) [28] | - | 33 | - | Accidental | Oral (pills) |
| Kerrigan et al. (2005) [29] | 192 mg/L | 39 | F | Accidental | Intravenous |
| | 567 mg/L | 29 | M | Accidental | Oral (pills) |
| Takeuchi et al. (2007) [30] | - | - | - | Accidental | Oral (pills) |
| Rudolph et al. (2010) [31] | - | 21 | F | Suicide | Oral (pills) |
| Thelander et al. (2010) [32] | 90 mg/L | 43 | M | Uncertain | Not reported |
| | 105 mg/L | 53 | M | Suicide | Not reported |
| | 170 mg/L | 47 | M | Uncertain | Not reported |
| | 86 mg/L | 26 | F | Uncertain | Not reported |
| | 210 mg/L | 25 | F | Suicide | Not reported |
| | 230 mg/L | 40 | F | Uncertain | Not reported |
| | 210 mg/L | 21 | M | Suicide | Not reported |
| | 153 mg/L | 31 | M | Suicide | Not reported |
| | 173 mg/L | 54 | M | Uncertain | Not reported |
| | 200 mg/L | 47 | F | Uncertain | Not reported |
| | 180 mg/L | 18 | F | Suicide | Not reported |
| | 166 mg/L | 20 | F | Suicide | Not reported |
| | 140 mg/L | 72 | F | Suicide | Not reported |
| | 80 mg/L | 24 | M | Suicide | Not reported |
| | 160 mg/L | 46 | F | Suicide | Not reported |
| | 113 mg/L | 73 | F | Uncertain | Not reported |
| | 138 mg/L | 66 | M | Accidental | Not reported |
| | 190 mg/L | 84 | M | Suicide | Not reported |
| | 192 mg/L | 79 | F | Suicide | Not reported |
| | 310 mg/L | 33 | F | Suicide | Not reported |
| Jabbar et al. (2013) [33] | 350 mg/L | 39 | M | Accidental | Oral (powder) |
| Jantos et al. (2013) [34] | 141 mg/L | 25 | F | Suicide | Oral (pills) |
| Poussel et al. (2013) [35] | 190 mg/L | 44 | M | Accidental | Oral (pills) |
| Bonsignore et al. (2014) [36] | 170 mg/L | 3 | M | Suicide | Oral (pills) |

**Table 1.** *Cont.*

| Author (Year) | Caffeine Blood Level (mg/L) | Age | Gender | Manner of Death | Route of Administration (Source) |
|---|---|---|---|---|---|
| | 320 mg/L | 50 | F | Uncertain | Oral (pills) |
| | 73 mg/L | 37 | F | Uncertain | Not reported |
| Banerjee et al. (2014) [37] | 320 mg/L | 43 | F | Suicide | Oral (pills) |
| | 74 mg/L | 44 | M | Uncertain | Oral (pills) |
| | 220 mg/L | 57 | M | Suicide | Oral (pills) |
| Eichner ER (2014) [38] | >70 mg/L | 18 | M | Accidental | Oral (powder) |
| Suzuki et al. (2014) [39] | 179 mg/L | 22 cases 20–90 years-old | - | 11 unknown 7 accidental 2 suicide 2 others | |
| Ishikawa et al. (2015) [40] | Blood 154.2 mg/L Bile 852.3 mg/L Stomach 197.5 mg/L | 20 | F | Suicide | Oral (pills) |
| Yamamoto et al. (2015) [41] | 290 mg/L | 18 | F | Suicide | Oral (pills) |
| Aknouche et al. (2017) [42] | 401 mg/L | 48 | M | Suicide | Oral (pills) |
| Magdalan et al. (2017) [43] | 140 mg/L | 27 | M | Accidental | Oral (pills) |
| | 613 mg/L | 20 | F | Uncertain | Oral (powder) |

## 3.3. Risk of Bias

This systematic review has a number of strengths that include the amount and breadth of the studies, which span the globe, the hand search and scan of reference lists for the identification of all relevant studies, and a flowchart that describe in detail the study selection process. It must be noted that this review includes studies that were published in a time frame of 59 years; thus, despite our efforts to fairly evaluate the existing literature, study results should be interpreted taking into account that the accuracy of the toxicological analyses, where reported, has changed over the years.

## 3.4. Caffeine-Related Fatalities

Despite the recent policy of sale restrictions of caffeine tablets, which was introduced in 2004 in several countries, we have identified an increase in caffeine-related deaths in the last years (Table 1).

Our study allowed us to identify the manner of death as suicide (36), accidental (27), intentional poisoning (2), and uncertain (27). Routes of administration of caffeine were: oral (pills, powder, liquid) in 46 cases, intravenous in three cases, and not reported in the remaining 43 cases.

Unintentional caffeine abuse due to excessive intake of caffeine is relatively frequent and responsible for classical clinical manifestations of overstimulation. However, death due to caffeine intoxication is rare and case reports of fatalities from caffeine toxicity are relatively infrequent. We have identified 28 cases (29%), among the 92 lethal cases described in the literature, in which death was attributed to accidental causes (Table 2). The majority of fatalities were related to the ingestion of a great amount of over-the-counter caffeine products. These tend to be weight loss supplements that are frequently used and perceived as safe, but that can be toxic and linked to serious health complications.

**Table 2.** Accidental causes among caffeine-related deaths.

| Causes | Cases |
|---|---|
| Not reported | 10 |
| Over-the-counter caffeine products | 9 |
| Errors in hospital medication | 3 |
| Drug abuse | 2 |
| Recreational use | 2 |
| Accidental ingestion by children | 1 |

As a result, the category of individuals consuming caffeine-containing products for dietary purposes represent a group at risk for severe intoxications, potentially leading to decease.

Despite cases where consumption of caffeine has accidentally lead to death and where caffeine was taken with suicidal purposes, we recognized three categories of individuals who have often been involved in caffeine-related deaths: athletes, psychiatric patients and infants.

In the latter group, the manner of death is linked to: intentional poisoning and child abuse; the low frequency of these categories in the other groups has encouraged us to emphasize these aspects.

### 3.5. Athletes

Five caffeine-related deaths (5%) among athletes have been described in the literature; these subjects were two amateur body builders, a basketball player and a wrestler [33,35,38]. The age ranged from 18 to 44. In all cases, the cause of death was attributed to cardiac arrest due to ventricular fibrillation.

Among these patients, body builders are well known to suffer from altered perception of body image often leading to unhealthy eating, heavy exercise habits, or even drug-taking, often with little regard to safety in spite of well publicized side effects [44]. In physiologically predisposed individuals, a combination of excessive ingestion of caffeine and strenuous physical activity can induce myocardial ischaemia by coronary vasospasm, with potentially fatal results.

### 3.6. Psychiatric Patients

Thirty-seven cases (39%) with a history of a psychiatric disorder have been identified; among the psychiatric disorders, depression is undoubtedly the most frequent (Table 3). The age ranged from 21 to 84 years-old.

The manner of death was undetermined in most of the reviewed cases, even if suicide has been recognized as the second most frequent manner. Many of these individuals have a history of past suicide attempts. A recent review on this specific topic, showed that caffeine was still a rare factor in a number of studies concerning its association with suicide attempts and death [45].

**Table 3.** Psychiatric disorders diagnosed before death.

| Disease * | Number |
|---|---|
| Depression | 20 |
| Alcohol dependence | 6 |
| Sleep disorders | 6 |
| Drug dependence | 4 |
| Eating disorder | 3 |
| Panic disorder | 2 |
| Schizophrenia | 2 |
| Not specified | 2 |
| Paranoid disorder | 1 |

* More than one disease may have been identified for each case.

### 3.7. Infants

Fatal caffeine poisoning in children is rarely described in the literature [9,12,20,23], with the few existing cases possibly being related to child abuse and neglect, as well as accidental causes. Among accidental fatalities, iatrogenic medication errors should be taken into account.

Only two cases of intentional poisoning by using caffeine, with concomitant child abuse were reported in literature. Morrow et al. described the case of a 14-month-old child who died for caffeine intoxication [20]. Although in this case it is unknown when or how this child ingested caffeine, the clear evidence of prolonged vomiting and the high blood level of theophylline attest to a long period of severe toxicity during which no medical help was sought. These facts, as well as the delay in weight gain, chronic iron deficiency anemia, thymic involution and severe trauma to the ribs and spleen are diagnostic of child neglect and abuse.

Rivenes et al. described the case of a 5-week-old boy admitted to the hospital for evaluation of persistent tachycardia [23]. During the examination, a preliminary drug screening was negative, but a comprehensive screen subsequently performed by gas chromatographic-mass spectrometry (GC/MS) revealed the presence of high levels of caffeine, ranging from 5–12 mg/L, which are incompatible with the therapeutic values for the boy's age. Since the source of caffeine remained unknown and its levels were far too high to be consistent with transfer from the breast milk, a referral to Child Protective Services was made. Three weeks after discharge the infant was readmitted with subarachnoid haemorrhages. He died a few dies later the admission. At the autopsy, signs of abuse, i.e., old and new rib fractures, a left spiral radial fracture, a right distal clavicular fracture, and cerebral contusion, were observed. The father admitted giving caffeine tablets to the infant "to see what it did".

With regard to the accidental causes, only two cases of fatal caffeine intoxication are reported in the medical literature. The first, described by Di Maio et al., concerned a 5-year-old girl who ingested about 40 diuretic tablets that she found in her mother room [12]. The other one, reported by Farago, regarded a 15-month-old child who underwent a test meal in a hospital [9]. Instead of receiving 90 mL of a 2% caffeine sodium benzoate solution, the child was given 90 mL of a 20% caffeine solution (about 18 g of caffeine). Despite the prompt treatment with gastric lavage, calcium hexobarbitone, and transfusions his condition deteriorated, and he died few hours later.

## 4. Discussion

The effects of caffeine on the cardiovascular system are the result of the direct and/or indirect action of caffeine on the neuroendocrine control systems of vascular resistance, cardiac function, and electrolyte balance.

Although cases of lethal intoxication have been mainly associated with the occurrence of arrhythmic events induced by caffeine, human studies provided scarce evidence to support the substance's ability to induce arrhythmic events in healthy subjects and in subjects predisposed to such events [2,46–49]. These findings, however, even if provided by studies differing in sample size and methods, should not be considered in disagreement with the conclusions of those studies reporting cases of lethal intoxication, as they take into account caffeine doses below the ones considered toxic for humans.

Furthermore, it should be considered that the concepts of toxic and lethal doses in humans are relative concepts, as doses below the toxic and/or lethal range may play a causal role in inducing intoxication or death. This could be due to:

- interactions with other substances with a synergistic effect when consumed with caffeine or able to increase caffeine's blood levels;
- individuals' pre-existing diseases and/or conditions capable of potentiating the effects of caffeine;
- inter-individual differences, mostly genetically determined, that can affect caffeine metabolism in both directions (i.e., increase or reduction), contributing to a different individual "sensibility" to the effects of the substance.

### 4.1. Caffeine and Athletes

Drug use among athletes, especially bodybuilders and weightlifters, has become a recognized problem in sports. Athletes may use drugs for therapeutic indications, for recreational or social reasons, as ergogenic aids or to mask the presence of other drugs during drug testing. Stimulants, such as caffeine, were some of the first drugs used and studied as ergogenic aids.

Sometimes, a psychiatric pathogenesis could represent the basis for excessive caffeine consumption in athletes. Indeed, some disorders are typically linked to recreational and professional athletes who consume caffeine to face fatigue and intense workouts. An example is muscle dysmorphia.

This condition, also known as "reverse anorexia" or "Adonis complex", is a subtype of body dysmorphic disorder generally affecting men, with its onset in adolescence or early adulthood, characterized by obsessiveness and compulsivity directed toward achieving a lean and muscular physique, even at the expense of health. This raises the issue of whether caffeine use causes these disorders in athletes, by inducing neuroadaptive changes within the reward neural circuit and affecting mechanisms of resilience to stress, or, vice versa, athletes with pre-morbid abnormal personalities or a history of psychiatric disorders are attracted to caffeine use, encouraged by an extrinsic motivation for exercise focused on appearance and weight control. Further studies on this topic are necessary for a full comprehension of this phenomenon.

Prior to 2004, caffeine was included in the World Anti-Doping Agency (WADA) Prohibited List of substances and methods; it was then removed, allowing athletes who compete in sports compliant with the WADA code to consume caffeine within their usual diets or for specific purposes of performance [50]. This revision was based on the acknowledgment that caffeine enhances performance at doses that are impossible to differentiate from daily caffeine use and that the practice of monitoring caffeine use via urinary concentration is not completely reliable. Despite this premise, WADA continues to measure caffeine levels through urinary concentration testing within its Monitoring Program, in order to investigate patterns of misuse of substances in sport. Differently from the WADA, the National Collegiate Athletic Association (NCAA), a non-profit association that regulates the athletes of over 1000 American institutions and associations, has a urinary concentration limit of 15 µg/mL; thus, athletes in the NCAA have to take into account that caffeine is still on the list of controlled substances.

## 4.2. Caffeine and Psychiatric Patients

Psychiatric disorders have been related to large amounts and long-term use of caffeine [51]. Furthermore, it has been suggested that caffeine may act as a trigger of psychiatric symptoms, from anxiety to depression and even psychosis [52].

In the past years, many studies have highlighted the relationship between caffeine intake and specific psychiatric disorders, in particular, bipolar [53], anxiety [54], eating disorders [55], and psychoses [56]. In addition to causing or worsening psychiatric symptoms [57], caffeine use has been investigated for its potential to interact with many psychiatric medications [58]. Caffeine is metabolized by the CYP1A2 enzyme and also acts as a competitive inhibitor of this enzyme, being able to interact with a wide range of psychiatric medications, including antidepressant, antipsychotic, antimanic, antianxiety and sedative agents. These interactions may lead to caffeine-related or medication-related side effects that may complicate psychiatric treatment, and in the most severe cases, lead to death.

With regards to alcohol use disorder (AUD), behavioural and genetic associations indicate that there is a significant link between caffeine and alcohol intake [59]. Regarding caffeine abuse by alcoholics, individuals with AUD consume approximately 30% more caffeine daily, compared to non-alcoholic individuals [60]. Besides, reports suggest that detoxified alcoholics consume large quantities of coffee following cessation of alcohol drinking, compared to their prior intake [61]. This could be a serious concern for treatment-seeking alcoholics. For example, using caffeine intake as a substitute stimulus for alcohol consumption could interfere with psychological and physiological efforts to overcome addiction-related behaviours. In addition, it is uncertain what impact a history of alcohol drinking could have on caffeine's pharmacokinetics and metabolism profile, and whether this could affect the caffeine levels consumed by actively drinking and detoxified individuals. In conclusion, public health concern over caffeinated alcohol drinks is justified, although the nature of the caffeine/alcohol relationship is yet to be fully elucidated.

## 4.3. Caffeine and Infants

Poisoning is a severe and potentially lethal form of child abuse, and case reports have become increasingly frequent. Multiple agents have been used to poison children, including salt, water, narcotics, laxatives, diuretics, salicylates, phenothiazines, tricyclic antidepressants, insulin, sedatives and others [62–64].

These cases can be difficult to identify because of the clinical presentation and misleading histories. Indeed, many patients present at ages or with histories incompatible with "accidental" ingestion, others may even present with histories of recurrent illnesses suggesting previous undiagnosed poisonings.

Child abuse is considered to occur in several clinical patterns, including child neglect and physical/sexual abuse.

Generally speaking, when child neglect and abuse are carried out, other signs could be evident. For example, delay in normal weight gain and unexplained trauma are typical. In particular, intentional poisoning may be associated with other forms of abuse; approximately 20% of poisoned children may have evidence of physical abuse [65,66].

Other cases of child abuse poisoning are reported in the medical literature. Some of these are reports of deliberate parental poisoning of children and could represents the evidence of a Munchausen syndrome by proxy (MSBP), carried out by the caregiver of the child [67]. In these cases the most common mode of disease instigation involved poisoning through beverage/food contamination or subcutaneous injection [68].

The mortality rate among children diagnosed with MSBP is 9% and the most frequent causes of death are suffocation and poisoning [69]. For this reason, when child poisoning occurs, the eventual role of the caregiver as cause of poisoning must be taken into account.

Fatalities from accidental poisoning are, still nowadays, frequent in literature.

Some of these cases, as the aforementioned one, involve iatrogenic medication errors, particularly in neonatal intensive care unit where caffeine is routinely used for the treatment of the apnea [70,71]. These errors are related, in the majority of cases, to drug weighing processes.

Rivenes et al. reviewed cases of pediatric caffeine overdose and reported that the majority of cases occurred because of iatrogenic medication errors. Authors also indicated the blood levels of caffeine and highlighted that even high blood concentration of caffeine in infants can be successfully treated, thus preventing the death of the patient [23].

In conclusion when unexplained ingestions occur in children, these must be treated as non-accidental poisonings until proven otherwise. These cases required full evaluations of the social situations and sometimes required the involvement of Child Protective Services. However, caffeine toxicity could be missed because this drug is frequently not reported on routine toxicological analysis.

## 5. Conclusions

This paper represents a comprehensive review of fatal cases due to caffeine intoxication that can be found in the literature. Athletes, psychiatric patients, and infants should receive particular attention with regard to their caffeine consumption. Indeed, athletes seems to consume high quantities of caffeine as performing and image enhancing aids; at the same time, caffeine use in psychiatric patients must be considered as an important risk factor for possible intoxications because of the synergic action of caffeine with many psychiatric drugs. Finally, infants have been recognized as a last category of patients in which the use of caffeine should be completely avoided.

Indeed, previous authors have conducted systematic reviews of this topic, but they focused on specific aspects of forensic toxicology or, vice versa, took into consideration more general clinical-epidemiological issues [6,7]. Recently, a review focused on caffeine concentrations in postmortem blood in fatal cases attributed to overdose from the compound [6]. Again, a systematic review regarded the adverse effects of caffeine in pregnant women, adolescents, and children [7]. It is interesting to note how the authors hope for a change in methodology in the field of research dedicated to the use of caffeine. The topic is stressed in order to characterize the inter-individual trends, unhealthy populations, co-exposures, and outcomes, so to have a roadmap about the risk regarding caffeine-related adverse events [7].

The dangers of caffeine are related to the wide diffusion of the substance, which results in a partially conscious high consumption, due to the difficulty of ascertaining the actual amount of caffeine ingested daily and the inability to predict specific effects with regard to the "trigger role" that caffeine can have—even at "safe" doses—on underlying and not necessarily known cardiovascular conditions.

Caffeine, like alcohol and tobacco, is legally used, but, unlike the last two, its sale in the form of high concentration (e.g., powder or tablets) is not controlled or restricted.

Accidental deaths from the consumption of over-the-count and/or dietary caffeine products represent the most common cause of death in our study. The high frequency of use, the uncontrolled sales of these products, and the potentially triggering action of caffeine on cardiovascular system pose a serious risk to the health and safety of consumers.

The findings of our paper underline the importance of a fundamental principle of prevention strategy put forth by the eminent British epidemiologist Geoffrey Rose: "A large number of people exposed to a low risk is likely to produce more cases than a small number of people exposed to a high risk."

We sincerely hope that information given about the frequency and the categories at a higher risk for caffeine intoxications may be useful for both clinicians and pathologists for a better understanding of the potentially fatal complications of coffee consumption.

**Author Contributions:** S.C. and M.A. conceived and designed the study; D.P. performed the study; V.F. and P.F. analyzed the data; L.C. contributed reagents/materials/analysis tools; S.C. and D.P. wrote the paper.

**Acknowledgments:** This study is part of the FIRB project code RBFR12LD0W_002 and has been funded by a grant of the Italian Ministry of Research.

**Conflicts of Interest:** The authors declare no conflict of interest.

## References

1. Jokela, S.; Varliainen, A. Caffeine poisoning. *Acta Pharmacol. Toxicol.* **1959**, *15*, 331–334. [CrossRef]
2. Cappelletti, S.; Piacentino, D.; Sani, G.; Aromatario, M. Caffeine: Cognitive and physical performance enhancer or psychoactive drug? *Curr. Neuropharmacol.* **2015**, *13*, 71–88. [CrossRef] [PubMed]
3. Nawrot, P.; Jordan, S.; Eastwood, J.; Rotstein, J.; Hugenholtz, A.; Feeley, M. Effects of caffeine on human health. *Food Addit. Contam.* **2003**, *20*, 1–30. [CrossRef] [PubMed]
4. Higdon, J.V.; Frei, B. Coffee and health: A review of recent human research. *Crit. Rev. Food Sci. Nutr.* **2006**, *46*, 101–123. [CrossRef] [PubMed]
5. Musgrave, I.F.; Farrington, R.L.; Hoban, C.; Byard, R.W. Caffeine toxicity in forensic practice: Possible effects and under-appreciated sources. *Forensic Sci. Med. Pathol.* **2016**, *12*, 299–303. [CrossRef] [PubMed]
6. Jones, A.W. Review of caffeine-related fatalities along with postmortem blood concentrations in 51 poisoning deaths. *J. Anal. Toxicol.* **2017**, *41*, 167–172. [CrossRef] [PubMed]
7. Wikoff, D.; Welsh, B.T.; Henderson, R.; Brorby, G.P.; Britt, J.; Myers, E.; Goldberger, J.; Lieberman, H.R.; O'Brien, C.; Peck, J.; et al. Systematic review of the potential adverse effects of caffeine consumption in healthy adults, pregnant women, adolescents, and children. *Food Chem. Toxicol.* **2017**, *109*, 585–648. [CrossRef] [PubMed]
8. Moher, D.; Liberati, A.; Tetzlaff, J.; Altman, D.G. PRISMA Group. Preferred reporting items for systematic reviews and meta-analyses: The PRISMA statement. *J. Clin. Epidemiol.* **2009**, *62*, 1006–1012. [CrossRef] [PubMed]
9. Farago, A. Fatal accidental caffeine poisoning of a child. *Bull. Int. Assoc. Forensic Toxicol.* **1968**, *5*, 2–3.
10. Alstott, R.L.; Miller, A.J.; Forney, R.B. Report of a human fatality due to caffeine. *J. Forensic Sci.* **1973**, *18*, 135–137. [CrossRef] [PubMed]
11. Grusz-Hardy, E. Lethal caffeine poisoning. *Bull. Int. Assoc. Forensic Toxicol.* **1973**, *4*, 6–7.
12. Dimaio, V.J.M.; Garriott, J.C. Lethal caffeine poisoning in a child. *Forensic Sci.* **1974**, *3*, 275–278. [CrossRef]
13. Turner, J.E.; Cravey, R.H. A fatal ingestion of caffeine. *Clin. Toxicol.* **1977**, *10*, 341–344. [CrossRef] [PubMed]
14. McGee, M.B. Caffeine poisoning in a 19-year-old female. *J. Forensic Sci.* **1980**, *25*, 29–32. [CrossRef] [PubMed]
15. Bryant, J. Suicide by ingestion of caffeine. *Arch. Pathol. Lab. Med.* **1981**, *105*, 685–686. [PubMed]

16. Chaturvedi, A.K.; Rao, N.G.; McCoy, F.E. A multi-chemical death involving caffeine, nicotine and malathion. *Forensic Sci. Int.* **1983**, *23*, 265–275. [CrossRef]

17. Garriott, J.C.; Simmons, L.M.; Poklis, A.; Mackell, M.A. Five cases of fatal overdose from caffeine containing look alike drugs. *J. Anal. Toxicol.* **1985**, *9*, 141–143. [CrossRef] [PubMed]

18. Winek, C.L.; Wahba, W.; Williams, K.; Blenko, J.; Janssen, J. Caffeine fatality: A case report. *Forensic Sci. Int.* **1985**, *29*, 207–211. [CrossRef]

19. Hanzlick, R.; Gowitt, G.T.; Wall, W. Deaths due to caffeine in "look-alike drugs". *J. Anal. Toxicol.* **1986**, *10*, 126. [CrossRef] [PubMed]

20. Morrow, P.L. Caffeine toxicity: A case of child abuse by drug ingestion. *J. Forensic Sci.* **1987**, *32*, 1801–1805. [CrossRef] [PubMed]

21. Mrvos, R.M.; Reilly, P.E.; Dean, B.S.; Krenzelok, E.P. Massive caffeine ingestion resulting in death. *Vet. Hum. Toxicol.* **1989**, *31*, 571–572. [PubMed]

22. Takayasu, T.; Nishigami, J.; Ohshima, T.; Lin, Z.; Kondo, T.; Nakaya, T.; Sawaguchi, T.; Nagano, T. A fatal case due to intoxication with seven drugs detected by GC-MS and TDx methods. *Nihon Hoigaku Zasshi* **1993**, *47*, 63–71. [PubMed]

23. Rivenes, S.M.; Bakerman, P.R.; Miller, M.B. Intentional caffeine poisoning in an infant. *Pediatrics* **1997**, *99*, 736–738. [CrossRef] [PubMed]

24. Shum, S.; Seale, C.; Hathaway, D.; Chucovich, V.; Beard, D. Acute caffeine ingestion fatalities: Management issues. *Vet. Hum. Toxicol.* **1997**, *39*, 228–230. [PubMed]

25. Riesselmann, B.; Rosenbaum, F.; Roscher, S.; Schneider, V. Fatal caffeine intoxication. *Forensic Sci. Int.* **1999**, *103*, 49–52. [CrossRef]

26. Watson, W.A.; Litovitz, T.L.; Klein-Schwartz, W.; Rodgers, G.C., Jr.; Youniss, J.; Reid, N.; Rouse, W.G.; Rembert, R.S.; Borys, D. 2003 annual report of the American Association of Poison Control Centers Toxic Exposure Surveillance System. *Am. J. Emerg. Med.* **2004**, *22*, 335–404. [CrossRef] [PubMed]

27. Holmgren, P.; Nordén-Pettersson, L.; Ahlner, J. Caffeine fatalities—Four case reports. *Forensic Sci. Int.* **2004**, *139*, 71–73. [CrossRef] [PubMed]

28. Watson, W.A.; Litovitz, T.L.; Rodgers, G.C., Jr.; Klein-Schwartz, W.; Reid, N.; Youniss, J.; Flanagan, A.; Wruk, K.M. 2004 Annual report of the American Association of Poison Control Centers Toxic Exposure Surveillance System. *Am. J. Emerg. Med.* **2005**, *23*, 589–666. [CrossRef] [PubMed]

29. Kerrigan, S.; Lindsey, T. Fatal caffeine overdose: Two case reports. *Forensic Sci. Int.* **2005**, *153*, 67–69. [CrossRef] [PubMed]

30. Takeuchi, S.; Homma, M.; Inoue, J.; Kato, H.; Murata, K.; Ogasawara, T. Case of intractable ventricular fibrillation by a multicomponent dietary supplement containing ephedra and caffeine overdose. *Jpn. J. Toxicol.* **2007**, *20*, 269–271.

31. Rudolph, T.; Knudsen, K. A case of fatal caffeine poisoning. *Acta Anaesthesiol. Scand.* **2010**, *54*, 521–523. [CrossRef] [PubMed]

32. Thelander, G.; Jönsson, A.K.; Personne, M.; Forsberg, G.S.; Lundqvist, K.M.; Ahlner, J. Caffeine fatalities— Do sales restrictions prevent intentional intoxications? *Clin. Toxicol.* **2010**, *48*, 354–358. [CrossRef] [PubMed]

33. Jabbar, S.B.; Hanly, M.G. Fatal caffeine overdose: A case report and review of literature. *Am. J. Forensic Med. Pathol.* **2013**, *34*, 321–324. [CrossRef] [PubMed]

34. Jantos, R.; Stein, K.M.; Flechtenmacher, C.; Skopp, G. A fatal case involving a caffeine-containing fat burner. *Drug Test. Anal.* **2013**, *5*, 773–776. [CrossRef] [PubMed]

35. Poussel, M.; Kimmoun, A.; Levy, B.; Gambier, N.; Dudek, F.; Puskarczyk, E.; Poussel, J.F.; Chenuel, B. Fatal cardiac arrhythmia following voluntary caffeine overdose in an amateur body-builder athlete. *Int. J. Cardiol.* **2013**, *166*, e41–e42. [CrossRef] [PubMed]

36. Bonsignore, A.; Sblano, S.; Pozzi, F.; Ventura, F.; Dell'Erba, A.; Palmiere, C. A case of suicide by ingestion of caffeine. *Forensic Sci. Med. Pathol.* **2014**, *10*, 448–451. [CrossRef] [PubMed]

37. Banerjee, P.; Ali, Z.; Levine, B.; Fowler, D.R. Fatal caffeine intoxication: A series of eight cases from 1999 to 2009. *J. Forensic Sci.* **2014**, *59*, 865–868. [CrossRef] [PubMed]

38. Eichner, E.R. Fatal caffeine overdose and other risks from dietary supplements. *Curr. Sports Med. Rep.* **2014**, *13*, 353–354. [CrossRef] [PubMed]

39. Suzuki, H.; Tanifuji, T.; Abe, N.; Maeda, M.; Kato, Y.; Shibata, M.; Fukunaga, T. Characteristics of caffeine intoxication-related death in Tokyo, Japan, between 2008 and 2013. *Nihon Arukoru Yakubutsu Igakkai Zasshi* **2014**, *49*, 270–277. [PubMed]

40. Ishikawa, T.; Yuasa, I.; Endoh, M. Non specific drug distribution in an autopsy case report of fatal caffeine intoxication. *Leg. Med.* **2015**, *17*, 535–538. [CrossRef] [PubMed]

41. Yamamoto, T.; Yoshizawa, K.; Kubo, S.; Emoto, Y.; Hara, K.; Waters, B.; Umehara, T.; Murase, T.; Ikematsu, K. Autopsy report for a caffeine intoxication case and review of the current literature. *J. Toxicol. Pathol.* **2015**, *28*, 33–36. [CrossRef] [PubMed]

42. Aknouche, F.; Guibert, E.; Tessier, A.; Eibel, A.; Kintz, P. Suicide by ingestion of caffeine. *Egypt. J. Forensic Sci.* **2017**, *7*, 6. [CrossRef] [PubMed]

43. Magdalan, J.; Zawadzki, M.; Skowronek, R.; Czuba, M.; Porębska, B.; Sozański, T.; Szpot, P. Nonfatal and fatal intoxications with pure caffeine—Report of three different cases. *Forensic Sci. Med. Pathol.* **2017**, *13*, 355–358. [CrossRef] [PubMed]

44. FitzSimmons, C.R.; Kidner, N. Caffeine toxicity in a bodybuilder. *J. Accid. Emerg. Med.* **1998**, *15*, 196–197. [CrossRef] [PubMed]

45. Silva, A.C.; de Oliveira Ribeiro, N.P.; de Mello Schier, A.R.; Pereira, V.M.; Vilarim, M.M.; Pessoa, T.M.; Arias-Carrion, O.; Machado, S.; Nardi, A.E. Caffeine and suicide: A systematic review. *CNS Neurol. Disord. Drug Targets* **2014**, *13*, 937–944. [CrossRef] [PubMed]

46. Ciszowski, K.; Biedron, W.; Gomolka, E. Acute caffeine poisoning resulting in atrial fibrillation after guarana extract overdose. *Prz. Lek.* **2014**, *71*, 495–498. [PubMed]

47. Bryczkowski, C.; Geib, A.J. Combined butalbital/acetaminophen/caffeine overdose: Case files of the Robert Wood Johnson Medical School Toxicology Service. *J. Med. Toxicol.* **2012**, *8*, 424–431. [CrossRef] [PubMed]

48. Berger, A.J.; Alford, K. Cardiac arrest in a young man following excess consumption of caffeinated "energy drinks". *Med. J. Aust.* **2009**, *190*, 41–43. [PubMed]

49. Rhidian, R. Running a risk? Sport supplement toxicity with ephedrine in an amateur marathon runner, with subsequent rhabdomyolysis. *BMJ Case Rep.* **2011**. [CrossRef] [PubMed]

50. World Anti-Doping Agency. The 2009 Monitoring Program. Available online: https://www.wada-ama.org/sites/default/files/resources/files/WADA_Monitoring_Program_2009_EN.pdf (accessed on 21 March 2018).

51. Ogawa, N.; Ueki, H. Clinical importance of caffeine dependence and abuse. *Psychiatry Clin. Neurosci.* **2007**, *61*, 263–268. [CrossRef] [PubMed]

52. Broderick, P.; Benjamin, A.B. Caffeine and psychiatric symptoms: A review. *J. Okla. State Med. Assoc.* **2004**, *97*, 538–542. [PubMed]

53. Ogawa, N.; Ueki, H. Secondary mania caused by caffeine. *Gen. Hosp. Psychiatry* **2003**, *25*, 138–139. [CrossRef]

54. Childs, E.; Hohoff, C.; Deckert, J.; Xu, K.; Badner, J.; de Wit, H. Association between ADORA2A and DRD2 polymorphisms and caffeine-induced anxiety. *Neuropsychopharmacology* **2008**, *33*, 2791–2800. [CrossRef] [PubMed]

55. Stock, S.L.; Goldberg, E.; Corbett, S.; Katzman, D.K. Substance use in female adolescents with eating disorders. *J. Adolesc. Health* **2002**, *31*, 176–182. [CrossRef]

56. Caykoylu, A.; Ekinci, O.; Kuloglu, M. Improvement from treatment-resistant schizoaffective disorder, manic type after stopping heavy caffeine intake: A case report. *Prog. Neuropsychopharmacol. Biol. Psychiatry* **2008**, *32*, 1349–1350. [CrossRef] [PubMed]

57. Hedges, D.W.; Woon, F.L.; Hoopes, S.P. Caffeine-induced psychosis. *CNS Spectr.* **2009**, *14*, 127–129. [CrossRef] [PubMed]

58. Broderick, P.J.; Benjamin, A.B.; Dennis, L.W. Caffeine and psychiatric medication interactions: A review. *J. Okla. State Med. Assoc.* **2005**, *98*, 380–384. [PubMed]

59. Kendler, K.S.; Prescott, C.A.; Myers, J.; Neale, M.C. The structure of genetic and environmental risk factors for common psychiatric and substance use disorders in men and women. *Arch. Gen. Psychiatry* **2003**, *60*, 929–937. [CrossRef] [PubMed]

60. Zeiner, A.R.; Stanitis, T.; Spurgeon, M.; Nichols, N. Treatment of alcoholism and concomitant drugs of abuse. *Alcohol* **1985**, *2*, 555–559. [CrossRef]

61. Aubin, H.J.; Laureaux, C.; Tilikete, S.; Barrucand, D. Changes in cigarette smoking and coffee drinking after alcohol detoxification in alcoholics. *Addiction* **1999**, *94*, 411–416. [CrossRef] [PubMed]

62. Fischler, R.S. Poisoning: A syndrome of child abuse. *Am. Fam. Physician* **1983**, *28*, 103–108. [PubMed]

63. Tenenbein, M. Recent advancements in pediatric toxicology. *Pediatr. Clin. N. Am.* **1999**, *46*, 1179–1188. [CrossRef]

64. Meadow, R. ABC of child abuse. Poisoning. *Br. Med. J.* **1989**, *27*, 1445–1446. [CrossRef]

65. Henretig, F.M.; Paschall, R.T.; Donaruma-Kwoh, M.M. Child abuse by poisoning. In *Child Abuse: Medical Diagnosis and Management*; Reece, R.M., Christian, C.W., Eds.; American Academy of Pediatrics: Farmington Hills, MI, USA, 2009; pp. 549–599.

66. Dine, M.S.; McGovern, M.E. Intentional poisoning of children—An overlooked category of child abuse: Report of seven cases and review of the literature. *Pediatrics* **1982**, *70*, 32–35. [CrossRef] [PubMed]

67. Lee, J.C.; Lin, K.L.; Lin, J.J.; Hsia, S.H.; Wu, C.T. Non-accidental chlorpyrifos poisoning-an unusual cause of profound unconsciousness. *Eur. J. Pediatr.* **2010**, *169*, 509–511. [CrossRef] [PubMed]

68. Ayoub, C.C.; Alexander, R.; Beck, D.; Bursch, B.; Feldman, K.W.; Libow, J.; Sanders, M.J.; Schreier, H.A.; Yorker, B. APSAC Taskforce on Munchausen by Proxy, Definitions Working Group. Position paper: Definitional issues in Munchausen by proxy. *Child Maltreat* **2002**, *7*, 105–111. [CrossRef] [PubMed]

69. Moldavsky, M.; Stein, D. Munchausen Syndrome by Proxy: Two case reports and an update of the literature. *Int. J. Psychiatry Med.* **2003**, *33*, 411–423. [CrossRef] [PubMed]

70. Anderson, B.J.; Gunn, T.R.; Holford, N.H.; Johnson, R. Caffeine overdose in a premature infant: Clinical course and pharmacokinetics. *Anaesth. Intensive Care* **1999**, *27*, 307–311. [PubMed]

71. Ergenekon, E.; Dalgiç, N.; Aksoy, E.; Koç, E.; Atalay, Y. Caffeine intoxication in a premature neonate. *Paediatr. Anaesth.* **2001**, *11*, 737–739. [CrossRef] [PubMed]

*nutrients*

MDPI

*Reply*

# Response to "Are There Non-Responders to the Ergogenic 3 Effects of Caffeine Ingestion on Exercise Performance?"

Kyle Southward [1], Kay Rutherfurd-Markwick [2,3], Claire Badenhorst [1,3] and Ajmol Ali [1,3,*]

[1]  School of Sport, Exercise and Nutrition, Massey University, North Shore Mail Centre, Private Bag 102 904, Auckland 0745, New Zealand; K.A.Southward@massey.ac.nz (K.S.); C.Badenhorst@massey.ac.nz (C.B.)
[2]  School of Health Sciences, Massey University, Auckland 0745, New Zealand; K.J.Rutherfurd@massey.ac.nz
[3]  Centre for Metabolic Health Research, Massey University, Auckland 0745, New Zealand
*   Correspondence: a.ali@massey.ac.nz; Tel.: +64-(0)9-213-6414

Received: 8 November 2018; Accepted: 9 November 2018; Published: 13 November 2018

In response to "Letter: are there non-responders to the ergogenic effects of caffeine ingestion on exercise performance" by Grgic [1], we welcome the additional context that this letter provides to our paper [2]. We agree with the sentiment that responders and non-responders are misleading to readers and thus avoided using these terms in our publication [2] as much as possible. As stated by Grgic [1], an individual may perform well in one test and not another following caffeine ingestion, likewise the individual may perform better or worse on different days given the same caffeine supplementation due to multiple external factors (as mentioned in our paper [2]) and variation in performance.

With regards to the study design of future research, while it may be beneficial to use multiple exercise modes to determine the ergogenicity of caffeine, it is quite often not realistic to do so within the same study. Most studies investigating the ergogenic benefits of supplements use a specific exercise modality to answer a specific research question, for example exploring the effects of caffeine intake on endurance time-trial performance [3–6]. Including multiple exercise modalities within the same study would greatly increase the participant burden, financial costs and time to carry out the study. However, we agree that researchers should still be encouraged to use a variety of valid exercise modalities to gain a comprehensive understanding of a particular supplement. The recommendations put forward by Grgic [1] are welcomed and should be applied where applicable, particularly the reporting of individual data in response to caffeine supplementation as well as when drawing conclusions from the results.

**Funding:** No sources of funding were used in the preparation of this article.

**Conflicts of Interest:** The authors declare no conflict of interest.

## References

1.  Jozo, G. Are There Non-Responders to the Ergogenic Effects of Caffeine Ingestion on Exercise Performance? *Nutrients* **2018**, *10*, 1736. [CrossRef]
2.  Southward, K.; Rutherfurd-Markwick, K.; Badenhorst, C.; Ali, A.; Southward, K.; Rutherfurd-Markwick, K.; Badenhorst, C.; Ali, A. The Role of Genetics in Moderating the Inter-Individual Differences in the Ergogenicity of Caffeine. *Nutrients* **2018**, *10*, 1352. [CrossRef] [PubMed]
3.  Astorino, T.A.; Cottrell, T.; Lozano, A.T.; Aburto-Pratt, K.; Duhon, J. Ergogenic Effects of Caffeine on Simulated Time-Trial Performance Are Independent of Fitness Level. *J. Caffeine Res.* **2011**, *1*, 179–185. [CrossRef]
4.  Astorino, T.A.; Cottrell, T.; Lozano, A.T.; Aburto-Pratt, K.; Duhon, J. Increases in cycling performance in response to caffeine ingestion are repeatable. *Nutr. Res.* **2012**, *32*, 78–84. [CrossRef] [PubMed]

5.   Desbrow, B.; Barrett, C.M.; Minahan, C.L.; Grant, G.D.; Leveritt, M.D. Caffeine, Cycling Performance, and Exogenous CHO Oxidation. *Med. Sci. Sports Exerc.* **2009**, *41*, 1744–1751. [CrossRef] [PubMed]
6.   Bortolotti, H.; Altimari, L.R.; Vitor-Costa, M.; Cyrino, E.S. Performance during a 20-km cycling time-trial after caffeine ingestion. *J. Int. Soc. Sports Nutr.* **2014**, *11*, 45. [CrossRef] [PubMed]

![nutrients logo] *nutrients*

MDPI

*Letter*

# Are There Non-Responders to the Ergogenic Effects of Caffeine Ingestion on Exercise Performance?

Jozo Grgic

Institute for Health and Sport (IHES), Victoria University, Melbourne 3001, Australia; jozo.grgic@live.vu.edu.au

Received: 24 October 2018; Accepted: 9 November 2018; Published: 12 November 2018

**Keywords:** individual responses; ergogenic aid; supplement; did not respond; responders

I have read with interest the recent review paper by Southward and colleagues [1]. While acknowledging that this was not the main focus of the paper, the authors attempted to estimate the average number of non-responders to caffeine ingestion in studies that investigated the effects of caffeine on time-trial performance [1]. Southward and colleagues [1] suggested that there might up to 33% of those who do not enhance performance following caffeine ingestion (i.e., non-responders). The authors came to this estimate by examining the change in performance following caffeine and placebo ingestion from the individual responses in several studies that reported these data. They used an approach where each participant that did not perform better on caffeine (as compared to placebo) was deemed as a non-responder. However, the authors did not consider that some of these individual differences between the caffeine and placebo conditions might have been merely an error of the measurement of the performance tests and not a true lack of response. Therefore, I believe that some additional discussion is needed to avoid confusion on this topic and to clarify the interpretation of these results.

## 1. Reliability of the Exercise Protocol

Reliability refers to the reproducibility of values of a given test [2]. In sport and exercise science, reliability is commonly determined by the error of measurement using the coefficient of variation (expressed as the percentage of the mean) [2]. If we examine the same set of studies as Southward et al. [1], while factoring in the coefficient of variation for the performance tests (for studies that provided these data), it becomes clear that the percentage of those that did not enhance performance following caffeine ingestion reduces from the initially suggested 33% to only 5% (Table 1) [3–21]. While such an amount is low, I would further question if there are really non-responders to the ergogenic effects of caffeine ingestion on exercise performance or if such inferences are an over-extrapolation of the results from the current body of evidence.

**Table 1.** Revised analysis of the prevalence of non-responders to the ergogenic effects of caffeine ingestion on exercise performance (based on the review by Southward et al. [1]).

| Reference | Number of Non-Responders as Classified by Southward et al. [1] | Coefficient of Variation for the Performance Test | Number of Non-Responder While Factoring in the Range of the Error of the Measurement (for Studies That Provided These Data) |
|---|---|---|---|
| Acker-Hewitt et al. [3] | 2/10 | 1.4% | 0/10 |
| Astorino et al. [4] | 3/16 | 1.5% | 1/16 |
| Astorino et al. [5] | 1/9 | 2.5% | 0/9 |
| Beaumont et al. [6] | 1/8 | Not reported | Unable to determine |
| Christensen et al. [7] | 4/12 | Not reported | Unable to determine |
| Church et al. [8] | 8/20 | Not reported | Unable to determine |
| Desbrow et al. [9] | 3/9 | Not reported | Unable to determine |
| Desbrow et al. [10] | 4/16 | Not reported | Unable to determine |
| Gonçalves et al. [11] | 6/40 | 2.9% | 4/40 |
| Graham-Paulson et al. [12] | 1/11 | Not reported | Unable to determine |
| Guest et al. [13] 2 mg/kg | 38/101 | Not reported | Unable to determine |
| Guest et al. [13] 4 mg/kg | 32/101 | Not reported | Unable to determine |
| O'Rourke et al. [14] | 3/30 | Not reported | Unable to determine |
| Pitchford et al. [15] | 2/9 | Not reported | Unable to determine |
| Roelands et al. [16] | 4/8 | Not reported | Unable to determine |
| De Alcantara Santos et al. [17] | 2/8 | 0.9% | 1/8 |
| Skinner et al. [18] | 1/14 | 0.9% | 0/14 |
| Stadheim et al. [19] | 2/10 | 3.2% | 0/10 |
| Stadheim et al. [20] | 4/13 | 2% | 0/13 |
| Womack et al. [21] | 3/35 | Not reported | Unable to determine |
| Pooled number of participants and the number of non-responders | | | 6/120 (5%) |

## 2. Using Multiple Exercise Tests When Examining the Effects of Caffeine

We have reported that caffeine ingestion in the dose of 6 mg/kg enhanced lower-body one-repetition maximum (1RM) strength and upper-body ballistic performance [22]. A scrutiny of the individual data from our study shows the problems when classifying responders and non-responders solely based on the results from one test. The figures provided in that paper indicate that participant number 8 experienced a 7% decrease in 1RM strength following caffeine ingestion (as compared to placebo). In contrast to the results for strength, the same participant experienced a 4% increase in upper-body ballistic performance following the ingestion of caffeine. If we were to present findings from only one test of performance the same participant can be classified as a non-responder to caffeine (based on the strength data) or as a responder (based on the ballistic performance data). Therefore, it becomes clear that caffeine might not enhance performance in one test while being effective in another. Classifying an individual as a non-responder to caffeine while focusing on the results from only one performance test may undermine the effects of caffeine that the same individual might experience in a different exercise task. These concepts can be juxtaposed with the findings by Churchward-Venne et al. [23] who reported no non-responders to a resistance training program when using several different test of performance (e.g., strength assessment in different exercise, chair-raise time, changes in lean body mass) given that each participant improved in at least some of the employed tests.

## 3. Using the Same Exercise Test with Different Doses of Caffeine

Jenkins et al. [24] tested the effects of 1, 2, and 3 mg/kg of caffeine on cycling performance. On average, their results indicated that only a 2 mg/kg dose of caffeine was effective for acute increases in cycling performance. However, the individual data presented in this study provided some very insightful findings. For instance, participant number 7 had a 6% decrease in performance following the ingestion of 1 mg/kg of caffeine. If Jenkins et al. [24] only used this dose of caffeine, this individual would be classified as a non-responder. However, in the 3 mg/kg caffeine condition, this participant improved cycling performance by +10%, and, if the researchers used only this dose of caffeine, the same participant would be considered as a high-responder to caffeine. While there are participants from the study by Jenkins et al. [24] that did not improve performance with any of the three caffeine doses, it is possible that higher doses of caffeine (e.g., 4–6 mg/kg) would elicit an acute improvement in performance even in these individuals. Some of these initial observations suggest that if an individual does not respond to a specific dose of caffeine, it would be erroneous to classify him as a non-responder given that a different dose (higher or lower) might be highly effective even in the same exercise task.

## 4. Repeated Testing Using the Same Exercise Test and the Same Dose of Caffeine

Astorino et al. [25] tested the same group of participants, using the same exercise test (cycling time-trial), and the same caffeine dose (5 mg/kg) on two different occasions. The results obtained by Astorino et al. [25] suggested that the effects of caffeine are repeatable in the majority of the participants as most tended to improve performance on both caffeine conditions. More importantly, the individual data presented in that study suggests that one participant improved cycling performance only in the second administration of the 5 mg/kg caffeine dose. Accordingly, if the study reported only the results from the first administration of caffeine, this participant would be considered as a non-responder. By contrast, if only the results from the second testing were reported, the same participant would be classified as a responder to caffeine. While working with limited data, these initial results imply a possible 'learning effect' for caffeine which needs to be considered when interpreting individual data.

## 5. Conclusions

As discussed herein, the estimate by Southward et al. [1] that there might be up to 33% of those that do not respond to caffeine ingestion might be an over-extrapolation of the current data. In fact, the number of those that do not respond to caffeine might be minimal given that using a different test of performance, changing the dose of caffeine, conducting repeated measures with the same test and the same dose of caffeine on different occasions (i.e., providing a 'learning effect' period), or possibly even adjusting the timing of caffeine ingestion based on genotype (as already nicely highlighted by Southward et al. [1]) might change a response to caffeine ingestion from a negative to a positive. To expand our current knowledge on the variation in responses to caffeine ingestion future studies should consider presenting the individual responses, and the interpretation of these responses should reflect some of the information presented herein.

**Funding:** No sources of funding were used to assist in the preparation of this article.

**Conflicts of Interest:** The author declares no conflict of interest.

## References

1. Southward, K.; Rutherfurd-Markwick, K.; Badenhorst, C.; Ali, A. The role of genetics in moderating the inter-individual differences in the ergogenicity of caffeine. *Nutrients* **2018**, *10*, 1352. [CrossRef] [PubMed]
2. Hopkins, W.G. Measures of reliability in sports medicine and science. *Sports Med.* **2000**, *30*, 1–15. [CrossRef] [PubMed]
3. Acker-Hewitt, T.L.; Shafer, B.M.; Saunders, M.J.; Goh, Q.; Luden, N.D. Independent and combined effects of carbohydrate and caffeine ingestion on aerobic cycling performance in the fed state. *Appl. Physiol. Nutr. Metab.* **2012**, *37*, 276–283. [CrossRef] [PubMed]
4. Astorino, T.A.; Cottrell, T.; Lozano, A.T.; Aburto-Pratt, K.; Duhon, J. Ergogenic effects of caffeine on simulated time-trial performance are independent of fitness level. *J. Caffeine Res.* **2011**, *1*, 179–185. [CrossRef]
5. Astorino, T.A.; Roupoli, L.R.; Valdivieso, B.R. Caffeine does not alter RPE or pain perception during intense exercise in active women. *Appetite* **2012**, *59*, 585–590. [CrossRef] [PubMed]
6. Beaumont, R.E.; James, L.J. Effect of a moderate caffeine dose on endurance cycle performance and thermoregulation during prolonged exercise in the heat. *J. Sci. Med. Sport* **2016**, *20*, 1024–1028. [CrossRef] [PubMed]
7. Christensen, P.M.; Petersen, M.H.; Friis, S.N.; Bangsbo, J. Caffeine, but not bicarbonate, improves 6 min maximal performance in elite rowers. *Appl. Physiol. Nutr. Metab.* **2014**, *39*, 1058–1063. [CrossRef] [PubMed]
8. Church, D.D.; Hoffman, J.R.; LaMonica, M.B.; Riffe, J.J.; Hoffman, M.W.; Baker, K.M.; Varanoske, A.N.; Wells, A.J.; Fukuda, D.H.; Stout, J.R. The effect of an acute ingestion of Turkish coffee on reaction time and time trial performance. *J. Int. Soc. Sports Nutr.* **2015**, *12*, 37. [CrossRef] [PubMed]
9. Desbrow, B.; Barrett, C.M.; Minahan, C.L.; Grant, G.D.; Leveritt, M.D. Caffeine, cycling performance and exogenous, CHO oxidation: A dose-response study. *Med. Sci. Sports Exerc.* **2009**, *41*, 1744–1751. [CrossRef] [PubMed]
10. Desbrow, B.; Biddulph, C.; Devlin, B.; Grant, G.D.; Anoopkumar-Dukie, S.; Leveritt, M.D. The effects of different doses of caffeine on endurance cycling time trial performance. *J. Sports Sci.* **2012**, *30*, 115–120. [CrossRef] [PubMed]
11. Gonçalves, L.S.; Painelli, V.S.; Yamaguchi, G.; de Oliveira, L.F.; Saunders, B.; da Silva, R.P.; Maciel, E.; Artioli, G.G.; Roschel, H.; Gualano, B. Dispelling the myth that habitual caffeine consumption influences the performance response to acute caffeine supplementation. *J. Appl. Physiol.* **2017**, *123*, 213–220. [CrossRef] [PubMed]
12. Graham-Paulson, T.S.; Perret, C.; Watson, P.; Goosey-Tolfrey, V.L. Improvement of sprint performance in wheelchair sportsmen with caffeine supplementation. *Int. J. Sports Physiol. Perform.* **2016**, *11*, 214–220. [CrossRef] [PubMed]
13. Guest, N.; Corey, P.; Vescovi, J.; El-Sohemy, A. Caffeine, CYP1A2 genotype, and endurance performance in athletes. *Med. Sci. Sports Exerc.* **2018**, *50*, 1570–1578. [CrossRef] [PubMed]

14. O'Rourke, M.P.; O'Brien, B.J.; Knez, W.; Paton, C.D. Caffeine has a small effect on 5-km running performance of well-trained and recreational runners. *J. Sci. Med. Sport* **2008**, *11*, 231–233. [CrossRef] [PubMed]

15. Pitchford, N.W.; Fell, J.W.; Leveritt, M.D.; Desbrow, B.; Shing, C.M. Effect of caffeine on cycling time-trial performance in the heat. *J. Sci. Med. Sport* **2014**, *17*, 445–449. [CrossRef] [PubMed]

16. Roelands, B.; Buyse, L.; Pauwels, F.; Delbeke, F.; Deventer, K.; Meeusen, R. No effect of caffeine on exercise performance in high ambient temperature. *Eur. J. Appl. Physiol.* **2011**, *111*, 3089–3095. [CrossRef] [PubMed]

17. Santos Rde, A.; Kiss, M.A.; Silva-Cavalcante, M.D.; Correia-Oliveira, C.R.; Bertuzzi, R.; Bishop, D.J.; Lima-Silva, A.E. Caffeine alters anaerobic distribution and pacing during a 4000-m cycling time trial. *PLoS ONE* **2013**, *8*, e75399. [CrossRef] [PubMed]

18. Skinner, T.L.; Jenkins, D.G.; Taaffe, D.R.; Leveritt, M.D.; Coombes, J.S. Coinciding exercise with peak serum caffeine does not improve cycling performance. *J. Sci. Med. Sport* **2013**, *16*, 54–59. [CrossRef] [PubMed]

19. Stadheim, H.K.; Kvamme, B.; Olsen, R.; Drevon, C.A.; Ivy, J.L.; Jensen, J. Caffeine increases performance in cross-country double-poling time trial exercise. *Med. Sci. Sports Exerc.* **2013**, *45*, 2175–2183. [CrossRef] [PubMed]

20. Stadheim, H.K.; Nossum, E.M.; Olsen, R.; Spencer, M.; Jensen, J. Caffeine improves performance in double poling during acute exposure to 2000-m altitude. *J. Appl. Physiol.* **2015**, *119*, 1501–1509. [CrossRef] [PubMed]

21. Womack, C.J.; Saunders, M.J.; Bechtel, M.K.; Bolton, D.J.; Martin, M.; Luden, N.D.; Dunham, W.; Hancock, M. The influence of a CYP1A2 polymorphism on the ergogenic effects of caffeine. *J. Int. Soc. Sports Nutr.* **2012**, *9*, 7. [CrossRef] [PubMed]

22. Grgic, J.; Mikulic, P. Caffeine ingestion acutely enhances muscular strength and power but not muscular endurance in resistance-trained men. *Eur. J. Sport Sci.* **2017**, *17*, 1029–1036. [CrossRef] [PubMed]

23. Churchward-Venne, T.A.; Tieland, M.; Verdijk, L.B.; Leenders, M.; Dirks, M.L.; de Groot, L.C.; van Loon, L.J. There are no nonresponders to resistance-type exercise training in older men and women. *J. Am. Med. Dir. Assoc.* **2015**, *16*, 400–411. [CrossRef] [PubMed]

24. Jenkins, N.T.; Trilk, J.L.; Singhal, A.; O'Connor, P.J.; Cureton, K.J. Ergogenic effects of low doses of caffeine on cycling performance. *Int. J. Sport Nutr. Exerc. Metab.* **2008**, *18*, 328–342. [CrossRef] [PubMed]

25. Astorino, T.A.; Cottrell, T.; Lozano, A.T.; Aburto-Pratt, K.; Duhon, J. Increases in cycling performance in response to caffeine ingestion are repeatable. *Nutr. Res.* **2012**, *32*, 78–84. [CrossRef] [PubMed]

MDPI

St. Alban-Anlage 66

4052 Basel

Switzerland

Tel. +41 61 683 77 34

Fax +41 61 302 89 18

www.mdpi.com

*Nutrients* Editorial Office

E-mail: nutrients@mdpi.com

www.mdpi.com/journal/nutrients

www.ingramcontent.com/pod-product-compliance
Lightning Source LLC
Chambersburg PA
CBHW051714210326
41597CB00032B/5482